Advanced Materials for Wastewater Treatment and Desalination

Advanced Materials for Wastewater Treatment and Desalination: Fundamentals to Applications offers a comprehensive overview of current progress in the development of advanced materials used in wastewater treatment and desalination. The book is divided into two major sections, covering both fundamentals and applications.

This book:

- Describes the synthesis and modification of advanced materials, including metal oxides, carbonaceous materials, perovskite-based materials, polymer-based materials, and advanced nanocomposites
- Examines relevant synthesis routes and mechanisms as well as correlates materials properties with their characterization
- Details new fabrication techniques including green synthesis, solvent-free, and energy-saving synthesis approaches
- Highlights various applications, such as removal of organic contaminants, discoloration of dye wastewater, petrochemical wastewater treatment, and electrochemically-enhanced water treatment.

With chapters written by leading researchers from around the world, this book will be of interest to chemical, materials, and environmental engineers working on progressing materials applications to improve water treatment technologies.

Emerging Materials and Technologies

Series Editor:
Boris I. Kharissov

2D Monoelemental Materials (Xenes) and Related Technologies: Beyond Graphene
Zongyu Huang, Xiang Qi, Jianxin Zhong

Atomic Force Microscopy for Energy Research
Cai Shen

Self-Healing Cementitious Materials: Technologies, Evaluation Methods, and Applications
Ghasan Fahim Huseien, Iman Faridmehr, Mohammad Hajmohammadian Baghban

Thin Film Coatings: Properties, Deposition, and Applications
Fredrick Madaraka Mwema, Tien-Chien Jen, and Lin Zhu

Biosensors: Fundamentals, Emerging Technologies, and Applications
Sibel A. Ozkan, Bengi Uslu, and Mustafa Kemal Sezgintürk

Error-Tolerant Biochemical Sample Preparation with Microfluidic Lab-on-Chip
Sudip Poddar and Bhargab B. Bhattacharya

Geopolymers as Sustainable Surface Concrete Repair Materials
Ghasan Fahim Huseien, Abdul Rahman Mohd Sam, and Mahmood Md. Tahir

Nanomaterials in Manufacturing Processes
Dhiraj Sud, Anil Kumar Singla, Munish Kumar Gupta

Advanced Materials for Wastewater Treatment and Desalination: Fundamentals to Applications
Ahmad Fauzi Ismail, Pei Sean Goh, Hasrinah Hasbullah, and Farhana Aziz

Green Synthesized Iron-Based Nanomaterials: Application and Potential Risk
Piyal Mondal and Mihir Kumar Purkait

For more information about this series, please visit: https://www.routledge.com/Emerging-Materials-and-Technologies/book-series/CRCEMT

Advanced Materials for Wastewater Treatment and Desalination

Fundamentals to Applications

Edited by
Ahmad Fauzi Ismail, Pei Sean Goh,
Hasrinah Hasbullah, and Farhana Aziz

CRC Press
Taylor & Francis Group
Boca Raton London New York

CRC Press is an imprint of the
Taylor & Francis Group, an informa business

First edition published 2023
by CRC Press
6000 Broken Sound Parkway NW, Suite 300, Boca Raton, FL 33487-2742

and by CRC Press
4 Park Square, Milton Park, Abingdon, Oxon, OX14 4RN

CRC Press is an imprint of Taylor & Francis Group, LLC

ISBN: 978-0-367-76516-3 (hbk)
ISBN: 978-0-367-76517-0 (pbk)
ISBN: 978-1-003-16732-7 (ebk)

DOI: 10.1201/9781003167327

Typeset in Times
by codeMantra

Contents

Preface

Water is a key component of living organisms. With more human activities, industrialization and urbanization, the demand for clean water is also increasing. On the other hand, the reduction of reliable water sources and increasing water pollution are known as one of the key global environmental issues of the 21st century. Water reclamation is an important strategy to resolve the water shortage issue. Wastewater treatment and desalination have shown great potential to provide sustainable clean water to meet water requirements. In the past few decades, a wide variety of treatment technologies and materials have been studied and applied for wastewater treatment and desalination. Tremendous efforts have been made in these aspects, and this book aims to make a compilation in the state-of-the-art progress made in the development of advanced materials for wastewater treatment and desalination application. This book, entitled *Advanced Materials for Wastewater Treatment: Fundamental to Application*, aims to bring together the ideas of researchers working in this field. Through contributions from leading experts from around the world, the book offers a detailed overview of the principles and applications of advanced materials in wastewater treatment.

This edited book is divided into two major sections: (a) Fundamentals and (b) Applications. The first part encompasses the synthesis and modification of advanced materials to eliminate any pollutants from wastewater and for desalination purpose. This includes the revolutionary material synthesis, modification, and characterization techniques. Advanced materials synthesized in different dimensions such as metal oxide, carbon-based materials, perovskite-based materials, polymer-based composite materials, and advanced nanocomposites are discussed. New fabrication techniques, including green synthesis, solvent-free, energy-saving synthesis approaches, are elaborated. The relevant synthesis route and mechanisms as well as the correlation of materials properties with their characterization are also included in the discussions. The second section of this book highlights the potential applications by advanced materials in water treatment technologies and desalination. The applications of a wide spectrum of functional and advanced materials in the removal of organic contaminant, discoloration of dye wastewater and agricultural wastewater reclamation, just to name a few, are discussed in this section. With further advancement, the innovations made in material advancement are expected to fulfill today's wastewater treatment demand with better quality.

It is expected that functional materials will continue to flourish in the field of water reclamation. This edited book targets material scientists, graduate students, post-doctoral researchers, professors and young researchers who work in this interesting field. It is hoped that this edited book serves as a reference for this targeted group to provide useful information on the progresses made in this field.

Editors,
Ahmad Fauzi Ismail
Pei Sean Goh
Hasrinah Hasbullah
Farhana Aziz
Advanced Membrane Technology Research Centre
School of Chemical and Energy Engineering,
Faculty of Engineering
Universiti Teknologi Malaysia

Editors

Professor Dr. Ahmad Fauzi Ismail is the seventh Vice-Chancellor of Universiti Teknologi Malaysia. Ahmad Fauzi Ismail graduated with a B.Eng. (Petroleum Engineering) and an M.Sc. (Chemical Engineering) from Universiti Teknologi Malaysia (UTM). He was awarded the Commonwealth Academic Staff Scholarship to pursue his Ph.D. in Chemical and Process Engineering at the University of Strathclyde, Glasgow, UK, specializing in Membrane Technology. Ahmad Fauzi Ismail's outstanding achievement in research has made him an inspirational role model for academic staff and post-graduate students. He champions research on the development of polymeric, inorganic, and mixed matrix membranes for water desalination, wastewater treatment, gas separation processes, membrane for fuel cell applications, palm oil refining, hemodialysis membrane, and smart optical fiber for tracking migration of oil flow. He owns 11 patents, with 20 more pending for approval. As a prolific author, his compendium of publications ranges from over 1000 papers in refereed journals, over 50 book chapters and six books. He is a Member of the Malaysia Science Council, Member of Higher Education Advisory Committee, Ministry of Higher Education, Chairman of Malaysia Membrane Society (MyMembrane), Chairman for Selection Committee of the14th National Academic Award and Chairman of Malaysia Research University Network (MRUN). He serves as a Chief Editor and Editorial Board Members for several reputable international journals.

Associate Professor Dr. Pei Sean Goh is an Associate Professor in the School of Chemical and Energy Engineering, Faculty of Engineering, Universiti Teknologi Malaysia (UTM). Pei Sean is a research fellow of the Advanced Membrane Research Technology Research Centre (AMTEC), UTM. She is also the Head of Nanostructured Materials Research Group in UTM. Her research interests focus on the synthesis of a wide range of nanostructured materials and their composites for membrane-based separation processes. One of the main focuses of her research is the applications of two-dimensional nanomaterials and polymeric nanocomposite membranes for acidic gas removal as well as desalination and wastewater treatment. Pei Sean has authored or co-authored more than 200 ISI-indexed research articles. She is an Associate Member of Academy of Science of Malaysia and Institute of Chemistry Malaysia, and a Member of prestigious Global Young Academy. She is also the recipient of L'Oreal-UNESCO for Women in Science 2020 award.

Dr. Hasrinah Hasbullah is a senior lecturer at the Department of Energy Engineering, Universiti Teknologi Malaysia (UTM) and a research associate of Advanced Membrane Technology Research Center (AMTEC). She has been in various administration posts including the Director of Energy Engineering Department. Hasrinah completed her PhD in Chemical Engineering specialization in Membrane Technology from Imperial College (IC) London, UK and prior to that her Bachelor and Master of Chemical Engineering from Universiti Teknologi Malaysia. She is an active researcher as the principal investigator and research member of more than 60 research grants including Sakura Exchange Program in Science Grant that she was awarded with Kobe University in 2017 by Japan Science and Technology Agency (JICA). Her research interest is mainly in the areas of polymer synthesis and membrane science and technology, in addition to various fields involving membrane such as gas separation, energy related and water as well as waste water treatment. She is currently involved in developing the first ever home-grown membrane for hemodialysis. Dr. Hasrinah has a good collaboration with academicians from local and international institutions, with whom she has successfully published numerous publications on material and membrane science and technology as well as energy and separation.

Dr. Farhana Aziz is currently a Senior Lecturer in the Department of Energy Engineering, School of Chemical and Energy Engineering, and Associate Research Fellow of Advanced Membrane Technology Research Center (AMTEC), Universiti Teknologi Malaysia (UTM). She graduated with a B.Eng. (Chemical Engineering), an M.Eng. (Gas Engineering), and a PhD (Gas Engineering) from Universiti Teknologi Malaysia (UTM) in 2007, 2010, and 2015, respectively. Her passion and dedication toward her research are reflected in her numerous scientific publications in high-impact international refereed journals and academic books. To date, she has been awarded with 12 research grants and has been member of more than 50 research projects. Her specialized research area includes photocatalysts and nanomaterials synthesis and characterizations for gas separation, wastewater treatment, and energy applications. Recently, Dr. Farhana has received a Special Program for Research Against COVID-19 grant from the Japan International Cooperation Agency (JICA) project for AUN/SEED-Net. She has also obtained a fellowship from JICA under AUN/SEED-Net Short-Term Research Program in Japan in Hokkaido University, Japan in 2017.

List of Contributors

Abdul Latif Ahmad
School of Chemical Engineering
Campus of Engineering, Universiti Sains Malaysia
Pulau Pinang, Malaysia

Syed Ossama Ali Ahmad
Solar Cell Applications Research Lab, Department of Physics
Government College University Lahore
Lahore, Pakistan

Salamat Ali
Department of Physics
Riphah Institute of Computing and Applied Sciences (RICAS), Riphah International University
Lahore, Pakistan

Aliah
Chemical Engineering Department, Faculty of Engineering
Lambung Mangkurat University
Banjarbaru, Indonesia
and
Material and Membrane Research Group (M2ReG)
Lambung Mangkurat University
Banjarbaru, Indonesia

Gloria Amo-Duodu
Green Engineering and Sustainability Research Group, Department of Chemical Engineering
Durban University of Technology
Durban, South Africa

Edward K. Armah
Green Engineering and Sustainability Research Group, Department of Chemical Engineering
Durban University of Technology
Durban, South Africa

Dennis Asante-Sackey
Kumasi Technical University
Kumasi, Ghana

Atif Ashfaq
Solar Cell Applications Research Lab, Department of Physics
Government College University Lahore
Lahore, Pakistan

Farhana Aziz
Advanced Membrane Technology Research Centre (AMTEC)
Universiti Teknologi Malaysia
Skudai, Malaysia
and
School of Chemical and Energy Engineering, Faculty of Engineering
Universiti Teknologi Malaysia
Skudai, Malaysia

Paria Banino
Department of Biology
Science and Research Branch, Islamic Azad University
Tehran, Iran

Saeed Bazgir
Department of Polymer Engineering, Petroleum and Chemical Engineering Faculty
Science and Research Branch, Islamic Azad University
Tehran, Iran

Muthia Elma
Chemical Engineering Department, Faculty of Engineering
Lambung Mangkurat University
Banjarbaru, Indonesia
and
Material and Membrane Research Group (M2ReG)
Lambung Mangkurat University
Banjarbaru, Indonesia
and
Wetland-Based Materials Research Center
Lambung Mangkurat University
Banjarbaru, Indonesia

Elorm Obotey Ezugbe
Green Engineering and Sustainability Research Group, Department of Chemical Engineering
Durban University of Technology
Durban, South Africa

Nur Fajrina
Advanced Membrane Technology Research Centre (AMTEC)
Universiti Teknologi Malaysia
Skudai, Malaysia
and
School of Chemical and Energy Engineering, Faculty of Engineering
Universiti Teknologi Malaysia
Skudai, Malaysia

Pei Sean Goh
Universiti Teknologi Malaysia,
Johor, Malaysia

Jahanzeb Hassan
Department of Physics
Riphah Institute of Computing and Applied Sciences (RICAS), Riphah International University
Lahore, Pakistan

Nurul Huda
Chemical Engineering Department, Faculty of Engineering
Lambung Mangkurat University
Banjarbaru, Indonesia
and
Material and Membrane Research Group (M2ReG)
Lambung Mangkurat University
Banjarbaru, Indonesia

Muhammad Ikram
Solar Cell Application Research Lab, Department of Physics
Government College University Lahore
Lahore, Pakistan

Ahmad Fauzi Ismail
Advanced Membrane Technology Research Centre (AMTEC), School of Chemical and Energy Engineering
Universiti Teknologi Malaysia
Johor, Malaysia

Nur Azizah Johari
Advanced Membrane Technology Research Centre (AMTEC)
Universiti Teknologi Malaysia
Skudai, Malaysia
and

School of Chemical and Energy Engineering, Faculty of Engineering
Universiti Teknologi Malaysia
Skudai, Malaysia

Wan Zulaisa Amira Wan Jusoh
Department of Chemical Engineering
College of Engineering, Universiti Malaysia Pahang, Lebuhraya Tun Razak
Kuantan, Malaysia

Mohammadreza Kafi
Chemical Engineering Department
Faculty of Engineering
University of Kashan
Kashan, Iran

Tutuk Djoko Kusworo
Chemical Engineering Department, Faculty of Engineering
Universitas Diponegoro
Semarang, Indonesia

Wan Mohd Asyraf Wan Mahmood
Faculty of Science
Universiti Teknologi Malaysia
Skudai, Malaysia

S. Martini
Universitas Muhammadiyah
Palembang, Indonesia

Nadzirah Mohd Mokhtar
Faculty of Civil Engineering Technology
Universiti Malaysia Pahang, Lebuhraya Tun Razak
Kuantan, Malaysia

Fitri Ria Mustalifah
Chemical Engineering Department, Faculty of Engineering
Lambung Mangkurat University
Banjarbaru, Indonesia
and
Material and Membrane Research Group (M2ReG)
Lambung Mangkurat University
Banjarbaru, Indonesia

Nur Hidayati Othman
Oil and Gas Engineering Programme, School of Chemical Engineering
Universiti Teknologi MARA
Selangor, Malaysia

Usman Qumar
Department of Physics
Riphah Institute of Computing and Applied Sciences (RICAS), Riphah International University
Lahore, Pakistan

Aulia Rahma
Material and Membrane Research Group (M2ReG)
Lambung Mangkurat University
Banjarbaru, Indonesia

Sunarti Abdul Rahman
Department of Chemical Engineering
College of Engineering, Universiti Malaysia Pahang, Lebuhraya Tun Razak
Kuantan, Malaysia

Erdina Lulu Atika Rampun
Material and Membrane Research Group (M2ReG)
Lambung Mangkurat University
Banjarbaru, Indonesia

Sudesh Rathilal
Green Engineering and Sustainability Research Group, Department of Chemical Engineering
Durban University of Technology
Durban, South Africa

Ali Raza
Department of Physics
University of Sialkot, Pakistan

Nur Aqilah Mohd Razali
Advanced Membrane Technology Research Centre (AMTEC)
Universiti Teknologi Malaysia
Skudai, Malaysia
and
School of Chemical and Energy Engineering, Faculty of Engineering
Universiti Teknologi Malaysia
Skudai, Malaysia

Wan Norharyati Wan Salleh
Advanced Membrane Technology Research Centre (AMTEC)
Universiti Teknologi Malaysia
Skudai, Malaysia
and
School of Chemical and Energy Engineering, Faculty of Engineering
Universiti Teknologi Malaysia
Skudai, Malaysia

Hamidreza Sanaeepur
Department of Chemical Engineering, Faculty of Engineering
Arak University
Arak, Iran

Mohammad Mahdi A. Shirazi
Chemical Engineering Department, Faculty of Engineering
University of Kashan
Kashan, Iran
and
Membrane Industry Development Institute
Tehran, Iran
and
Green-MEMTECH Team, Incubation Center
Chemistry and Chemical Engineering Research Center of Iran (CCERCI)
Tehran, Iran
and
Department of Polymer Engineering, Petroleum and Chemical Engineering Faculty
Science and Research Branch, Islamic Azad University
Tehran, Iran

Emmanuel K. Tetteh
Green Engineering and Sustainability Research Group, Department of Chemical Engineering
Durban University of Technology
Durban, South Africa

Anwar Ul-Hamid
Core Research Facilities
King Fahd University of Petroleum & Minerals
Dhahran, Saudi Arabia

Dani Puji Utomo
Chemical Engineering Department, Faculty of Engineering
Universitas Diponegoro
Semarang, Indonesia

Erna Yuliwati
Universitas Muhammadiyah
Palembang, Indonesia

Norhaniza Yusof
Advanced Membrane Technology Research Centre (AMTEC)
Universiti Teknologi Malaysia
Skudai, Malaysia

Section 1

Fundamentals

1 Graphitic Carbon Nitride ($g\text{-}C_3N_4$)-Based Photocatalysts for Wastewater Treatment

Nur Aqilah Mohd Razali,
Wan Norharyati Wan Salleh,
Farhana Aziz, Ahmad Fauzi Ismail, and
Wan Mohd Asyraf Wan Mahmood
Universiti Teknologi Malaysia

CONTENTS

1.1 INTRODUCTION

For decades, one of the most popular ideas in the literature of water studies is the idea that revealed the industrial demand for water has been increasing along with population growth, economic development, and socio-economic factors (Mao et al. 2021; World 2018). The three important components, namely, social, environmental, and economic aspects, are influential toward a water-secure world and/or global change, which include urbanization, population growth, socio-economic change, evolving energy needs, and climate change (Mishra et al. 2021). In addition, the world's water-stressed population will increase rapidly to about 3.3 billion in 2050 if there is no efficient implementation for each aspect (Hayashi, Akimoto, and Tomoda 2013). As a result, the impacts of insufficient safe water would contribute to poor health,

DOI: 10.1201/9781003167327-2

destruction of livelihood, and unnecessary suffering for the poor globally (Hanjra and Qureshi 2010). Furthermore, severe environmental contamination could occur due to the different hazardous effluents coming from various industries (Liu et al. 2016).

A challenging problem that arises in this domain is the water pollutants that are mainly found in industrial effluents, such as textile (dyes), cosmetics, paint (heavy metal ions), automobile manufacturing, and electrical industries (Gopinath et al. 2020). Industrial wastewater effluents are the second largest contributor of pollution, and have acute (direct) and chronic (long-term) impacts on human health (Reghizzi 2021). Unfortunately, the main contributor of pollution results in industrial effluent problems related to the production of emerging pollutants (EPs) which are divided into four categories: (a) persistent organic pollutants (POPs), (b) pharmaceuticals and personal care products (PPCPs), (c) endocrine disrupting chemicals (EDCs), and (d) agricultural chemicals (e.g., pesticides, herbicides) (Gasperi et al. 2014). Similarly, another issue that should be highlighted is the persistency and unchanged concentration of chemical pollutants in wastewater even after undergoing conventional biological treatments (Chen, Ngo, and Guo 2012).

In addition, the adsorption and coagulation process can only transform the pollutants into another phase, which is incapable of complete degradation (Crini and Lichtfouse 2019). Ibrahim and Halim (2008) also studied the limitations of sedimentation, filtration, chemical, and membrane technologies, which are expensive and produce toxic secondary pollutants when released into the ecosystem. Therefore, photocatalytic technology was introduced (Fujishima and Honda 1972). The photocatalytic technology, classified as advanced oxidation processes (AOPs), can degrade bio-recalcitrant compounds including phenols, pesticides, pharmaceuticals, dyes, and petrochemicals (Zhu and Zhou 2019). This technology can also be applied to solve the aforementioned problems due to its (a) ambient operating temperature and pressure, (b) complete mineralization of parents and intermediate compounds without secondary pollution, and (c) inexpensiveness (Chong et al. 2010).

In addition, semiconductor photocatalysis is an easy treatment that can utilize energy by natural sunlight or artificial indoor illumination (Tong et al. 2012). Nowdays, TiO_2 still numerously used as a semiconductor photocatalyst that has been progressively used to split water into hydrogen as its photoanode (Fujishima and Honda 1972). However, TiO_2 is usually restricted by unsatisfactory photocatalytic efficiency due to its broad band gaps (~3.2 eV) (Jiang, Yuan, Pan, Liang, and Zeng 2017). Hence, it is necessary to establish ways to increase the activity for its catalytic applications. Compared with other photocatalysts, graphitic carbon nitride (g-C_3N_4) is considered a promising photocatalyst with a narrow band gap (2.7 eV) and the ability for enhancing visible light absorption.

Therefore, g-C_3N_4 has a clear advantage in photocatalytic technology. It has unique physicochemical and photophysical properties and high thermal and chemical stability. It can be readily doped or chemically functionalized on large scale. It has a large surface area, is low priced, and exhibits long service life, ease of synthesis, controllable band gap properties, low toxicity, and high photocatalytic activity (Raghava et al. 2019; Kumar, Karthikeyan, and Lee 2018). These special properties make g-C_3N_4 application suitable for photocatalytic degradation, photocatalytic sterilization, photocatalytic H_2 generation, and CO_2 reduction. However, the rapid recombination of photoinduced charges can still occur for pristine g-C_3N_4. Most studies of tailored problems are from various modifications such as morphological control (Malik and Tomer 2021), doping elements (Jiang, Yuan, Pan, Liang, Zeng, et al. 2017), deposition of noble metals (Kavitha, Nithya, and Kumar 2020), construction of heterojunctions (Huang et al. 2018), and two-dimensional (2D) nanosheets (Ong 2017).

This chapter overviews understanding of the development of g-C_3N_4 photocatalyst, specifically in water-treatment technology, from the history to the fundamentals of catalyst and its unique properties and role in photocatalytic degradation. Then, the synthesis method of g-C_3N_4 photocatalyst is briefly summarized in Section 1.3 which covers "top-down" and "bottom-up" techniques. Subsequently, the photocatalytic principles and mechanism of Z-scheme and heterojunction are well explained in order to promote a better understanding of the g-C_3N_4 photocatalyst. Finally, an outline of feasible applications of g-C_3N_4 toward various pollutants is presented and explained.

1.2 ROAD MAP OF g-C_3N_4 PHOTOCATALYSTS FOR PHOTOCATALYTIC DEGRADATION

The development history of C_3N_4 originates from Berzelius and Liebig's (1834) embryonic form of melon as the oldest synthetic polymer, which is interconnected with tri-s-triazine using nitrogen (Liebig 1834a). Another review by Fakhrul, Samsudin, and Bacho (2018) stated that in 1922, the composition of melon has been found by Franklin after heating treatment, while the tri-s-triazine in C_3N_4 structure had already been studied by Pauling and Sturdivant in 1937. In 1989, Cohen and Liu had investigated the diamond that acts as an ultrahard material (Liebig 1834a). The investigation was carried out by classifying the hardness into three categories—hard, superhard, and ultrahard—on several materials such as oxides, borides, nitrides and carbides of metals, cermets, carbon nitrides, cubic boron nitride (c-BN), and diamond which have high hardness, high incompressibility, and chemical inertness (Kanyanta 2016). The β-Si_3N_4 structure was selected due to its local density states and the first pseudopotential approximation method. Subsequently, Liu and Cohen introduced β-C_3N_4 by pulsed laser ablation, which is an amorphous and crystalline solid and has high hardness and high thermal stability (Liu and Cohen 1989). Next, Si was substituted for C to form β-C_3N_4. Additionally, Rodeman and Lucas (1940) discovered that the melon has a graphite structure as one of the phases of g-C_3N_4 during synthesis.

According to Teter and Hemley, the calculation methods were used to determine that g-C_3N_4 has five structural types: α-g-C_3N_4 phase (5.49 eV), β-g-C_3N_4 (4.85 eV), a cubic phase (4.13 eV), pseudo-cubic phase (4.30 eV), g-o-triazine (0.93 eV), g-h-triazine (2.97 eV), and g-h-heptazine (2.88 eV) (Teter and Hemley 1996; Chan, Liu, and Hsiao 2019). The g-C_3N_4 has extremely high hardness values and was considered the most stable allotrope at ambient conditions (Wang et al. 2017; Fox and Dulay 1993). Furthermore, the basic tectonic units are triazine (C_3N_4) and tri-s-triazine/heptazine (C_6N_7) rings (Salman et al. 2019). Wang et al. (2015) also mentioned the tri-s-triazine-based g-C_3N_4 as tri-s-triazine rings that are cross-linked by trigonal nitrogen atoms, making it the most favorable and energetically stable phase at ambient conditions. The application of carbon nitride began in 2006, specifically in the field of heterogeneous catalysis (Goettmann et al. 2006). Schlo et al. (2008) also mentioned that g-C_3N_4 has a set of diverse carbon and nitrogen-rich starting compounds with a C/N molar ratio close to 0.75. Hence, the historical development of g-C_3N_4-based photocatalyst showed a huge achievement, which makes it the best material for photocatalytic application due to its unique properties.

1.3 UNIQUE PROPERTIES OF g-C_3N_4 PHOTOCATALYSTS

Accordingly, numerous studies on visible-light-driven (VLD) photocatalysts have focused not only on air pollutants and photocatalytic hydrogen production but also on wastewater treatment and disinfection (Le et al. 2020; Zhang et al. 2019). Among the emerging VLD photocatalysts, g-C_3N_4 is deemed an important catalyst because of its appropriate band gap and unique light-harvesting ability to absorb visible light of less than 450 nm (Mamba and Mishra 2016). g-C_3N_4 is composed of earth-abundant elements with strong covalent bonds between carbon and nitrogen atoms, which delocalizes conjugated structure containing graphitic stacking of carbon nitride (C_3N_4) layers that are interconnected via tertiary amines (Cao and Yu 2014; Thomas et al. 2008). Their unique properties trigger their properties of good biocompatibility, high wear resistance, good catalyst carriers, high chemical and thermal stability, and good electronic conductivity (Ismael 2020; Thomas, Fischer, Goettmann, et al. 2008). Furthermore, the stacking of layers is due to the chemical resistance from Van der Waals' forces, which makes them insoluble in toluene, diethyl ether, water, ethanol, and tetrahydrofuran (Zhou, Hou, and Chen 2018; Cheng et al. 2013).

Tectonic units constitute potential allotropes of g-C_3N_4 consisting of triazine (C_3N_3) and tri-s-triazine/heptazine (C_6N_7) rings. Meanwhile, the C_6N_7-based g-C_3N_4 that is connected to the planar amino groups becomes the most energetic and stable phase under ambient conditions, as compared with other phases (Maeda et al. 2009; Xu and Gao 2012; Shen et al. 2018; Li et al. 2015). There are several steps in the transformation of $C_3H_6N_6$ into g-C_3N_4 (Sudhaik et al. 2018; Zhang, Mori, and Ye 2012). The melamine would undergo rearrangements that lead to the formation of tri-s-triazine unit products and elimination of ammonia above 350°C. Subsequently, the polycondensation of the structure completely converts into g-C_3N_4 at approximately 520°C. Moreover, a polymeric derivative had been found by Berzelius and termed as "melon" after it undergoes the formation of melam and melem (Liebig 1834b). It can be seen that g-C_3N_4 does not have a crystalline structure and consists of only carbon and nitrogen atoms with a C/N molar ratio of 0.75 and a small amount of H.

There are various applications of g-C_3N_4, including degradation of organic pollutants, water splitting, photoreduction, organic contaminants purification, catalytic organic synthesis, and fuel cells (Wang et al. 2008). g-C_3N_4 has been demonstrated to be an excellent visible-light-responsive photocatalyst, which is inexpensive, has ease of synthesis, and is a suitable electronic structure with outstanding photocatalytic performance (Reddy et al. 2019). Furthermore, the unique characteristic of g-C_3N_4 is its visible-light-response element contributed by their band gaps (Eg = 2.7 eV; CB = −1.1eV; VB = 1.6 eV) via normal hydrogen electrode (Darkwah and Ao 2018). Moreover, it should be noted that g-C_3N_4 has the advantages of hydrophilicity, large specific surface area, inertness, and is environmentally friendly (Wu et al. 2018).

On the other hand, Wang et al. (2017) reported that g-C_3N_4 exhibits a low specific surface area, irregular morphology, and a hydrophobic surface that often results in

TABLE 1.1
Advantages and Disadvantages of $g-C_3N_4$-Based Photocatalyst

Advantages	Disadvantages
• Low-cost	• Poor conductivity
• Good biocompatibility	• Poor contact and inhomogeneity
• Unique electronic	• Poor electron transfer
• Nontoxicity	
• High water dispersibility	
• Emit visible light (400–475 nm)	
• Low biotoxicity	

limitation in its photocatalytic performance. Also, $g-C_3N_4$ limits its extensive application, such as poor utilization of long-wavelength light, low quantum efficiency, and high recombination of photoexcited electron–hole pairs (Li et al. 2016; Zhang et al. 2014). The electron transportation in $g-C_3N_4$ showed dramatically fast recombination of photogenerated charge carriers because of the restriction from their layered structure and hydrophobicity (Zhou et al. 2016). Table 1.1 illustrates the summary of advantages and disadvantages of $g-C_3N_4$-based photocatalyst that were reviewed by several studies.

1.4 SYNTHESIS METHOD OF $g-C_3N_4$ PHOTOCATALYST

The synthesis method of $g-C_3N_4$-based photocatalyst is classified into two types: "top-down" and "bottom-up", which are able to transform structure phase into bulks, sheets, lines, and dots (Chan, Liu, and Hsiao 2019). High-temperature (825°C–950°C) method becomes a "top-down" synthesis method and leads to corrosion phenomenon. This happens because of the weak force between each g-CN layer and the change of bulk g-CNs into sheet form. Meanwhile, the "bottom-up" synthesis is a hydrothermal method with the presence of carbon and nitrogen precursors, such as sodium citric acid and urea. The advantage of this process is the presence of carboxyl group from sodium citric acid and urea for the maintenance of subsequent surface functional modification.

Figure 1.1 illustrates the various methods for the synthesis of $g-C_3N_4$-based photocatalyst. Examples of their synthesis methods are chemical vapor deposition, solid-state, solvothermal, and thermal annealing of nitrogen-rich precursor (Kumar et al. 2020; Ghosh and Pal 2019; Xu et al. 2018; Thomas, Fischer, Goettmann, Antonietti, et al. 2008). Chemical vapor deposition is involved in the deposition technique, in which a substrate surface is coated with the desired nanomaterial by heat treatment accompanied by chemical reaction and precursor gases. From this method, the N/C ratio of the $g-C_3N_4$-based photocatalyst would increase until it reaches 1.0 and is illustrated in their C–N single bond (Hidekazu et al. 2007). Another method is a solid reaction, which synthesizes at low temperature, produces high nitrogen-enriched graphitic carbon (nitrogen content > 50 atomic wt%) and has an excellent hydrogen storage capacity (0.34 wt%) at room temperature under 100 bar and low BET surface area ($10\,m^2g$) (Jae et al. 2009).

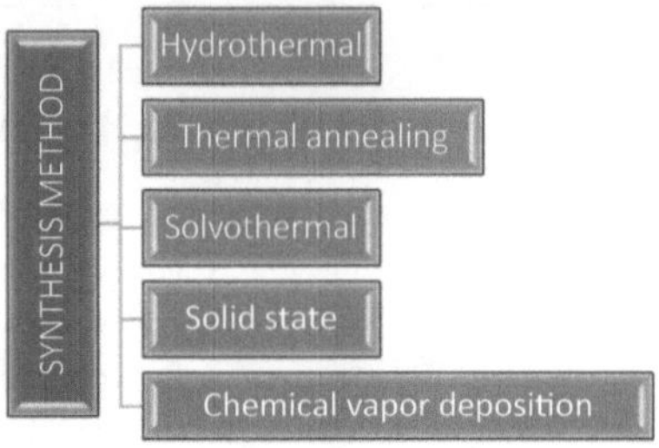

FIGURE 1.1 Types of g-C_3N_4 synthesis.

The solvothermal method that uses a solvent and high temperature is another option for synthesizing g-C_3N_4, which has several advantages, one being its easy and simple way of obtaining nanocrystalline powder. However, the synthesis of pure g-C_3N_4 is difficult due to the strong chemical affinity between hydrogen and nitrogen/carbon. Kojima and Ohfuji (2018) explained that $C_3N_5H_3$, which forms stacked layers of s-triazine ring units, showed a low degree of polymerization. The presence of high pressure and high temperature would disrupt the production, such that only hydrogen-containing $C_2N_2(NH)$ is obtained (Kojima and Ohfuji 2013). The thermal annealing of nitrogen-rich precursors, such as the calcination method, is extensively applied to the design of a two-phase contact interface (Jiang et al. 2018).

Chang et al. (2013) reported many reviews on g-C_3N_4 structure and preparation in the last few years, which has increased tremendously. g-C_3N_4 can be simply synthesized by inexpensive metal salt preparation, which is thermally poly-condensing cheap nitrogen-rich precursors and oxygen-free compounds containing C–N core structures, such as dicyanamide ($C_2H_4N_4$), cyanamide (CH_2N_2), melamine ($C_3H_6N_6$), thiourea (CH_4N_2S), and urea (CH_4N_2O) (Li et al. 2012; Thomas, Fischer, Goettmann, Antonietti, et al. 2008). Figure 1.2 shows the facile synthetic pathway to generate g-C_3N_4 including its intermediate step. It can be seen that g-C_3N_4-based photocatalyst using the calcination method can be achieved in an appropriate temperature range (Groenewolt et al. 2005; Sudhaik et al. 2018; Wang et al. 2015). It can be seen that there is a different temperature range for each precursor: dicyanamide and cyanamide (550°C), melamine (500°C–580°C), thiourea (450°C–650°C), and urea (520°C–550°C), all of which are for the generation of g-C_3N_4. Shen et al. (2018) also reported via the thermal gravimetric analysis (TGA) that g-C_3N_4 is non-volatile and is slightly unstable above 600°C due to the emergence of nitrogen and cyano fragments, and finally, its complete decomposition at 700°C.

1.5 PHOTOCATALYTIC PRINCIPLES AND MECHANISMS OVER g-C_3N_4 PHOTOCATALYSTS

In recent years, researchers have been devoted to establishing, harnessing, and enhancing the potential of unitary photocatalysts using different mechanisms. Figure 1.3 illustrates the types of photocatalytic mechanisms that have been involved in the synthesis of g-C_3N_4 photocatalyst. They are divided into two groups, namely, Z-scheme (direct, indirect, and heterojunction) and heterojunction (Type I, Type II,

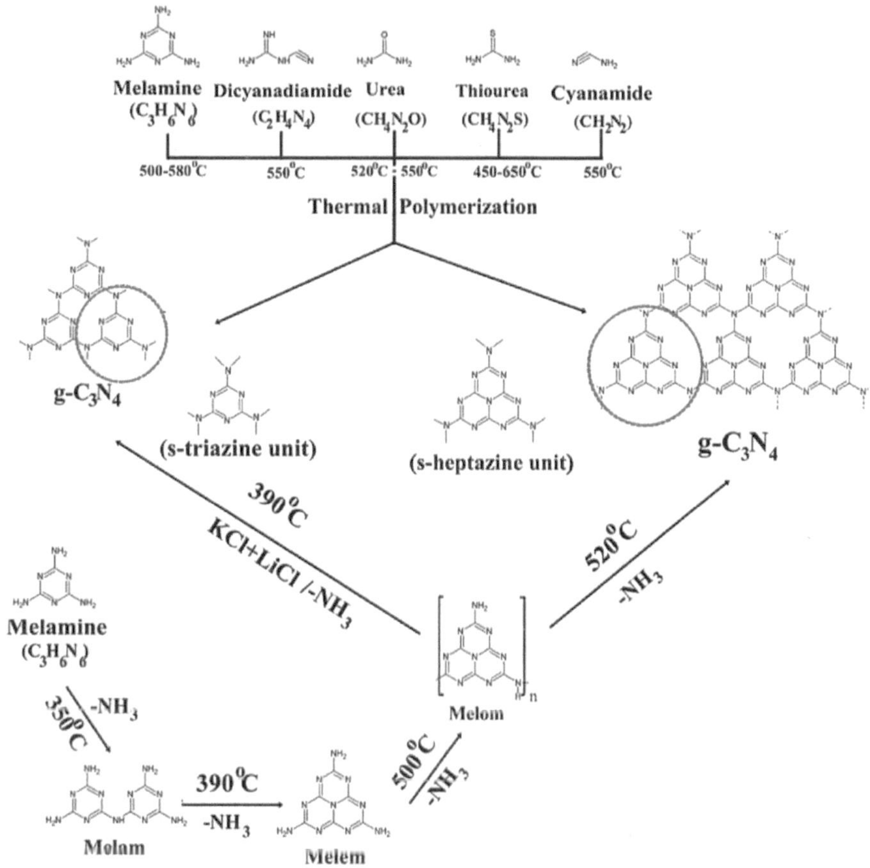

FIGURE 1.2 Thermal polymerization pathways of g-C_3N_4 (Sudhaik et al. 2018).

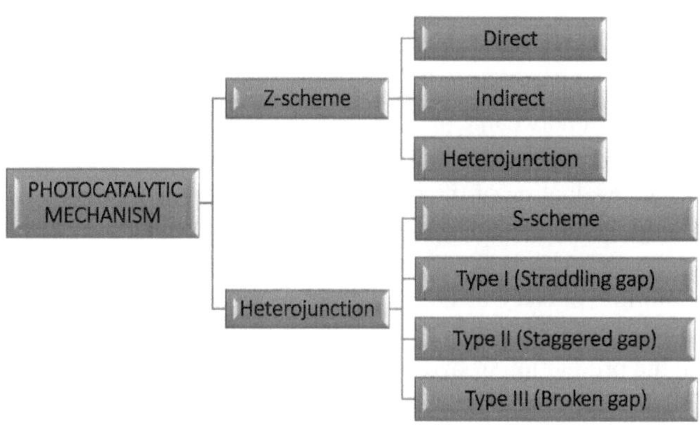

FIGURE 1.3 Schematic diagram showing the types of photocatalytic mechanism.

Type III, and S-scheme) mechanisms (Lin et al. 2021; Xu et al. 2018). The following are the brief introductions of each newly developed mechanism.

1.5.1 Z-Scheme Mechanism

According to Bard (1979), the Z-scheme is initially introduced once it forms the liquid phase by involving two semiconductors with an electron acceptor or donor pair as illustrated in Figure 1.4a. However, the disadvantage of this scheme is the redox mediator reversibility, and it is applicable only to the liquid phase. Consequently, the investigation of Z-scheme photocatalyst had been improved by replacing the electron mediator with the combination of two different semiconductors with a noble metal nanoparticle (NP) through all-solid-state Z-scheme photocatalysts (Tada et al. 2006), as shown in Figure 1.4b. In addition, the Z-scheme mechanism by two narrow band gap semiconductors could produce strong reduction and oxidation potentials, as well as increase the light-harvesting capacity (Figure 1.4c) (Ghosh and Pal 2019).

Most of the earlier studies, including the current work, have focused on the construction of Z-scheme toward metal oxides, metal sulfides, bismuth-based, and silver-based semiconductors, as well as carbon-based semiconductor. An example of a carbon-based semiconductor that is normally applied in the Z-scheme is g-C_3N_4, which increases light absorption, facilitates charge separation, promotes redox ability, and prolongs charge carrier lifetime. In comparison with other mechanisms, Z-scheme involves the transportation of photoexcited electrons from the CB to VB through an electron mediator, leaving the holes of another semiconductor with a strong redox ability. Jiang et al. (2018) summarized the criteria required by g-C_3N_4 in the Z-scheme mechanism: (a) it is a visible-light-driven photocatalyst; (b) it has high VB potential, high redox ability, and strong oxidability; and (c) it needs a strong anchoring support as electron mediator between two semiconductors. This has been previously assessed only to a very limited extent (a) because the mechanism and their electron–hole transfer process is yet to be established; (b) because of the suitable arrangement of the Fermi levels between electron mediator and photosystems; (c) because of the limitation of photoexcited electrons and holes in the heterojunction-type system under identical conditions; and (d) because of their doping, nanostructures fabrication, band gap engineering, facet control, and surface plasmon resonance (SPR) effect of noble metal.

In addition, Gu et al. (2019) reported that it is sufficient to highlight the Z-scheme mechanism between Co_9S_8 and g-C_3N_4 for the degradation of Cr(VI)/2,4-dichlorophenoxyacetic acid under visible light irradiation for 180 min, as illustrated in Figure 1.5. The evaluation of the mechanism proved that the potential of CB level has enhanced the reduction rate of Cr (VI) in all-solid-state Z-scheme systems. Furthermore, the production of active site prevented the recombination process due to the longer lifetime of electrons in the CB of Co_9S_8. This is applicable for the degradation of 2,4-D that was aided by their photoexcited holes located in the VB of g-C_3N_4, while the production of hydroxyl radical was generated during the reduction of Cr (VI) due to the photogenerated electrons in the CB of g-C_3N_4. Furthermore, the recombination of the electron–hole pairs on the surface of Co_9S_8/g-C_3N_4 could be observed for both mechanisms and in the degradation of pollutants.

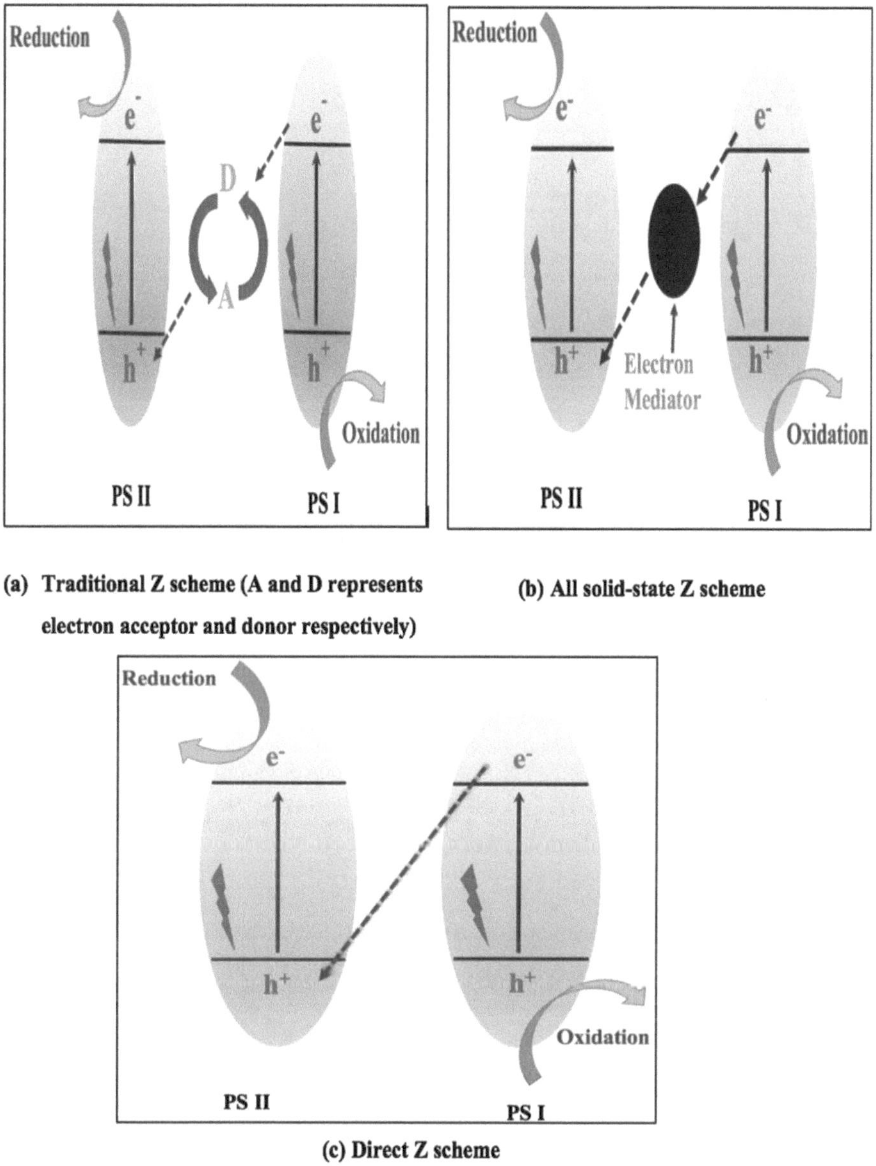

FIGURE 1.4 Schematic illustrations of (a) traditional, (b) all-solid-state, and (c) direct Z-scheme photocatalysts (Ghosh and Pal 2019).

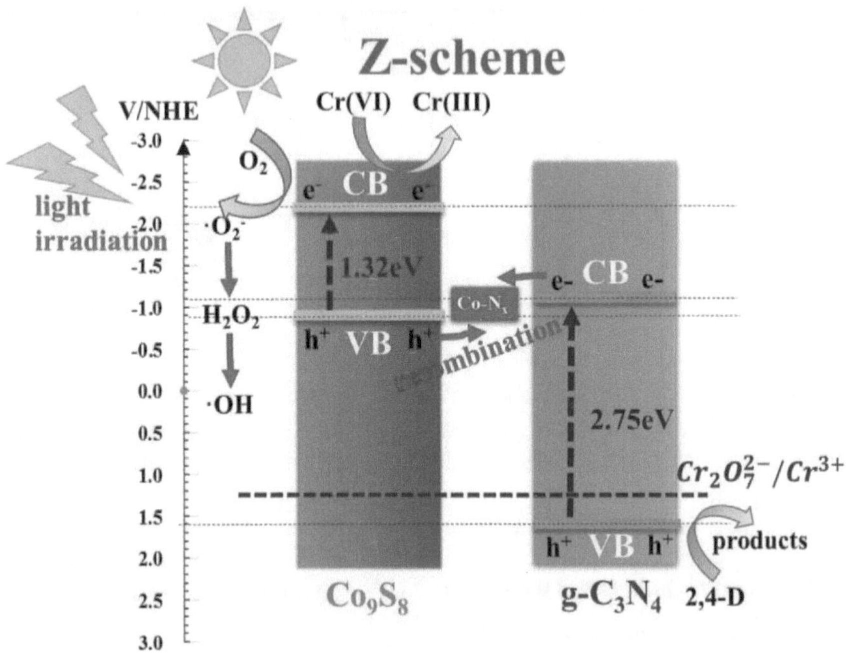

FIGURE 1.5 A schematic illustration of simultaneous reduction of Cr(VI) and degradation of 2,4-D over 15% Co_9S_8/g-C_3N_4 under simulated solar irradiation sites (Gu et al. 2019).

1.5.2 Heterojunction Mechanism

The heterojunction mechanism between g-C_3N_4 could be demonstrated in Type I, Type II, Type III, and Z/S-scheme. There are many advantages of the heterojunction mechanism, whereby it can enable the creation of an electric field from the space charge layer and enhance the separation between electrons and holes during irradiation (Shevlin, Martin, and Guo 2015).

Generally, the heterojunction mechanism involves a surface-based process that requires a combination of metal and semiconductor, which would create a high work function in metal as compared with the semiconductor (Li et al. 2020). Next, the electron transports from the semiconductor to the metal and forms a Schottky heterojunction. A Type II heterojunction would be formed if two conditions are met: (a) two photocatalysts, both n-type semiconductors with a staggered band structure, should be involved, and (b) the position of CB in photocatalyst A must be higher than the VB which provides a higher work function than photocatalyst B. The foremost problem in Type II heterojunction is the lower reduction and oxidation potentials that have a bad impact on their redox ability. Additionally, the electrostatic repulsion between electron and electron or hole and hole could influence their migration of electrons (Low et al. 2017).

Meanwhile, Z-scheme or S-scheme heterojunction has the same structure as Type II heterojunction but provide a smaller work function. Furthermore, the S-scheme heterojunction can increase the separation rate of photoinduced charges and retain the high redox ability of each component. Besides that, the mechanism that involved photocatalyst A (n-type) and photocatalyst B (p-type) is called the p–n heterojunction.

On the other hand, Wang et al. (2014) reported that a larger specific surface area, large electron-storage capacity, and the presence of metallic conductivity in $g\text{-}C_3N_4$ can create a Schottky barrier junction for boosting the photocatalytic degradation efficiency by accepting photon-excited electrons, as well as hindering the recombination. The proposed mechanism of heterojunction elucidates the enhancement of the photocatalytic properties of $C_3N_4\text{-}Cu_2O$ composites (Tian et al. 2014).

Comparatively, $g\text{-}C_3N_4$ (CB: −1.13 V; VB:1.57 V) is more typical n-type semiconductor than Cu_2O (CB: −0.7 V; VB:1.3 V), as its Fermi level is close to that of CB, while Cu_2O would be considered as a typical p-type semiconductor because its Fermi level is close to that of VB (Paracchino et al. 2011). The heterojunction mechanism cannot be constructed due to the straddling band structure of C_3N_4 and Cu_2O. However, the internal electric field was formed until the Fermi levels of C_3N_4 and Cu_2O reached equilibration. A high-energy photon excites an electron from C_3N_4 to Cu_2O and holes would diffuse from Cu_2O to C_3N_4 in their contacted interface when Cu_2O was deposited on the surface of C_3N_4. The photogenerated electrons moved to the positive field ($n\text{-}C_3N_4$) and holes moved to the negative field ($p\text{-}Cu_2O$). Moreover, the $C_3N_4\text{-}Cu_2O$ heterojunction has metallic Cu that acts as an acceptor to the photogenerated electrons from the CB of C_3N_4 and Cu_2O.

Among the heterojunction in $g\text{-}C_3N_4$, the $CoB/g\text{-}C_3N_4$ composite system represents a good example for illustrating the enhancement of photocatalytic properties that applied heterojunction mechanism. Guo et al. (2021) proposed a Schottky heterojunction mechanism to elucidate the enhancement of the photocatalytic properties of $CoB/g\text{-}C_3N_4$ composites, as shown in Figure 1.6. The $CoB/g\ C_3N_4$ composite highlights the Schottky junction that flow of electrons from CNs to CoB have different band gaps which are 4.69 and 5.34 eV, respectively. The interfacial Co-N bond

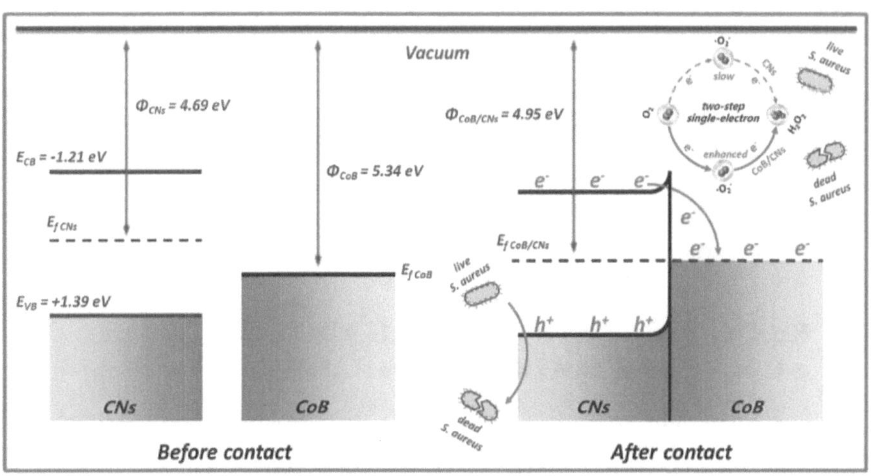

FIGURE 1.6 Schematic illustration of CoB/CNs-2 Schottky heterojunction under visible light irradiation (Guo et al. 2021).

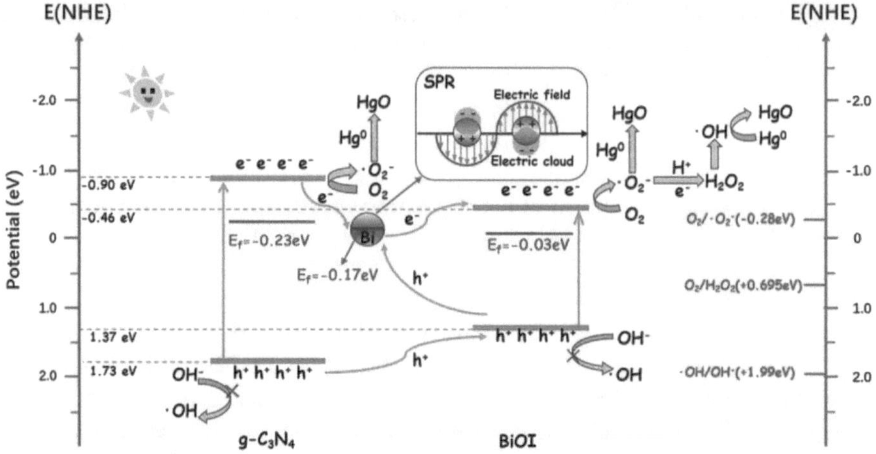

FIGURE 1.7 Mechanism of the electron transfer and photocatalysis for the indirect Z-scheme ternary 10%g-C_3N_4@Bi/BiOI under visible light irradiation (Wang et al. 2021).

has constructed high-energy electrons and holes at CNs, and the photogenerated electrons can be captured by CoB. Therefore, the existence of upward band bending helps in efficient spatial separation of photogenerated electron–hole pairs due to the inefficient electrons in CoB to flow back toward CNs. As a result, about 7×10^7 CFU mL of *Staphylococcus aureus (S. aureus)* had been inactivated after 125 min under visible light irradiation. Our results demonstrate that the ability of CoB could undergo the two-step single-electron process between O_2 molecules and reduce them to $\bullet O_2^-$ and H_2O_2.

A recent study by Wang et al. (2021) introduced the Z-scheme heterojunction with the appearance of the Schottky barrier between g-C_3N_4 and Bi/BiOI photocatalyst for mercury oxidation, as shown in Figure 1.7. The presence of metallic Bi as electron mediator is conceivable because the position of the Fermi level of g-C_3N_4 (–0.23 eV) is much more negative than that of Bi (–0.17 eV). The mechanism introduced by the authors has the advantage of Schottky barrier at the interface between g-C_3N_4 and Bi. A more comprehensive description can be found in Z-scheme heterojunction where the intrinsic surface plasmon resonance (SPR) effect could be developed by metallic Bi under visible irradiation. Besides, there are two paths involved by Bi and g-C_3N_4 which are VB of BiOI and CB of BiOI for production of holes and O_2 to form radical $\bullet O_2^-$, respectively.

1.6 RECENT PROGRESS IN THE APPLICATION OF g-C_3N_4 TOWARD VARIOUS POLLUTANTS

Data revealed significant application of g-C_3N_4-based photocatalysts, as tabulated in Table 1.2. There were various combinations of g-C_3N_4 with other photocatalysts to evaluate the degradation of various pollutants. Recent studies reviewed alternatives for improving g-C_3N_4 by doping with metal, non-metal, co-doping, and heterojunction for enhancing the light absorption, facilitating the charge separation, and

TABLE 1.2
Recent Key Advances in g-C_3N_4-Based Photocatalysts

Photocatalyst	Method	Pollutants	Precursor	Performances	Mechanism	References
BWO-OCN	Hydrothermal	Tetracycline	Dicyandiamide	0.047 min^{-1}; 60 min;4th	Z-scheme	Huo et al. (2021)
SnS/g-C_3N_4	Chemical deposition	RhB	Urea	91.8 %;90 min; 3th	Z-scheme	Jia et al. (2020)
AgBr/P-g-C_3N_4	Thermal polymerization; deposition-precipitation	Ephedrine	Melamine	99.9 %; 60 min	Z-scheme; heterojunction	Chen et al. (2020)
CdS/C_3N_4	Thermal polymerization, co-precipitation	RhB	Melamine	75%–62%; min; 5th	Z-scheme	You et al. (2021)
g-C_3N_4/CeO_2	Hydrothermal	Clofibric acid	Melamine	98.5%; 60 min; 3th	Heterojunction	Lin et al. (2021)
g-C_3N_4/Bi_2S_3/In_2S_3	Hydrothermal	RhB	Urea	90.9%; 60 min; 4th	Heterojunction	Zhao et al. (2021)
g-C_3N_4 nanosheet/$MgBi_2O_6$	Hydrothermal	MO, RhB, CR, MB, and phenol	Melamine	RhB;1337.84 × 10^{-4} min^{-1}; 120 min	Z-scheme	Nguyen et al. (2021)
Pt/g-C_3N_4/$SrTiO_3$	Calcination	Acid Red 1	Urea	93% for dye; 471 μmol $h^{-1}g^{-1}$	Z-scheme	Tan et al. (2020)
CdS/CQDs/g-C_3N_4	Calcination	RhB, MB and phenol	Melamine	RhB; 100%; 20 min	Z-scheme	Feng et al. (2020)
CTFNS/CNNS	Chemical exfoliation	Sulfamethazine	Melamine	95.8%; 180 min	Heterojunction	Cao et al. (2020)
$ZnWO_4$/g-C_3N_4	Hydrothermal	RhB and 4-chlorophenol	Melamine	99%; 100 min	Heterojunctions	Rathi, Panneerselvam, and Sathiyapriya (2020)

transportation, and prolonging the charge carrier lifetime (Ajiboye, Kuvarega, and Onwudiwe 2020; Dong et al. 2014; Liang et al. 2021; Jiang, Yuan, Pan, Liang, and Zeng 2017). As the band gap of g-C_3N_4-based photocatalysts is 2.7 eV, their CB is −1.1eV and VB is 1.6 eV; as such, their electronegativity could enhance the electron transport to another potential photocatalyst.

As a unique example, a BWO-OV/OCN composite with a black body nature and oxygen-rich structure of OCN was reported to have formed a Z-scheme due to its oxygen vacancies (Huo et al. 2021). Similarly, Jia et al. (2020) studied the migration of electrons from g-C_3N_4 from their VB would recombine with the holes and their excess of electrons on their CB in SnS/g-C_3N_4 hybrids for reduction of dissolved oxygen. Besides that, the analysis from VB-XPS and Mott–Schottky proved that the combination of g-C_3N_4 with $MgBi_2O_6$ would improve the lifetime of the photoinduced charges, as well as their photo-ability (Nguyen et al. 2021). Feng et al. (2020) mentioned that the holes in the VB of the UCN could construct the Z-scheme mechanism with STO photocatalyst (Feng et al. 2020). However, in line with the Z-scheme mechanism, it can be concluded that VB of g-C_3N_4 could generate electrons, as well as holes.

When comparing Z-scheme to heterojunction mechanisms in g-C_3N_4, it must be highlighted that the electronegativity of g-C_3N_4 could ease their photogenerated electrons to move onto the CB of CeO_2 in the heterojunction of g-C_3N_4/CeO_2 (Lin et al. 2021). The same mechanism had been involved in g-C_3N_4/Bi_2S_3/In_2S_3, and the electron at CB of g-C_3N_4 would receive the electron from CBs In_2S_3 and Bi_2S_3 as well as the transferring hole from VB of g-C_3N_4 to the VBs of In_2S_3 and Bi_2S_3 (Zhao et al. 2021). Furthermore, Rathi, Panneerselvam, and Sathiyapriya (2020) also studied the generation electrostatic induction from different charges where g-C_3N_4 (negative) and $ZnWO_4$ (positive) would form heterojunction after reaching the electrical equilibrium. Cao et al. (2020) concluded that the requirement of a large interface contact area and short charge transport distance in the heterojunction mechanism would greatly enhance the transfer of excited-state electrons. This can be done by surface modification between carbon-based materials and their functional groups that constructed a homogeneous heterojunction. Overall, these findings are in accordance with Z-scheme of g-C_3N_4 to those heterojunction mechanisms should be considered the electronegativity of g-C_3N_4.

1.7 CONCLUSION AND CHALLENGES

This chapter has shown important findings that are imperative for understanding the effect of g-C_3N_4-based photocatalyst on visible light-driven photocatalytic applications. The g-C_3N_4-based photocatalyst offers an abundant scope for designing novel photocatalysts, which gives rise to a wide range of applications. Although there are some reports on the photocatalytic activity of g-C_3N_4-based photocatalyst, detailed studies on these materials remain limited. More efforts are needed to understand the fundamentals through their unique properties, as well as their synthesis methods. Among the photocatalytic mechanism, Z-scheme and heterojunction are normally applied for g-C_3N_4-based photocatalyst. The strategy of adopting the Z-scheme and heterojunction in narrow band gap photocatalysts is widely used to induce visible light absorption and subsequently, to enable photocatalytic activities. However, very limited knowledge is available on the g-C_3N_4-based photocatalyst on their photophysical properties of the compounds; this should be properly investigated in the future. Research works on g-C_3N_4-based photocatalysts show that these photocatalysts offer distinct advantages and disadvantages for wastewater treatment.

ACKNOWLEDGMENTS

The authors would like to thank the Ministry of Education and Universiti Teknologi Malaysia for the financial support provided under Fundamental Research Grant Scheme (Project Number: R.J130000.7851.5F369) and UTM High Impact Research Grant (Project Number: Q.J130000.2451.08G36) in completing this work. N. A. M. Razali would like to acknowledge the support from Universiti Teknologi Malaysia for the ZAMALAH scholarship.

REFERENCES

Ajiboye, Timothy O, Alex T Kuvarega, and Damian C Onwudiwe. 2020. "Graphitic Carbon Nitride-Based Catalysts and Their Applications : A Review." *Nano-Structures & Nano-Objects* 24: 100577. doi:10.1016/j.nanoso.2020.100577.

Bard, Allen J. 1979. "Photoelectrochemistry and Heterogenous Photo-Catalysis at Semiconductor." *Journal of Photochemistry* 10 (August): 59–75.

Cao, Shaowen, and Jiaguo Yu. 2014. "G-C_3N_4-Based Photocatalysts for Hydrogen Generation." *The Journal of Physical Chemistry Letter* 5: 2101–2107.

Cao, Shihai, Yu Zhang, Nannan He, Jing Wang, Huan Chen, and Fang Jiang. 2020. "Metal-Free 2D/2D Heterojunction of Covalent Triazine-Based Frameworks/Graphitic Carbon Nitride with Enhanced Interfacial Charge Separation for Highly Efficient Photocatalytic Elimination of Antibiotic Pollutants." *Journal of Hazardous Materials* 391 (October): 122204. doi:10.1016/j.jhazmat.2020.122204.

Chan, Ming-Hsien, Ru-Shi Liu, and Michael Hsiao. 2019. "Graphitic Carbon Nitride-Based Nanocomposites and Their Biological Applications: A Review." *Nanoscale* 11: 14993–15003. doi:10.1039/c9nr04568f.

Chang, Fei, Yunchao Xie, Chenlu Li, Juan Chen, Jieru Luo, Xuefeng Hu, and Jiaowen Shen. 2013. "A Facile Modification of G-C_3N_4 with Enhanced Photocatalytic Activity for Degradation of Methylene Blue." *Applied Surface Science* 280: 967–974.

Chen, Miao, Changsheng Guo, Song Hou, Jiapei Lv, Yan Zhang, Heng Zhang, and Jian Xu. 2020. "A Novel Z-Scheme AgBr/P-g-C3N4 Heterojunction Photocatalyst: Excellent Photocatalytic Performance and Photocatalytic Mechanism for Ephedrine Degradation." *Applied Catalysis B: Environmental* 266 (January): 118614. doi:10.1016/j.apcatb.2020.118614.

Chen, Zhuo, Huu Hao Ngo, and Wenshan Guo. 2012. "A Critical Review on Sustainability Assessment of Recycled Water Schemes." *Science of the Total Environment* 426: 13–31. doi:10.1016/j.scitotenv.2012.03.055.

Cheng, Alfred, Hon Huan, Tze Chien Sum, Rajeev Ahuja, and Zhong Chen. 2013. "Defect Engineered G-C_3N_4 for Efficient Visible Light Photocatalytic Hydrogen Production." *ACS Applied Materials & Interfaces* 5 (C): 10317–10324.

Chong, Meng Nan, Bo Jin, Christopher WK Chow, and Chris Saint. 2010. "Recent Developments in Photocatalytic Water Treatment Technology: A Review." *Water Research.* doi:10.1016/j.watres.2010.02.039.

Crini, Grégorio, and Eric Lichtfouse. 2019. "Advantages and Disadvantages of Techniques Used for Wastewater Treatment." *Environmental Chemistry Letters* 17 (1): 145–155. doi:10.1007/s10311-018-0785-9.

Darkwah, Williams Kweku, and Yanhui Ao. 2018. "Mini Review on the Structure and Properties (Photocatalysis), and Preparation Techniques of Graphitic Carbon Nitride Nano-Based Particle, and Its Applications." *Nanoscale Research Letters* 13. Nanoscale Research Letters. doi:10.1186/s11671-018-2702–3.

Dong, Guoping, Yuanhao Zhang, Qiwen Pan, and Jianrong Qiu. 2014. "A Fantastic Graphitic Carbon Nitride (g-C_3N_4) Material: Electronic Structure, Photocatalytic and Photoelectronic Properties." *Journal of Photochemistry and Photobiology C: Photochemistry Reviews* 20: 33–50.

Fakhrul, Mohamad, Ridhwan Samsudin, and Nurfatien Bacho. 2018. "Recent Development of Graphitic Carbon Nitride-Based Photocatalyst for Environmental Pollution Remediation." *Nanocatalysts* (2003): 1–15.

Feng, Shuting, Tian Chen, Zhichao Liu, Jianhui Shi, Xiuping Yue, and Yuzhen Li. 2020. "Z-Scheme CdS/CQDs/g-C3N4 Composites with Visible-near-Infrared Light Response for Efficient Photocatalytic Organic Pollutant Degradation." *Science of the Total Environment* 704: 135404. doi:10.1016/j.scitotenv.2019.135404.

Fox, MA, and MT Dulay. 1993. "Heterogenous Photocatalysis." *Chemical Reviews* 93: 341–357.

Fujishima, Akira, and Kenichi Honda. 1972. "Electrochemical Photolysis of Water at a Semiconductor Electrode." *Nature* 238: 37–38.

Gasperi, J, C Sebastian, V Ruban, M Delamain, S Percot, L Wiest, C Mirande, et al. 2014. "Micropollutants in Urban Stormwater: Occurrence, Concentrations, and Atmospheric Contributions for a Wide Range of Contaminants in Three French Catchments." *Environmental Science and Pollution Research* 21 (8): 5267–5281. doi:10.1007/s11356-013-2396-0.

Ghosh, Utpal, and Anjali Pal. 2019a. "Graphitic Carbon Nitride Based Z Scheme Photocatalysts: Design Considerations, Synthesis, Characterization and Applications." *Journal of Industrial and Engineering Chemistry* 79: 383–408. The Korean Society of Industrial and Engineering Chemistry. doi:10.1016/j.jiec.2019.07.014.

Goettmann, Fischer F., M. A., Antonietti, and A. Thomas. 2006. "Metal-Free Catalysis of Sustainable Friedel–Crafts Reactions: Direct Activation of Benzene by Carbon Nitrides to Avoid the Use of Metal Chlorides and Halogenated Compounds." *Chemical Communications* 43: 4530–4532.

Groenewolt, By Matthijs, and Markus Antonietti. 2005. "Synthesis of G-C_3N_4 Nanoparticles in Mesoporous Silica Host Matrices." *Advance Materials* 17 (14): 1789–1792. doi:10.1002/adma.200401756.

Gu, Jiayu, Huan Chen, Fang Jiang, Xin Wang, and Liyuan Li. 2019. "All-Solid-State Z-Scheme Co_9S_8/Graphitic Carbon Nitride Photocatalysts for Simultaneous Reduction of Cr(VI) and Oxidation of 2,4-Dichlorophenoxyacetic Acid under Simulated Solar Irradiation." *Chemical Engineering Journal* 360 (September): 1188–1198. doi:10.1016/j.cej.2018.10.137.

Guo, Hai, Cheng Gang Niu, Ya Ya Yang, Chao Liang, Huai Yuan Niu, Hui Yun Liu, Lu Li, and Ning Tang. 2021. "Interfacial Co-N Bond Bridged CoB/g-C3N4 Schottky Junction with Modulated Charge Transfer Dynamics for Highly Efficient Photocatalytic Staphylococcus Aureus Inactivation." *Chemical Engineering Journal* 422 (April). doi:10.1016/j.cej.2021.130029.

Hanjra, Munir A, and M Ejaz Qureshi. 2010. "Global Water Crisis and Future Food Security in an Era of Climate Change." *Food Policy* 35 (5): 365–377. doi:10.1016/j.foodpol.2010.05.006.

Hayashi, Ayami, Keigo Akimoto, and Toshimasa Tomoda. 2013. "Global Evaluation of the Effects of Agriculture and Water Management Adaptations on the Water-Stressed Population." *Mitig Adapt Strateg Glob Change* 18: 591–618. doi:10.1007/s11027-012-9377-3.

Hidekazu, T., M. Sougawa, K. Takarabe, S. Sato, and O. Ariyada. 2007. "Use of Nitrogen Atmospheric Pressure Plasma for Synthesizing Carbon Nitride." *Japanese Journal of Applied Physics* 46 (4R): 1596.

Huang, Danlian, Xuelei Yan, Ming Yan, Guangming Zeng, Chengyun Zhou, and Jia Wan. 2018. "Graphitic Carbon Nitride-Based Heterojunction Photoactive Nanocomposites : Applications and Mechanism Insight." *ACS Applied Materials & Interfaces* 10: 21035–21055. doi:10.1021/acsami.8b03620.

Huo, Xiuqin, Yang Yang, Qiuya Niu, Yuan Zhu, Guangming Zeng, Cui Lai, Huan Yi, et al. 2021. "A Direct Z-Scheme Oxygen Vacant BWO/Oxygen-Enriched Graphitic Carbon Nitride Polymer Heterojunction with Enhanced Photocatalytic Activity." *Chemical Engineering Journal* 403 (June): 126363. doi:10.1016/j.cej.2020.126363.

Ibrahim, Umar, and Abdul Halim. 2008. "Heterogeneous Photocatalytic Degradation of Organic Contaminants over Titanium Dioxide : A Review of Fundamentals, Progress and Problems." *Journal of Photochemistry and Photobiology C : Photochemistry Reviews* 9: 1–12. doi:10.1016/j.jphotochemrev.2007.12.003.

Ismael, Mohammed. 2020. "A Review on Graphitic Carbon Nitride (g-C_3N_4) Based Nanocomposites : Synthesis, Categories, and Their Application in Photocatalysis." *Journal of Alloys and Compounds* 846: 156446. doi:10.1016/j.jallcom.2020.156446.

Jae, Seung, Jung Hyun, Gyu Hwan, Kee Suk, and Chong Rae. 2009. "Easy Synthesis of Highly Nitrogen-Enriched Graphitic Carbon with a High Hydrogen Storage Capacity at Room Temperature." *Carbon* 47 (6): 1585–1591. doi:10.1016/j.carbon.2009.02.010.

Jia, Tiekun, Fang Fu, Jili Li, Zhao Deng, Fei Long, Dongsheng Yu, Qi Cui, and Weimin Wang. 2020. "Rational Construction of Direct Z-Scheme SnS/g-C3N4 Hybrid Photocatalyst for Significant Enhancement of Visible-Light Photocatalytic Activity." *Applied Surface Science* 499 (September): 143941. doi:10.1016/j.apsusc.2019.143941.

Jiang, Longbo, Xingzhong Yuan, Yang Pan, Jie Liang, and Guangming Zeng. 2017. "Doping of Graphitic Carbon Nitride for Photocatalysis : A Reveiw." *Applied Catalysis B : Environmental* 217: 388–406.

Jiang, Longbo, Xingzhong Yuan, Yang Pan, Jie Liang, Guangming Zeng, Zhibin Wu, and Hou Wang. 2017. "Doping of Graphitic Carbon Nitride for Photocatalysis: A Reveiw." *Applied Catalysis B: Environmental* 217: 388–406. doi:10.1016/j.apcatb.2017.06.003.

Jiang, Longbo, Xingzhong Yuan, Guangming Zeng, Jie Liang, Zhibin Wu, and Hou Wang. 2018. "Construction of an All-Solid-State Z-Scheme Photocatalyst Based on Graphite Carbon Nitride and Its Enhancement to Catalytic Activity." *Environmental Science: Nano* 5 (3): 599–615. doi:10.1039/c7en01031a.

Kanyanta, Valentine. 2016. Microstructure-Property Correlations for Hard, Superhard, and Ultrahard Materials, 1–239. Springer First Edition. doi:10.1007/978--3--319--29291--5.

Kavitha, R, PM Nithya, and S Girish Kumar. 2020. "Noble Metal Deposited Graphitic Carbon Nitride Based Heterojunction Photocatalysts." *Applied Surface Science* 508 (November): 145142. doi:10.1016/j.apsusc.2019.145142.

Kojima, Yohei, and Hiroaki Ohfuji. 2013. "Structure and Stability of Carbon Nitride under High Pressure and High Temperature up to 125 GPa and 3000 K." *Diamond & Related Materials* 39. Elsevier B.V.: 1–7. doi:10.1016/j.diamond.2013.07.006.

Kojima, Yohei, and Hiroaki Ohfuji. 2018. "Reexamination of Solvothermal Synthesis of Layered Carbon Nitride." *Journal of Materials* (2).

Kumar, Abhinandan, Pankaj Raizada, Pardeep Singh, Reena V Saini, Adesh K Saini, and Ahmad Hosseini-Bandegharaei. 2020. "Perspective and Status of Polymeric Graphitic Carbon Nitride Based Z-Scheme Photocatalytic Systems for Sustainable Photocatalytic Water Purification." Chemical Engineering Journal 391. doi:10.1016/j.cej.2019.123496.

Kumar, Santosh, Sekar Karthikeyan, and Adam F Lee. 2018. "G-C3N4-Based Nanomaterials for Visible Light-Driven Photocatalysis." *Catalysts* 8 (74). doi:10.3390/catal8020074.

Le, Yao, Lifang Qi, Chao Wang, and Shaoxian Song. 2020. "Hierarchical Pt/WO_3 Nanoflakes Assembled Hollow Microspheres for Room-Temperature Formaldehyde Oxidation Activity." *Applied Surface Science* 512 (February). doi:10.1016/j.apsusc.2020.145763.

Li, Jianghua, Biao Shen, Zhenhua Hong, Bizhou Lin, Bifen Gaoa and Yilin Chen. 2012. "A Facile Approach to Synthesize Novel Oxygen-Doped g-C3N4 with Superior Visible-Light Photoreactivity." *Chemical Communications* 48: 12017–12019.

Li, Shuna, Guohui Dong, Reshalaiti Hailili, Liping Yang, Yingxuan Li, Fu Wang, Yubin Zeng, and Chuanyi Wang. 2016. "Effective Photocatalytic H_2O_2 Production under Visible Light Irradiation at g-C_3N_4 Modulated by Carbon Vacancies." *Applied Catalysis B:Environmental* 190: 26–35. doi:10.1016/j.apcatb.2016.03.004.

Li, Xiaobo, Gareth Hartley, Antony J. Ward, Pamela A. Young, Anthony F. Masters, and Thomas Maschmeyer. 2015. "Hydrogenated Defects in Graphitic Carbon Nitride Nanosheets for Improved Photocatalytic Hydrogen Evolution." *Journal of Physical Chemistry C* 119 (27): 14938–14946. doi:10.1021/acs.jpcc.5b03538.

Li, Yunfeng, Minghua Zhou, Bei Cheng, and Yan Shao. 2020. "Recent Advances in G-C3N4-Based Heterojunction Photocatalysts." *Journal of Materials Science & Technology* 56: 1–17. doi:10.1016/j.jmst.2020.04.028.

Liang, Qinghua, Binbin Shao, Shehua Tong, Zhifeng Liu, Lin Tang, Yang Liu, Min Cheng, et al. 2021. "Recent Advances of Melamine Self-Assembled Graphitic Carbon Nitride-Based Materials : Design, Synthesis and Application in Energy and Environment." *Chemical Engineering Journal* 405 (August): 126951. doi:10.1016/j.cej.2020.126951.

Liebig, J. 1834a. "Analyse Der Harnsäure, Ann." *Archiv der Pharmazie* 10: 47–48.

Liebig, JV. 1834b. "About Some Nitrogen Compounds." *Annales pharmaceutiques françaises* 10 (10): 10.

Lin, Heng, Xin Tang, Jing Wang, Qingyuan Zeng, Hanxiao Chen, Wei Ren, Jie Sun, and Hui Zhang. 2021. "Enhanced Visible-Light Photocatalysis of Clofibric Acid Using Graphitic Carbon Nitride Modified by Cerium Oxide Nanoparticles." *Journal of Hazardous Materials* 405 (January): 124204. doi:10.1016/j.jhazmat.2020.124204.

Lin, Jingkai, Wenjie Tian, Huayang Zhang, Xiaoguang Duan, Hongqi Sun, and Shaobin Wang. 2021. "Graphitic Carbon Nitride-Based Z-Scheme Structure for Photocatalytic CO_2 Reduction." *Energy and Fuels* 35 (1): 7–24. doi:10.1021/acs.energyfuels.0c03048.

Liu, AY, and ML Cohen. 1989. "Prediction of New Low Compressibility Solids." *Science* 245: 841–842.

Liu, Bailin, Xinwang Ma, Shiwei Ai, Saiyong Zhu, and Wenya Zhang. 2016. "Spatial Distribution and Source Identification of Heavy Metals in Soils under Different Land Uses in a Sewage Irrigation Region, Northwest China." *Journal of Soils and Sediments* 16: 1547–1556. doi:10.1007/s11368-016-1351-3.

Low, Jingxiang, Jiaguo Yu, Mietek Jaroniec, Swelm Wageh, and Ahmed A Al-ghamdi. 2017. "Heterojunction Photocatalysts." *Advanced Materials* 1601694: 1–20. doi:10.1002/adma.201601694.

Maeda, Kazuhiko, Xinchen Wang, Yasushi Nishihara, Daling Lu, Markus Antonietti, and Kazunari Domen. 2009. "Photocatalytic Activities of Graphitic Carbon Nitride Powder for Water Reduction and Oxidation under Visible Light." *Journal of Physical Chemistry A* 113: 4940–4947.

Malik, Ritu, and Vijay K Tomer. 2021. "State-of-the-Art Review of Morphological Advancements in Graphitic Carbon Nitride (g-CN) for Sustainable Hydrogen Production." *Renewable and Sustainable Energy Reviews* 135 (August): 110235. doi:10.1016/j.rser.2020.110235.

Mamba, G, and AK Mishra. 2016. "Graphitic Carbon Nitride (g-C_3N_4) Nanocomposites : A New and Exciting Generation of Visible Light Driven Photocatalysts for Environmental Pollution Remediation." *Applied Catalysis B, Environmental* 198: 347–377. doi:10.1016/j.apcatb.2016.05.052.

Mao, Guozhu, Haoqiong Hu, Xi Liu, John Crittenden, and Ning Huang. 2021. "A Bibliometric Analysis of Industrial Wastewater Treatments from 1998." *Environmental Pollution* 275: 115785. doi:10.1016/j.envpol.2020.115785.

Mishra, Binaya Kumar, Pankaj Kumar, Chitresh Saraswat, Shamik Chakraborty, and Arjun Gautam. 2021. "Water Security in a Changing Environment : Concept, Challenges and Solutions." *Water* 13 (490). https://doi.org/10.3390/w13040490

Nguyen, Van Huy, Mitra Mousavi, Jahan B Ghasemi, Seyed Ali Delbari, Quyet Van Le, Abbas Sabahi Namini, Mehdi Shahedi Asl, Mohammadreza Shokouhimehr, Yashar Azizian-Kalandaragh, and Mohsen Mohammadi. 2021. "Z-Scheme g-C3N4 Nanosheet/MgBi2O6 Systems with the Visible Light Response for Impressive Photocatalytic Organic Contaminants Degradation." *Journal of Photochemistry and Photobiology A: Chemistry* 406 (August): 113023. doi:10.1016/j.jphotochem.2020.113023.

Ong, Wee Jun. 2017. "2D/2D Graphitic Carbon Nitride (g-C3N4) Heterojunction Nanocomposites for Photocatalysis: Why Does Face-to-Face Interface Matter?" *Frontiers in Materials* 4 (April): 1–10. doi:10.3389/fmats.2017.00011.

Paracchino, A, V Laporte, K Sivula, M Grätzel, and E Thimsen. 2011. "Highly Active Oxide Photocathode for Photoelectrochemical Water Reduction." *Nature Materials* 10 (6): 456–461.

Raghava, Kakarla, CH Venkata, Mallikarjuna N Nadagouda, Nagaraj P Shetti, Shim Jaesool, and Tejraj M Aminabhavi. 2019. "Polymeric Graphitic Carbon Nitride (g-C3N4) -Based Semiconducting Nanostructured Materials : Synthesis Methods, Properties and Photocatalytic Applications." *Journal of Environmental Management* 238 (March): 25–40. doi:10.1016/j.jenvman.2019.02.075.

Rathi, V, A. Panneerselvam, and R Sathiyapriya. 2020. "Graphitic Carbon Nitride (g-C3N4) Decorated ZnWO4 Heterojunctions Architecture Synthesis, Characterization and Photocatalytic Activity Evaluation." *Diamond and Related Materials* 108 (June): 107981. doi:10.1016/j.diamond.2020.107981.

Reddy, Kakarla Raghava, CH Venkata Reddy, Mallikarjuna N. Nadagouda, Nagaraj P. Shetti, Shim Jaesool, and Tejraj M. Aminabhavi. 2019. "Polymeric Graphitic Carbon Nitride (g-C3N4)-Based Semiconducting Nanostructured Materials: Synthesis Methods, Properties and Photocatalytic Applications." *Journal of Environmental Management* 238 (March): 25–40. doi:10.1016/j.jenvman.2019.02.075.

Reghizzi, OC. 2021. "Chapter 19-Reducing Pollution from Industrial Wastewater in Developing and Emerging Countries." In *Sustainable Industrial Water Use: Perspectives, Incentives, and Tools*, 216–223. IWA Publishing.

Salman, Muhammad, Guorui Yang, Iqra Ayub, Silan Wang, and Ling Wang. 2019. "Recent Development in Graphitic Carbon Nitride Based Photocatalysis for Hydrogen Generation." *Applied Catalysis B : Environmental* 257 (March). doi:10.1016/j.apcatb.2019.117855.

Shen, Rongchen, Jun Xie, Pinyu Guo, Leshi Chen, Xiaobo Chen, and Xin Li. 2018. "Bridging the G-C_3N_4 Nanosheets and Robust CuS Cocatalysts by Metallic Acetylene Black Interface Mediators for Active and Durable Photocatalytic H_2 Production." *ACS Applied Energy Materials* 1 (5): 2232–2241. doi:10.1021/acsaem.8b00311.

Shevlin, Stephen, David Martin, and Z Xiao Guo. 2015. "Visible-Light Driven Heterojunction Photocatalysts for Water Splitting-a Critical Review." *Energy & Environmental Science*, March. Royal Society of Chemistry. doi:10.1039/C4EE03271C.

Sudhaik, Anita, Pankaj Raizada, Pooja Shandilya, Dae-yong Jeong, and Ji-ho Lim. 2018. "Review on Fabrication of Graphitic Carbon Nitride Based Efficient Nanocomposites for Photodegradation of Aqueous Phase Organic Pollutants." *Journal of Industrial and Engineering Chemistry* 67: 28–51. The Korean Society of Industrial and Engineering Chemistry. doi:10.1016/j.jiec.2018.07.007.

Tada, Hiroaki, Tomohiro Mitsui, Tomokazu Kiyonaga, Tomoki Akita, and Koji Tanaka. 2006. "All-Solid-State Z-Scheme in CdS-Au-TiO2 Three-Component Nanojunction System." *Nature Materials* 5 (10): 782–786. doi:10.1038/nmat1734.

Tan, Chung En, Ju Ting Lee, En Chin Su, and Ming Yen Wey. 2020. "Facile Approach for Z-Scheme Type Pt/g-C3N4/SrTiO3 Heterojunction Semiconductor Synthesis via Low-Temperature Process for Simultaneous Dyes Degradation and Hydrogen Production." *International Journal of Hydrogen Energy* 45 (24): 13330–13339. doi:10.1016/j.ijhydene.2020.03.034.

Teter, D.M., and R.J. Hemley. 1996. "Low-Compressibility Carbon Nitrides." *Science* 271: 53–55.

Thomas, Arne, Anna Fischer, Frederic Goettmann, Markus Antonietti, Jens-Oliver Muller, Robert Schlogl, and Johan M Carlsson. 2008. "Graphitic Carbon Nitride Materials : Variation of Structure and Morphology and Their Use as Metal-Free Catalysts." *Journal of Materials Chemistry* 18: 4893–4908. doi:10.1039/b800274f.

Tian, Yanlong, Binbin Chang, Jie Fu, Baocheng Zhou, Jiyang Liu, Fengna Xi, and Xiaoping Dong. 2014. "Graphitic Carbon Nitride/Cu2O Heterojunctions: Preparation, Characterization, and Enhanced Photocatalytic Activity under Visible Light." *Journal of Solid State Chemistry* 212: 1–6. doi:10.1016/j.jssc.2014.01.011.

Tong, H., S. Ouyang, Y. Bi, N. Umezawa, M. Oshikiri, and J Ye. 2012. "Nano-Photocatalytic Materials: Possibilities and Challenges." *Advanced Materials* 24: 229–251.

Wang, Aiwu, Chundong Wang, Li Fu, and Winnie Wong-ng Yucheng. 2017. "Recent Advances of Graphitic Carbon Nitride-Based Structures and Applications in Catalyst, Sensing, Imaging and LEDs." *Nano-Micro Letters* 9 (47). doi:10.1007/s40820-017-0148-2.

Wang, Huanli, Lisha Zhang, Zhigang Chen, Junqing Hu, Shijie Li, Zhaohui Wang, Jianshe Liu, and Xinchen Wang. 2014. "Semiconductor Heterojunction Photocatalysts: Design, Construction, and Photocatalytic Performances." *Chemical Society Reviews* 43 (15): 5234–5244. doi:10.1039/c4cs00126e.

Wang, Run, Jiang Wu, Xu Mao, Jianmin Wang, Qizhen Liu, Yongfeng Qi, Ping He, Xuemei Qi, Guolong Liu, and Yu Guan. 2021. "Bi Spheres Decorated G-C3N4/BiOI Z-Scheme Heterojunction with SPR Effect for Efficient Photocatalytic Removal Elemental Mercury." *Applied Surface Science* 556 (November): 149804. doi:10.1016/j.apsusc.2021.149804.

Wang, Wei, Jiaojiao Fang, Shaofeng Shao, Min Lai, and Chunhua Lu. 2017. "Compact and Uniform TiO_2@g-C_3N_4 Core-Shell Quantum Heterojunction for Photocatalytic Degradation of Tetracycline Antibiotics." *Applied Catalysis B: Environmental* 217: 57–64. doi:10.1016/j.apcatb.2017.05.037.

Wang, X, K Maeda, A Thomas, K Takanabe, G Xin, JM Carlsson, K Domen, and M Antonietti. 2008. "A Metal-Free Polymeric Photocatalyst for Hydrogen Production from Water under Visible Light." *Nature Materials*, 8 (1), 76–80.

Wang, X, Yun Zheng, Lihua Lin, Bo Wang, and Xinchen Wang. 2015. "Graphitic Carbon Nitride Polymers toward Sustainable Photoredox Catalysis." *Angewandte Chemical International Edition* 54 (44): 12868–12884. doi:10.1002/anie.201501788.

World, United Nations. 2018. "UN World Water Development Report." *UN World Water Development Report.*

Wu, Yan, Hou Wang, Wenguang Tu, Yue Liu, Yong Zen Tan, Xingzhong Yuan, and Jia Wei Chew. 2018. "Quasi-Polymeric Construction of Stable Perovskite-Type $LaFeO_3$/g-C_3N_4 Heterostructured Photocatalyst for Improved Z-Scheme Photocatalytic Activity via Solid p-n Heterojunction Interfacial Effect." *Journal of Hazardous Materials* 347 (September): 412–422. doi:10.1016/j.jhazmat.2018.01.025.

Xu, Bentuo, Mohammad Boshir, John L. Zhou, Ali Altaee, Gang Xu, and Minghong Wu. 2018. "Graphitic Carbon Nitride Based Nanocomposites for the Photocatalysis of Organic Contaminants under Visible Irradiation : Progress, Limitations and Future Directions." *Science of the Total Environment* 633: 546–559. doi:10.1016/j.scitotenv.2018.03.206.

Xu, Quanlong, Liuyang Zhang, Jiaguo Yu, Swelm Wageh, Ahmed A Al-ghamdi, and Mietek Jaroniec. 2018. "Direct Z-Scheme Photocatalysts : Principles, Synthesis, and Applications." *Materials Today* 21 (10): 1042–1063. doi:10.1016/j.mattod.2018.04.008.

Xu, Yuan, and Shang-peng Gao. 2012. "Band Gap of C3N4 in the GW Approximation." *International of Journal Hydrogen Energy* 7: 8. doi:10.1016/j.ijhydene.2012.04.138.

You, Junhua, Wanting Bao, Lu Wang, Aiguo Yan, and Rui Guo. 2021. "Preparation, Visible Light-Driven Photocatalytic Activity, and Mechanism of Multiphase CdS/C3N4 Inorganic-Organic Hybrid Heterojunction." *Journal of Alloys and Compounds* 866: 158921. doi:10.1016/j.jallcom.2021.158921.

Zhang, Fubao, Xianming Wang, Haonan Liu, Chunli Liu, Yong Wan, and Yunze Long. 2019. "Recent Advances and Applications of Semiconductor Photocatalytic Technology." *Applied Science* 9: 2489. doi:10.3390/app9122489.

Zhang, Mo, Jing Xu, Ruilong Zong, and Yongfa Zhu. 2014. "Enhancement of Visible Light Photocatalytic Activities via Porous Structure of G-C3N4." *Applied Catalysis B: Environental* 147: 229–235. doi:10.1016/j.apcatb.2013.09.002.

Zhang, Yuanjian, Toshiyuki Mori, and Jinhua Ye. 2012. "Polymeric Carbon Nitrides : Semiconducting Properties and Emerging Applications in Photocatalysis and Photoelectrochemical Energy Conversion." *Science of Advanced Materials* 4: 282–291. doi:10.1166/sam.2012.1283.

Zhao, Teng, Xiaofeng Zhu, Yufan Huang, and Zijun Wang. 2021. "One-Step Hydrothermal Synthesis of a Ternary Heterojunction g-C3N4/Bi2S3/In2S3 Photocatalyst and Its Enhanced Photocatalytic Performance." *RSC Advances* 11 (17): 9788–9796. doi:10.1039/d1ra00729g.

Zhou, Lingyu, Yuxing Xu, Wei Yu, Xin Guo, Shuwen Yu, Jian Zhang, and Can Li. 2016. "Ultrathin Two-Dimensional Graphitic Carbon Nitride as a Solution-Processed Cathode Interfacial Layer for Inverted Polymer Solar Cells." *Journal of Materials Chemistry A* 4 (21): 8000–8004. doi:10.1039/c6ta01894g.

Zhou, Minjie, Zhaohui Hou, and Xiaobo Chen. 2018. "The Effects of Hydrogenation on Graphitic C_3N_4 Nanosheets for Enhanced Photocatalytic Activity." *Particle and Particle Systems Characterization* 35 (1): 1–9. doi:10.1002/ppsc.201700038.

Zhu, Dandan, and Qixing Zhou. 2019. "Action and Mechanism of Semiconductor Photocatalysis on Degradation of Organic Pollutants in Water Treatment : A Review." *Environmental Nanotechnology, Monitoring & Management* 12 (June). Elsevier: 100255. doi:10.1016/j.enmm.2019.100255.

2 Metal-Organic Frameworks for Wastewater Treatment

Nur Azizah Johari, Nur Fajrina, and Norhaniza Yusof
Universiti Teknologi Malaysia

CONTENTS

DOI: 10.1201/9781003167327-3

2.1 INTRODUCTION

According to the 2020 edition of the World Water Development Report, 1.6 billion people lack access to clean water (UNESCO 2019). Moreover, to fulfill the ever-increasing demand for numerous essential commodities, the industrialization process is constantly advancing, resulting in the production of a vast volume of toxic waste. Hazardous wastewater is dumped into the environment without being treated, causing serious contamination. In recent years, various pollutants such as metal ions (Abdullah et al. 2017), pharmaceuticals and personal care products (PPCPs) (Wu et al. 2014), herbicides (Hasan and Jhung 2015), dyes (Johari et al. 2021), oil (Banerjee et al. 2012), and aromatics/organic substances (Yan et al. 2015) have been detected at a critical level in wastewater, ground and surface water. Because these impurities constitute a serious hazard to human life, appropriate separation/removal technologies are urgently needed. So far, various technologies have been developed and used to remove harmful pollutants from water, such as the advanced oxidation process (Hu et al. 2019), adsorption (Arora et al. 2019), biological oxidation (Pavithra et al. 2019), chemical treatment (Chen et al. 2017), and membrane (Johari et al. 2021). Nevertheless, several factors need to be considered in remediation technologies including the effectiveness of materials, economic value, design/operation flexibility and possible reversibility, and feasibility.

The discovery of metal-organic frameworks (MOFs) opened the door for the development of porous materials for water and wastewater treatment. MOFs have received a lot of attention in the last 10 years as promising adsorbent materials for removing contaminants from water bodies. Figure 2.1 presents the number of articles published between 2010 and 2020, broken down by contaminant types through the MOF adsorption process. It can be readily seen that publications were predominantly on adsorptions compared to photocatalysis and membrane applications. Dyes were also popular topics of research applications. This is not surprising as studies on the removal, degradation, and rejection of dyes are easy to be executed using spectrophotometers, and the effects can be seen with naked eyes. Studies on non-chromogenic

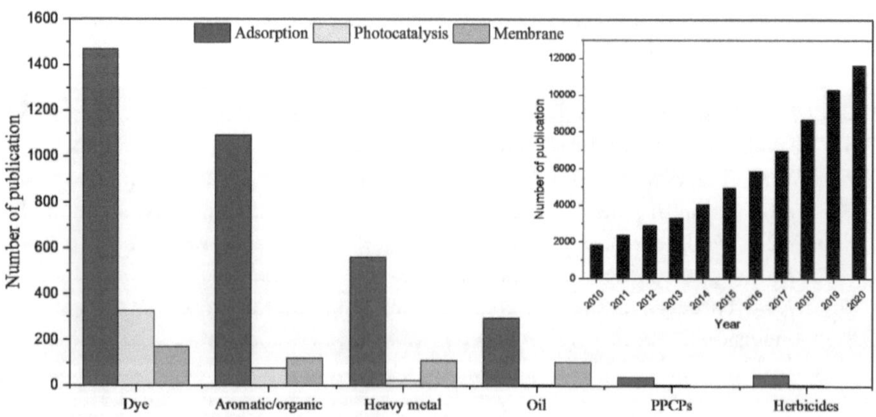

FIGURE 2.1 Number of publications related to MOFs for water treatment from 2010 to 2020. (Data extracted from Web of Science.)

pollutants, on the other hand, will require instruments that are less easily available, such as AAS, HPLC, or HPLC-MS. Nonetheless, investigations involving MOFs for various types of contaminants are projected to increase dramatically in the future years.

Due to the importance of treating wastewater from the contaminant, it is necessary to overview their adsorptive removal by promising MOFs structure. This chapter provides an overview of MOFs modification to improve its efficiency in adsorption, recyclability, and stability for the removal of dyes, heavy metals, and PPCPs from wastewater. The SWOT analysis of MOFs' practicality has also been summarized here.

2.2 HISTORY AND SYNTHESIS OF MOFs

Metal-organic frameworks (MOFs) are a crystalline subset of microporous materials that have attracted the attention of many scientists around the world in recent decades (Johari et al. 2020; Johari et al. 2021; Johari et al. 2021; Sarker et al. 2019; Ang et al. 2015; Jang et al. 2020). Due to the lack of an accepted standard definition during the development of this new family of hybrid materials (Li et al. 2012), they are also known as porous coordination polymers (PCPs) (Kitagawa et al. 2004), porous polymer networks (PPNs) (W. Lu et al. 2010), microporous coordination polymers (Cychosz et al. 2008), iso-reticular MOFs (Eddaoudi et al. 2002), metal peptide frameworks (MPF) (Mantion et al. 2008), and zeolite-like MOFs (Liu et al. 2006).

Despite the fact that some scientists believe that "until the mid-1990s, there were basically two types of porous materials, namely inorganic and carbon-based materials" (Kitagawa et al. 2004), Tomic (1965) was the first to introduce porous coordination polymers, which are now known as MOFs. Various organic–inorganic hybrid materials were then developed in the early 1990s and the scope of research on these porous structures evolved significantly in the years that followed. Then, Hoskins and Robson (1990) developed scaffolding-like materials, a novel type of PCPs that was employed through the mechanism of anion-exchange process.

Fujita et al. (1994) published a two-dimensional square network material for catalytic applications 4 years later. Following that, Yaghi et al. (1995) and Li et al. (1999) were the pioneers to define the novel idea of MOFs and conduct systematic surveys and research. In 1995, they reported the synthesis of a novel MOF designed to bind aromatic guest molecules selectively (Yaghi et al. 1995). They synthesized the first robust and highly porous structure of MOF-5 in late 1999 and analyzed x-ray single crystal structure and pore size of the material (Li et al. 1999). Other articles on porous polymer frameworks were published during these years, such as the study reported by Kitagawa's group in 1997, for gas adsorption applications such as CH_4, N_2, and O_2 (Kondo et al. 1997). They discovered that the porous structure they created could reversibly adsorb these gasses at pressures ranging from 1 to 36 atm without affecting its crystal framework or deforming. Nowadays, various relevant and updated reviews in the disciplines of PCP and MOFs have been published, which describe the rapidly developing research-related activities. Kitagawa et al. (2004), Rowsell and Yaghi (2004), and Férey (2008) are among the most comprehensive reviews and studies.

MOFs are made up of two inorganic metal ions/clusters and two organic building units, such as carboxylates or other organic anions (phosphonate, sulfonate, and heterocyclic compounds), which are bound together by coordination bonds (Férey 2008; Lu et al. 2014). They have some exceptional and unique properties that distinguish them from other classic porous solids, including simple synthesis, extremely high internal surface area, high thermal and mechanical stabilities, low density, large micropore volumes, permanent porosities, and tailorable frameworks (Robles-García et al. 2016). Metal ions or metal clusters (inorganic component) and bridging ligands or linkers (organic unit) are used to integrate MOF materials via coordination bonding (Kuppler et al. 2009; Furukawa et al. 2014; McGuire and Forgan 2015). This is accomplished through self-assembly, which results in one-, two-, or three-dimensional networks with extremely varied configurations (i.e., surface areas, crystal structures, pore sizes) (Cook et al. 2013; Wu and Navrotsky 2015). By tailoring the structures and functionalities of MOFs, their characteristics can be readily systematically regulated to attain the required target or improve performance. For example, adjusting the inorganic moiety's coherence and the organic linker's composition can change the surface area, shape, and pore size (Jhung et al. 2012; Lu et al. 2014), or modifying the organic linker can increase the surface area (Cavka et al. 2008). Several similar or separate topological frameworks with various physico-chemical properties have been discovered due to the diversity of metal ions or metal clusters, as well as the variety of organic linkers (Yaghi et al. 2003; Cavka et al. 2008; Jhung et al. 2012; Moorhouse et al. 2016). The timeline of the development of MOFs is shown in Figure 2.2.

Many various mechanical or chemical processes can be used to create these highly structured materials. Such a wide range of manufacturing techniques can be divided into six categories: [1] The solvo-thermal approach is the most common and widely utilized method, and it has been used to make a variety of well-known MOFs such as MOF-177 (Li et al. 2007), MIL-101 (Chowdhury et al. 2009), and MIL-53 (Fe) (Haque et al. 2010); [2] mass transfer techniques to form inorganic salts crystals through layering of solutions, or evaporation of a precursor solution, or slow diffusion of one material into a solution containing metal salts or organic acid as invented by Li et al. (1999); [3] synthesizing MOFs via irradiation

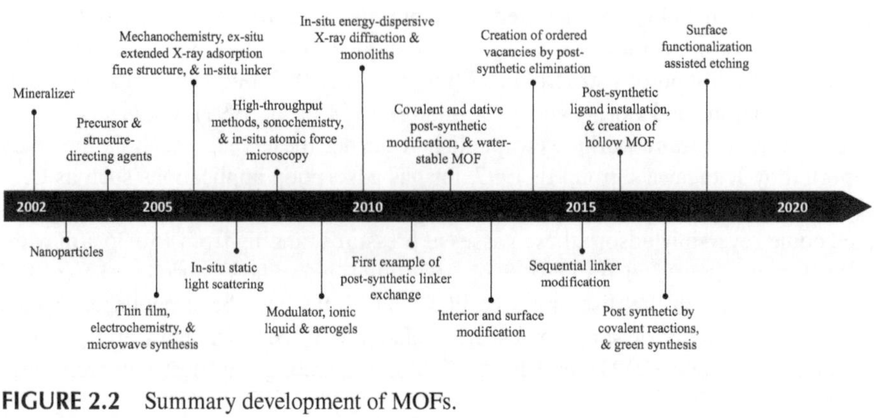

FIGURE 2.2 Summary development of MOFs.

of ultrasound and microwave that consume short synthesis time (Safarifard and Morsali 2015); [4] the electrochemical procedure was first introduced by Mueller et al. (2006) for the synthesis of HKUST-1; [5] the mechano-chemical method (Lee et al. 2013; Fri and Abi 2009); and [6] the free-solvent synthesis technique was introduced by Pichon et al. (2006), and Yang et al. (2011) used this method to prepare $Cu_3(BTC)_2$. Other less traditional and reliable approaches, such as iono-thermal, microfluidics, and dry-gel conversion, have been investigated in addition to these methods, which are widely utilized in the effective synthesis of MOFs (Lee et al. 2013).

2.3 RECENT MODIFICATION OF MOFs

Several modifications that have been done to enhance the properties of MOFs materials in water treatment are: (1) modifying the organic ligand with functionalized groups, (2) altering the pore size and structures of MOFs, and (3) compound MOFs with other material to create MOF-based composite, that not only limit the drawbacks of MOFs but also achieve synergistic properties.

2.3.1 Surface Functionalization

Introducing different functional groups such as amino ($-NH_2$) and hydroxyl (-OH) that have chelation/complexation with pollutants into MOFs is an effective strategy for the adsorption of pollutants. There are usually two methods for synthesizing functionalized MOFs, i.e., pre-synthesis modification and post-synthesis modification. In the pre-synthesized modification, the designed blocks can be purposefully assembled into an extended network with definite functions. Meanwhile, post-synthetic modification was done by introducing functional groups into MOFs after the framework synthesis.

2.3.1.1 NH_2-Functionalized

Researchers have created NH_2-MIL-101(Cr), NH_2-MIL-53(Al), and NH_2-MIL-125, which are all amine-functionalized MOF. Due to its amine pendant, relatively high surface areas, excellent moisture stability, and straightforward synthesis technique, NH_2-MIL-101(Cr) has received a lot of attention. Furthermore, MIL-101(Cr), its parent, was first described by Férey (2005) and quickly rose to prominence as one of the most prominent members of the MOFs family. It has mesopores, is extremely stable, and has open metal sites. The modified MOFs further enhanced the properties of parent MOF, thus improving the adsorption performance. Chen et al. (2011) also used the above-mentioned experimental technique to lower the synthesis temperature and add some base to create NH_2-MIL-101(Cr), which has outstanding stability and a large surface area (up to $1{,}675\,m^2g^{-1}$) and pore volume (up to $1.67\,cm^3g^{-1}$). Then, Jiang et al. (2012) discovered a post-synthesis approach to NH_2-MIL-101(Cr). Another amine-functionalized MOF in the MIL family was introduced by Fu et al. (2012), which is NH_2-MIL-125. The MOF is made up of H_2BDC and titanium oxo-hydroxo clusters, to form a quasi-cubic tetragonal framework. Owing to its metal cluster, the MOF is usually used for the removal of dye (Hu et al. 2017).

2.3.1.2 OH-Functionalized

The introduction of a hydroxyl (-OH) group in the organic linker of MOFs was critical for gaining key acid–base and hydrogen-bond interaction sites. In 2016, hydroxyl-functionalized MOFs, MIL-101(OH) and MIL-101$(OH)_2$ were synthesized and used for the adsorption of PPCPs from water (Seo et al. 2016). The favorable contributions of -OH and -OH_2 on MIL-101 in the increased adsorption of ibuprofen and oxybenzone (especially based on porosity) confirmed the importance of H-bonding mechanism. In another study, hydroxyl-functionalized MIL-101s such as MIL-101$(OH)_3$ and MIL-101(OH) were prepared by grafting amino-alcohols on pristine MIL-101 (Sarker et al. 2017). The highly porous MIL-1010 bearing –$(OH)_3$ groups exhibited remarkable organic arsenic acids adsorption capacities with the help of a hydrogen bonding mechanism. Modification of MOFs by tannic acid (TA), which contains a vast -OH group, has attracted significant attention recently because of its coating and surface functionalization abilities. Hu et al. (2016) modified ZIF-8 with TA via a simple post-synthesis method by controlling its reaction time and concentration of acid, resulting in MOFs with a high degree of hydrophilicity. Another study reported the same trend with a different type of MOF like MIL-100(Fe) (Johari et al. 2021). Importantly, the extra -OH group that was attached to the organic linker of MOFs can be a necessary component to form hydrogen-bond acceptors and donors between pollutants and MOFs, respectively, which is very useful for dyes, heavy metals, and PPCPs adsorption.

2.3.2 Hollow-structure MOFs

Taking advantage of the unique properties of MOFs, researchers started to create novel hollow-structure MOFs and at the same time overcome the limitations of the MOF itself (e.g., smaller intrinsic open channels hinder the possibility for MOFs to encapsulate guest species with size more than 2 nm). Designing MOFs with hollow structure can be done by two main strategies, which are the bottom-up approach (e.g., sacrificial template or Ostwald ripening) and the top-down approach (e.g., etching).

2.3.2.1 Sacrificial Template

The required size of the open channel of MOFs can be constructed based on the size of the sacrificial template. As illustrated in Figure 2.3a, in-situ methods have been developed to fabricate MOFs with voids by removing the sacrificial template *via* self-assembly, layer-by-layer growth, or stepwise synthetic approach. Ejima and co-workers reported Fe-MOF crystal through a one-step assembly on the polystyrene template (Ejima et al. 2013). Meanwhile, Zhang et al. (2014) synthesized nanocages MOF-5 by using layer-by-layer growth on a poly(vinylpyrrolidone) template. Kim and co-workers used a copper-based metal-organic polyhedron crystal as a sacrificial hard template to create a hollow MOF (Kim et al. 2015). Although this technique is facile, rapid, and straightforward, this method has a limitation on creating the void for single-crystalline MOFs. Usually, the formed MOFs by this technique are either a composite of the smaller non-hollow MOF crystals and the polymer template or an aggregate of smaller non-hollow MOF crystals.

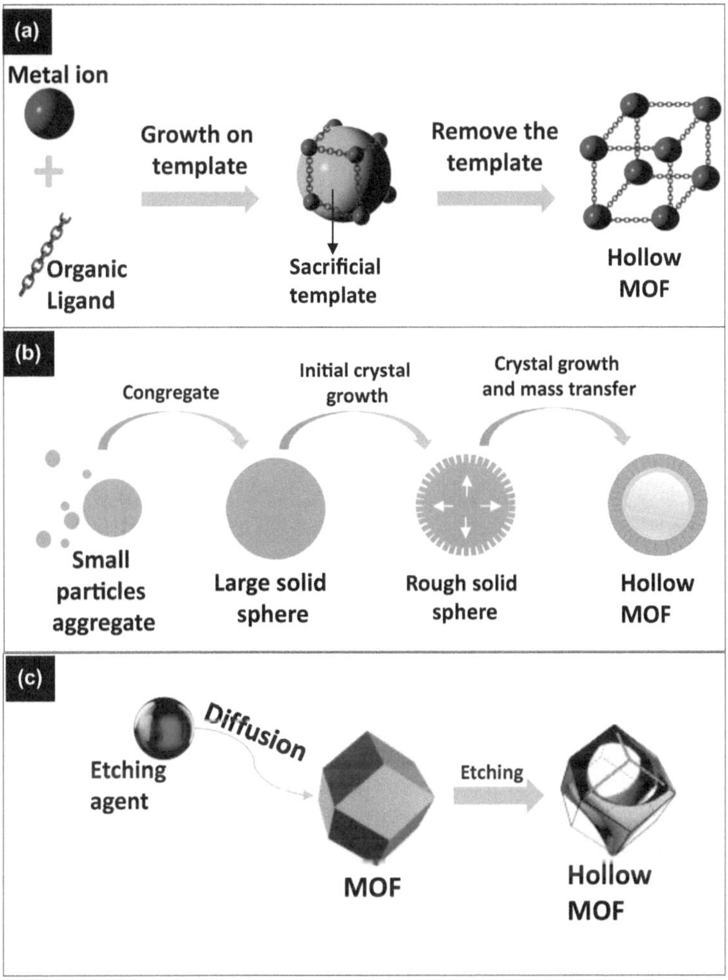

FIGURE 2.3 Schematic diagram of illustration to construct a hollow-structure MOFs by (a) sacrificial template, (b) Ostwald ripening, and (c) etching methods.

2.3.2.2 Ostwald Ripening

Ostwald ripening is a bottom-up approach for obtaining hollow materials with dispersed morphology by a one-pot hydrothermal reaction without any additional template, also known as a template-free strategy (Hao Yang et al. 2018), as illustrated in Figure 2.3b. Feng et al. (2019) described that the formation of internal void MOFs was due to a surface-energy-driven mechanism and mass transfer during the secondary growth. This technique is found to be effective in some cases. For instance, Huo et al. (2010) produced hollow ferrocenyl with microporous coordination polymer shells. Meanwhile, Liu et al. (2019) reported a uniform and controlled shell thickness of a Zr-MOF, which has potential in photodynamic cancer therapy due to its capability to generate more singlet oxygen compared to other MOF. In contrast, several studies claimed this strategy could not create a void in the robust open framework like HKUST or MIL-64 due to its difficulty in determining optimum conditions for Ostwald ripening (Shen et al. 2014; Ke, Yuan, et al. 2011; Tu et al. 2014), which limit its potential in creating a hollow structure in various MOF.

2.3.2.3 Etching

The most well-known top-down approach to creating a void inside the crystals is the etching technique, where the void is created by the diffusion of the etching agent to the MOFs while maintaining its outer shell, as illustrated in Figure 2.3c. For instance, Liu et al. (2017) synthesized multi-shelled hollow MIL-101 by using glacial acetic acid and regulating the reactant concentration and the etching time. Meanwhile, some researchers showed a green approach to construct hollow MOF by using water as the etching agent. For example, Avci et al. (2015) introduced ZIF-8 to water as an etching agent, but it leads to the formation of non-uniform voids at some unstable parts of ZIF-8. A couple of studies suggested that water can be applied only to MOFs that have hydrophilic surfaces (Gonzalez et al. 2011; Yin et al. 2004). So, it is impractical to apply to the MOFs with hydrophobic surfaces. Moreover, a problem like uncontrolled etching leads to the frame-like structure and hindered the development of this technique (Zhang et al. 2008; Campagnol et al. 2013). Although the construction of MOFs with hollow structures is less well studied but with further development the hollow MOFs will have high potential in various areas.

2.3.3 Composite

MOFs exhibit promising chemical and physical properties for many adsorption applications. Besides modification during MOFs synthesis, MOFs can further improve properties by the alteration of physical and structural characteristics by combining with other nanocomposite materials. Recently, MOFs composite is rare, and it is essential to find suitable complementary materials to synthesize composite based on MOF. Among the most suitable materials for the synthesis of MOF, composites are layered double hydroxide (LDH) and magnetite MOF.

2.3.3.1 LDH

LDH has a high surface area with a large pore diameter, which is significant for obtaining excellent adsorption properties for water pollutants. LDHs are widely used as adsorbents to treat water contaminants with poor adsorption capacity. The composite nanoparticle of MOF with LDH is expected to enhance the adsorption performance. Wang et al. (2018) successfully fabricated an LDH nanoflower lantern derived from ZIF-67 nanocrystal and reported Congo red (CR) and Rhodamine B (RhB) adsorption capacities of 328.77 and 112.35 mg g^{-1}, respectively. Nazir et al. (2020) showed a sandwich-like structure of the in-situ growth of ZIF-67 on LDH to form porous composites to remove methylene blue (MB) and methyl orange (MO). The finding showed that ZIF-67@CoAl-LDH has a good adsorption capacity of 180.5 mg g^{-1} for MO than LDH and ZIF-67. LDH with a thin layered unique structure has a large surface area and this has provided a high amount of adsorption site for dye. Therefore, it can be deduced that the composite filler of LDH/MOF can enhance the removal of dye for wastewater.

2.3.3.2 Magnetite

Magnetite MOF is developed to induce magnetic properties on the MOF for enhancing adsorption on heavy metals removal. The new adsorbents combine the advantages of MOFs and magnetic nanoparticles and have a large capacity, low cost, fast removal, and easy separation of the solid phase, which makes them an excellent sorbent for wastewaters treatment (Zhao et al. 2015). Also, magnetic MOF can overcome the restriction such as poor separation ability from solid–liquid phase.

Bao et al. (2018) synthesized amino-functionalization magnetic multi-metal-organic framework material as an effective adsorbent for methyl orange (MO). It showed that the magnetic MOF not only has higher removal efficiency toward MO but also had higher stability in an aqueous solution. Huang et al. (2015) successfully prepared a magnetic MOF composite of Fe_3O_4@SiO_2@HKUST-1 core–shell nanostructures, and it has been successfully employed for the selective removal of Hg^{2+} from water. Ke et al. (2017) synthesized thiol-functionalized magnetic MOF adsorbents and exhibited highly selective removal of Hg^{2+} and Pb^{2+} from water. This adsorbent can be easily recycled because of the presence of the magnetic Fe_3O_4 core.

The introduction of magnetic nature to MOF can enhance the removal of charge pollutants such as heavy metal as well as a dye which also come with a charge. Importantly, magnetite MOF causes the cost of adsorption application more effective as it can be used repeatedly. Therefore, it can be concluded that the magnetite MOF enhances the adsorbent performance and reduce the operation process.

2.4 REMOVAL MECHANISMS BY MOFs

The possibility to synthesize hundreds of frameworks from various clusters of metal ions with organic linkers gives rise to an unlimited number of crystalline MOFs with microporous or mesoporous structures. Additionally, different functional groups in the organic linkers and metal node serves as adsorption centers for various types of organic contaminants (Hasan and Jhung 2015). MOFs also offer selective adsorption of organic molecules due to the functionalities of the organic linkers, possibly forcing inclusion complexes with the guest adsorbate molecules. The mode of adsorption interactions is usually through electrostatic interaction, hydrogen bonding, π–π interaction/stacking, hydrophobic interaction, and acid–base interaction, as illustrated in Figure 2.4. Molecular modeling has shown that when the pore sizes of the MOF are bigger than the pollutant molecule the guest molecule preferably resides in the pores of MOFs (Gaikwad et al. 2020). Alternatively, the guest molecule is adsorbed on the outside if it is bigger than the pores of the MOFs. Thus, choosing the MOF for the adsorption of an analyte is important to optimize the adsorption (Petit 2018; Gaikwad et al. 2020; Johari et al. 2020). MOFs with promising adsorption properties have been selectively used for the removal of contaminants in water.

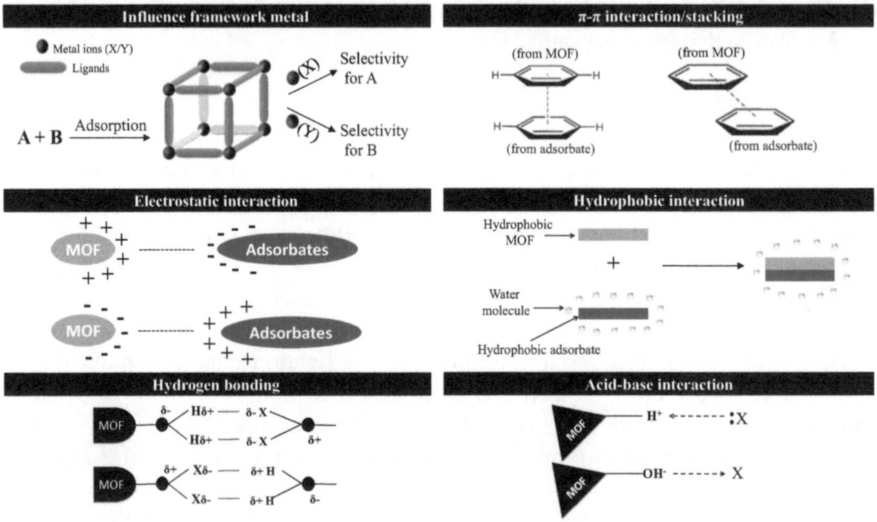

FIGURE 2.4 Possible mechanisms between MOFs and pollutants.

2.5 ADSORPTION OF POLLUTANTS FROM WASTEWATER BY MOFs

2.5.1 Removal of Dye

Azo and organic dyes are a kind of chemicals widely used in the food, paper, and textile industries. In order to meet industrial needs, a large number of dyes are produced and applied. The majority of industrial dyes are poisonous, carcinogenic, and teratogenic, and their release into the ecosystem is hazardous (Yagub et al. 2014). Moreover, throughout the manufacturing process, more than 15% of the colors generated each year are lost (Zhao et al. 2015), resulting in the discharge of the dyes containing wastewater into the environment. Various types of MOFs, including chromium, alumina, ferric, zinc, copper, and cobalt-based MOF, have been used for dye removal applications. Given the intrinsic micro-mesopores and tunable organic ligands with various hybrid atoms such as O, N, or S, which may exhibit specific affinity to dye molecules, their performance in water treatment has been explored by scholars around the world.

The first research on the adsorption of dyes by MOFs was conducted by Haque et al. (2010), in which two highly porous chromium-based MOFs (MIL-53-Cr and MIL-101-Cr) were used for the adsorption of methyl orange (MO) from aqueous solutions. The study shows MIL-101 to have higher adsorption capacity and adsorption kinetic constant than MIL-53. They also highlighted that both MOFs have higher adsorption capacity than activated carbon during the MO removal.

Compared to other materials, the distinguishing feature of MOFs is their flexibility to modulate their pore size without affecting the intrinsic properties (Hasan and Jhung 2015). Due to this unique property, Wang et al. (2013) studied the size-selective adsorption performance of copper-based MOFs with different organic linkers (such as pyrazine, 1,2-di(pyridine-4-yl)ethane, and 4,4-bipyridine(bpy)) for the removal of different size azo dyes from water. All the MOFs were capable to adsorb MB and MO; however, they were unable to adsorb RhB because of their smaller pore size than the size of dye, which restricted the passage of the pollutant.

Molavi et al. (2018) also prepared a partially positive charged UiO-66 applied as an energetic adsorbent for the selective removal of ionic dyes in water. Zaki et al. (2018) compared nickel and cobalt-based MOF for methyl orange (MO), malachite green (MG), and MB adsorption and reported that nickel-based MOF adsorbed only MB from water with a saturated adsorption capacity of 216 mg g^{-1}, while the cobalt-based MOFs adsorb MO and MG from water with saturated adsorption capacities of 671 and 405 mg g^{-1}, respectively. These studies showed that different metal ion of MOFs will only selectively remove dye from the solution.

Several studies reported that some MOFs have the capability to adsorb more than one dye simultaneously, although the simultaneous removal is not easily achieved. A MOF-235 can be synthesized from [Fe_3O(terephthalate)$_3$(DMF)$_3$] and [$FeCl_4$] for the MO and MB dyes adsorption (Haque et al. 2011). It is non-toxic and readily available at low temperatures. In the liquid phase, the MOF-235 can rapidly adsorb both cationic and anionic dyes, despite limited nitrogen adsorption at low temperatures. It is noteworthy that the MOF-235 has a higher adsorption capacity than activated carbon, for MO removal. Nevertheless, the major role behind MO/MB adsorption on the active site of MOF is by entropy effect instead of an enthalpy change (Sudik et al. 2005).

Moreover, another MOF like MIL-100(Fe) could also efficiently capture large amounts of both MO (1,045.2 mg g^{-1}) and MB (736.2 mg g^{-1}) simultaneously associated with rapid kinetics, whereas MIL-100(Cr) could only selectively adsorb MB (645.3 mg g^{-1}) from an equal mass MO–MB mixture (Tong et al. 2013). In another study, zirconium-metalloporyphyrin (PCN-22) was used for mixed dye solution (MB and MO) adsorption from an aqueous solution (Li et al. 2017). They demonstrated that the MOF can mutually adsorb dyes by 36.8% (1,239 mg g^{-1}) for MB and 73.5% (1,022 mg g^{-1}) for MO.

Charu et al. (2019) reported that 94.74% (9.72 mg g^{-1}) methylene blue (MB) removal was attained by 25 mg of Fe-BDC after 24 h. They highlighted that the adsorption process is spontaneous and endothermic in nature and proceeds through chemisorption, and the interaction behavior of the MOF surface with the MB dye followed the pseudo-second-order model. Mahmoodi et al. (2018) prepared composite MOF of MIL-68(Al) with silica SBA-15 and reported that the composite MOF adsorbed 100% of MG solution with a maximum adsorption capacity of 68.9 mg g^{-1} and followed pseudo-second-order and the Langmuir isotherm models. MOFs show great removal efficiency of dye in wastewater, owing to their higher surface area and much more surface-active sites than bulk materials.

2.5.2 Removal of Heavy Metal

Oxyhydroxides can adsorb heavy metals transported by water and deposited on the aquatic plants like algae, which can disturb their food web due to the excess bioaccumulation of contaminants. Even though many heavy metals are required by the biological system (e.g., copper and zinc), exceeding the presumptive maximum tolerable daily intake might induce toxicity and damage the system. The most common heavy metal ions studied in recent years are arsenic, cadmium, chromium, lead, and mercury.

The Fe-BTC synthesized by Zhu et al. (2012) using the hydrothermal method was used for arsenic (V) adsorption. This MOF was based on iron nodes and tricarboxylic acid linkers, and they discovered that the removal efficiency of As(V) using the MOF was more than 96% when the solution was between pH 2 and 10. The MOF was reported to be 7 times more efficient than commercially available powdered Fe_2O_3 and 37 times more effective than iron oxide. EDX confirmed the As(V) adsorption on the internal surface of the MOF, which supported the appearance of a new IR band at 824 cm^{-1} corresponding with Fe-O-As groups (Zhu et al. 2012).

Several MOFs were reported to be a possible cadmium(II) adsorbents such as AMOF-1, UiO-66-NHC(S)NHMe, and HS-mSi@MOF-5. Chakraborty et al. (2016) prepared AMOF-1 from zinc(II) nodes and tetracarboxylate linkers and reported that 41 mg g^{-1} of Cd(II) was adsorbed 24 h after the MOF achieved equilibrium. However, it only exhibit partial selectivity in the adsorption of Cd(II) ions. In another study by Saleem et al. (2016), UiO-66-NHC(S)NHMe was capable to reach a maximum adsorption capacity of up to 49 mg g^{-1}. Meanwhile, HS-mSi@MOF-5, post-modified MOF-5 by Zhang et al. (2016), has achieved 2.25 times higher adsorption capacity than its parent MOF-5.

At an optimal pH of 7, a magnetic Cu-terephthalate MOF was assessed as effective manganese(II), iron(III), and zinc(II) adsorbent that achieved equilibrium in 120 minutes for all three with maximum adsorption capacities of around 150, 175, and 115 mg g^{-1}, respectively (Rahimi and Mohaghegh 2016). In another study by Li et al. (2017), a silver-triazolato MOF1NO_3^- was used in the ion-exchange process to adsorb chromium(VI) (in the form of $Cr_2O_{72}^-$). The chromium adsorption was significant and took around 4 hours to reach equilibrium. Interestingly, the MOF was discovered to be reusable and preserved 93% of its original efficacy even after four sequential adsorption–desorption cycles. However, the MOF displayed lower adsorption over other anions such as SO_{42}^-, NO_3^-, ClO_4^-, or Cl^-.

Apart from efficient Cd(II) adsorption, MOF-5 also displayed excellent lead(II) removal from aqueous solution with an adsorption capacity of up to 659 mg g^{-1} at 45°C. The results of uptake value fluctuated with changes in temperature, with no consistent rise or decrease tendencies, and conclude that the adsorption by the MOF is temperature independent. The adsorption value at pH 5 was recorded to be 450 mg g^{-1}, but due to the unique structure of the MOF that has both acidic and basic active sites, the adsorption value increases with both decreasing and increasing pH of the solution to 660 mg g^{-1} (pH 6) and 750 mg g^{-1} (pH 4) (Rivera et al. 2016).

Next, the most common heavy metal adsorption is of mercury(II). A study by Abbasi et al. (2015) synthesized Co(II)-MOF for Hg(II) adsorption at several concentrations from 10 to 40 ppm. The MOF was able to remove 70% of the Hg(II) ions at an optimal pH of 6 within 100 mins. Interestingly, the structure of MOF showed no severe damage after adsorption, which was confirmed by PXRD analysis. Therefore, from previous studies, it is proved that MOFs could be beneficial and effective adsorbents in the removal of heavy metal ions from polluted water, by achieving high maximum adsorption capacities while maintaining their structure in various conditions.

2.5.3 Removal of PPCPs

PPCPs are emerging pollutants that may cause potential harm to the environment and human health. PPCPs are widely used in various industries and in day-to-day life. PPCPs may often remain in the environment even after they have been consumed completely (Bulloch et al. 2015; Bu et al. 2021; C. Jung et al. 2021), because PPCPs usually have long shelf lives to meet customers' demands, and some PPCPs are inadvertently dumped into the environment. For example, various PPCPs have recently been found in surface water, ground water, and even in the tissues of fishes and vegetables (Gadipelly et al. 2014; X. Wu et al. 2014); therefore, PPCPs are typical examples of so-called emerging contaminants. It is reported that PPCPs may cause endocrine disruptions that can change hormonal actions (Bulloch et al. 2015; C. Jung et al. 2021), although the adverse impact of PPCPs on human health and the environment is still not fully understood. Therefore, the removal of PPCPs from surface/ground water and aquatic systems has recently been attracting much attention, even though PPCPs in the environment have not been regulated explicitly so far.

Pharmaceuticals waste removal from aqueous solution by using MOF was first investigated by Férey et al. (2005). Water-stable MOF – MIL-100(Cr) – demonstrated its capability to adsorb furosemide and sulfasalazine up to 11.8 and 6.2 mg g^{-1}, respectively. Generally, these two drugs are used in the treatment of diuretic and inflammatory bowel diseases. The studies reported that the results of adsorption experiments were comparable with HKUST-1 (<1 mg g^{-1}) and zeolite (no adsorption). Because of the strong trinuclear chromium clusters in its structure, MIL-100(Cr) could be an effective adsorbent for pharmaceuticals removal from contaminated water.

Tetracycline and oxytetracycline hydrochloride are most commonly used for treating infectious diseases by bacteria. The pharmaceuticals waste reached the water bodies directly through human feces and urine during the biological metabolism process, causing harm to the aquatic environment, triggering the resistant genes which can results in long-term contamination (Hasan et al. 2015). The adsorption of tetracycline was conducted by Yuan et al. (2019) using three types of ferric-based MOFs – MIL-100(Fe), MIL-53(Fe), and MIL-101(Fe), which then achieved adsorption capacity of up to 51.52, 12.51, and 43.41 mg g^{-1}, respectively.

Another study investigated the effect of adsorption properties with different MOFs (UiO-66, MOF-802, and MOF-808) for the removal of several pharmaceuticals waste (Lin et al. 2018). Findings from this study proved that MOFs had excellent adsorption effects on a wide range of drugs including ketoprofen, furosemide, ibuprofen, indomethacin, salicylic acid, and naproxen. Meanwhile, Chen et al. (2017) highlighted that the adsorption capacity of UiO-66 significantly increased even after it reached equilibrium. The experiments revealed that when one drug's adsorption approached adsorption equilibrium, introducing another substance increased the former's adsorption capacity while exerting minimal influence on its own adsorption.

Surprisingly, PPCPs with a diameter greater than the MOF framework could be adsorbed. For instance, tetracycline can be adsorbed by UiO-66 and ZF-8.

The occurrence of this phenomenon might be due to the framework flexibility of the MOF structure or defects in its window (Zhang et al. 2013; Fairen-Jimenez et al. 2011). The aperture size of ZIF-8 is 3.4 Å; nonetheless, the gate-opening in ZIF-8's 6-membered ring would cause big molecules to disperse throughout the structure. In comparison to traditional adsorbents such as zeolites and activated carbon, ZIF-8 has outstanding structural flexibility to be kinetically selective adsorbent (Zhang et al. 2016; Han et al. 2019). On the other hand, structural defects in UiO-66 frameworks unintentionally caused the ability to adsorb large-sized PPCPs. Controlling defects could be a strategical technique for improving PPCPs removal efficiency. Despite the fact that the number of MOFs used for PPCP removal is the least popular among 80,000 reported MOFs (Moghadam et al. 2017), the excellent adsorption efficiency (compared to traditional adsorbents) has demonstrated MOFs' ability to adsorb PPCPs pollutants from effluents.

2.6 RECYCLABILITY AND STABILITY OF MOFs

In terms of economic and environmental sustainability, recyclability is the crucial factor in determining the efficacy of MOFs for the removal of contaminants from water. A variety of solvents were applied for the desorption of pollutants from the MOFs and varying solvents affect differently in their degree of regeneration. The appropriate solvents play an important role in maintaining the framework of MOF prior washing step and subsequent adsorption studies. Thus, MOF stability also needs to be considered since an unstable structure after adsorption/desorption steps will impair the adsorption performance in the long run.

The majority of MIL-series MOFs could be easily regenerated after soaking with several solvents including methanol, aqueous or acidic ethanol, or acetone. It was reported that MIL-53(Al) removed the pollutants (bisphenol A and sulfamethoxazole) attached to its active sites after three cycles of methanol washing (Gao, Kang, et al. 2019; M. Zhou et al. 2013). Meanwhile, several studies of MIL-53(Al and Cr) discovered that the MOFs can be regenerated after several washing steps (four cycles of acetone and three cycles of ethanol soaking) to ensure that dimetridazole was totally detached from the MOF (Jung et al. 2013; Peng et al. 2018).

During p-arsanilic acid removal, MIL-100(Fe) demonstrated an intriguing phenomenon; employing acidic ethanol, it can recycle up to three cycles, but pure ethanol remarkably lowers the MOF performance (Wang et al. 2018). Meanwhile, Yuan et al. (2016) discovered that after washing MOFs [MIL-53(Fe), MIL-100(Fe), and MIL-101(Fe)] with aqueous ethanol, all three MOFs maintains their structure and retained high recyclability even after four cycles. On the other hand, organic arsenic acids such as phenylarsonic acid and p-arsanilic acid were adsorbed by MI-101$(OH)_3$ with no notable declining adsorption performance after four cycles and can be easily regenerated by acid ethanol treatment (Sarker et al. 2017).

Several studies demonstrated the use of washing agents such as methanol, acetone, or acidic medium to achieve high returnability of zirconium-based MOFs. For instance, carbamazepine deposited on the UiO-66 was desorbed with chloroform, and the regenerated MOF retained half of its maximum adsorption capacity after the fourth cycle (Jin et al. 2020). Other studies by Chen et al. (2017) highlighted that after the fifth cycle, the PCN-22 maintains excellent removal of bisphenol A up to 99%. The stability of PCN-22 was proved by comparing PXRD signals before and after washing treatment (Chen et al. 2017). Meanwhile, the effectiveness of NU-1000 in removing atrazine from the water was nearly unchanged after washing with acetone for up to three cycles, and the regenerated MOF maintained its framework stability (Meng et al. 2017). The same results were obtained with other MOFs such as BUT-12 and BUT-13 where they displayed excellent returnability after the sixth cycle (Gao et al. 2019).

Most of the reported MOFs displayed excellent adsorption capacity and have good stability and regeneration. Adsorbate desorption from MOFs necessitates the use of appropriate washing steps, especially the choice of solvents or a particular technique that takes into account the interaction between MOFs and adsorbates. Most importantly, desorption agents must be chosen with several considerations such as the structure of pollutants (dyes, metal ions, and pharmaceuticals), MOFs, and their effects on the structure. MOFs should be considered as materials that have the capability to purify water due to their efficacity and effectiveness.

2.7 SWOT ANALYSIS OF MOFs' PRACTICALITY IN WASTEWATER TREATMENT

2.7.1 Strength

MOFs have attracted the attention of many researchers because of their large surface area, high thermal stability, and huge porosity. High surface and porosity may provide the accessibility of the adsorption site and contaminant diffusion within the framework.

Water stability of MOF is essential for its effective use for heavy metal ions removal from water for wastewater treatment (Wang et al. 2016). The "weakest" point in MOFs is metal-ligand coordination bonding. The strength of coordination bonds is closely associated with the stability of MOFs in water (Canivet et al. 2014). The relations between MOFs and water can be counted as the competing coordination of metal nodes/metal ions between organic linkers and water molecules (Burtch et al. 2014). The water molecule is hard to replace the original bond if the metal-organic coordination connection is strong enough. All the MOFs with strong metal-organic coordination, therefore, have strong stability in water. Many other aspects influence water stability such as porosity, crystallinity, metal-ligand coordination geometry, and hydrophobic surface pores (Feng et al. 2018).

Also, modified MOF and MOF composite have the tendency to enhance simultaneous adsorption and removal of more than one pollutant. Boix et al. (2020) reported the corresponding adsorption capacities for MOF and inorganic nanoparticles into composite materials have removed multiple heavy metals such as As^{3+}, As^{5+}, Cd^{2+}, Cr^{3+}, Cr^{6+}, Cu^{2+}, Pb^{2+}, and Hg^{2+} from water simultaneously. Molavi et al. (2020)

utilized MOF composite nanoparticles adsorbent in contaminated water to remove cationic malachite green (MG) dye and anionic methyl red (MR) dye and found that prepared adsorbent simultaneously adsorbs 43% of MG and 59% of MR in a mixture of both dyes for 2 min. Lv et al. (2019) applied UiO-66-NH_2 to remove methyl orange (MO) and methylene blue (MB) simultaneously, and the results showed excellent adsorption ability with 148.4 mg g^{-1} for MO and 549.6 mg g^{-1} for MB mg g^{-1}.

2.7.2 Weakness

The small pores with diameters normally dropping within the micropore range are one of the disadvantages of using MOFs. The number of colorants that can be adsorbed into the system is thus restricted. This makes it necessary to synthesize pores in the mesoporous range (Adeyemo, Adeoye, and Bello 2012).

During the adsorption process, MOF tends to degrade and become hydrophobic MOF which reduces the performance of the adsorbent. It has been reported that MOFs deteriorate through two key processes, ligand displacement and hydrolysis (Low et al. 2009). These pathways were achieved by computational chemistry and experimentally tested. The displacement of a ligand happens as a water molecule is stuck into a metal-ligand M–O bond of the framework leading to the formation of a hydrated cation and the discharge of a free ligand. On the other hand, the metal-ligand bonds are detached, and water dissociates in a hydrolysis reaction to create a hydroxylated cation and free protonated ligand. Conversely, hydrophobic MOFs have water-resistant sites on their surfaces and limit the adsorption pollutant.

2.7.3 Opportunities

Having high specific surface areas beyond 6,000 $m^2 g^{-1}$, MOFs are considered as potential candidates for adsorption, separation, and catalysis due to the tunable pore sizes (Arora et al. 2019). MOF can be designed in multi-dimensional structures with diverse properties (Zhou et al. 2012). MOF pore size can be properly adjusted to accommodate a specific pollutant (Russo et al. 2020). MOF pores are strongly organized due to their crystalline structure. The MOF shape and size of the pores can be tuned by choosing the linkers used and the connectivity of the metal ions for heavy metal removal.

2.7.4 Threats

Although a few MOFs such as ZIF-8 and MIL-101, have shown excellent chemical stability, many other organic–inorganic hybrid materials are usually unstable in strong acidic conditions. Consequently, the recycling of MOF-like absorbents after heavy metal adsorption may be a challenge, as efficient absorbents regeneration (e.g., thiol-functional mesoporous, activated charcoal, and ion-exchange resin), should often be performed following heavy metal adsorption in strong acidic conditions (Ke et al. 2011).

Some of the surface molecules adsorbed by oil are not fully washed away, resulting in decreased adsorption area availability (Banerjee et al. 2012). The regeneration and recyclability are the concerning areas because the stability of sorbents implicitly

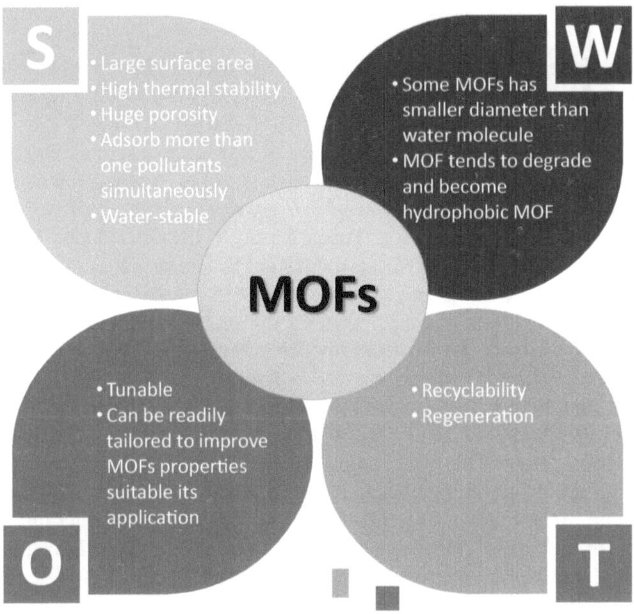

FIGURE 2.5 SWOT summary for MOFs toward wastewater treatment.

and thereafter reduces the net expense of adsorbents. More research is required to find improved regeneration methods for the absorbents used to make them economical. A Summary of the SWOT analysis is shown in Figure 2.5.

2.8 CONCLUSION

This chapter highlights the adsorption-based purification of water by MOFs. The application of MOFs and their composite in the field of wastewater treatment have been studied by numerous researchers, especially its adsorption capability on various contaminants such as dyes, heavy metals, and PPCPs. These unique hybrid nanoparticles can be modified with various functional groups, tuning their shape, and compound with other nanomaterials to control their hydrophilicity, stability, and enhanced adsorption properties which are suitable for various applications. In contrast to conventional adsorbents (which usually rely primally on the unspecific Van der Waals force), MOFs adsorption has been linked to the simultaneous usage of several interactions, including cationic–anionic, hydrogen bonding, Van der Waals interaction, hydrophobic attraction, redox reaction, and π–π stacking. Because of the orientation of their frameworks, MOFs can give better selectivity to wide range of contaminants than other traditional adsorbents. This is attributed to the enormous number of pores with uniform sizes. To a larger extent, composites of MOFs offer a great advantage over their pristine forms due to their multiple functionalities. Most of the reported MOFs demonstrated their good stability and returnability based on their outstanding adsorption capacity. Thus, MOFs have proven to be promising materials for the adsorption of different classes of pollutants and their practicality.

REFERENCES

Abbasi, Alireza, Tahereh Moradpour, and Kristof Van Hecke. 2015. "A New 3D Cobalt (II) Metal-Organic Framework Nanostructure for Heavy Metal Adsorption." *Inorganica Chimica Acta* 430 (May): 261–267. doi:10.1016/j.ica.2015.03.019.

Abd El Salam, H M, and T Zaki. 2018. "Removal of Hazardous Cationic Organic Dyes from Water Using Nickel-Based Metal-Organic Frameworks." *Inorganica Chimica Acta* 471 (February): 203–210. doi:10.1016/j.ica.2017.10.040.

Abdullah, Norfadhilatuadha, Norhaniza Yusof, Rasool J. Gohari, Ahmad F. Ismail, Juhana Jaafar, Woei J Lau, Nurasyikin Misdan, and Nur H H Hairom. 2017. "Characterizations of Polysulfone/Ferrihydrite Mixed Matrix Membranes for Water/Wastewater Treatment." *Water Environment Research* 90 (1): 64–73. doi:10.2175/106143017x15054988926541.

Adeyemo, Aderonke Ajibola, Idowu Olatunbosun Adeoye, and Olugbenga Solomon Bello. 2012. "Metal Organic Frameworks as Adsorbents for Dye Adsorption: Overview, Prospects and Future Challenges." *Toxicological & Environmental Chemistry* 10: 37–41. doi:10.1080/02772248.2012.744023.

Akpinar, Isil, Riki J Drout, Timur Islamoglu, Satoshi Kato, Jiafei Lyu, and Omar K Farha. 2019. "Exploiting π–π Interactions to Design an Efficient Sorbent for Atrazine Removal from Water." *ACS Applied Materials and Interfaces*. doi:10.1021/acsami.8b20355.

Ang, Wei Lun, Abdul Wahab Mohammad, Nidal Hilal, and Choe Peng Leo. 2015. "A Review on the Applicability of Integrated/Hybrid Membrane Processes in Water Treatment and Desalination Plants." *Desalination* 363: 2–18. doi:10.1016/j.desal.2014.03.008.

Arora, Charu, Sanju Soni, Suman Sahu, Jyoti Mittal, Pranaw Kumar, and P K Bajpai. 2019. "Iron Based Metal Organic Framework for Ef Fi Cient Removal of Methylene Blue Dye from Industrial Waste." *Journal of Molecular Liquids* 284: 343–352. doi:10.1016/j.molliq.2019.04.012.

Avci, Civan, Javier Ariñez-Soriano, Arnau Carné-Sánchez, Vincent Guillerm, Carlos Carbonell, Inhar Imaz, and Daniel Maspoch. 2015. "Post-Synthetic Anisotropic Wet-Chemical Etching of Colloidal Sodalite ZIF Crystals." *Angewandte Chemie International Edition* 54 (48): 14417–14421. doi:10.1002/anie.201507588.

Banerjee, Abhik, Rohan Gokhale, Sumit Bhatnagar, Jyoti Jog, Monika Bhardwaj, Benoit Lefez, Beatrice Hannoyer, and Satishchandra Ogale. 2012. "MOF Derived Porous Carbon-Fe 3O 4 Nanocomposite as a High Performance, Recyclable Environmental Superadsorbent." *Journal of Materials Chemistry* 22 (37): 19694–19699. doi:10.1039/c2jm33798c.

Boix, Gerard, Javier Troyano, Luis Garzón-Tovar, Ceren Camur, Natalia Bermejo, Amirali Yazdi, Jordi Piella, et al. 2020. "MOF-Beads Containing Inorganic Nanoparticles for the Simultaneous Removal of Multiple Heavy Metals from Water." *ACS Applied Materials and Interfaces* 12 (9). 10554–10562. doi:10.1021/acsami.9b23206.

Bu, Q, B Wang, J Huang, S Deng, G Yu 2013. "Pharmaceuticals and Personal Care Products in the Aquatic Environment in China: A Review." *Journal of Hazardous Materials*. Elsevier. Accessed April 29. https://www.sciencedirect.com/science/article/pii/S0304389413006067.

Bulloch, Daryl N, Eric D Nelson, Steve A Carr, Chris R Wissman, Jeffrey L Armstrong, Daniel Schlenk, and Cynthia K Larive. 2015. "Occurrence of Halogenated Transformation Products of Selected Pharmaceuticals and Personal Care Products in Secondary and Tertiary Treated Wastewaters from Southern California." *Environmental Science and Technology* 49 (4): 2044–2051. doi:10.1021/es504565n.

Burtch, Nicholas C, Himanshu Jasuja, and Krista S Walton. 2014. "Water Stability and Adsorption in Metal-Organic Frameworks." *Chemical Reviews* 114(20):10575–10612. doi:10.1021/cr5002589.

Campagnol, Nicolò, Tom Van Assche, Tom Boudewijns, Joeri Denayer, Koen Binnemans, Dirk De Vos, and Jan Fransaer. 2013. "High Pressure, High Temperature Electrochemical Synthesis of Metal Organic Frameworks: Films of MIL-100 (Fe) and HKUST-1 in Different Morphologies." *Journal of Materials Chemistry A* 1 (19): 5827–5830. doi:10.1039/c3ta10419b.

Canivet, Jérôme, Alexandra Fateeva, Youmin Guo, Benoit Coasne, and David Farrusseng. 2014. "Water Adsorption in MOFs: Fundamentals and Applications." *Chemical Society Reviews* 43: 5594–5617. Royal Society of Chemistry. doi:10.1039/c4cs00078a.

Cavka, Jasmina Hafizovic, Søren Jakobsen, Unni Olsbye, Nathalie Guillou, Carlo Lamberti, Silvia Bordiga, and Karl Petter Lillerud. 2008. "A New Zirconium Inorganic Building Brick Forming Metal Organic Frameworks with Exceptional Stability." *Journal of the American Chemical Society* 130 (42): 13850–13851. doi:10.1021/ja8057953.

Chakraborty, Anindita, Sohini Bhattacharyya, Arpan Hazra, Ashta Chandra Ghosh, and Tapas Kumar Maji. 2016. "Post-Synthetic Metalation in an Anionic MOF for Efficient Catalytic Activity and Removal of Heavy Metal Ions from Aqueous Solution." *Chemical Communications* 52 (13): 2831–2834. doi:10.1039/c5cc09814a.

Chen, Caiqin, Dezhi Chen, Shasha Xie, Hongying Quan, Xubiao Luo, and Lin Guo. 2017. "Adsorption Behaviors of Organic Micropollutants on Zirconium Metal-Organic Framework UiO-66: Analysis of Surface Interactions." *ACS Applied Materials and Interfaces* 9 (46): 41043–41054. doi:10.1021/acsami.7b13443.

Chowdhury, Pradip, Chaitanya Bikkina, and Sasidhar Gumma. 2009. "Gas Adsorption Properties of the Chromium-Based Metal Organic Framework MIL-101." *Journal of Physical Chemistry C* 113 (16): 6616–6621. doi:10.1021/jp811418r.

Cook, Timothy R., Yao Rong Zheng, and Peter J Stang. 2013. "Metal-Organic Frameworks and Self-Assembled Supramolecular Coordination Complexes: Comparing and Contrasting the Design, Synthesis, and Functionality of Metal-Organic Materials." *Chemical Reviews*. doi:10.1021/cr3002824.

Cychosz, Katie A., Antek G Wong-Foy, and Adam J Matzger. 2008. "Liquid Phase Adsorption by Microporous Coordination Polymers: Removal of Organosulfur Compounds." *Journal of the American Chemical Society* 130 (22): 6938–6939. doi:10.1021/ja802121u.

Dan-Hardi, Meenakshi, Christian Serre, Théo Frot, Laurence Rozes, Guillaume Maurin, Clément Sanchez, and Gérard Férey. 2009. "A New Photoactive Crystalline Highly Porous Titanium(IV) Dicarboxylate." *Journal of the American Chemical Society* 131 (31): 10857–10859. doi:10.1021/ja903726m.

Eddaoudi, Mohamed, Jaheon Kim, Nathaniel Rosi, David Vodak, Joseph Wachter, Michael O'Keeffe, and Omar M Yaghi. 2002. "Systematic Design of Pore Size and Functionality in Isoreticular MOFs and Their Application in Methane Storage." *Science* 295 (5554): 469–472. doi:10.1126/science.1067208.

Ejima, Hirotaka, Joseph J Richardson, Kang Liang, James P Best, Martin P van Koeverden, Georgina K Such, Jiwei Cui, and Frank Caruso. 2013. "One-Step Assembly of Coordination Complexxes for Versatile Film and Particle Engineering." *Science* 154: 341. doi:10.1126/science.1237265.

Fairen-Jimenez, D, S A Moggach, M T Wharmby, P A Wright, S Parsons, and T Düren. 2011. "Opening the Gate: Framework Flexibility in ZIF-8 Explored by Experiments and Simulations." *Journal of the American Chemical Society* 133 (23): 8900–8902. doi:10.1021/ja202154j.

Feng, Liang, Kun Yu Wang, Joshua Powell, and Hong Cai Zhou. 2019. "Controllable Synthesis of Metal Organic Frameworks and Their Hierarchical Assemblies." *Matter* 1 (4): 801–824. doi:10.1016/j.matt.2019.08.022.

Feng, Mingbao, Peng Zhang, Hong Cai Zhou, and Virender K Sharma. 2018. "Water-Stable Metal-Organic Frameworks for Aqueous Removal of Heavy Metals and Radionuclides: A Review." *Chemosphere*. Elsevier Ltd. doi:10.1016/j.chemosphere.2018.06.114.

Férey, C., C Mellot-Draznieks, C Serre, F Millange, J Dutour, S Surblé, and I Margiolaki. 2005. "Chemistry: A Chromium Terephthalate-Based Solid with Unusually Large Pore Volumes and Surface Area." *Science* 309 (5743): 2040–2042. doi:10.1126/science.1116275.

Férey, G. 2005. “Erratum: A Chromium Terephthalate-Based Solid with Unusually Large Pore Volumes and Surface Area (Science (September 23) (2040)).” *Science*. American Association for the Advancement of Science. doi:10.1126/science.310.5751.1119.

Férey, Gérard. 2008. “Hybrid Porous Solids: Past, Present, Future.” *Chemical Society Reviews* 37 (1): 191–214. doi:10.1039/B618320B.

Fri, Tomislav, and F Abi. 2009. “Mechanochemical Conversion of a Metal Oxide into Coordination Polymers and Porous Frameworks Using Liquid-Assisted Grinding (LAG) †.” *CrystEngComm* 11: 743–745. doi:10.1039/b822934c.

Fu, Yanghe, Dengrong Sun, Yongjuan Chen, Renkun Huang, Zhengxin Ding, Xianzhi Fu, and Zhaohui Li. 2012. “An Amine-Functionalized Titanium Metal-Organic Framework Photocatalyst with Visible-Light-Induced Activity for CO_2 Reduction.” *Angewandte Chemie International Edition* 51 (14): 3364–3367. doi:10.1002/anie.201108357.

Fujita, Makoto, Satoru Washizu, Katsuyuki Ogura, and Yoon Jung Kwon. 1994. “Preparation, Clathration Ability, and Catalysis of a Two-Dimensional Square Network Material Composed of Cadmium(II) and 4, 4′-Bipyridine.” *Journal of the American Chemical Society* 116 (3): 1151–1152. doi:10.1021/ja00082a055.

Furukawa, Shuhei, Julien Reboul, Stéphane Diring, Kenji Sumida, and Susumu Kitagawa. 2014. “Structuring of Metal-Organic Frameworks at the Mesoscopic/Macroscopic Scale.” *Chemical Society Reviews*. Royal Society of Chemistry. doi:10.1039/c4cs00106k.

Gadipelly, Chandrakanth, Antía Pérezpérez-Gonzaíez, Ganapati D Yadav, Inmaculada Ortiz, Raquel Ibáñ Ibáñ Ez, Virendra K Rathod, and Kumudini V Marathe. 2014. “Pharmaceutical Industry Wastewater: Review of the Technologies for Water Treatment and Reuse.” *ACS Publications* 53 (29): 11571–11592. doi:10.1021/ie501210j.

Gaikwad, Sanjit, Seok Jhin Kim, and Sangil Han. 2020. “Novel Metal–Organic Framework of UTSA-16 (Zn) Synthesized by a Microwave Method: Outstanding Performance for CO_2 Capture with Improved Stability to Acid Gases.” *Journal of Industrial and Engineering Chemistry* 87 (July): 250–263. doi:10.1016/j.jiec.2020.04.015.

Gao, Yanxin, Ruoxi Kang, Jing Xia, Gang Yu, and Shubo Deng. 2019. “Understanding the Adsorption of Sulfonamide Antibiotics on MIL-53s: Metal Dependence of Breathing Effect and Adsorptive Performance in Aqueous Solution.” Journal of Colloid and Interface Science 535 (February): 159–168. doi:10.1016/j.jcis.2018.09.090.

Gao, Yanxin, Jing Xia, Dengchao Liu, Ruoxi Kang, Gang Yu, and Shubo Deng. 2019. “Synthesis of Mixed-Linker Zr-MOFs for Emerging Contaminant Adsorption and Photodegradation under Visible Light.” *Chemical Engineering Journal* 378 (December). doi:10.1016/j.cej.2019.122118.

Gonzalez, E, Jordi Arbiol, and Víctor F Puntes. 2011. “Carving at the Nanoscale: Sequential Galvanic Exchange and Kirkendall Growth at Room Temperature.” *Science* 334 (6061): 1377–13780. doi:10.1126/science.1212822.

Han, Rebecca, Nina Tymińska, J R Schmidt, and David S Sholl. 2019. “Propagation of Degradation-Induced Defects in Zeolitic Imidazolate Frameworks.” *Journal of Physical Chemistry C* 123 (11): 6655–6666. doi:10.1021/acs.jpcc.9b00304.

Haque, Enamul, Jong Won Jun, and Sung Hwa Jhung. 2011. “Adsorptive Removal of Methyl Orange and Methylene Blue from Aqueous Solution with a Metal-Organic Framework Material, Iron Terephthalate (MOF-235).” *Journal of Hazardous Materials* 185 (1): 507–511. doi:10.1016/j.jhazmat.2010.09.035.

Haque, Enamul, Nazmul Abedin Khan, Hwa Jung Park, and Sung Hwa Jhung. 2010. “Synthesis of a Metal-Organic Framework Material, Iron Terephthalate, by Ultrasound, Microwave, and Conventional Electric Heating: A Kinetic Study.” *Chemistry – A European Journal* 16 (3): 1046–10452. doi:10.1002/chem.200902382.

Hasan, Zubair, and Sung Hwa Jhung. 2015. “Removal of Hazardous Organics from Water Using Metal-Organic Frameworks (MOFs): Plausible Mechanisms for Selective Adsorptions.” *Journal of Hazardous Materials* 283: 329–339. doi:10.1016/j.jhazmat.2014.09.046.

Hasan, Zubair, Jon Won Jun, and Sung Hwa Jhung. 2015. "Sulfonic Acid-Functionalized MIL-101(Cr): An Efficient Catalyst for Esterification of Oleic Acid and Vapor-Phase Dehydration of Butanol." *Chemical Engineering Journal* 278: 265–271. doi:10.1016/j.cej.2014.09.025.

Hoskins, B F, and R Robson. 1990. "Design and Construction of a New Class of Scaffolding-like Materials Comprising Infinite Polymeric Frameworks of 3D-Linked Molecular Rods. A Reappraisal of the Zn(CN)2 and Cd(CN)2 Structures and the Synthesis and Structure of the Diamond-Related Framework." *Journal of the American Chemical Society* 112 (4): 1546–1554. doi:10.1021/ja00160a038.

Hu, Han, Haixuan Zhang, Yujia Chen, and Huase Ou. 2019. "Enhanced Photocatalysis Using Metal–Organic Framework MIL-101(Fe) for Organophosphate Degradation in Water." *Environmental Science and Pollution Research* 101. doi:10.1007/s11356-019-05649-2.

Hu, Ming, Yi Ju, Kang Liang, Tomoya Suma, Jiwei Cui, and Frank Caruso. 2016. "Void Engineering in Metal Organic Frameworks via Synergistic Etching and Surface Functionalization." *Advanced Functional Materials* 26: 5827–5834. doi:10.1002/adfm.201601193.

Hu, Shen, Min Liu, Keyan Li, Chunshan Song, Guoliang Zhang, and Xinwen Guo. 2017. "Surfactant-Assisted Synthesis of Hierarchical NH2-MIL-125 for the Removal of Organic Dyes." *RSC Advances* 7 (1): 581–587. doi:10.1039/c6ra25745c.

Huang, Lijin, Man He, Beibei Chen, and Bin Hu. 2015. "A Designable Magnetic MOF Composite and Facile Coordination-Based Post-Synthetic Strategy for the Enhanced Removal of Hg2+ from Water." *Journal of Materials Chemistry A* 3 (21): 11587–11595. doi:10.1039/c5ta01484k.

Huo, Jia, Li Wang, Elisabeth Irran, Haojie Yu, Jingming Gao, Dengsen Fan, Bao Li, et al. 2010. "Hollow Ferrocenyl Coordination Polymer Microspheres with Micropores in Shells Prepared by Ostwald Ripening." *Angewandte Chemie International Edition* 49 (48): 9237–9241. doi:10.1002/anie.201004745.

Jang, Sanha, Sehwan Song, Ji Hwan Lim, Han Seong Kim, Bach Thang Phan, Ki Tae Ha, Sungkyun Park, and Kang Hyun Park. 2020. "Application of Various Metal-Organic Frameworks (Mofs) as Catalysts for Air and Water Pollution Environmental Remediation." *Catalysts* 10(2), 195. doi.10.3390/catal10020195.

Jhung, Sung Hwa, Nazmul Abedin Khan, and Zubair Hasan. 2012. "Analogous Porous Metal-Organic Frameworks: Synthesis, Stability and Application in Adsorption." *CrystEngComm* 14 (21): 7099–7109. doi:10.1039/c2ce25760b.

Jiang, Dongmei, Luke L Keenan, Andrew D Burrows, and Karen J Edler. 2012. "Synthesis and Post-Synthetic Modification of MIL-101(Cr)-NH2via a Tandem Diazotisation Process." *Chemical Communications* 48 (99): 12053–12055. doi:10.1039/c2cc36344c.

Jin, Eunji, Soochan Lee, Eunyoung Kang, Yeongjin Kim, and Wonyoung Choe. 2020. "Metal-Organic Frameworks as Advanced Adsorbents for Pharmaceutical and Personal Care Products." *Coordination Chemistry Reviews*. doi:10.1016/j.ccr.2020.213526.

Johari, Nur Azizah, Norhaniza Yusof, and Ahmad Fauzi Ismail. 2021. "Fabrication and Characterizations of Hybrid Membrane Containing Tannin-Modified Metal-Organic Framework for Water Treatment." *Materials Today: Proceedings*. Elsevier. doi:10.1016/j.matpr.2021.02.245.

Johari, Nur Azizah, Norhaniza Yusof, Ahmad Fauzi Ismail, Farhana Aziz, Wan Norharyati Wan Salleh, Juhana Jaafar, Nur Hanis Hayati Hairon, and Nurasyikin Misdan. 2020. "The Application of Ferric-Metal-Organic Framework for Dye Removal: A Mini Review." *Journal of Advanced Research in Fluid Mechanics and Thermal Sciences* 75 (1): 68–80. doi:10.37934/ARFMTS.75.1.6880.

Johari, Nur Azizah, Norhaniza Yusof, Woei Jye Lau, Norfadhilatuladha Abdullah, Wan Norharyati Wan Salleh, Juhana Jaafar, Farhana Aziz, and Ahmad Fauzi Ismail. 2021. "Polyethersulfone Ultrafiltration Membrane Incorporated with Ferric-Based

Metal-Organic Framework for Textile Wastewater Treatment." *Separation and Purification Technology* 270 (September): 118819. doi:10.1016/j.seppur.2021.118819.

Jung, Beom K., Zubair Hasan, and Sung Hwa Jhung. 2013. "Adsorptive Removal of 2,4-Dichlorophenoxyacetic Acid (2,4-D) from Water with a Metal-Organic Framework." *Chemical Engineering Journal* 234 (September): 99–105. doi:10.1016/j.cej.2013.08.110.

Jung, C, A Son, N Her, KD Zoh, J Cho, Y Yoon 2015. "Removal of Endocrine Disrupting Compounds, Pharmaceuticals, and Personal Care Products in Water Using Carbon Nanotubes: A Review." *Journal of Industrial and Engineering Chemistry* 27: 1–11. Accessed April 29. https://www.sciencedirect.com/science/article/pii/S1226086X14007047.

Ke, Fei, Jing Jiang, Yizhi Li, Jing Liang, Xiaochun Wan, and Sanghoon Ko. 2017. "Highly Selective Removal of Hg 2+ and Pb 2+ by Thiol-Functionalized Fe 3 O 4 @metal-Organic Framework Core-Shell Magnetic Microspheres." Applied Surface Science 413 (August): 266–274. doi:10.1016/j.apsusc.2017.03.303.

Ke, Fei, Ling Guang Qiu, Yu Peng Yuan, Fu Min Peng, Xia Jiang, An Jian Xie, Yu Hua Shen, and Jun Fa Zhu. 2011. "Thiol-Functionalization of Metal-Organic Framework by a Facile Coordination-Based Postsynthetic Strategy and Enhanced Removal of Hg 2+ from Water." *Journal of Hazardous Materials* 196 (November): 36–43. doi:10.1016/j.jhazmat.2011.08.069.

Ke, Fei, Yu Peng Yuan, Ling Guang Qiu, Yu Hua Shen, An Jian Xie, Jun Fa Zhu, Xing You Tian, and Li De Zhang. 2011. "Facile Fabrication of Magnetic Metal Organic Framework Nanocomposites for Potential Targeted Drug Delivery." *Journal of Materials Chemistry* 21 (11): 3843–3848. doi:10.1039/c0jm01770a.

Kim, Hyehyun, Minhak Oh, Dongwook Kim, Jeongin Park, Junmo Seong, Sang Kyu Kwak, and Myoung Soo Lah. 2015. "Single Crystalline Hollow Metal Organic Frameworks: A Metal Organic Polyhedron Single Crystal as a Sacrificial Template." *Chemical Communications* 51 (17): 3678–3681. doi:10.1039/C4CC10051D.

Kitagawa, Susumu, Ryo Kitaura, and Shin Ichiro Noro. 2004. "Functional Porous Coordination Polymers." *Angewandte Chemie – International Edition.* doi:10.1002/anie.200300610.

Kondo, Mitsuru, Tomomichi Yoshitomi, Kenji Seki, Hiroyuki Matsuzaka, and Susumu Kitagawa. 1997. "Three-Dimensional Framework with Channeling Cavities for Small Molecules: {[M2(4,4′-Bpy)3(NO3) 4]·xH2O}n (M = Co, Ni, Zn)." *Angewandte Chemie (International Edition in English)* 36 (16): 1725–1727. doi:10.1002/anie.199717251.

Kuppler, Ryan J., Daren J Timmons, Qian Rong Fang, Jian Rong Li, Trevor A Makal, Mark D Young, Daqiang Yuan, Dan Zhao, Wenjuan Zhuang, and Hong Cai Zhou. 2009. "Potential Applications of Metal-Organic Frameworks." *Coordination Chemistry Reviews.* doi:10.1016/j.ccr.2009.05.019.

Lee, Yu Ri, Jun Kim, and Wha Seung Ahn. 2013. "Synthesis of Metal-Organic Frameworks: A Mini Review." *Korean Journal of Chemical Engineering.* doi:10.1007/s11814-013-0140-6.

Li, Haichao, Xinyu Cao, Chuang Zhang, Qing Yu, Zijian Zhao, Xuedun Niu, Xiaodong Sun, Yunling Liu, Li Ma, and Zhengqiang Li. 2017. "Enhanced Adsorptive Removal of Anionic and Cationic Dyes from Single or Mixed Dye Solutions Using MOF PCN-222." *RSC Advances* 7 (27): 16273–16281. doi:10.1039/c7ra01647f.

Li, Hallian, Mohamed Eddaoudi, M O'Keeffe, and O M Yaghi. 1999. "Design and Synthesis of an Exceptionally Stable and Highly Porous Metal- Organic Framework." *Nature* 402 (6759): 276–279. doi:10.1038/46248.

Li, Jian Rong, Julian Sculley, and Hong Cai Zhou. 2012. "Metal-Organic Frameworks for Separations." *Chemical Reviews.* doi:10.1021/cr200190s.

Li, Li Li, Xiao Quan Feng, Run Ping Han, Shuang Quan Zang, and Guang Yang. 2017. "Cr(-VI) Removal via Anion Exchange on a Silver-Triazolate MOF." *Journal of Hazardous Materials* 321 (January): 622–628. doi:10.1016/j.jhazmat.2016.09.029.

Li, Yingwei, and Ralph T Yang. 2007. "Gas Adsorption and Storage in Metal-Organic Framework MOF-177." *Langmuir* 23 (26): 12937–12944. doi:10.1021/la702466d.

Lin, Shuo, Yufeng Zhao, and Yeoung Sang Yun. 2018. "Highly Effective Removal of Nonsteroidal Anti-Inflammatory Pharmaceuticals from Water by Zr(IV)-Based Metal-Organic Framework: Adsorption Performance and Mechanisms." *ACS Applied Materials and Interfaces* 10 (33): 28076–28085. doi:10.1021/acsami.8b08596.

Liu, Lieju, T M Yang, Wolfgang Liedtke, and S A Simon. 2006. "Chronic IL-1β Signaling Potentiates Voltage-Dependent Sodium Currents in Trigeminal Nociceptive Neurons." *Journal of Neurophysiology* 95 (3): 1478–1490. doi:10.1152/jn.00509.2005.

Liu, Wenxian, Jijiang Huang, Qiu Yang, Shiji Wang, Xiaoming Sun, Weina Zhang, Junfeng Liu, and Fengwei Huo. 2017. "Multi-Shelled Hollow Metal Organic Frameworks." *Angewandte Chemie International Edition* 56 (20): 5512–5516. doi:10.1002/anie.201701604.

Liu, Yuan, Yu Yang, Yujia Sun, Jibin Song, Nicholas G Rudawski, Xiaoyuan Chen, and Weihong Tan. 2019. "Ostwald Ripening-Mediated Grafting of Metal–Organic Frameworks on a Single Colloidal Nanocrystal to Form Uniform and Controllable MXF." *Journal of the* American Chemical Society 141 (18): 7407–7413. doi:10.1021/jacs.9b01563.

Low, John J., Annabelle I Benin, Paulina Jakubczak, Jennifer F Abrahamian, Syed A Faheem, and Richard R Willis. 2009. "Virtual High Throughput Screening Confirmed Experimentally: Porous Coordination Polymer Hydration." *Journal of the American Chemical Society* 131 (43): 15834–15842. doi:10.1021/ja9061344.

Lu, Chun Mei, Jun Liu, Kefeng Xiao, and Andrew T Harris. 2010. "Microwave Enhanced Synthesis of MOF-5 and Its CO_2 Capture Ability at Moderate Temperatures across Multiple Capture and Release Cycles." *Chemical Engineering Journal* 156 (2): 465–470. doi:10.1016/j.cej.2009.10.067.

Lu, Weigang, Zhangwen Wei, Zhi Yuan Gu, Tian Fu Liu, Jinhee Park, Jihye Park, Jian Tian, et al. 2014. "Tuning the Structure and Function of Metal-Organic Frameworks via Linker Design." *Chemical Society Reviews*. Royal Society of Chemistry. doi:10.1039/c4cs00003j.

Lu, Weigang, Daqiang Yuan, Dan Zhao, Christine Inge Schilling, Oliver Plietzsch, Thierry Muller, Stefan Bräse, et al. 2010. "Porous Polymer Networks: Synthesis, Porosity, and Applications in Gas Storage/Separation." *Chemistry of Materials* 22 (21): 5964–5972. doi:10.1021/cm1021068.

Lv, Shi Wen, Jing Min Liu, Hui Ma, Zhi Hao Wang, Chun Yang Li, Ning Zhao, and Shuo Wang. 2019. "Simultaneous Adsorption of Methyl Orange and Methylene Blue from Aqueous Solution Using Amino Functionalized Zr-Based MOFs." *Microporous and Mesoporous Materials* 282 (July): 179–187. doi:10.1016/j.micromeso.2019.03.017.

Mahmoodi, Niyaz Mohammad, Jafar Abdi, Mina Oveisi, Mokhtar Alinia Asli, and Manouchehr Vossoughi. 2018. "Metal-Organic Framework (MIL-100 (Fe)): Synthesis, Detailed Photocatalytic Dye Degradation Ability in Colored Textile Wastewater and Recycling." *Materials Research Bulletin* 100: 357–366. doi:10.1016/j.materresbull.2017.12.033.

Mantion, Alexandre, Lars Massüger, Pierre Rabu, Cornelia Palivan, Lynne B McCusker, and Andreas Taubert. 2008. "Metal-Peptide Frameworks (MPFs): 'Bioinspired' Metal Organic Frameworks." *Journal of the American Chemical Society* 130 (8): 2517–2526. doi:10.1021/ja0762588.

McGuire, Christina V, and Ross S Forgan. 2015. "The Surface Chemistry of Metal-Organic Frameworks." *Chemical Communications* 51 (25): 5199–5217. doi:10.1039/c4cc04458d.

Meng, Ai Na, Ling Xiao Chaihu, Huan Huan Chen, and Zhi Yuan Gu. 2017. "Ultrahigh Adsorption and Singlet-Oxygen Mediated Degradation for Efficient Synergetic Removal of Bisphenol A by a Stable Zirconium-Porphyrin Metal-Organic Framework." *Scientific Reports* 7 (1). Nature Publishing Group. doi:10.1038/s41598-017-06194-z.

Moghadam, Peyman Z, Aurelia Li, Seth B Wiggin, Andi Tao, Andrew G P Maloney, Peter A Wood, Suzanna C Ward, and David Fairen-Jimenez. 2017. "Development of a Cambridge

Structural Database Subset: A Collection of Metal-Organic Frameworks for Past, Present, and Future." *Chemistry of Materials*. doi:10.1021/acs.chemmater.7b00441.

Molavi, Hossein, Alireza Hakimian, Akbar Shojaei, and Milad Raeiszadeh. 2018. "Selective Dye Adsorption by Highly Water Stable Metal-Organic Framework: Long Term Stability Analysis in Aqueous Media." *Applied Surface Science* 445 (July): 424–436. doi:10.1016/j.apsusc.2018.03.189.

Molavi, Hossein, Milad Neshastehgar, Akbar Shojaei, and Hossein Ghashghaeinejad. 2020. "Ultrafast and Simultaneous Removal of Anionic and Cationic Dyes by Nanodiamond/UiO-66 Hybrid Nanocomposite." *Chemosphere* 247 (May): 125882. doi:10.1016/j.chemosphere.2020.125882.

Moorhouse, Saul J., Yue Wu, and Dermot O'Hare. 2016. "An in Situ Study of Resin-Assisted Solvothermal Metal-Organic Framework Synthesis." *Journal of Solid State Chemistry* 236 (April): 209–214. doi:10.1016/j.jssc.2015.07.035.

Mueller, U, M Schubert, F Teich, H Puetter, K Schierle-Arndt, and J Pastré. 2006. "Metal-Organic Frameworks - Prospective Industrial Applications." *Journal of Materials Chemistry*, 16: 626–636. doi:10.1039/b511962f.

Nazir, Muhammad Altaf, Naseem Ahmad Khan, Chao Cheng, Syed Shoaib Ahmad Shah, Tayyaba Najam, Muhammad Arshad, Ahsan Sharif, Shahbaz Akhtar, and Azizur Rehman. 2020. "Surface Induced Growth of ZIF-67 at Co-Layered Double Hydroxide: Removal of Methylene Blue and Methyl Orange from Water." *Applied Clay Science* 190 (June): 105564. doi:10.1016/j.clay.2020.105564.

Pavithra, K Grace, P Senthil Kumar, V Jaikumar, and P Sundar Rajan. 2019. "Removal of Colorants from Wastewater: A Review on Sources and Treatment Strategies." *Journal of Industrial and Engineering Chemistry* 75: 1–19. doi:10.1016/j.jiec.2019.02.011.

Peng, Yaguang, Yuxi Zhang, Hongliang Huang, and Chongli Zhong. 2018. "Flexibility Induced High-Performance MOF-Based Adsorbent for Nitroimidazole Antibiotics Capture." *Chemical Engineering Journal* 333 (February): 678–685. doi:10.1016/j.cej.2017.09.138.

Petit, Camille. 2018. "Present and Future of MOF Research in the Field of Adsorption and Molecular Separation." *Current Opinion in Chemical Engineering*. doi:10.1016/j.coche.2018.04.004.

Pichon, Anne, Ana Lazuen-Garay, and Stuart L James. 2006. "Solvent-Free Synthesis of a Microporous Metal-Organic Framework." *CrystEngComm* 8 (3): 211–214. doi:10.1039/b513750k.

Rahimi, Esmaeil, and Neda Mohaghegh. 2016. "Entfernung von Toxischen Metallionen Aus Sauerwasser Mittels Mordenit, Graphen-Nanoschichten Sowie Einer Neuartigen Metallorganischen Gerüstverbindung." *Mine Water and the Environment* 35 (1): 18–28. doi:10.1007/s10230-015-0327-7.

Rivera, José Mariá, Susana Rincón, Cherif Ben Youssef, and Alejandro Zepeda. 2016. "Highly Efficient Adsorption of Aqueous Pb(II) with Mesoporous Metal-Organic Framework-5: An Equilibrium and Kinetic Study." *Journal of Nanomaterials* 2016. doi:10.1155/2016/8095737.

Robles-García, Miguel Angel, Francisco Rodríguez-Félix, Enrique Márquez-Ríos, José Antonio Aguilar, Arturo Barrera-Rodríguez, Jacobo Aguilar, Saúl Ruiz-Cruz, and Carmen Lizette Del-Toro-Sánchez. 2016. "Applications of Nanotechnology in the Agriculture, Food, and Pharmaceuticals." *Journal of Nanoscience and Nanotechnology* 16 (8): 8188–8207. doi:10.1166/jnn.2016.12925.

Rowsell, Jesse LC, and Omar M Yaghi. 2004. "Metal-Organic Frameworks: A New Class of Porous Materials." *Microporous and Mesoporous Materials*. doi:10.1016/j.micromeso.2004.03.034.

Russo, Vincenzo, Maryam Hmoudah, Francesco Broccoli, Maria Rosaria Iesce, Ok-Sang Jung, and Martino Di Serio. 2020. "Applications of Metal Organic Frameworks in

Wastewater Treatment: A Review on Adsorption and Photodegradation." *Frontiers in Chemical Engineering* 2 (October). doi:10.3389/fceng.2020.581487.

Safarifard, Vahid, and Ali Morsali. 2015. "Applications of Ultrasound to the Synthesis of Nanoscale Metal-Organic Coordination Polymers." *Coordination Chemistry Reviews*. doi:10.1016/j.ccr.2015.02.014.

Saleem, Hira, Uzaira Rafique, and Robert P Davies. 2016. "Investigations on Post-Synthetically Modified UiO-66-NH2 for the Adsorptive Removal of Heavy Metal Ions from Aqueous Solution." *Microporous and Mesoporous Materials* 221 (February): 238–244. doi:10.1016/j.micromeso.2015.09.043.

Sarker, Mithun, Subin Shin, Jong Hwa Jeong, and Sung Hwa Jhung. 2019. "Mesoporous Metal-Organic Framework PCN-222(Fe): Promising Adsorbent for Removal of Big Anionic and Cationic Dyes from Water." *Chemical Engineering Journal* 371 (March): 252–259. doi:10.1016/j.cej.2019.04.039.

Sarker, Mithun, Ji Yoon Song, and Sung Hwa Jhung. 2017. "Adsorption of Organic Arsenic Acids from Water over Functionalized Metal-Organic Frameworks." *Journal of Hazardous Materials* 335 (August): 162–169. doi:10.1016/j.jhazmat.2017.04.044.

Seo, Pill Won, Biswa Nath Bhadra, Imteaz Ahmed, Nazmul Abedin Khan, and Sung Hwa Jhung. 2016. "Adsorptive Removal of Pharmaceuticals and Personal Care Products from Water with Functionalized Metal-Organic Frameworks: Remarkable Adsorbents with Hydrogen-Bonding Abilities." *Scientific Reports* 6 (1): 1–11. doi:10.1038/srep34462.

Shen, Zhurui, Ji Liu, Fangyun Hu, Song Liu, Ning Cao, Ying Sui, Qingdao Zeng, and Yongtao Shen. 2014. "Bottom-up Synthesis of Cerium-Citric Acid Coordination Polymers Hollow Microspheres with Tunable Shell Thickness and Their Corresponding Porous CeO2 Hollow Spheres for Pt-Based Electrocatalysts." *CrystEngComm* 16 (16): 3387–3394. doi:10.1039/c3ce42400f.

Sudik, Andrea C., Adrien P Côté, and Omar M Yaghi. 2005. "Metal-Organic Frameworks Based on Trigonal Prismatic Building Blocks and the New 'Acs' Topology." *Inorganic Chemistry* 44 (9): 2998–3000. doi:10.1021/ic050064g.

Tomic, E A. 1965. "Thermal Stability of Coordination Polymers." *Journal of Applied Polymer Science* 9 (11): 3745–3752. doi:10.1002/app.1965.070091121.

Tong, Minman, Dahuan Liu, Qingyuan Yang, Sabine Devautour-Vinot, Guillaume Maurin, and Chongli Zhong. 2013. "Influence of Framework Metal Ions on the Dye Capture Behavior of MIL-100 (Fe, Cr) MOF Type Solids." *Journal of Materials Chemistry A* 1 (30): 8534–8537. doi:10.1039/c3ta11807j.

Tu, Binbin, Qingqing Pang, Doufeng Wu, Yuna Song, Linhong Weng, and Qiaowei Li. 2014. "Ordered Vacancies and Their Chemistry in Metal Organic Frameworks." *Journal of the* American Chemical Society 136 (41): 14465–14471. doi:10.1021/ja5063423.

Wang, Bin, Xiu Liang Lv, Dawei Feng, Lin Hua Xie, Jian Zhang, Ming Li, Yabo Xie, Jian Rong Li, and Hong Cai Zhou. 2016. "Highly Stable Zr(IV)-Based Metal-Organic Frameworks for the Detection and Removal of Antibiotics and Organic Explosives in Water." *Journal of the American Chemical Society* 138 (19): 6204–6216. doi:10.1021/jacs.6b01663.

Wang, Dongbo, Feiyue Jia, Hou Wang, Fei Chen, Ying Fang, Wenbo Dong, Guangming Zeng, Xiaoming Li, Qi Yang, and Xingzhong Yuan. 2018. "Simultaneously Efficient Adsorption and Photocatalytic Degradation of Tetracycline by Fe-Based MOFs." *Journal of Colloid and Interface Science* 519 (June): 273–284. doi:10.1016/j.jcis.2018.02.067.

Wang, Hai Ning, Fu Hong Liu, Xin Long Wang, Kui Zhan Shao, and Zhong Min Su. 2013. "Three Neutral Metal-Organic Frameworks with Micro- and Meso-Pores for Adsorption and Separation of Dyes." *Journal of Materials Chemistry A* 1 (42): 13060–13063. doi:10.1039/c3ta13242k.

Wang, Qian, Xiaofei Wang, and Chunlei Shi. 2018. "LDH Nanoflower Lantern Derived from ZIF-67 and Its Application for Adsorptive Removal of Organics from Water."

Industrial and Engineering Chemistry Research 57 (37): 12478–12484. doi:10.1021/acs.iecr.8b01324.

Wu, Di, and Alexandra Navrotsky. 2015. "Thermodynamics of Metal-Organic Frameworks." *Journal of Solid State Chemistry* 223: 53–58. doi:10.1016/j.jssc.2014.06.015.

Wu, Xiaoqin, Jeremy L Conkle, Frederick Ernst, and Jay Gan. 2014. "Treated Wastewater Irrigation: Uptake of Pharmaceutical and Personal Care Products by Common Vegetables under Field Conditions." *ACS Publications* 48 (19): 11286–11293. doi:10.1021/es502868k.

Yaghi, O M, Guangming Li, and Hailian Li. 1995. "Selective Binding and Removal of Guests in a Microporous Metal–Organic Framework." *Nature* 378 (6558): 703–706. doi:10.1038/378703a0.

Yaghi, Omar M, Michael O'Keeffe, Nathan W Ockwig, Hee K Chae, Mohamed Eddaoudi, and Jaheon Kim. 2003. "Reticular Synthesis and the Design of New Materials." *Nature*. doi:10.1038/nature01650.

Yagub, Mustafa T, Tushar Kanti Sen, Sharmeen Afroze, and H M Ang. 2014. "Dye and Its Removal from Aqueous Solution by Adsorption: A Review." *Advances in Colloid and Interface Science*. doi:10.1016/j.cis.2014.04.002.

Yan, Han, Hu Wu, Kun Li, Yawen Wang, Xue Tao, Hu Yang, Aimin Li, and Rongshi Cheng. 2015. "Influence of the Surface Structure of Graphene Oxide on the Adsorption of Aromatic Organic Compounds from Water." *ACS Applied Materials and Interfaces* 7 (12): 6690–6697. doi:10.1021/acsami.5b00053.

Yang, Hao, Xiaopo Cheng, Xuanxuan Cheng, Fusheng Pan, Hong Wu, Guanhua Liu, Yimeng Song, Xingzhong Cao, and Zhongyi Jiang. 2018. "Highly Water-Selective Membranes Based on Hollow Covalent Organic Frameworks with Fast Transport Pathways." *Journal of Membrane Science* 565 (November): 331–341. doi:10.1016/j.memsci.2018.08.043.

Yang, Hongwei, Samuel Orefuwa, and Andrew Goudy. 2011. "Study of Mechanochemical Synthesis in the Formation of the Metal-Organic Framework Cu3(BTC)2 for Hydrogen Storage." *Microporous and Mesoporous Materials* 143 (1): 37–45. doi:10.1016/j.micromeso.2011.02.003.

Yin, Yadong, Robert M Rioux, Can K Erdonmez, Steven Hughes, Gabor A Somorjal, and A Paul Alivisatos. 2004. "Formation of Hollow Nanocrystals Through the Nanoscale Kirkendall Effect." *Science* 304 (5671): 711–714. doi:10.1126/science.1096566.

Yong, Jae, Choi Jeo Kim, Hwa Jhung, Hye-Kyoung Kim, Jong-San Chang, and Hee K Chae. 2006. "Microwave Synthesis of a Porous Metal-Organic Framework, Zinc Terephthalate MOF-5." *Communications to the Editor Bulletin of the Korean Chemical Society* 27.

Yuan, Binqin, Xun Wang, Xin Zhou, Jing Xiao, and Zhong Li. 2019. "Novel Room-Temperature Synthesis of MIL-100(Fe) and Its Excellent Adsorption Performances for Separation of Light Hydrocarbons." *Chemical Engineering Journal* 355 (May): 679–686. doi:10.1016/j.cej.2018.08.201.

Yuan, Shuai, Ying Pin Chen, Jun Sheng Qin, Weigang Lu, Lanfang Zou, Qiang Zhang, Xuan Wang, Xing Sun, and Hong Cai Zhou. 2016. "Linker Installation: Engineering Pore Environment with Precisely Placed Functionalities in Zirconium MOFs." *Journal of the American Chemical Society* 138 (28): 8912–8919. doi:10.1021/jacs.6b04501.

Zhang, Chenyang, Chu Han, David S Sholl, and J R Schmidt. 2016. "Computational Characterization of Defects in Metal-Organic Frameworks: Spontaneous and Water-Induced Point Defects in ZIF-8." *Journal of Physical Chemistry Letters* 7 (3): 459–464. doi:10.1021/acs.jpclett.5b02683.

Zhang, Jinmiao, Zhenhu Xiong, Chen Li, and Chunsheng Wu. 2016. "Exploring a Thiol-Functionalized MOF for Elimination of Lead and Cadmium from Aqueous Solution." *Journal of Molecular Liquids* 221 (September): 43–50. doi:10.1016/j.molliq.2016.05.054.

Zhang, Ke, Ryan P Lively, Chen Zhang, Ronald R Chance, William J Koros, David S Sholl, and Sankar Nair. 2013. "Exploring the Framework Hydrophobicity and Flexibility of Zif-8: From Biofuel Recovery to Hydrocarbon Separations." *Journal of Physical Chemistry Letters* 4 (21): 3618–3622. doi:10.1021/jz402019d.

Zhang, Tierui, Jianping Ge, Yongxing Hu, Qiao Zhang, Shaul Aloni, and Yadong Yin. 2008. "Formation of Hollow Silica Colloids through a Spontaneous Dissolution-Regrowth Process." *Angewandte Chemie - International Edition* 47 (31): 5806–5811. doi:10.1002/anie.200800927.

Zhang, Zhicheng, Yifeng Chen, Xiaobin Xu, Jingchao Zhang, Guolei Xiang, Wei He, and Xun Wang. 2014. "Well-Defined Metal Organic Framework Hollow Nanocages." *Angewandte Chemie - International Edition* 53 (2): 429–433. doi:10.1002/anie.201308589.

Zhao, Xiaoli, Shuangliu Liu, Zhi Tang, Hongyun Niu, Yaqi Cai, Wei Meng, Fengchang Wu, and John P Giesy. 2015. "Synthesis of Magnetic Metal-Organic Framework (MOF) for Efficient Removal of Organic Dyes from Water." *Scientific Reports* 5 (1): 1–10. doi:10.1038/srep11849.

Zhou, Hong Cai, Jeffrey R Long, and Omar M Yaghi. 2012. "Introduction to Metal-Organic Frameworks." *Chemical Reviews*. American Chemical Society. doi:10.1021/cr300014x.

Zhou, Meimei, Yi nan Wu, Junlian Qiao, Jing Zhang, Amanda McDonald, Guangtao Li, and Fengting Li. 2013. "The Removal of Bisphenol A from Aqueous Solutions by MIL-53(Al) and Mesostructured MIL-53(Al)." *Journal of Colloid and Interface Science* 405 (September): 157–163. doi:10.1016/j.jcis.2013.05.024.

Zhu, Bang Jing, Xin Yao Yu, Yong Jia, Fu Min Peng, Bai Sun, Mei Yun Zhang, Tao Luo, Jin Huai Liu, and Xing Jiu Huang. 2012. "Iron and 1,3,5-Benzenetricarboxylic Metal-Organic Coordination Polymers Prepared by Solvothermal Method and Their Application in Efficient as(V) Removal from Aqueous Solutions." *Journal of Physical Chemistry C* 116 (15): 8601–8607. doi:10.1021/jp212514a.

3 Impact of Metal Oxide Nanoparticles on Adsorptive and Photocatalytic Schemes

Muhammad Ikram
Government College University Lahore

Usman Qumar and Salamat Ali
Riphah International University

Anwar Ul-Hamid
King Fahd University of Petroleum & Minerals

CONTENTS

DOI: 10.1201/9781003167327-4

3.1 INTRODUCTION

Rapid industrialization, expanding urbanization, increasing population, widespread use of non-viable means, and unbalanced use of natural resources are creating dangerous, irreversible, and severe environmental problems. The major threat to aquatic life lies under rivers, ponds, oceans, lakes, and so on. A number of industries such as paper, textiles chemicals, pulp, microorganisms, fertilizers, metal-plating, pharmaceutical, refineries, and food processing sectors release millions of gallons of waste in water every day, which contaminate our atmosphere, terrestrial land, and water species. Wastewater contains a wide range of contaminants, including environmental (dyes, fertilizers, organo-halides, microorganisms, medicinal ingredients, phenol, surfactants, and so on) and inorganic (MO, heavy metal ions, salts, metal compounds, and so on) pollutants, bacteria, industrial and mineral pollution, and so on. Organic toxins are much important due to their (a) diverse uses and full release to aquatic life and terrestrial lands, (b) increased resolve, (c) highly resistive, and (d) synergistic impact on the ecosystem and living beings. Persistent organic pollutants (POPs) are greatly resistant to decay and have serious consequences for humans and other living things [1–4]. It is believed that POPs affect the growth of babies and children, produce tumors, immune dysfunctions, congenital incapacities, and multiplicative processes. Organic contaminants are complex compounds that enhance the color, texture, or composition of materials or substances, and some of them are resistant to heat and detergents [5–7]. Dyes are the major source of industrial procedures for a broad range of goods. Presently, >100,000 different kinds of pollutants are available at a commercial scale with an annual fabrication of over 7×10^5 tons [8,9]. Since a large number of dyes are soluble in water, they have a lower biodegradability and are difficult to be detected at low concentrations. Soluble dyes are difficult to detect due to the release of organic compounds into the hydrosphere, which reduces the entry of sunlight and comprises biological and photochemical of aquatic bodies. Some thyroids are mutagenic, carcinogenic, and teratogenic and have a poisonous effect [3,10]. Dyes contaminated water can cause serious health issues and damage human organs like kidneys, liver, brain, central nervous systems, etc., and maybe allergic to the skin [11,12]. Pesticides in water bodies are another major source of organic toxins. Pesticides are found in aquatic bodies mostly as a result of (a) agricultural use, where they are mostly used for pest control to protect crops from pest destruction, (b) pesticide effluents manufacturing industries, and (c) domestic use. Rodenticides, insecticides, herbicides, fungicides, and plant growth regulators are all examples of pesticides. Pollution of water (agricultural runoff) caused by unbalanced and excessive use of microorganisms has serious consequences for living beings. Apart from aldrin, dieldrin, heptachlor, and heptachlor epoxide, which are allowable under a 0.3 ppm limit, according to the World Health Organization, per pesticide (0.1 ppm) and total pesticides (0.5 ppm) are the highest permissible restrictions in fresh water [13,14]. The pesticides outside allowable limits impart dermatological, neurological, gastrointestinal, carcinogenic respiratory, reproduction, and endocrine impacts on living beings [15–17]. Water demand has risen as a result of rapid industrialization, urbanization, low-cost growth, dynamic shifts in everyday life, growing population, and changing attitudes toward water used [18–21]. Overall, people face the danger of water shortage and are unable to satisfy their basic needs. Water is so essential for life. Hence, efforts are being made to combat shortages of water, improve water quality, harvest rainwater, and conserve water for a long time [6,19,22]. In near future, Asia and Africa are the main continents that may face a shortage of fresh water. However, an extensive struggle has been undertaken to preserve water and remove pollutants from wastewater in order to make it convenient for agricultural and domestic uses. Environmental remediation, in particular various kinds of organic pollutants removal with economically possible and environment friendly techniques, is attaining significant interest. In this part, for their effective dye degradation in affordable ways, the science of interface among organic pollutants in water and compounds to be used is becoming significant for technological deployment [16,23–26].

3.2 ADVANCES IN WATER REMEDIATION APPROACHES

Organic contaminants are therefore important for a diversified area of research and are thus of considerable importance for their additional use, environmental disposal, and management without proper protection. However, the removal of organic contaminants from water has become a big challenge in the world. Numerous efforts have been made for degradation of organic contaminants from water. Figure 3.1 indicates the minimization and removal of organic pollutants and their degraded goods from aquatic bodies using several physical, chemical, and biological approaches. These techniques are either independently used or in conjunction with others to

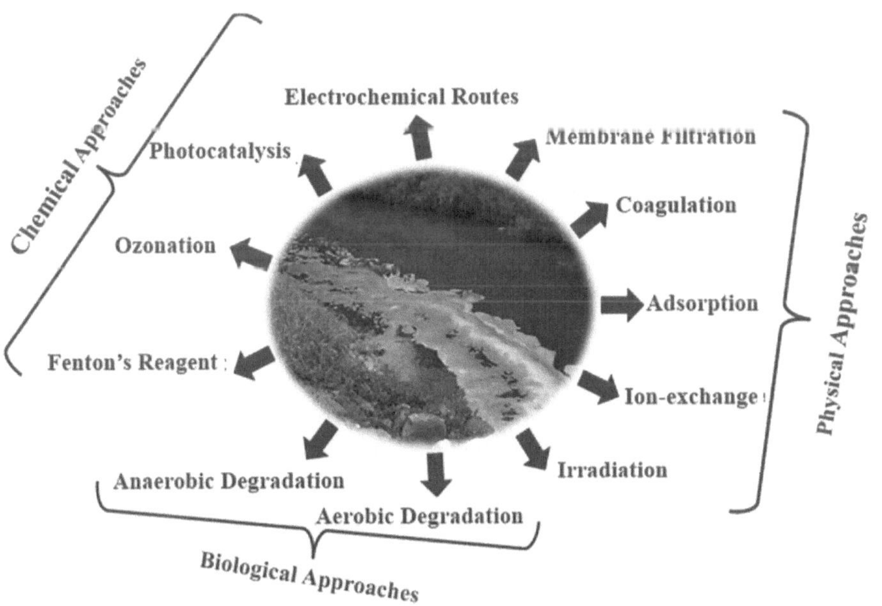

FIGURE 3.1 Biological, chemical, and physical techniques for removal of organic contaminants or degradation in wastewater. (Reproduced with permission from Ref. [4] Copyright 2019 Elsevier B.V.)

purify contaminated water. Some of the generally used techniques for purification of water are adsorption, biological and photocatalytic response, membrane filtration, liquid – liquid extraction, reverse osmosis, oxidation, nanofiltrations, UV-irradiation, etc [27–31]. As a result, we have chosen to emphasize adsorption and photocatalysis degradation approaches for the reduction of organic pollutants from polluted water because they are well recognized as efficient and economically scalable techniques.

3.3 METAL OXIDE-NANOSTRUCTURED PHOTOCATALYSTS

A vital requisite for photocatalysis is its resistance to the reaction that takes place at solid–liquid interface, which may be due to its physiochemical properties. Therefore, it is hard to produce a good photocatalyst that possesses all properties such as physiochemical capability, resistive to corrosion, collecting visible light efficiently, and having a suitable band gap (BG) that could increase photocatalytic activity as a feasible alternative. Conveniently, current research in the nanoscience and nanotechnology field has enhanced the achievable photocatalysts alteration and the exploration and improvement of young materials [32,33]. The rapidly growing scientific publications in bulk have established energetic bibliographical confirmation for consequence of this current subject matter. Many studies deliberated the diverse effect of NPs on their efficiency as photocatalysts, as their energy conversion efficiency is generally investigated through nanoscopic features. In this framework, metal oxide (MO) semiconductors showed increased scientific importance in the field of electronic and environmental industry as they can generate electrons when disturbed with necessary energy expanse. The inspiring electronic arrangement organization, effective high absorption abilities, and maximum interesting charge transport types of MOs have promisingly concluded its use as photocatalyst [34]. However, there are a few shortcomings in semiconductor photocatalysts, for instance, wide BG and rapid electron–hole pair recombination which allows the probability to fabricate young nanocomposite (NC) photocatalysts. The most significant aim of engineering composite photocatalyst is to alter the semiconductor NPs photo-electrochemical belongings. Classification of metal–semiconductor and composite semiconductor nanomaterials has been explained to relax charge refinement in the MO semiconducting nanostructured [33]. The noble metal may act as a sink for degenerated charge carriers and therefore favors interfacial charge carrier process in these NC photocatalysts [35]. In a recent report, a research effort was made to involve noble metal ions to incorporate the titania photo-response keen on UV-irradiation [36].

3.4 METAL–METAL OXIDE HETEROSTRUCTURE PHOTOCATALYSTS

Heterogeneous NPs consisting of two or more different functional units are of huge research interest in epitaxial conjugation since they contain inclusive physiochemical, optical, electronic, magnetic, catalysis, and photocatalysis reactions

[37–40]. Typically, these properties can be obtained by their nanoscopic junction properties lying in the structure. It is considered that for heterojunction among the metal–MO NP heterostructure, the material features similar to the interface at both ends are changed from their bulk counterpart. This may be prompted due to many influences, consisting of surface reconstruction, lattice dissimilar convinced crystal strain, and transfer/interaction of electrons at the interface [41,42]. Generally, the MO nanostructured semiconductor photocatalysts may be differentiated into two types. Firstly, the material must be photon-energetic and during excitation, the excited electrons are transferred to the next part of the material, which therefore stimulates the catalytic response. The plasmonic nanostructured (e.g., Au, Ag, Pt) groups of high band gap semiconductors such as titania exist in this class, where the plasmon electrons are transferred during photoexcitation from gold to TiO_2 semiconducting material used for initiating catalytic response [43–45]. These are named plasmonic photocatalysts that were described and reported completely in the literature. A probability of electron removal from the gold surface plasmonic state to the semiconductor nanomaterials is shown in Figure 3.2 matched with the present literature [46–48]. The second step is that both metal and semiconductor materials are photosensitized and can stimulate solar light. The material class where the gold plasmonic element is connected with the semiconductor has small band gap; for instance, Au-CdS, Au-CdSe, and Au-PbS are mentioned in this class [49–51]. Interestingly, it is observed that the transporting of electron from both ends is possible, e.g., either semiconductor to metal or inverse exposed to the arrangement of BG using excitation basis. In a certain situation, if metal is excited only, then electron transference could trail alike to the scheme as illustrated in Figure 3.2a, offering a plasmonic surface

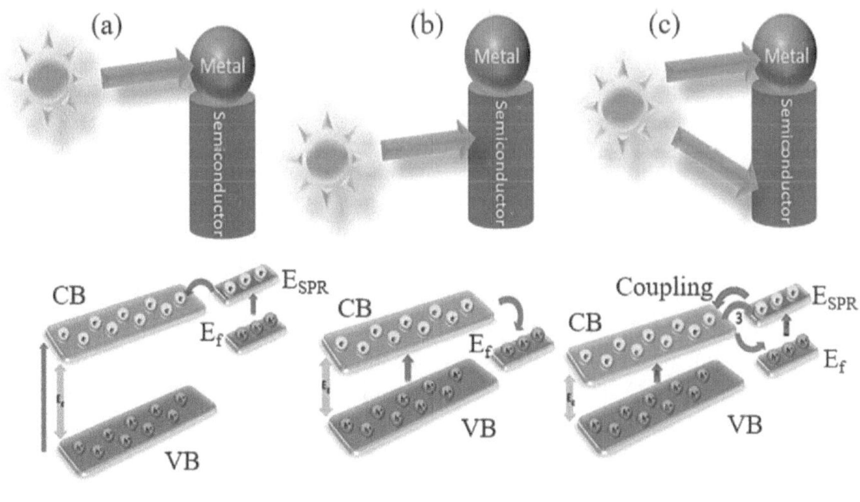

FIGURE 3.2 Electron transfer illustration in a metal–semiconductor heterostructure: (a) transfer of electrons from VB to CB, (b) transfer of electron to Fermi level, and (c) coupling phenomenon. (Reproduced with permission from Ref. [33] Copyright 2016 Royal Society of Chemistry.)

state rests above the semiconductor conduction band (CB). In the same manner, selective semiconductor excitation may relax the movement of electrons from gold CB to the electronic band, and a schematic illustration of such electronic excitation is shown in Figure 3.2b. But then, the situation turns out to be more complex when both metal and semiconductors are excited simultaneously (as shown in Figure 3.2c). The coupling of photoexcited electron is also possible among the semiconductor and metal plasmon [51,52]. Conversely, it is reported that when both metal and semiconductor are excited, together they could attract solar radiation and generate more excitons than the rest of the two types. Even though in most cases, besides gold, it is demonstrated for metal–semiconductor heterostructures, more metals can also be coupled with several semiconductors and applied in proficient photocatalysis. For instance, the platinum plasmon can be demonstrated even more efficiently in some circumstances [53,54]. But normally, the Au metallic is utilized for metal–semiconductor photocatalysis as it contains surface plasmon resonance (SPR) that drops in the strong solar spectrum vicinity.

3.5 METAL OXIDE CORE–SHELL NANOSTRUCTURED PHOTOCATALYSTS

In recent times, remarkable developments have occurred in the engineering of core–shell metal nanostructure photocatalysts [55–60]. Till now, the effort was made to use these core–shell semiconductor photocatalysts in the photoexcitation process, e.g., hydrogen generation, splitting of water, it is imperious to understand the metal core–shell effect on the photocatalytic properties of the exterior MO core shell (displayed in high-resolution transmission electron microscope images of an individual Au–Cu_2O core–metal shell NC along with its selected area electron diffraction patterns in Figure 3.3a and b, respectively) [61]. As a distinct kind of young metal–core shell photocatalysts, both metal core and semiconductor shell nanostructures are able to demonstrate remarkable advantages as a potential applicant in heterogeneous photocatalysts. Initially, inside a MO semiconductor, by encapsulating metal NPs outer shell, their stability can be potentially increased compared to agglomeration and favor freeing of undesirable photo-corrosion or suspension in catalytic reaction process in real uses [62,63]. Secondly, the lower Fermi energy level of these materials may act as a group of photoexcited electrons and enhance the lifetime of photoexcited charge carriers. However, perhaps this reaction enhances the complete photocatalytic efficiency [64–70]. These properties have been described well by Kamat and his colleagues, which might be described realistically through a semiconductor metal nanostructure system as indicated [71]. Additionally, the core–shell nanomaterials design provides a 3D close contact between semiconductor shell and core metal which earns the maximum benefit of metal-support in-contact. In this way, it can ease the interfacial charge transport process. Finally, MO core–shell NC configuration offers a homogeneous photocatalytic procedure for continuous reactions (Figure 3.3c) [72,73]. Essentially, these remarkable advantages accommodating MO and core–shell semiconductor nanostructure photocatalysts suggest that they might be applied

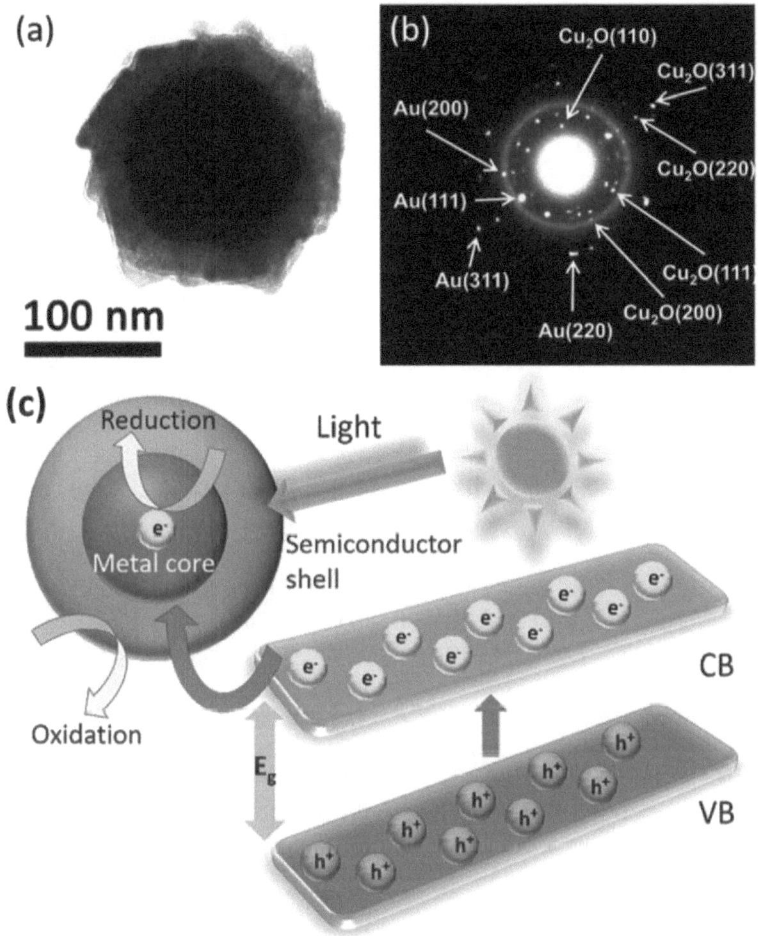

FIGURE 3.3 (a) TEM images of separated Au–Cu_2O core–shell NP through a compact metal oxide shell and (b) SAED pattern exhibiting specific crystal planes. (Reproduced with permission from Ref. [61]. Copyright 2012 American Chemical Society.) (c) Scheme of a light-generated charge separation phenomena in a critical metal–metal oxide semiconductor core–shell in a photocatalytic phenomenon. (Reproduced with permission from Ref. [33]. Copyright 2016 Royal Society of Chemistry.)

as a pioneering technique for light collecting in heterogeneous photocatalytic activity.

3.6 PHOTOCATALYTIC POTENTIAL OF VARIOUS METAL OXIDES

3.6.1 TiO_2

Titanium dioxide (TiO_2) is known as titanium (IV) oxide. Titania is an outstanding photocatalyst due to its durable physiochemical stability and is significantly

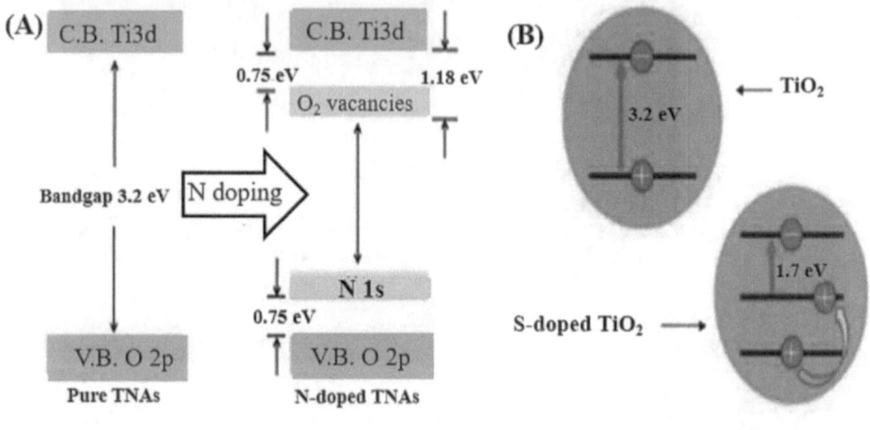

FIGURE 3.4 Schematic representation of the energy band diagram (A) TiO_2 and N-doped TiO_2, (B) TiO_2 and S-doped TiO_2. (Reproduced with permission from Refs. [76] and [77], Copyright 2014 and 2015, Elsevier B.V.)

economical. It agrees to band gap (BG) energy of the order 3.0–3.2 eV, which lies in the UV portion [74,75]. There exist three TiO_2 polymorphs nominated as anatase, brookite, and rutile; among these three crystallographic phases, the rutile phase is regarded as the most stable, while the remaining two belong to the metastable state of stability. Other phases could be modified into the most stable rutile phase at specific temperatures irreversibly. In addition, TiO_2 photocatalyst has important merits in antibacterial, lithium-ion batteries, sensors, anticancer, and significantly in catalysis as well as wastewater remediation. Due to these factors, the synthesis route of TiO_2 photocatalyst, its transformation from anatase to rutile, and the photocatalytic route are elucidated in the following paragraphs.

Hou et al. [76] explored the influence using different ratios of deionized water (DI) in solutions and aqueous-NH_3 on the morphological structures of N-doped TiO_2 nanotubes that were synthesized through a hydrothermal route. The N-doped TiO_2 nanotubes dissolved in NH_3 yielded advanced adsorption ability compared with water. Moreover, N-dopant reduced the BG energy from 3.2 to 2.84 eV and hence boost the degradation response of MO dye as displayed in Figure 3.4a. Correspondingly, McManamon investigated the impact of S-doping inside TiO_2 crystal structure and found a remarkable drop of BG energy from 3.2 to 1.7 eV as visualized in Figure 3.4b. The absorption of S-dopant managed to magnify photocatalytic progress for malachite green degradation [77].

Later, Bakar et al. [78] demonstrated the template-free oxygen peroxide practice followed by a hydrothermal process for S-doped TiO_2 crystallization and synthesis. In this scheme, oxygen atoms were exchanged with S anion that prolongs the optical absorption spectra from UV to visible portion and lowers BG energy as shown in Figure 3.4a. Additionally, S-dopant causes oxygen vacancies on the surface that assist to trap electrons and reduces the excitons recombination rate which surely lifts the degradation performance of MO dye. Wu et al. examined the C-doped TiO_2 through solvothermal route which displays same features compared with S-doped

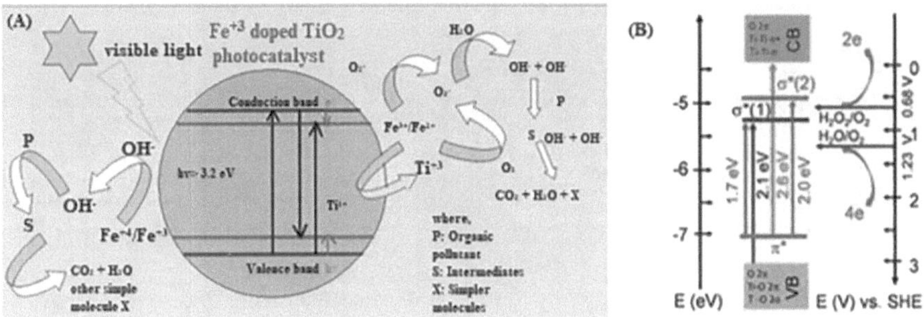

FIGURE 3.5 Schematic presentation of the band gaps diagram and photocatalytic routes (A) in Fe-doped TiO_2 and (B) Cu-doped TiO_2. (Reproduced with permission from Refs. [80] and [81], Copyright 2015, Elsevier B.V. and 2014, American Chemical Society.)

TiO_2 using superior photocatalytic response for MO dye and nitric oxide oxidation [79].

Sood et al. [80] prepared Fe-doped TiO_2 with an ultrasonic-assisted hydrothermal strategy in which Ti is substituted by Fe instead of O which turns into a lowering of BG energy from 3.2 to 2.9 eV and as a result increases the photocatalytic progress toward para nitrophenol degradation. Its mechanism is displayed in Figure 3.5a. Diverse behavior was seen for Cu-doped TiO_2 which was synthesized by a hydrothermal process. The d-orbital splitting elucidated the formation of the intermediate state between VB and CB for Cu-doped TiO_2. Sub-orbitals included in valence band were Ti–O 2σ, Ti–O 2π, and O 2π, for conduction band (bottom): Ti–Ti σ, Ti–Ti σ*, Ti–Ti π, Ti–Ti π*, and Ti–O π* while Ti –O σ* for conduction band (top), as shown in Figure 3.5b. Moreover, an atomic number of transition metals arose due to fluctuation of π* orbitals energy level to CMB that reduced BG energy between σ* (1) and σ*(2). These orbitals were found adjacent to peroxide oxidation sites and H_2O; therefore, they tend to respond to σ orbitals of transitional species and fall potential activity [81].

Sol–gel synthesis of indium (In)-doped TiO_2 exhibits significant prospects including a shift of absorption band edge headed for shorter wavelength, inferior recombination rate, increase in surface-active sites, and large BG energy [82]. These belongings cause to produce CH_4 in larger concentration by choosing In-doped TiO_2 to lift photocatalytic progress for CO_2 reduction. Furthermore, the photocatalytic mechanism of bare and In-doped TiO_2 is displayed in Figure 3.6A and B.

Kőrösi et al. [83] fabricated PF co-doped anatase TiO_2 powder using hydrothermal method by adopting hexafluorophosphoric acid (HPF_6) as a dopant by selecting several PF concentrations. The morphological qualities of NPs were influenced by dopants concentration. Larger concentrations of HPF_6 stemmed into round-shaped particles, while at lower concentrations polymeric NPs in spherical and rod-like shapes were identified. PF co-doped TiO_2 proved excellent photocatalytic progress compared with bear, and doped TiO_2 (F-TiO_2, P-TiO_2). To pay critical significance to surface configuration, the PF ratio should be changed effectively to obtain maximum progress of photocatalysis with PF-TiO_2.

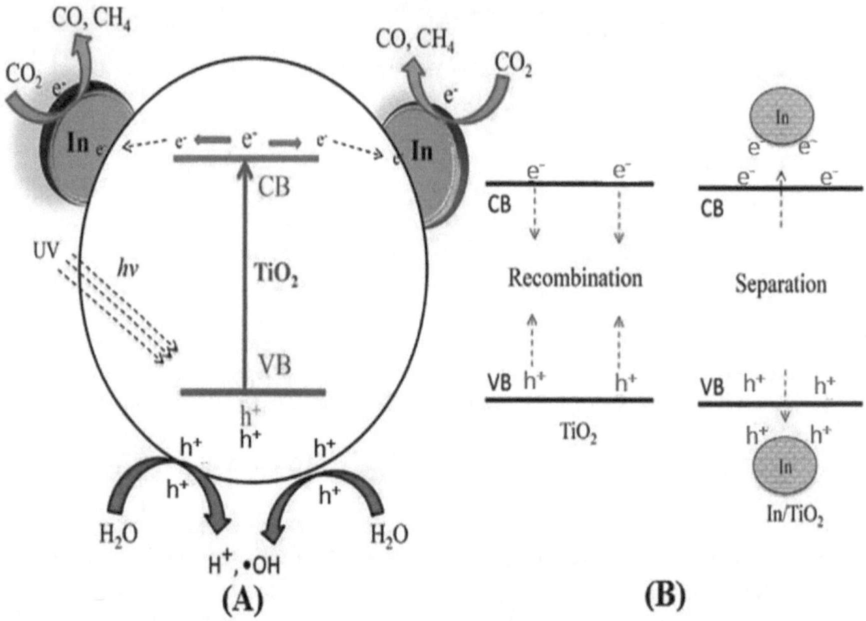

FIGURE 3.6 Schematic diagram for possible reaction phenomena of photocatalytic CO_2 reduction through water: (A) CO_2 photocatalytic transformation and redox responses over In-doped TiO_2 and (B) photoexcited electrons recombination and separation. (Reproduced with permission from Ref. [82], Copyright 2015, Elsevier B.V.)

3.7 ZnO

Zinc oxide (ZnO) is water-insoluble, and it is extracted from the mineral zincite in white powder form by synthetic approaches. It belongs to the n-type semiconductors family with wide BG energy and resides in the II–IV group of the periodic table. It is an economical compound having chemical stability as well as toxic-free and copiously utilized in different industries, for instance, ceramics, plastic, glass, cement, and rubber as well. ZnO holds vast radiation hardness, substantial optical absorption (i.e., in the UV region), efficacious transparency, and excellent thermal properties. These belongings make ZnO useful in numerous applications including solar cells, antimicrobial agents, medicine, optoelectronics, catalytic activity, and ultimately photocatalysis.

Pawar et al. [84] generated polar surface microstructure of Cu-doped ZnO via a single-stage chemical technique. FESEM images disclosed the rod-like ZnO nature into spheres, whereas the absorption of doping disk-like structure was discovered. Enormous absorption of Cu above ZnO stemmed into the photocatalysis enhancement for both rhodamine B (RhB) and methylene blue (MB) dyes degradation. In addition, the presence of polar surfaces, CuO production, and the introduction of O_2 vacancies were the parameters for improved photocatalysis. Yin et al. [85] prepared hierarchical Ni-doped ZnO nanostructures with solvothermal method. Transformation of solid into the empty spherical structure was carried out via Ostwald ripening strategy. The RhB dye degradation progress was obtained by utilizing 1 mol% doping of Ni in the ZnO nanostructure.

Utilizing 10 mol% Ni-doping causes to diminish energy and increases the recombination of the photogenerated electrons, hence resulting in photocatalysis dominance.

3.8 CuO

Copper oxide (CuO) is a black solid inorganic material and is conventionally nominated as tenorite mineral which is also termed as cupric oxide or copper (II) oxide. CuO belongs to the p-type semiconductor family and is related to antiferromagnetic properties, having low BG energy in the order of 1.2 eV and low electrical resistance. The electrical conductivity falls when it is visible to species of reduction gas. It owns $3d^9$ electronic configuration and suggests the monoclinic crystal system. CuO is an effective material owing to its applications in gas sensing, heterogenous catalysis, solar cells, and lithium electrode materials. Several synthesis approaches might be utilized to synthesize CuO and its literature-related photocatalytic activity is discussed below.

Ardekani et al. [86] deposited thin nanofilms of N-doped ZnO–CuO NC on the glass substrate via spray pyrolysis method. Copper acetate, zinc acetate, and ammonium acetate were used as precursors in many molar ratios. Field effect scanning electron microscope examination shows nano-dimensioned particles of crumpled shape. The elemental valuation was inspected via energy dispersive x-rays exploration which confirmed the presence of copper, oxygen, nitrogen, and zinc. Further, x-ray diffraction studies guaranteed the existence of crystalline phases of CuO and ZnO with the grain size of ~18 nm. Degradation response of MO dye indicates high progress of NC. Altered materials may be used as dopants with MO through such a technique.

3.9 METAL OXIDES AS ADSORBATES

3.9.1 Removal of Heavy Metallic Ions from Water

Various treatment technologies are used for the removal of heavy metal from wastewater. Conventional treatment methods include co-precipitation [87,88], membrane separation [89], ion exchange [90], electrocoagulation [91], reverse osmosis [92], filtration/ultrafiltration [93], etc. with each of them having particular advantage. For instance, chemical precipitation (CP) results in the removal of effective metal. The CP method is very simple to carry out without the use of expensive apparatus [94,95]. Ion exchange, electrochemical, and electro-dialysis have the capability for pollutants selectivity removal from a mixture. In addition, photocatalysis offers the capability to eliminate both organic and metal contaminants simultaneously. Each technology also has a few inherent disadvantages [96]. For instance, adsorption technology is an established method for the removal of heavy metals from water [97–100]. Specifically in developing countries where financial means is a major issue, this method provides a simple way to eliminate heavy metal ions from wastewater. Thus, adsorption proficiency depends on the adsorbents type. The significant and fluently used activated carbon is very expensive, making it unfavorable for a wide range of applications (Figure 3.7) [101–103]. So, the quest for young and efficient adsorbent material was always an active research field. It is notable to set experimental data for the most proper mathematical equation for a large range of experimental situations to explore

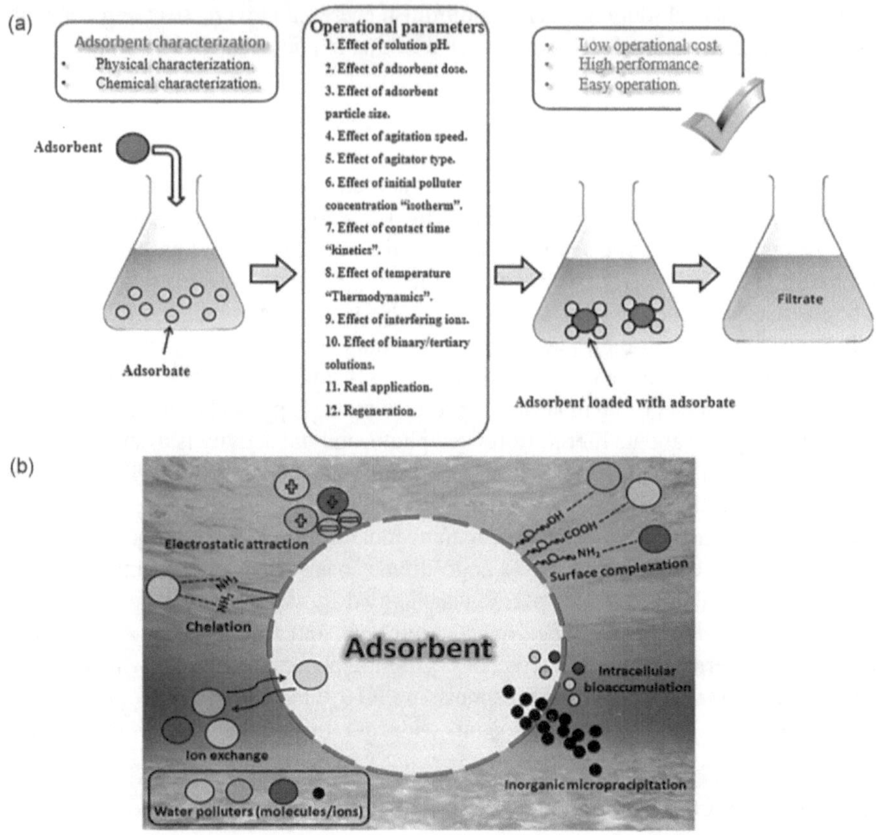

FIGURE 3.7 (a) Schematic diagram of batch adsorption and (b) various mechanisms occurring in the adsorption process. (Reproduced with permission from Ref. [103]. Copyright 2020 Royal Society of Chemistry.)

a distinct adsorbent for an adsorption technique. Normally, adsorption isotherm models, e.g., Langmuir or Freundlich models, are used broadly [104]. In general, adsorption isotherms elucidate how variable adsorbates make interaction with an adsorbent and are typical choices for optimizing adsorption phenomena, expression of surface belongings, adsorbents tendency, and efficient design of adsorption systems [105]. Usually, an adsorption isotherm is a vital curve elucidating mechanism leading to retention or substance mobility from an aqueous system to a solid phase with persistent temperature and pH [106].

3.10 ADSORPTIVE REMOVAL BY VARIOUS METAL OXIDES

3.10.1 Iron Oxides

On Earth's surface, iron is one of the rich substances. It is extremely reactive and can exist in a variety of oxidation states. The chemical composition and iron oxidation state of iron oxide nanoparticles influence their crystalline and structural properties. Magnetite (Fe_3O_4), hematite ($–Fe_2O_3$), and maghemite ($–Fe_2O_3$) are the most common oxide types of iron [4,107]. Generally used magnetite iron oxide NPs contain two kinds of iron with variable oxidation states (i.e., Fe^{+2} and Fe^{+3}). The high-temperature process (400°C) of magnetite and spinel-structured maghemite yields hematite of corundum structure. Iron oxide particle with nano dimension offers a large specific surface-to-volume ratio and super-paramagnetism [4]. Iron oxide NPs have received significant attention for several applications due to their easy synthesis, surface modification, excess accessibility, noteworthy belongings at nano-scale, magnetic segregation, and small toxicity [108,109]. The size, textual qualities, structural and magnetic properties, and iron oxide NPs functional groups may be modified with synthesis methods and reaction precursor's selection [110]. Iron oxide NPs in their undoped, doped, or composite practices have been used widely for organic pollutants adsorption. Contaminants separation from the aqueous phase after adsorption of organic pollutants was a great challenge and it needed a complete energy process, i.e., centrifugation. Therefore, in the presence of an external magnetic field, after adsorption of organic contaminants, spent iron-oxide-based adsorbents could be simply separated focusing on their inherent magnetism [111]. Subsequently, iron oxide NPs and their NCs have gained growing research interest for photocatalytic applications. Iron oxide (magnetite) NPs have been utilized for proficient chlorinated pesticides removal, for instance, 2,4,5-trichlorophenoxyacetic acid (2,4,5-T), dieldrin, 2,4-dichlorophenoxyacetic acid (2,4-D), hexachlorocyclohexane (α-HCH and γ-HCH), lindane (1,2,3,4,5,6-hexachlorocyclohexane), etc. from polluted water. The chemical formation affects their adsorption on iron oxide NPs. The organic dyes, viz. Erichrome Black T, Bromophenol Blue, Bromocresol Green, and Fluorescein taking hydroxyl ions functional groups, have been adsorbed preferably on magnetite NPs as compared with non-hydroxylated dyes, e.g., methyl orange, methyl red, and MB as shown in Figure 3.8. The hydroxylated adsorption dyes were corresponded to their contact using hydroxyl groups on magnetic NPs surfaces through hydrogen linkages [112]. The empty structure of Fe_3O_4 nanospheres having large surface area and active edges competently removed Neutral Red dye and exhibited large adsorption ability [113]. The efficiency and selectivity of iron oxide NPs might be enhanced using their surface modification with surface-acting agents. In recent times, potential work was carried out on iron oxide NPs surface functionalization with a wide range of organic species, i.e., organosilane, polyacrylic acid, glutamic acid, and so on. The functionalization may assist organic dyes' adsorption efficiency [114]. The functionalized iron oxide NPs along with humic acid increased the adsorption rate of MB (dye) from wastewater as it made more reactions with humic acid in comparison to pure iron oxide NPs [115]. Iron oxide NPs may also be functionalized with other compounds such as ionic, aqueous, polymers, and chitosan to increase selectivity and adsorption capability. Goltz et al. have altered magnetite NPs using imidazolium-based ionic liquids and utilized them for removal efficiency of Reactive Red (120 dye) [116]. The grafting of β-cyclodextrin on magnetic iron oxide NPs increased the removal proficiency of azo dyes (Direct Blue 15, Evans Blue, and Chicago Sky Blue dyes) from contaminated water [117]. The γ-Fe_2O_3 NPs cross-linked through chitosan showed a dynamic larger adsorption tendency for methyl orange (dye) compared

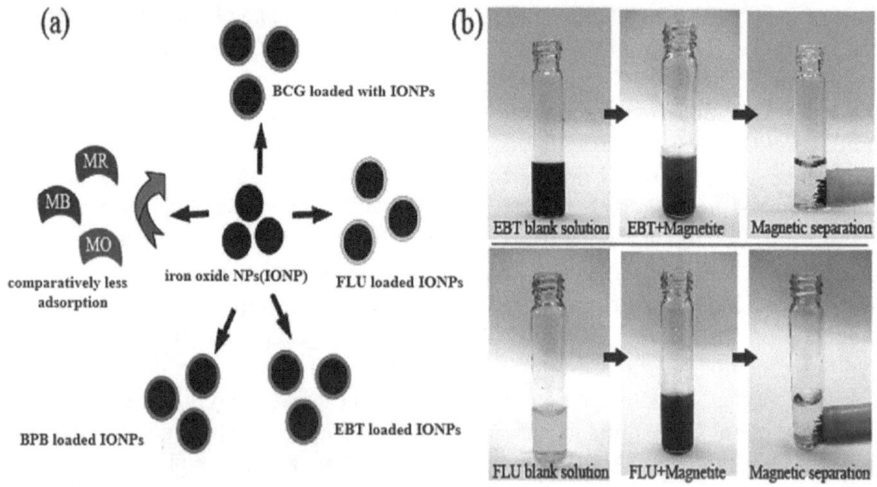

FIGURE 3.8 (a) Schematic visualization of desirable adsorption of variable dyes on iron oxide nanoparticles (IONPs) and (b) magnetic partition of EBT & FLU adsorbed iron nanoparticles via an external magnetic field. (Reproduced with permission from Ref. [112]. Copyright 2011 American Chemical Society.)

with standalone γ-Fe_2O_3 NPs and chitosan. The electrostatic contact between methyl orange (dye) and chitosan grafted on γ-Fe_2O_3 NPs increased adsorption ability [118]. A simple and viable scheme of spent adsorbent separation needs to be joined by adsorption characteristics. Magnetic mesoporous (Fe_3O_4@nSiO_2@mSiO_2) NPs consist of a super-paramagnetic core of Fe_3O_4, a sandwich between nonporous (nSiO_2) and outer mesoporous silica layer (mSiO_2) showed good adsorption rate and the ability for DDT insecticides removal [119]. The hierarchical core-level iron oxide@magnesium silicate (HIO@MgSi) can be fabricated with a hydrothermal route revealing rapid and effective adsorption rates of positively charged dyes.

3.10.2 Titanium Oxides

Titanium dioxide NCs have been investigated as an adsorbent due to their high surface reactivity, rich availability of precursors, easiness of synthesis, balanced structure and texture qualities, and desirable zero point charge (pHpze ranging from 6 to 6.8), which facilitates contact with organic contaminants [74,120]. Titania can also be used as a photocatalytic substance to convert hazardous organic materials into non-hazardous materials [75,121,122]. One of the basic requirements for photodegradation of organic pollutants is their adsorption on the surface of TiO_2. Adsorption of organic compounds promotes interaction with dye molecules and surface-active sites, such as photoexcited holes, and thus promotes photo-oxidation events [123–127]. However, photodegradation of organic contaminants with TiO_2 nanostructure governs through their adsorption performance. Fundamental studies focusing on initial and normally very fast photocatalytic reactions, induced with photon absorption, the energy of which increases BG energy of photocatalyst, inside and at semiconductor surface, are not frequently investigated. However, the knowledge of these methods is of utmost significance for understanding photocatalyst mechanism and for a significant scheme of photocatalytic reactions [128].

The particle size, shape, crystallinity, and phase configuration affect titania surface characteristics, subsequently adsorption and organic contaminants degradation. The hierarchical NCs such as TiO_2, i.e., TiO_2 1D nanorods, 3D-0D TiO_2 microspheres, and 3D-1D microspheres synthesized using hydrothermal route showed morphological characteristics dependent of phenol degradation. In hierarchical 3D NC titania, a greater scale of surface-active edges and favorable BG energy produced an efficient and rapid organic pollutants decomposition as compared with 1D-nanostructured analogous [129]. Hierarchical flower-like NCs of sodium titanate (TMF) consisting of 1D-nanoribbons exhibited excellent efficiency as an adsorbent for MB removal from wastewater. The electrostatic contact with the cationic charge of MB and anion titanate increased MB adsorption over the surface of sodium titanate. As compared to 1D-nanobelt (TNB) sodium-titanate structure, TMF hierarchical nanostructure showed greater MB adsorption, which was attributed to TMF's high active sites and specific surface area of 3D structural qualities. Because of the repulsive behavior of electrostatic interaction, the anion methyl orange demonstrated negligible adsorption on TMF hierarchical nanostructure [123]. The surface and structural properties of TiO_2 were modified morphologically and chemically, increasing their absorption ability for organic contaminants. Because of the large surface area and filled surface-active edges of mesoporous TiO_2, adsorption events are simplified, resulting in quick and efficient organic dye adsorption [130].

3.10.3 Zinc Oxides

ZnO is a favorable adsorbent for water treatment applications because of its non-toxicity, easy operation, economical, and controlled textural and surface features. Chitosan modified-ZnO with economic and biocompatible can adsorb Direct Blue (78) and Acid Black (26) dyes effectively [131]. When chitosan was grafted on ZnO NPs the adsorption performance of permethrin organisms through chitosan had enhanced from 49% to 99% [132]. The electrostatic interaction among cationic amino groups of chitosan and anionic functionalities of organic contaminants, microstructural features, ZnO NPs active surface sites, and large surface area increased adsorption for wastewater remediation applications. The Cr-doped ZnO NPs with different contents of Cr^{+3} have been described as potential adsorbents for the removal of anionic dyes from water. The incorporation of Cr has improved surface defects on ZnO NPs, which make hydroxyl groups for protonation; as a result, a positively charged particle is created on the surface. Cationic ZnO NPs adsorbed anionic dyes selectively driven with electrostatic interaction [133]. ZnO-ZIF-8 [Zn(-MeIM$_2$)] hollow microsphere composites were explained for selective dye adsorption, further increased through the composite core–shell configuration [134]. The introduction of $NiFe_2O_4$ in ZnO NPs minimizes photo-corrosion and enhances the Congo red dye adsorption suggesting to high photocatalytic activity. Thus, before photocatalytic performance, the primary step is efficient Congo red adsorption on $NiFe_2O_4$/ZnO surface which was then facilitated through electrostatic contact between $NiFe_2O_4$/ZnO and Congo red elements [135]. High surface area and hierarchical mesoporous microsphere pore volume of ZnO nanosheets showed excellent large adsorption capacity (334 mg g^{-1}) for Congo red as compared to commercially available ZnO residue [136].

3.11 SUMMARY AND OUTLOOK

Water is a vital source of life for all living things. The amount of wastewater generated has rapidly risen over time, and it has emerged as a major global problem. Various techniques for wastewater disposal and water conservation have been used. Most experimental solutions are focused on natural toxins in the water. Adsorption and photocatalytic methods for removing organic chemicals from polluted water are reviewed in this chapter using different types of MOs and their NCs as photocatalytic and adsorptive materials. MO nanomaterials of portable crystalline, morphological and textural features, chemical nature, and surface characteristics display exceptional ability for organic contaminants adsorption determined by various adsorptive pathways viz. physiosorption, chemiosorption, charge-induced interact, and so on. The applications of MO-based nanomaterials and their NCs for organic contaminants removal, especially dyes and microorganisms, from contaminated water are thoroughly described. In this context, a wide variety of MOs such as oxides of copper, iron, zinc, tungsten, metal oxides titanium oxides, nanostructured, and graphene-based metal oxides nanomaterials, among others, are analyzed, with an emphasis on existing problems, extension, and prospects for water treatment. Adsorption has been a recognized method with different materials to remove or degrade a diversified range of pollutants involving organic contaminants. Consequently, secondary pollutants are deficient, whereas organic contaminants are distributed from one phase, i.e., soluble to other phases(spent adsorbent). Though adsorptive composites have been used in a variety of industries, they still require a significant amount of innovation to address recyclability issues, decomposition of adsorbed contaminants in a greener procedure, adsorption capacity enhancement, and so on. In recent times, a broad range of MO nanomaterials and their NCs have drawn cumulative consideration as photocatalytic materials to decompose organic contaminants or transform them into environment friendly products. The metal oxides such as titania, CuO, ZnO, MOs nanomaterials, and graphene-based MOs NCs have been proven as effective and recyclable photocatalysts for degrading organic pollutants under light irradiation. Thus, at the industrial scale, the applicability of MO nanomaterials with high proficiency, economically feasible technique for their high degree of construction, use of environment friendly and non-toxic precursors for their production, and well-controlled materials fabrication that is largely significant under sunlight irradiation were some of the serious problems encountered. Systematic interventions are necessary for building future practical materials for wastewater treatment through zero discharge. Smart materials engineering with large surface area and crystalline behavior through economically viable and environment friendly process are the need of the hour. The reproduction of adsorbents/photocatalysts is also a significant parameter that aids economic viability, restricts toxic disposal, keeps stability among wastewater management and secondary waste management, and the process sustainability. Secondary waste management is a seriously discussed topic. Secondary waste (spent adsorbent) could be utilized in various applications such as catalysis. However, the possibility and mitigation of contaminants during the synthesis of MO nanomaterials and later their applications in environmental remediation, their occupation for another ambition, can assist to offer a greener platform. The faster, greener, economically possible, and effective approaches for organic contaminants removal or degradation in a viable way and resurgence of fresh and hygienic water for the community and living beings need substantial development.

REFERENCES

1. G.L. Dotto, G. McKay, Current scenario and challenges in adsorption for water treatment, *Journal of Environmental Chemical Engineering*, 8 (2020) 103988.
2. H. Wang, X. Mi, Y. Li, S. Zhan, 3D graphene-based macrostructures for water treatment, *Advanced Materials*, 32 (2020) 1806843.
3. J. Qian, X. Gao, B. Pan, Nanoconfinement-mediated water treatment: From fundamental to application, *Environmental Science & Technology*, 54 (2020) 8509–8526.
4. R. Gusain, K. Gupta, P. Joshi, O.P. Khatri, Adsorptive removal and photocatalytic degradation of organic pollutants using metal oxides and their composites: A comprehensive review, *Advances in Colloid and Interface Science*, 272 (2019) 102009.
5. A. Yusuf, A. Sodiq, A. Giwa, J. Eke, O. Pikuda, G. De Luca, J.L. Di Salvo, S. Chakraborty, A review of emerging trends in membrane science and technology for sustainable water treatment, *Journal of Cleaner Production*, 266 (2020) 121867.
6. M. Selvaraj, A. Hai, F. Banat, M.A. Haija, Application and prospects of carbon nanostructured materials in water treatment: A review, *Journal of Water Process Engineering*, 33 (2020) 100996.
7. P. Raizada, V. Soni, A. Kumar, P. Singh, A.A. Parwaz Khan, A.M. Asiri, V.K. Thakur, V.-H. Nguyen, Surface defect engineering of metal oxides photocatalyst for energy application and water treatment, *Journal of Materiomics*, 7 (2021) 388–418.
8. T.K. Sen, S. Afroze, H.M. Ang, Equilibrium, kinetics and mechanism of removal of methylene blue from aqueous solution by adsorption onto pine cone biomass of Pinus radiata, *Water, Air, & Soil Pollution*, 218 (2011) 499–515.
9. S. Gautam, H. Agrawal, M. Thakur, A. Akbari, H. Sharda, R. Kaur, M. Amini, Metal oxides and metal organic frameworks for the photocatalytic degradation: A review, *Journal of Environmental Chemical Engineering*, 8 (2020) 103726.
10. L. Wang, C. Shi, L. Wang, L. Pan, X. Zhang, J.-J. Zou, Rational design, synthesis, adsorption principles and applications of metal oxide adsorbents: A review, *Nanoscale*, 12 (2020) 4790–4815.
11. K. Kannan, D. Radhika, K.K. Sadasivuni, K.R. Reddy, A.V. Raghu, Nanostructured metal oxides and its hybrids for photocatalytic and biomedical applications, *Advances in Colloid and Interface Science*, 281 (2020) 102178.
12. H. Saleem, S.J. Zaidi, Developments in the application of nanomaterials for water treatment and their impact on the environment, *Nanomaterials*, 10 (2020) 1764.
13. K. Shaheen, H. Suo, T. Arshad, Z. Shah, S.A. Khan, S.B. Khan, M.N. Khan, M. Liu, L. Ma, J. Cui, Y.T. Ji, Y. Wang, Metal oxides nanomaterials for the photocatalytic mineralization of toxic water wastes under solar light illumination, *Journal of Water Process Engineering*, 34 (2020) 101138.
14. C. Karthikeyan, P. Arunachalam, K. Ramachandran, A.M. Al-Mayouf, S. Karuppuchamy, Recent advances in semiconductor metal oxides with enhanced methods for solar photocatalytic applications, *Journal of Alloys and Compounds*, 828 (2020) 154281.

15. C. Ros, T. Andreu, J.R. Morante, Photoelectrochemical water splitting: A road from stable metal oxides to protected thin film solar cells, *Journal of Materials Chemistry A*, 8 (2020) 10625–10669.
16. T. Velempini, E. Prabakaran, K. Pillay, Recent developments in the use of metal oxides for photocatalytic degradation of pharmaceutical pollutants in water: A review, *Materials Today Chemistry*, 19 (2021) 100380.
17. M. Ikram, J. Hassan, M. Imran, J. Haider, A. Ul-Hamid, I. Shahzadi, M. Ikram, A. Raza, U. Qumar, S. Ali, 2D chemically exfoliated hexagonal boron nitride (hBN) nanosheets doped with Ni: Synthesis, properties and catalytic application for the treatment of industrial wastewater, *Applied Nanoscience*, 10 (2020) 3525–3528.
18. M.S. Danish, L.L. Estrella, I.M.A. Alemaida, A. Lisin, N. Moiseev, M. Ahmadi, M. Nazari, M. Wali, H. Zaheb, T. Senjyu, Photocatalytic applications of metal oxides for sustainable environmental remediation, *Metals*, 11 (2021) 1764.
19. Q. Zhu, K. Ye, W. Zhu, W. Xu, C. Zou, L. Song, E. Sharman, L. Wang, S. Jin, G. Zhang, Y. Luo, J. Jiang, A hydrogenated metal oxide with full solar spectrum absorption for highly efficient photothermal water evaporation, *The Journal of Physical Chemistry Letters*, 11 (2020) 2502–2509.
20. A. Raza, M. Ikram, M. Aqeel, M. Imran, A. Ul-Hamid, K.N. Riaz, S. Ali, Enhanced industrial dye degradation using Co doped in chemically exfoliated MoS2 nanosheets, *Applied Nanoscience*, 10 (2020) 1535–1544.
21. A. Raza, U. Qumar, A. Haider, S. Naz, J. Haider, A. Ul-Hamid, M. Ikram, S. Ali, S. Goumri-Said, M. Binali Kanoun, Liquid-phase exfoliated MoS_2 nanosheets doped with p-type transition metals: A comparative analysis of photocatalytic and antimicrobial potential combined with density functional theory, *Dalton Transactions*, 50 (2021) 6598–6619.
22. K.K. Chenab, B. Sohrabi, A. Jafari, S. Ramakrishna, Water treatment: Functional nanomaterials and applications from adsorption to photodegradation, *Materials Today Chemistry*, 16 (2020) 100262.
23. I. Rabani, C. Bathula, R. Zafar, G.Z. Rabani, S. Hussain, S.A. Patil, Y.S. Seo, Morphologically engineered metal oxides for the enhanced removal of multiple pollutants from water with degradation mechanism, *Journal of Environmental Chemical Engineering*, 9 (2021) 104852.
24. Z. Fallah, E.N. Zare, M. Ghomi, F. Ahmadijokani, M. Amini, M. Tajbakhsh, M. Arjmand, G. Sharma, H. Ali, A. Ahmad, P. Makvandi, E. Lichtfouse, M. Sillanpää, R.S. Varma, Toxicity and remediation of pharmaceuticals and pesticides using metal oxides and carbon nanomaterials, *Chemosphere*, 275 (2021) 130055.
25. M. Ikram, A. Raza, M. Imran, A. Ul-Hamid, A. Shahbaz, S. Ali, Hydrothermal synthesis of silver decorated reduced graphene oxide (rGO) nanoflakes with effective photocatalytic activity for wastewater treatment, *Nanoscale Research Letters*, 15 (2020) 95.
26. A. Raza, U. Qumar, J. Hassan, M. Ikram, A. Ul-Hamid, J. Haider, M. Imran, S. Ali, A comparative study of dirac 2D materials, TMDCs and 2D insulators with regard to their structures and photocatalytic/sonophotocatalytic behavior, *Applied Nanoscience*, 10 (2020) 3875–3899.
27. M. Rani, U. Shanker, V. Jassal, Recent strategies for removal and degradation of persistent & toxic organochlorine pesticides using nanoparticles: A review, *Journal of Environmental Management*, 190 (2017) 208–222.
28. R. Rosal, A. Rodríguez, J.A. Perdigón-Melón, A. Petre, E. García-Calvo, M.J. Gómez, A. Agüera, A.R. Fernández-Alba, Occurrence of emerging pollutants in urban wastewater and their removal through biological treatment followed by ozonation, *Water Research*, 44 (2010) 578–588.
29. A. Pérez-González, A.M. Urtiaga, R. Ibáñez, I. Ortiz, State of the art and review on the treatment technologies of water reverse osmosis concentrates, *Water Research*, 46 (2012) 267–283.

30. M. Ikram, M.I. Khan, A. Raza, M. Imran, A. Ul-Hamid, S. Ali, Outstanding performance of silver-decorated MoS2 nanopetals used as nanocatalyst for synthetic dye degradation, *Physica E-: Low-dimensional Systems and Nanostructures*, 124 (2020) 114246.
31. M. Ikram, R. Tabassum, U. Qumar, S. Ali, A. Ul-Hamid, A. Haider, A. Raza, M. Imran, S. Ali, Promising performance of chemically exfoliated Zr-doped MoS2 nanosheets for catalytic and antibacterial applications, *RSC Advances*, 10 (2020) 20559–20571.
32. M. Ahmed, G. Xinxin, A review of metal oxynitrides for photocatalysis, *Inorganic Chemistry Frontiers*, 3 (2016) 578–590.
33. K. Mondal, A. Sharma, Recent advances in the synthesis and application of photocatalytic metal–metal oxide core–shell nanoparticles for environmental remediation and their recycling process, *RSC Advances*, 6 (2016) 83589–83612.
34. S. Bano, S. Sultana, S. Sabir, Metal oxide nanostructured materials for water treatment: Prospectives and challenges, in: M. Oves, M.O. Ansari, M. Zain Khan, M. Shahadat, I.M.I. Ismail (Eds.) *Modern Age Waste Water Problems: Solutions Using Applied Nanotechnology*, Springer International Publishing, Cham, 2020, pp. 213–231.
35. T. Jiang, X. Qin, Y. Sun, M. Yu, UV photocatalytic activity of Au@ZnO core–shell nanostructure with enhanced UV emission, *RSC Advances*, 5 (2015) 65595–65599.
36. L. Zang, W. Macyk, C. Lange, W.F. Maier, C. Antonius, D. Meissner, H. Kisch, Visible-light detoxification and charge generation by transition metal chloride modified titania, *Chemistry – A European Journal*, 6 (2000) 379–384.
37. H. Yu, M. Chen, P.M. Rice, S.X. Wang, R.L. White, S. Sun, Dumbbell-like bifunctional Au–Fe3O4 nanoparticles, *Nano Letters*, 5 (2005) 379–382.
38. C. Xia, H. Wang, J.K. Kim, J. Wang, Rational design of metal oxide-based heterostructure for efficient photocatalytic and photoelectrochemical systems, *Advanced Functional Materials*, 31 (2021) 2008247.
39. J. Theerthagiri, S. Chandrasekaran, S. Salla, V. Elakkiya, R.A. Senthil, P. Nithyadharseni, T. Maiyalagan, K. Micheal, A. Ayeshamariam, M.V. Arasu, N.A. Al-Dhabi, H.-S. Kim, Recent developments of metal oxide based heterostructures for photocatalytic applications towards environmental remediation, *Journal of Solid State Chemistry*, 267 (2018) 35–52.
40. K. Afroz, M. Moniruddin, N. Bakranov, S. Kudaibergenov, N. Nuraje, A heterojunction strategy to improve the visible light sensitive water splitting performance of photocatalytic materials, *Journal of Materials Chemistry A*, 6 (2018) 21696–21718.
41. Z.-P. Liu, X.-Q. Gong, J. Kohanoff, C. Sanchez, P. Hu, Catalytic role of metal oxides in gold-based catalysts: A first principles study of CO oxidation on TiO_2 supported Au, *Physical Review Letters*, 91 (2003) 266102.
42. W.J. Huang, R. Sun, J. Tao, L.D. Menard, R.G. Nuzzo, J.M. Zuo, Coordination dependent surface atomic contraction in nanocrystals revealed by coherent diffraction, *Nature Materials*, 7 (2008) 308–313.
43. Z. Bian, T. Tachikawa, P. Zhang, M. Fujitsuka, T. Majima, Au/TiO_2 superstructure-based plasmonic photocatalysts exhibiting efficient charge separation and unprecedented activity, *Journal of the American Chemical Society*, 136 (2014) 458–465.
44. S. Linic, P. Christopher, D.B. Ingram, Plasmonic-metal nanostructures for efficient conversion of solar to chemical energy, *Nature Materials*, 10 (2011) 911–921.
45. F. Wang, Y. Jiang, A. Gautam, Y. Li, R. Amal, Exploring the origin of enhanced activity and reaction pathway for photocatalytic H_2 production on Au/B-TiO_2 catalysts, *ACS Catalysis*, 4 (2014) 1451–1457.
46. Z.W. Seh, S. Liu, M. Low, S.-Y. Zhang, Z. Liu, A. Mlayah, M.-Y. Han, Janus Au-TiO_2 photocatalysts with strong localization of plasmonic near-fields for efficient visible-light hydrogen generation, *Advanced Materials*, 24 (2012) 2310–2314.
47. S.K. Dutta, S.K. Mehetor, N. Pradhan, Metal semiconductor heterostructures for photocatalytic conversion of light energy, *The Journal of Physical Chemistry Letters*, 6 (2015) 936–944.

48. A. Furube, L. Du, K. Hara, R. Katoh, M. Tachiya, Ultrafast plasmon-induced electron transfer from gold nanodots into TiO_2 nanoparticles, *Journal of the American Chemical Society*, 129 (2007) 14852–14853.
49. J.-S. Lee, E.V. Shevchenko, D.V. Talapin, Au–PbS core–shell nanocrystals: Plasmonic absorption enhancement and electrical doping via intra-particle charge transfer, *Journal of the American Chemical Society*, 130 (2008) 9673–9675.
50. M.T. Sheldon, P.-E. Trudeau, T. Mokari, L.-W. Wang, A.P. Alivisatos, Enhanced semiconductor nanocrystal conductance via solution grown contacts, *Nano Letters*, 9 (2009) 3676–3682.
51. J.W. Ha, T.P.A. Ruberu, R. Han, B. Dong, J. Vela, N. Fang, Super-resolution mapping of photogenerated electron and hole separation in single metal–semiconductor nanocatalysts, *Journal of the American Chemical Society*, 136 (2014) 1398–1408.
52. J. Zhang, Y. Tang, K. Lee, M. Ouyang, Tailoring light–matter–spin interactions in colloidal hetero-nanostructures, *Nature*, 466 (2010) 91–95.
53. L. Amirav, A.P. Alivisatos, Photocatalytic hydrogen production with tunable nanorod heterostructures, *The Journal of Physical Chemistry Letters*, 1 (2010) 1051–1054.
54. X. Yu, A. Shavel, X. An, Z. Luo, M. Ibáñez, A. Cabot, Cu2ZnSnS4-Pt and Cu2ZnSnS4-Au heterostructured nanoparticles for photocatalytic water splitting and pollutant degradation, *Journal of the American Chemical Society*, 136 (2014) 9236–9239.
55. S.-F. Hung, Y.-C. Yu, N.-T. Suen, G.-Q. Tzeng, C.-W. Tung, Y.-Y. Hsu, C.-S. Hsu, C.-K. Chang, T.-S. Chan, H.-S. Sheu, J.-F. Lee, H.M. Chen, The synergistic effect of a well-defined Au@Pt core–shell nanostructure toward photocatalytic hydrogen generation: Interface engineering to improve the Schottky barrier and hydrogen-evolved kinetics, *Chemical Communications*, 52 (2016) 1567–1570.
56. A.N. Kadam, D.P. Bhopate, V.V. Kondalkar, S.M. Majhi, C.D. Bathula, A.-V. Tran, S.-W. Lee, Facile synthesis of Ag-ZnO core–shell nanostructures with enhanced photocatalytic activity, *Journal of Industrial and Engineering Chemistry*, 61 (2018) 78–86.
57. W. Gao, Q. Liu, S. Zhang, Y. Yang, X. Zhang, H. Zhao, W. Qin, W. Zhou, X. Wang, H. Liu, Y. Sang, Electromagnetic induction derived micro-electric potential in metal-semiconductor core-shell hybrid nanostructure enhancing charge separation for high performance photocatalysis, *Nano Energy*, 71 (2020) 104624.
58. J. Liu, J. Feng, J. Gui, T. Chen, M. Xu, H. Wang, H. Dong, H. Chen, X. Li, L. Wang, Z. Chen, Z. Yang, J. Liu, W. Hao, Y. Yao, L. Gu, Y. Weng, Y. Huang, X. Duan, J. Zhang, Y. Li, Metal@semiconductor core-shell nanocrystals with atomically organized interfaces for efficient hot electron-mediated photocatalysis, *Nano Energy*, 48 (2018) 44–52.
59. B. Babu, V.V.N. Harish, R. Koutavarapu, J. Shim, K. Yoo, Enhanced visible-light-active photocatalytic performance using CdS nanorods decorated with colloidal SnO_2 quantum dots: Optimization of core–shell nanostructure, *Journal of Industrial and Engineering Chemistry*, 76 (2019) 476–487.
60. D. Gao, W. Liu, Y. Xu, P. Wang, J. Fan, H. Yu, Core-shell Ag@Ni cocatalyst on the TiO_2 photocatalyst: One-step photoinduced deposition and its improved H_2-evolution activity, *Applied Catalysis B: Environmental*, 260 (2020) 118190.
61. L. Zhang, H. Jing, G. Boisvert, J.Z. He, H. Wang, Geometry control and optical tunability of metal–cuprous oxide core–shell nanoparticles, *ACS Nano*, 6 (2012) 3514–3527.
62. Q. Zhang, T. Zhang, J. Ge, Y. Yin, Permeable silica shell through surface-protected etching, *Nano Letters*, 8 (2008) 2867–2871.
63. I. Lee, J.B. Joo, Y. Yin, F. Zaera, A Yolk@Shell nanoarchitecture for Au/TiO_2 catalysts, *Angewandte Chemie International Edition*, 50 (2011) 10208–10211.
64. V. Hasija, P. Raizada, A. Sudhaik, K. Sharma, A. Kumar, P. Singh, S.B. Jonnalagadda, V.K. Thakur, Recent advances in noble metal free doped graphitic carbon nitride based nanohybrids for photocatalysis of organic contaminants in water: A review, *Applied Materials Today*, 15 (2019) 494–524.

65. A. Zada, P. Muhammad, W. Ahmad, Z. Hussain, S. Ali, M. Khan, Q. Khan, M. Maqbool, Surface plasmonic-assisted photocatalysis and optoelectronic devices with noble metal nanocrystals: Design, synthesis, and applications, *Advanced Functional Materials*, 30 (2020) 1906744.
66. N. Singh, J. Prakash, R.K. Gupta, Design and engineering of high-performance photocatalytic systems based on metal oxide–graphene–noble metal nanocomposites, *Molecular Systems Design & Engineering*, 2 (2017) 422–439.
67. W. Chen, Y. Wang, S. Liu, L. Gao, L. Mao, Z. Fan, W. Shangguan, Z. Jiang, Non-noble metal Cu as a cocatalyst on TiO_2 nanorod for highly efficient photocatalytic hydrogen production, *Applied Surface Science*, 445 (2018) 527–534.
68. Y. Zhang, Y. Ma, L. Wang, Q. Sun, F. Zhang, J. Shi, Facile one-step hydrothermal synthesis of noble-metal-free hetero-structural ternary composites and their application in photocatalytic water purification, *RSC Advances*, 7 (2017) 50701–50712.
69. M.A. Islam, J. Church, C. Han, H.-S. Chung, E. Ji, J.H. Kim, N. Choudhary, G.-H. Lee, W.H. Lee, Y. Jung, Noble metal-coated MoS_2 nanofilms with vertically-aligned 2D layers for visible light-driven photocatalytic degradation of emerging water contaminants, *Scientific Reports*, 7 (2017) 14944.
70. P. Ribao, J. Corredor, M.J. Rivero, I. Ortiz, Role of reactive oxygen species on the activity of noble metal-doped TiO_2 photocatalysts, *Journal of Hazardous Materials*, 372 (2019) 45–51.
71. T. Hirakawa, P.V. Kamat, Charge separation and catalytic activity of Ag@TiO_2 core–shell composite clusters under UV–irradiation, *Journal of the American Chemical Society*, 127 (2005) 3928–3934.
72. N. Zhang, X. Fu, Y.-J. Xu, A facile and green approach to synthesize Pt@CeO_2 nanocomposite with tunable core-shell and yolk-shell structure and its application as a visible light photocatalyst, *Journal of Materials Chemistry*, 21 (2011) 8152–8158.
73. N. Zhang, S. Liu, X. Fu, Y.-J. Xu, A simple strategy for fabrication of "plum-pudding" type Pd@CeO_2 semiconductor nanocomposite as a visible light-driven photocatalyst for selective oxidation, *The Journal of Physical Chemistry C*, 115 (2011) 22901–22909.
74. M. Ikram, E. Umar, A. Raza, A. Haider, S. Naz, A. Ul-Hamid, J. Haider, I. Shahzadi, J. Hassan, S. Ali, Dye degradation performance, bactericidal behavior and molecular docking analysis of Cu-doped TiO_2 nanoparticles, *RSC Advances*, 10 (2020) 24215–24233.
75. M. Ikram, J. Hassan, A. Raza, A. Haider, S. Naz, A. Ul-Hamid, J. Haider, I. Shahzadi, U. Qamar, S. Ali, Photocatalytic and bactericidal properties and molecular docking analysis of TiO_2 nanoparticles conjugated with Zr for environmental remediation, *RSC Advances*, 10 (2020) 30007–30024.
76. X. Hou, C.-W. Wang, W.-D. Zhu, X.-Q. Wang, Y. Li, J. Wang, J.-B. Chen, T. Gan, H.-Y. Hu, F. Zhou, Preparation of nitrogen-doped anatase TiO_2 nanoworm/nanotube hierarchical structures and its photocatalytic effect, *Solid State Sciences*, 29 (2014) 27–33.
77. C. McManamon, J. O'Connell, P. Delaney, S. Rasappa, J.D. Holmes, M.A. Morris, A facile route to synthesis of S-doped TiO_2 nanoparticles for photocatalytic activity, *Journal of Molecular Catalysis A: Chemical*, 406 (2015) 51–57.
78. S.A. Bakar, C.J.R. Ribeiro, Rapid and morphology controlled synthesis of anionic S-doped TiO_2 photocatalysts for the visible-light-driven photodegradation of organic pollutants, *RSC Advances*, 6 (2016) 36516–36527.
79. X. Wu, S. Yin, Q. Dong, T.J.P.C.C.P. Sato, Preparation and visible light induced photocatalytic activity of C-$NaTaO_3$ and C-$NaTaO_3$–Cl-TiO_2 composite, *Physical Chemistry Chemical Physics*, 15 (2013) 20633–20640.
80. S. Sood, A. Umar, S.K. Mehta, S.K.J.J. Kansal, Highly effective Fe-doped TiO2 nanoparticles photocatalysts for visible-light driven photocatalytic degradation of toxic organic compounds, *Journal of Colloid and Interface Science*, 450 (2015) 213–223.

81. N. Roy, Y. Sohn, K.T. Leung, D. Pradhan, Engineered electronic states of transition metal doped TiO2 nanocrystals for low overpotential oxygen evolution reaction, *The Journal of Physical Chemistry C*, 118 (2014) 29499–29506.
82. M. Tahir, N.S.J.A.C.B.E. Amin, Indium-doped TiO_2 nanoparticles for photocatalytic CO_2 reduction with H_2O vapors to CH_4, 162 (2015) 98–109.
83. L. Kőrösi, M. Prato, A. Scarpellini, A. Riedinger, J. Kovács, M. Kus, V. Meynen, S. Papp, Hydrothermal synthesis, structure and photocatalytic activity of PF-co-doped TiO_2, *Materials Science in Semiconductor Processing*, 30 (2015) 442–450.
84. R.C. Pawar, D.-H. Choi, J.-S. Lee, C.S.J.M.C. Lee, Physics, formation of polar surfaces in microstructured ZnO by doping with Cu and applications in photocatalysis using visible light, *Materials Chemistry and Physics*, 151 (2015) 167–180.
85. Q. Yin, R. Qiao, Z. Li, X.L. Zhang, L.J.J. Zhu, Hierarchical nanostructures of nickel-doped zinc oxide: Morphology controlled synthesis and enhanced visible-light photocatalytic activity, *Journal of Alloys and Compounds*, 618 (2015) 318–325.
86. S.R. Ardekani, A.S. Rouhaghdam, M.J.C.P.L. Nazari, N-doped ZnO-CuO nanocomposite prepared by one-step ultrasonic spray pyrolysis and its photocatalytic activity, *Chemical Physics Letters*, 705 (2018) 19–22.
87. T. Kameda, Y. Suzuki, T. Yoshioka, Removal of arsenic from an aqueous solution by coprecipitation with manganese oxide, *Journal of Environmental Chemical Engineering*, 2 (2014) 2045–2049.
88. G.K. Sarma, S. Sen Gupta, K.G. Bhattacharyya, Nanomaterials as versatile adsorbents for heavy metal ions in water: A review, *Environmental Science and Pollution Research*, 26 (2019) 6245–6278.
89. K. Sunil, G. Karunakaran, S. Yadav, M. Padaki, V. Zadorozhnyy, R.K. Pai, Al-Ti2O6 a mixed metal oxide based composite membrane: A unique membrane for removal of heavy metals, *Chemical Engineering Journal*, 348 (2018) 678–684.
90. A.F. Tag El-Din, M.E. El-Khouly, E.A. Elshehy, A.A. Atia, W.A. El-Said, Cellulose acetate assisted synthesis of worm-shaped mesopores of MgP ion-exchanger for cesium ions removal from seawater, *Microporous and Mesoporous Materials*, 265 (2018) 211–218.
91. P.V. Nidheesh, T.S.A. Singh, Arsenic removal by electrocoagulation process: Recent trends and removal mechanism, *Chemosphere*, 181 (2017) 418–432.
92. A. Abejón, A. Garea, A. Irabien, Arsenic removal from drinking water by reverse osmosis: Minimization of costs and energy consumption, *Separation and Purification Technology*, 144 (2015) 46–53.
93. B. Lam, S. Déon, N. Morin-Crini, G. Crini, P. Fievet, Polymer-enhanced ultrafiltration for heavy metal removal: Influence of chitosan and carboxymethyl cellulose on filtration performances, *Journal of Cleaner Production*, 171 (2018) 927–933.
94. M.A. Barakat, New trends in removing heavy metals from industrial wastewater, *Arabian Journal of Chemistry*, 4 (2011) 361–377.
95. D. Magrì, G. Caputo, G. Perotto, A. Scarpellini, E. Colusso, F. Drago, A. Martucci, A. Athanassiou, D. Fragouli, Titanate fibroin nanocomposites: A novel approach for the removal of heavy-metal ions from water, *ACS Applied Materials & Interfaces*, 10 (2018) 651–659.
96. A.T. Le, S.-Y. Pung, S. Sreekantan, A. Matsuda, D.P. Huynh, Mechanisms of removal of heavy metal ions by ZnO particles, *Heliyon*, 5 (2019) e01440.
97. S.H. Khorzughy, T. Eslamkish, F.D. Ardejani, M.R. Heydartaemeh, Cadmium removal from aqueous solutions by pumice and nano-pumice, *Korean Journal of Chemical Engineering*, 32 (2015) 88–96.
98. A.K. Shukla, J. Alam, M. Alhoshan, L. Arockiasamy Dass, F.A.A. Ali, M. M. R, U. Mishra, M.A. Ansari, Removal of heavy metal ions using a carboxylated graphene oxide-incorporated polyphenylsulfone nanofiltration membrane, *Environmental Science: Water Research & Technology*, 4 (2018) 438–448.

99. H. Chen, Y. Meng, S. Jia, W. Hua, Y. Cheng, J. Lu, H. Wang, Graphene oxide modified waste newspaper for removal of heavy metal ions and its application in industrial wastewater, *Materials Chemistry and Physics*, 244 (2020) 122692.
100. F. Ahmadijokani, S. Tajahmadi, A. Bahi, H. Molavi, M. Rezakazemi, F. Ko, T.M. Aminabhavi, M. Arjmand, Ethylenediamine-functionalized Zr-based MOF for efficient removal of heavy metal ions from water, *Chemosphere*, 264 (2021) 128466.
101. J. Wei, Z. Yang, Y. Sun, C. Wang, J. Fan, G. Kang, R. Zhang, X. Dong, Y. Li, Nanocellulose-based magnetic hybrid aerogel for adsorption of heavy metal ions from water, *Journal of Materials Science*, 54 (2019) 6709–6718.
102. M. Sharma, J. Singh, S. Hazra, S. Basu, Adsorption of heavy metal ions by mesoporous ZnO and TiO2@ZnO monoliths: Adsorption and kinetic studies, *Microchemical Journal*, 145 (2019) 105–112.
103. K.Z. Elwakeel, A.M. Elgarahy, Z.A. Khan, M.S. Almughamisi, A.S. Al-Bogami, Perspectives regarding metal/mineral-incorporating materials for water purification: With special focus on Cr(vi) removal, *Materials Advances*, 1 (2020) 1546–1574.
104. P. Zhang, L. Wang, Extended Langmuir equation for correlating multilayer adsorption equilibrium data, *Separation and Purification Technology*, 70 (2010) 367–371.
105. M.I. El-Khaiary, Least-squares regression of adsorption equilibrium data: Comparing the options, *Journal of Hazardous Materials*, 158 (2008) 73–87.
106. G. Limousin, J.P. Gaudet, L. Charlet, S. Szenknect, V. Barthès, M. Krimissa, Sorption isotherms: A review on physical bases, modeling and measurement, *Applied Geochemistry*, 22 (2007) 249–275.
107. M.E. McHenry, D.E. Laughlin, Nano-scale materials development for future magnetic applications, *Acta Materialia*, 48 (2000) 223–238.
108. M. Bystrzejewski, K. Pyrzyńska, A. Huczko, H. Lange, Carbon-encapsulated magnetic nanoparticles as separable and mobile sorbents of heavy metal ions from aqueous solutions, *Carbon*, 47 (2009) 1201–1204.
109. S.T. Selvan, T.T.Y. Tan, D.K. Yi, N.R. Jana, Functional and multifunctional nanoparticles for bioimaging and biosensing, *Langmuir*, 26 (2010) 11631–11641.
110. U. Jeong, X. Teng, Y. Wang, H. Yang, Y. Xia, Superparamagnetic colloids: Controlled synthesis and niche applications, *Advanced Materials*, 19 (2007) 33–60.
111. R.D. Ambashta, M. Sillanpää, Water purification using magnetic assistance: A review, *Journal of Hazardous Materials*, 180 (2010) 38–49.
112. B. Saha, S. Das, J. Saikia, G. Das, Preferential and enhanced adsorption of different dyes on iron oxide nanoparticles: A comparative study, *The Journal of Physical Chemistry C*, 115 (2011) 8024–8033.
113. M. Iram, C. Guo, Y. Guan, A. Ishfaq, H. Liu, Adsorption and magnetic removal of neutral red dye from aqueous solution using Fe_3O_4 hollow nanospheres, *Journal of Hazardous Materials*, 181 (2010) 1039–1050.
114. Z. Wu, J. Wu, H. Xiang, M.-S. Chun, K. Lee, Organosilane-functionalized Fe_3O_4 composite particles as effective magnetic assisted adsorbents, *Colloids and Surfaces A: Physicochemical and Engineering Aspects*, 279 (2006) 167–174.
115. X. Zhang, P. Zhang, Z. Wu, L. Zhang, G. Zeng, C. Zhou, Adsorption of methylene blue onto humic acid-coated Fe_3O_4 nanoparticles, *Colloids and Surfaces A: Physicochemical and Engineering Aspects*, 435 (2013) 85–90.
116. G. Absalan, M. Asadi, S. Kamran, L. Sheikhian, D.M. Goltz, Removal of reactive red-120 and 4-(2-pyridylazo) resorcinol from aqueous samples by Fe_3O_4 magnetic nanoparticles using ionic liquid as modifier, *Journal of Hazardous Materials*, 192 (2011) 476–484.
117. M. Arslan, S. Sayin, M. Yilmaz, Removal of carcinogenic azo dyes from water by new cyclodextrin-immobilized iron oxide magnetic nanoparticles, *Water, Air, & Soil Pollution*, 224 (2013) 1527.

118. H.-Y. Zhu, R. Jiang, L. Xiao, W. Li, A novel magnetically separable γ-Fe_2O_3/crosslinked chitosan adsorbent: Preparation, characterization and adsorption application for removal of hazardous azo dye, *Journal of Hazardous Materials*, 179 (2010) 251–257.
119. F. Liu, H. Tian, J. He, Adsorptive performance and catalytic activity of superparamagnetic Fe_3O_4@nSiO_2@mSiO_2 core–shell microspheres towards DDT, *Journal of Colloid and Interface Science*, 419 (2014) 68–72.
120. I.K. Konstantinou, T.A. Albanis, TiO_2-assisted photocatalytic degradation of azo dyes in aqueous solution: Kinetic and mechanistic investigations: A review, *Applied Catalysis B: Environmental*, 49 (2004) 1–14.
121. A. Ajmal, I. Majeed, R.N. Malik, H. Idriss, M.A. Nadeem, Principles and mechanisms of photocatalytic dye degradation on TiO_2 based photocatalysts: A comparative overview, *RSC Advances*, 4 (2014) 37003–37026.
122. U. Qumar, J. Hassan, S. Naz, A. Haider, A. Raza, A. Ul-Hamid, J. Haider, I. Shahzadi, I. Ahmad, M. Ikram, Silver decorated 2D nanosheets of GO and MoS_2 serve as nanocatalyst for water treatment and antimicrobial applications as ascertained with molecular docking evaluation, *Nanotechnology*, 32 (2021) 255704.
123. U.G. Akpan, B.H. Hameed, Parameters affecting the photocatalytic degradation of dyes using TiO2-based photocatalysts: A review, *Journal of Hazardous Materials*, 170 (2009) 520–529.
124. K. Nakata, A. Fujishima, TiO_2 photocatalysis: Design and applications, *Journal of Photochemistry and Photobiology C: Photochemistry Reviews*, 13 (2012) 169–189.
125. Q. Guo, C. Zhou, Z. Ma, X. Yang, Fundamentals of TiO_2 photocatalysis: Concepts, mechanisms, and challenges, *Advanced Materials*, 31 (2019) 1901997.
126. M.R. Al-Mamun, S. Kader, M.S. Islam, M.Z.H. Khan, Photocatalytic activity improvement and application of UV-TiO_2 photocatalysis in textile wastewater treatment: A review, *Journal of Environmental Chemical Engineering*, 7 (2019) 103248.
127. R. Qian, H. Zong, J. Schneider, G. Zhou, T. Zhao, Y. Li, J. Yang, D.W. Bahnemann, J.H. Pan, Charge carrier trapping, recombination and transfer during TiO_2 photocatalysis: An overview, *Catalysis Today*, 335 (2019) 78–90.
128. J. Schneider, M. Matsuoka, M. Takeuchi, J. Zhang, Y. Horiuchi, M. Anpo, D.W. Bahnemann, Understanding TiO_2 photocatalysis: Mechanisms and materials, *Chemical Reviews*, 114 (2014) 9919–9986.
129. L. Liu, H. Liu, Y.-P. Zhao, Y. Wang, Y. Duan, G. Gao, M. Ge, W. Chen, Directed synthesis of hierarchical nanostructured TiO_2 catalysts and their morphology-dependent photocatalysis for phenol degradation, *Environmental Science & Technology*, 42 (2008) 2342–2348.
130. L. Xiong, Y. Yang, J. Mai, W. Sun, C. Zhang, D. Wei, Q. Chen, J. Ni, Adsorption behavior of methylene blue onto titanate nanotubes, *Chemical Engineering Journal*, 156 (2010) 313–320.
131. R. Salehi, M. Arami, N.M. Mahmoodi, H. Bahrami, S. Khorramfar, Novel biocompatible composite (Chitosan–zinc oxide nanoparticle): Preparation, characterization and dye adsorption properties, *Colloids and Surfaces B: Biointerfaces*, 80 (2010) 86–93.
132. S. Moradi Dehaghi, B. Rahmanifar, A.M. Moradi, P.A. Azar, Removal of permethrin pesticide from water by chitosan–zinc oxide nanoparticles composite as an adsorbent, *Journal of Saudi Chemical Society*, 18 (2014) 348–355.
133. A. Meng, J. Xing, Z. Li, Q. Li, Cr-Doped ZnO nanoparticles: Synthesis, characterization, adsorption property, and recyclability, *ACS Applied Materials & Interfaces*, 7 (2015) 27449–27457.
134. S. El-Hankari, J. Aguilera-Sigalat, D. Bradshaw, Surfactant-assisted ZnO processing as a versatile route to ZIF composites and hollow architectures with enhanced dye adsorption, *Journal of Materials Chemistry A*, 4 (2016) 13509–13518.

135. H.-Y. Zhu, R. Jiang, Y.-Q. Fu, R.-R. Li, J. Yao, S.-T. Jiang, Novel multifunctional NiFe2O4/ZnO hybrids for dye removal by adsorption, photocatalysis and magnetic separation, *Applied Surface Science*, 369 (2016) 1–10.
136. C. Lei, M. Pi, C. Jiang, B. Cheng, J. Yu, Synthesis of hierarchical porous zinc oxide (ZnO) microspheres with highly efficient adsorption of Congo red, *Journal of Colloid and Interface Science*, 490 (2017) 242–251.

4 2D Nanostructures for Membrane-Enabled Water Desalination

Graphene and Beyond

Muhammad Ikram
Solar Cell Applications Research Lab
Department of Physics
Government College University Lahore

Anwar Ul-Hamid
King Fahd University of Petroleum & Minerals

Ali Raza
Department of Physics
University of Sialkot

CONTENTS

4.1 INTRODUCTION

Molecular separation methods are important techniques in industry. Nowadays, these separation methods are extremely in demand, but they also provide prospects for the fabrication of low-energy as well as low-carbon materials. Aside from that, up to 80% of industry still uses conventional thermal separation methods, which are more effective in purifying hydrocarbons despite being thermodynamically less

DOI: 10.1201/9781003167327-5

favorable [1–4]. However, since the 1970s, a non-thermal process—membrane-based reverse osmosis (RO)—has been used for seawater desalination, and it is still a globally accepted method [5,6]. The primary reason for preferring RO over thermal separation is that the former is economical, i.e., 5 to 10 times less expensive relative to thermal separation techniques for instance multistage flash [7]. This innovative membrane-based process is found to be very efficient in various industrial sectors. These membrane-based separation approaches rely substantially on the production of these kinds of responsive polymer filters that really can discriminate molecules depending on particle size [8,9]. At present, with regards to selectivity and permeability, new membranes have been fabricated in low dimensions, specifically in the 2D structure where professional architectures have brilliantly merged both molecular specificity and flux while texturing the permeable membranes.

Separation membranes with a lower thickness perform better in terms of selective molecule separation. The final viable refinement within substance thickness is discovered to be a one-atom-thick sheet, which is accomplished by 2D materials. In the recent past, numerous analytical methods including interfacial polymerization (polyamide membrane thickness < 10 nm) [10], atomic-layer deposition (ZnO layer thickness of around 200 nm and effective ZIF-8 membrane thickness < 200 nm) [11], and 3D printing (polyamide membrane thickness < 20–80 nm) have been employed to fabricate ultrathin membranes [12]. This chapter discusses the use of 2D materials as a basic building block of membrane structures, highlights recent advances in 2D-enabled membranes, and sketches of 2D-supported membranes future. 2D material-based membranes with more refined micropores, including inorganic membranes, tend to recognize same-sized particles. These advanced membranes meet separation needs in a variety of industries, including pharmaceuticals, biofuel processing, semiconductor manufacturing, and petrochemical separation. To deal with more difficult molecular mixtures, scientists and engineers are constantly looking for ways to improve and have stronger, thinner, and more selective membrane materials [13–15]. Especially, few-atom-thick 2D material membranes with finer micropores can more precisely separate even identical-sized molecules. By altering fundamental pores and intrinsic planes, many chemical substances can refine and improve the molecular specificity of such membranes. These variations lead to many trials on the permeability and molecular selectivity of 2D-enabled membranes. These challenges, distinct trends, and the development of materials used for 2D-enabled membranes construction are thoroughly discussed in this chapter. Separation of molecules can be further controlled by various 2D tools such as nanocomposites with well-engineered basal planes, which have been depicted in respective units.

4.2 GRAPHENE-BASED NANOPOROUS MEMBRANES

Graphene's one-atom thickness and enormous surface area might make it excellent for dividing membranes if it includes selectively permeable carbon planes. Graphene's pristine structure, on the other hand, is resistant to all forms of liquids and gases [16], even to very tiny molecules of helium [17]. In spite of this impervious characteristic of graphene, Suk and Aluru [18] tried to impart nanopores into carbon planes of graphene to make an ultimate desalination membrane (Figure 4.1) [19]. Later, at Manchester University, Nair and his coworkers [20] realized that GO membranes assist permeation of unconstrained water that was formerly discovered to

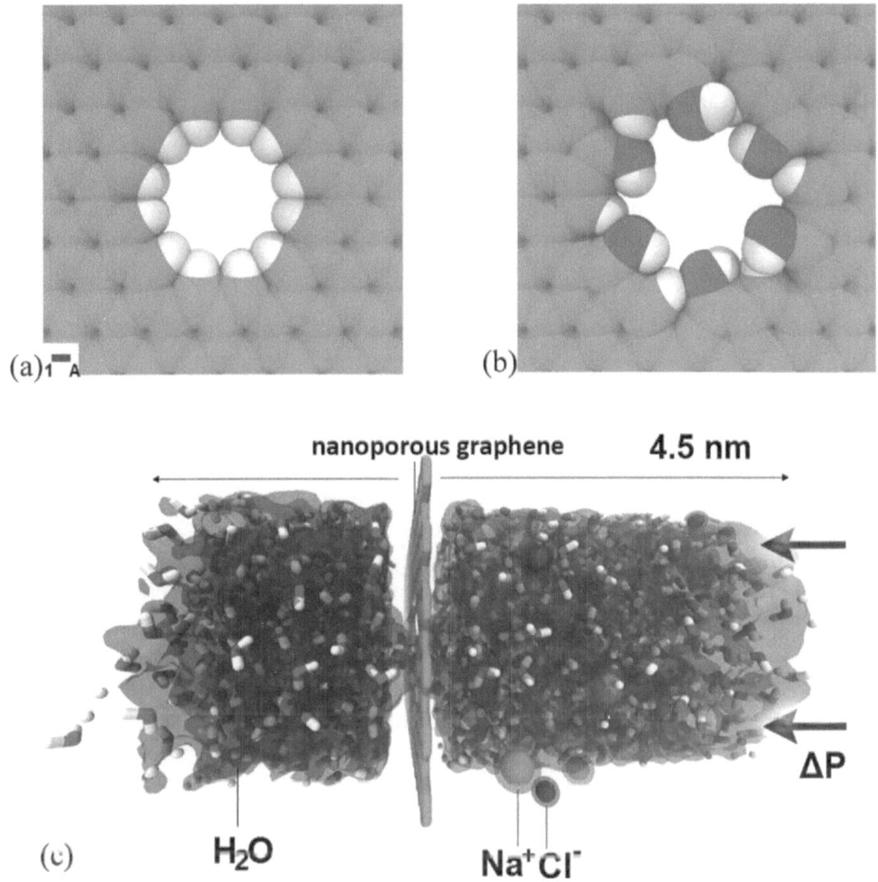

FIGURE 4.1 (a) Hydrogenated graphene nanoporous membrane, (b) hydroxylated graphene pores, and (c) side view of computational system. (Reproduced with permission from Ref. [19] Copyright 2012 American Chemical Society.)

be impervious to liquids, vapors, and gases. Water permeation via such an incomprehensible channel was ascribed to low-friction flow via 2D capillaries between membrane sheets. This unusual movement of molecules confirmed the filtration strategies and, moreover, it emphasized the further exploration of the layout of nanomaterials to a molecular degree.

Later in this interesting account, David and Jeffrey Grossman [19] put their efforts to observe the desalination proficiency of graphene membranes using various molecular dynamics and successfully revealed that the incorporation of nanopores into graphene sheets makes it more effective than conventional RO membranes that are used to filter NaCl salt from water. Pore size, chemical functionalization, and applied pressure are the major factors that influence membrane performance. From obtained data, it was suggested that graphene membranes can fully remove salt ions, while permitting water transport at a measurable rate of 10–100 L $cm^{-2}day^{-1}MPa^{-1}$ and atomic thickness of membrane and pore dimensions are the key elements that are responsible for this high water flux [21,22]. The study concluded that pore diameter essentially decides whether a membrane will allow water to pass through or will block salt passage. Cohen-Tanugi and Grossman investigated different chemical functional groups and their impact attached to graphene pores boundaries to aid in successful ion selectivity [19]. They analyzed that, besides pore size, salt removal is susceptible to pore chemistry. They found that the rate of water permeation becomes roughly double if hydrophilic hydroxyl groups come in contact with graphene membranes [23], and consequently, the development of synthetic nanopores of graphene membranes considerably stimulates the usage of graphene in desalination methods.

4.2.1 Nanopore Generation

Graphene membranes are primarily associated with precisely generated sub-nanometer pores existing on the surface area which is considered an interesting area of research [24–28]. Simulated and experimentally observed results have proved the existence of sub-nanometer pores residing in graphene sheets by employing various methods, for instance, oxidation associated with electron beam irradiation [29], ion bombardment technique [30], and doping methods, respectively (Figure 4.2) [31]. Moreover, O'Hern et al. [32] reported a detailed low-energy ion irradiation along with chemical-oxidation etching, generating nanoscale pores showing high density on monolayer graphene. Additionally, sub-nanometer pores appeared on large surface areas of graphene. Aforesaid creation was fabricated via chemical vapor deposition (CVD), thereby generating isolated defects that were prominently observed within single-layer sheets by employing ion bombardment. Furthermore, permeable membrane pores have been produced by effective oxidative etching. This body of literature clearly demonstrated that etching time varies accordingly, resulting in control over pore density. Variation in etching time from 0 to 120 min resulted in an increase in pore density from 0 to 6×10^{-12} cm^{-2}. However, pore diameter stability was discovered after a 60-minute etching cycle. Additionally, this regulatory capability

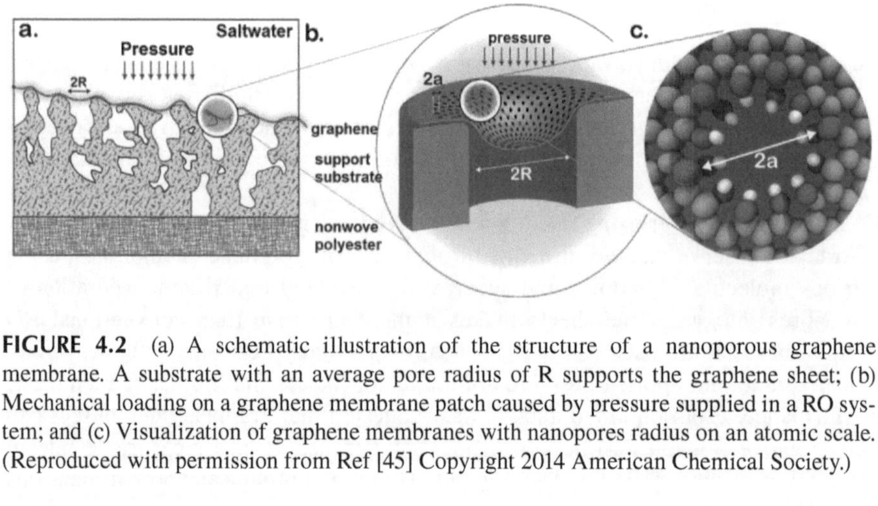

FIGURE 4.2 (a) A schematic illustration of the structure of a nanoporous graphene membrane. A substrate with an average pore radius of R supports the graphene sheet; (b) Mechanical loading on a graphene membrane patch caused by pressure supplied in a RO system; and (c) Visualization of graphene membranes with nanopores radius on an atomic scale. (Reproduced with permission from Ref [45] Copyright 2014 American Chemical Society.)

appears to induce ion and organic species rejection. The resulting ionic flow density has not been denied. The explanation for this was the presence of large pores, which were formed by permanganate etching as a naturally occurring defect inherited in graphene nanosheets (NS). Prominently, large area associated with defect-free and single-layer graphene sheets is still considered as challenging goal for developing graphene membranes for water filtration [26,33–36].

A great deal of effort has been invested into developing simple and dependable techniques for producing 2D-graphene membranes. Among these approaches, two-layer optimized CVD provided graphene with minimal defects and fine grain connectivity at approximately 49 pore punctures [37]. SiN_x specimen was observed to construct robust as well as 1 nm thick freestanding graphene layers [38]. However, nanopores were successfully framed through drilling techniques along with the association of FIB. Double-layer graphene contains a million pores capable to provide ultimate permeation that offered water transport rates connected with 2D natures. O'Hern and distinguished engineers at MIT, Oak Ridge National Laboratory, and Saudi Arabia's King Fahd University of Petroleum & Minerals have developed different techniques [39]. The aforementioned group constructed a precise defect-free (centimeter-scale) graphene membrane by employing multiscale leakage-sealing phenomenon. Copper based on graphene was used to build the membrane. Subsequently, synthesized membrane composite was disseminated to surface-modified membranes with 200 nm pores. Graphene sheet defects were integrated with hafnium oxide by the ALD technique. Holes and tears generated during the transformation of copper-substrate graphene were isolated using nylon-6 utilizing an interfacial polymerization technique [40–43].

Moreover, the defect repairing technique was rendered on nanometer-scale pores that were fabricated by ion bombardment. As a consequence, porous defect-free graphene membrane along with small molecules refused up to 90% of different ions; and water flow has been related to previously established molecular dynamic simulations. Similarly, salt was accomplished faster than water via membrane. Versatile outcomes supported potential contribution in graphene in the form of filtration membranes but still defect sealing techniques were not so advanced. Struggles pertaining to challengeable single pristine graphene sheets have been examined for purification of multilayer nanoporous graphene membranes [44]. Cohen-Tanugi et al. [45] have highlighted a bilayer system corresponding to NPG that offers a versatile strategy for active RO membrane. Further study has revealed the efficient performance of NPG membrane containing multilayer structure that has a direct effect on system design. The number of NPG layers have greatly affected the membrane properties that highlight the immense need for attractive design before declaring an efficient energy system resulting in an alternative cost-effective approach toward RO membranes [46]. However, an oxygen-plasma etching process was adopted for facile structure of tunable graphene nanopores [47]. Pore generation method has resulted in the form of suspended graphene along with tailored pores with desired dimensions, exhibiting water molecule adsorption/desorption over ionic solution. The salt rejection rate was evaluated as 100% additionally toward rapid water transport [47]. The outcomes again revealed the efficiency and potentiality in favor of nanoporous graphene application toward water filtration [25,48,49].

4.2.2 Flexibility of Nanoporous Graphene Membranes

Nanoporous graphene advancement has offered vast opportunities for desalination technologies. Super graphene has demonstrated strength not favorable; however, promising desalination material presents favorable flexibility that may be restricted by a water-drenched environment along with a high-pressure environment associated with RO desalination [50,51]. Moreover, to meet uncertainty, nanoporous graphene may come forward as a viable choice for maintaining mechanical integrity as suggested by Cohen-Tanugi and Jeffrey [45] who adopted famous approaches, viz. molecular dynamics or continuum fracture mechanics, to determine mechanical rigidity owing to nanoporous graphene from desalination technologies. RO desalination techniques often prepare thin-film-composite of membrane nature containing active layer, comprising polyamide indulging highly porous polysulfone substrate along with size in the range of 0.1–0.5 µm [45,51,52]. Polyamide nanoporous graphene provides mechanical support toward a favorable hydraulic load. Additionally, substrate choice is still challengeable subject to material availability for required information in favor of a reasonable polysulfone layer [45]. Cohen-Tanugi and Jeffrey's theoretical analysis emphasized system alignment using a variety of mechanisms, including applied pressure, elastic characteristics, fracture stress, and water impact. [45]. An evolutionary study has revealed the genuine role of nanoporous graphene in the form of hydraulic pressures with desalination technologies [45]. In addition, substrate choice is considered basic consideration toward membrane designing, therefore, favorable selected substrate having 1 µm cavity has offered graphene membranes providing resistance to pressures of 58 MPa [45]. Nanoporous graphene presents a unique filtration membrane, but still faces a variety of barriers that are attached to the large surface area of graphene membranes along with the creation of nanopores residing with graphene sheets. All results produced in favor of incongruous pores along with possible fracturing of membranes have been observed while applying high pressure [50].

4.3 GO MEMBRANES

A variety of structural designs for GO [53–61], particularly due to its hydrophilic nature, have been deeply studied. The mechanical strength with interesting properties has set a lot of future targets in favor of the material's application in purification strategies for water treatment. Nevertheless, GO may be produced in bulk form by employing well-known oxidation methods similar to pristine graphene [62,63]. Selective graphene membranes have offered a separation process via nanopore structure that is present within basal planes showing hexagonal crystalline structure. On the other hand, ions, as well as molecules, may selectively be transported by following interlayer spacing owing to many layers of 2D materials. The stacked form of GO sheets offers multilayered lamination that provides the required mechanical strength for follow-up of pressure-driven water purification technology [64–66]. Tight hydrogen bonding between isolated sheets causes the process to proceed. Oxygenated functional groups residing randomly at the

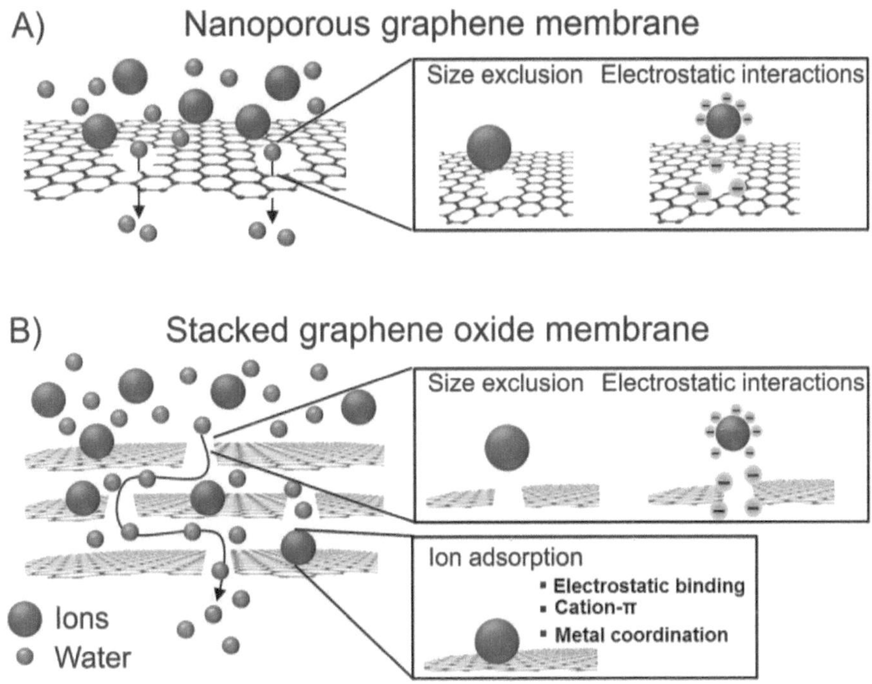

FIGURE 4.3 Two distinct forms of GO-based membranes are depicted schematically. (a) Nanoporous GO membranes are composed of a single layer of GO with defined-pore-size nanopores. Size exclusion and electrostatic repulsion between charged species and pores are used to create selectivity; (b) Membranes constructed entirely of stacked GO sheets. (Reproduced with permission from Ref. [24] Copyright 2015 Royal Society of Chemistry.)

edges of GO-NS spontaneously create appropriate interlayer spacing, while empty spaces occur among non-oxidized sites, resulting in the formation of a network of nano capillaries within the available film [67–72]. The aforementioned nanochannels have allowed water molecules saturation along with subsequent transportation toward hydrophobic non-oxidizing areas owing to membranes assisting the fast movement of water molecules [73]. However, other states of matter and vapors are strictly opposed (Figure 4.3) [20]. Oxygenated functional groups of soluble water molecules that occur on compound sheets show absorbance of water molecules, thereby diffusing between non-polar hydrocarbons owing to GO. Penetration of water molecules enlarges interlayer spacing among stacked form of GO-NS [74] that is generating water flux within nanochannels possessing high flow rate. A reported study has presented a theory that ions with less flux may penetrate GO membranes that present a greater role than that of a simple diffusion process [75]. This gives researchers an idea about the expansion of nanochannels of the hydrated state containing similar size ions. GO exhibits a hydrophilic nature with O_2 functional groups. GO's dispersion in various solvents and GO films' durability in water molecules have been vastly studied [66,75,76].

4.4 GRAPHENE ANALOGS FOR MEMBRANE TECHNOLOGY

Recently, tens of novel two-dimensional crystals were recovered from bulk materials, but only a few have been widely exploited for water purification. Due to its perfect features, such as 2D nanostructures and the ability to function as an effective membrane, graphene is projected to have a substantial impact on a wide variety of applications. Apart from GO, 2D nanostructured membranes exhibit very fast ion and molecule separation, according to the current research. In addition, 1D-BN nanotubes are very efficient substances for both desalination and water decontamination [77–81].

4.4.1 TMDCs

Among the Transition Metallic Dichalcogenide (TMDCs), MoS_2 is the most extensively used nanocomposite in a number of fields, and now it has been employed in separation techniques as well. MoS_2, a typical example of TMDCs, is comprised of hexagonal layers of Mo and S_2 atoms [82,83]. Generally, TMDCs are represented by a formula MX_2, and it includes over 40 different materials, comprised of a central transition metal (M) like Mo, W, Nb, Re, Ni, or V, surrounded by two chalcogens (X), for instance, S, Se, or Te [84]. An elastic separation membrane 'laminar', prepared from MoS_2 sheets, was reported to have 3–5 times higher water flux than GO and was able to remove 89% of Evans Blue and 98% of cytochrome C molecules [85].

Followed by lamellar separation membranes, Sun et al. [86] further probed semiconductor material proficiency, and found tungsten disulfide (WS_2), similar to MoS_2. Using the filtration process, bulk material was exfoliated to obtain a thin film of WS_2. The resultant membrane exhibited five times higher water flux than GO membranes and nearly twice more than MoS_2 laminar membranes and eliminated 90% Evans Blue molecules. Water penetration rate further increased from 450 to 930 $Lm^{-2}h^{-1}$ bar^{-1} by the inclusion of metal hydroxide nano strands. The occurrence of nano strands in between WS_2 sheets generated additional channels that allowed water to transport without rejection. On the other hand, in advance research for 2D nanostructured desalination membranes, Heiranian et al. [87] worked with a number of different materials possessing naturally hydrophilic sites in them as shown in Figure 4.4. Among the tested substances, i.e., $MoSe_2$, $MoTe_2$, WS_2, and WSe_2, MoS_2 showed paramount efficiency.

It has already been documented that the existence of hydroxyl groups causes hydrophilic attraction on the edge of graphene pores, thereby increasing water flux and transportation. The key challenge for researchers is to determine how defined functional groups can be added at the edges of nanopores, which prompted them to investigate the efficiency of various TMDCs membranes. To test the salt-removing capability of monolayer nanoporous MoS_2, the researchers created a molecular dynamic model consisting of a MoS_2 layer, H_2O, and ions, as well as a rigid graphene sheet that can withstand external pressure. The proficiency of MoS_2 membrane was investigated as a pore dimensions feature, chemistry, geometry, and applied hydrostatic strain. From the data, it is discovered that membrane efficiency was extremely

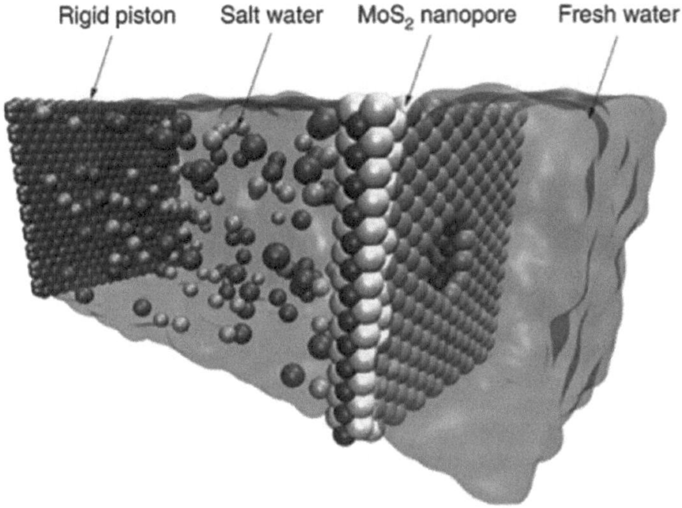

FIGURE 4.4 Simulation box comprising a MoS_2 sheet of water, ions, and a graphene sheet. (Reproduced with permission from Ref. [87] Copyright 2015 Springer Nature Limited.)

influenced by pore chemistry. Examination of Mo, S, and pore structure indicated that besides hydrophobic edges and hydrophilic center of Mo, pore dimensions were solely responsible for proficient ion rejection above 80%, and increased water flux by more than 70% compared to graphene membranes without aided nanopores [87,88].

4.4.2 MXenes

A new group of 2D materials known as 'MXenes', comprising single-atom-thick layers of TM carbides or carbonitrides, were utilized to examine the potential of separation membranes for ions and molecules removal from water [89]. These membranes were found to be more vigorous, elastic, and possess hydrophilic $Ti_3C_2T_x$ nature which improved the rate of water permeation up to 37.4 $Lm^{-2}h^{-1}$ bar^{-1} and showed better separation of higher charge cations compared to GO. In the industrial application as wastewater treatment, graphene with its natural porosity is the best separating material. According to molecular dynamics, tremendous water permeation can be achieved when graphene is used in nanoweb structure and thus, exhibits excellent rejection of heavy metal ions, hydrophobic organic chemicals, and monovalent salt ions [90,91]. The presence of natural nanopores of graphene, its ultrathin configuration, superior mechanical strength, and ability to tolerate high deformation stress in web-like membranes, impart amazing filtration performance.

Similar to other 2D materials, MXene nanoflakes with one-atom thickness can form nanocomposites with polymers. To fabricate 2D nanocomposite membranes, the hydrophilic nature of MXene-NS contributes an energetic part to the formation of highly stable concentrations in various organic aqueous (i.e., aprotic polar solvents) [92]. Stable dispersions produced in organic solvents are then converted into non-segregating polymer solutions for further processing to construct membranes.

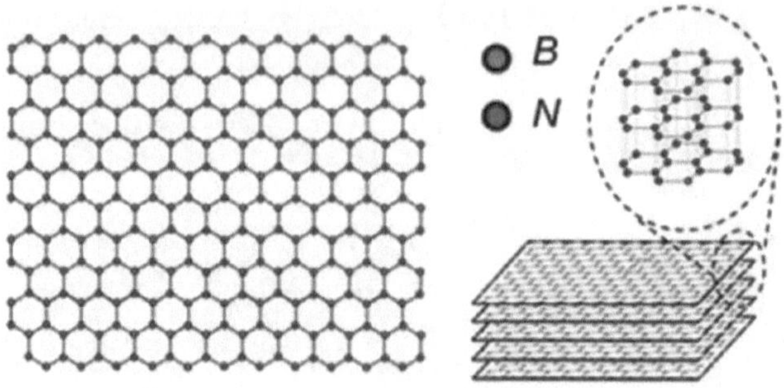

FIGURE 4.5 Schematic of preparation and separation of MXene/PES composite membrane. (Reproduced with permission from Ref. [94] Copyright 2017 Royal Society of Chemistry.)

Furthermore, 2D MXenes lamellar structure has an extremely narrow transport passage and a large number of nanochannels, and when these MXene-NS combine to form nanocomposite membranes, enhanced separation efficiency is revealed. Hence, mixed-matrix membranes (MMMs) that are made up of MXene-NS indicate very fast water permeance at the rate of greater than 1,000 $Lm^{-2}h^{-1} bar^{-1}$ with a promising dye removal rate of approximately 90% for Evans blue of molecular weight 960 and 97% for Cytochrome C of molecular weight 12,000 [93].

Han et al. [94] reported the formation of MMM based on 2D $Ti_3C_2T_x$ (where T indicates functional groups such as O, OH, and/or F) for forward osmosis process (Figure 4.5). Through etching and ultrasonic treatment of Ti_3AlC_2 structure, 2D MXene was prepared. Thus generated MXene-NS possesses a free lamellar structure with a favored transport channel for liquid molecules which affects the transport of water through fabricated MMM with polyethersulfone (PES). The best results were observed in PES/MXene membrane, as the rate of water flux reached up to ~ 115 $Lm^{-2}h^{-1}$, and about 92% of Congo red dye (molecular weight of 697) was removed. Further verifications of $Ti_3C_2T_x$ NS-based mixed-matrix membranes were carried out in dehydration of conventional organic solvents. Another mixed-matrix membrane with submicron-thick chitosan (Cs)/MXene has been prepared on porous polyacrylonitrile (PAN) substrate by spin-coating method.

The channels formed between layers of MXene membranes supplied a passage for quick and selective permeation of water molecules. Cs/MXene MMM exhibited an overall flux of 1.4–1.5 $kgm^{-2}h^{-1}$ and selectivity up to 1,400, 4,800, and 900 for ethanol dehydration, ethyl acetate, and dimethyl carbonate at 50°C, respectively. Han et al. [95] designed a crosslinked P84 copolyimide MMM based on MXene that sowed improved solvent resistance. A triethylenetetramine crosslinked P84/MXene mixed-matrix membrane was used to illustrate organic solvent nanofiltration. It was found that enhanced hydrophilicity of water channels in MXene is a major factor to improve the overall membrane performance. Finally, this membrane attained a maximum flux density of 268 $Lm^{-2}h^{-1}$ and 100% removal of gentian violet dye with a molecular weight of 408. Furthermore, its performance can sustain for 18 days if it is kept in acetone, methanol, and DMF.

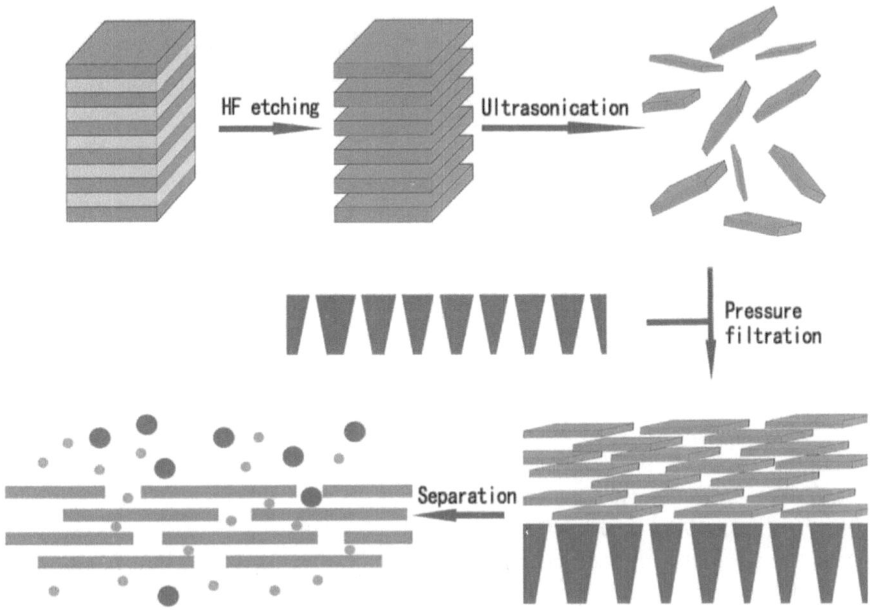

FIGURE 4.6 Front view of hexagonal structure of BN with indication of B and N, and edge structure of layers. (Reproduced with permission from Ref. [96] Copyright 2012 Royal Society of Chemistry.)

As the chemists have pinned their hopes on 2D materials they have explored BN-NS as a potential candidate for MXenes [97]. BN is a compound containing boron and nitride atoms with equal numbers, showing sp^2 hybridization in a honeycomb lattice (Figure 4.6) and exists in several crystallographic forms [96]. Hexagonal BN, commonly known as 'white graphene', is becoming popular and has excellent chemically stable insulating properties, which render it a good substrate for mono and bilayer graphene expedients enhancing graphene's electronic properties to a valuable extent [98,99]. BN-NS with the porous structure were fabricated through thermal treatment to eliminate organic solvents, oils, and dyes from water. The remarkable surface area of 1,427 $m^2 g^{-1}$ in superhydrophobic NS renders them to absorb more than 33 times the original weight. Moreover, owing to its excessive oxidative resistance, amazingly selective, porous, layered BN nanomaterial may be fabricated via burning/heating, which helps to reuse the NS.

4.5 CHALLENGES RELATED TO 2D WATER PURIFICATION MEMBRANES

Graphene and different 2D crystals expanded to industrial 2D nanostructure membranes are in their early stage for water treatment applications. Synthetic complexities and scalable processing challenges make their industrial usage applications costly and confined to small-scale devices owing to these materials' young nature.

Moreover, substantial investment is needed to replace traditional materials commonly utilized for treating wastewater with novel 2D nanostructure membrane materials. Challenges posed by 2D water treatment membranes react mostly to young processing procedures utilized to reach big areas and defect-free monolayers of these materials. High-quality graphene-like 2D nanostructure and pristine graphene remain a difficulty. Also making such ideal crystals on porous support poses issues. GO reduction rate gives a low-cost approach to creating graphene on a wide scale, but the reduction procedure requires additional investigation to restore mechanical and electrical characteristics similar to pure graphene. When proposing a realistic approach for manufacturing 2D materials, issues can still exist regarding pores-controlled creation and precise material sizes. Pores with precise graphene monolayer size may be created by focused electron beam irradiation, although this technology is not routinely exploited to generate vast regions of porous graphene sheets. Chemical or oxidative etching was an alternate source for large-scale manufacture, but still encountered some difficulties due to various flaws in graphene [19,100].

GO exploitation in water purification, rather than pure graphene, is another possibility for graphene-based water treatment. In comparison, oxidized graphene, similar to graphene, encounters problems for ease of use on an industrial scale. The continuous evolution of new technology and developed techniques has begun to encourage new phenomena and expand the use of 2D materials in the industry. Despite the fact that uncertainties surround issues such as graphene family health risks and environmental protection, graphene evolution is being studied in almost every sector. Researchers have implored valuation of material's toxicity [101–103]. The research explored graphene's cytotoxic effect and suggested that graphene toxicity is contingent on many parameters including chemical structure [104] and surface functionalization [105,106], dose and size [107–109], and exposure time. Moreover, graphene type, for example, GO, rGO, few layer of graphene and graphene NS, excess of layers, technique used to synthesize materials, the type of exposed cell, and in rare cases administration route have impacted the properties significantly [110].

Furthermore, the toxicity of graphene has begun to achieve great appeal to explain the brief examination of health hazards and environmental consequences of other 2D crystals owing to its unique nature. Examining their possible toxicity and environmental concerns is vital to assure safe design, production, and usage of 2D nanomembrane. To decrease the environmental and health concerns of 2D materials, it is vital to comprehend how nanostructures interact with tissues and cells by focusing specifically on exposure ways, such as lung and skin, as well as the nervous and immune systems. The substantial influence on terrestrial, aquatic, and microbiological lives should also be explored. Moreover, numerous groups emphasized the urgent need for graphene risk assessment and appropriate 2D materials [111–113]. Moreover, there seems to be little understanding of graphene and diverse 2D materials for post-recovery recycling and disposal treatment. If the right hypotheses for conservation and disposal of graphene materials are not used, an additional water crisis may arise. With the continual introduction of new nanomaterials and their use in nanotechnology domains, it is crucial for researchers to show interest in studying their impact on the environment and initiate appropriate disposal methods.

4.6 CONCLUSIONS AND SUMMARY

Conventional water treatment methods, like oxidation, distillation, sedimentation, and boiling, as well as solar disinfection and chemical disinfection, are unable to provide an affordable or satisfactory potable supply of water to the world. Improved technology should be accepted and industrialized to offer every distinct on the planet with elementary right to drink clean and fresh water. As a result, pressure-driven, water-treatment, and membrane-based methods like ultrafiltration, nanofiltration, and RO have developed encouraging and commercial-scale potable water production solutions. These technologies evade the utilization of most common chemical additives, spent media regeneration, and thermal inputs that causes to produce more operative wastewater treatment approaches. But, owing to the aforesaid shortcomings connected with these methods such as RO, poor selectivity, great energy demands, and membrane-based methods, high improvement is desired to dramatically meet the likely water calamity that is presently faced by humanity. Significant breakthroughs in cutting-edge nanomaterials were lately been employed to address the weaknesses of well-known membrane technology. The evolution of nanotechnology toward wastewater treatment has altered concentration from the use of non-porous conventional RO polymeric membranes to 2D-quasi nanomembrane. The novel properties of these complex nanosystems include distinct inorganic salts and also organic compounds with a low molecular weight depending on the size-exclusion process and surface charge interaction, enabling lower energy consumption and high flux rates related to conventional ultrafiltration or RO crystals. The performance of existing membrane separation methods may be increased in terms of water flow, salt rejection, and fouling resistance via noble features of such cutting-edge materials, thereby reducing the costs of several processes widely utilized in industry. Atomic thinness, higher surface area, and synthetic pores or naturally occurring 2D-based membrane nanochannels provide simple selective penetration, high water flow, and proficient release of contaminants without the high-pressure need of present technologies. Indeed, there are significant industrial-scale limits on 2D materials; yet, if handled, 2D nanomaterial membranes can give a solution to the future water issue by boosting the efficiency and financial feasibility of present wastewater treatment performance. As a result, subsequent research and development must focus on preceding these materials.

REFERENCES

1. R.P.L. David, S. Sholl, Seven chemical separations to change the world, *Nature News*, 532 (2016) 435–437.
2. T. Hyun, J. Jeong, A. Chae, Y.K. Kim, D.-Y. Koh, 2D-enabled membranes: Materials and beyond, *BMC Chemical Engineering*, 1 (2019) 12.
3. A. Raza, M. Ikram, M. Aqeel, M. Imran, A. Ul-Hamid, K.N. Riaz, S. Ali, Enhanced industrial dye degradation using Co doped in chemically exfoliated MoS_2 nanosheets, *Applied Nanoscience*, 10 (2020) 1535–1544.
4. Z. Xie, Y.-P. Peng, L. Yu, C. Xing, M. Qiu, J. Hu, H. Zhang, Solar-inspired water purification based on emerging 2D materials: Status and Challenges, *Solar RRL*, 4 (2020) 1900400.

5. L.F. Greenlee, D.F. Lawler, B.D. Freeman, B. Marrot, P. Moulin, Reverse osmosis desalination: Water sources, technology, and today's challenges, *Water Research*, 43 (2009) 2317–2348.
6. S. Remanan, N. Padmavathy, S. Ghosh, S. Mondal, S. Bose, N.C. Das, Porous graphene-based membranes: Preparation and properties of a unique two-dimensional nanomaterial membrane for water purification, *Separation & Purification Reviews*, 50 (2020) 1–21.
7. W.J. Koros, R.P. Lively, Water and beyond: Expanding the spectrum of large-scale energy efficient separation processes, *AIChE Journal*, 58 (2012) 2624–2633.
8. M. Ulbricht, Advanced functional polymer membranes, *Polymer*, 47 (2006) 2217–2262.
9. M. Ikram, A. Raza, M. Imran, A. Ul-Hamid, A. Shahbaz, S. Ali, Hydrothermal synthesis of silver decorated reduced graphene oxide (rGO) nanoflakes with effective photocatalytic activity for wastewater treatment, *Nanoscale Research Letters*, 15 (2020) 95.
10. S. Karan, Z. Jiang, A.G. Livingston, Sub–10nm polyamide nanofilms with ultrafast solvent transport for molecular separation, *Science*, 348 (2015) 1347.
11. X. Ma, P. Kumar, N. Mittal, A. Khlyustova, P. Daoutidis, K.A. Mkhoyan, M. Tsapatsis, Zeolitic imidazolate framework membranes made by ligand-induced permselectivation, *Science*, 361 (2018) 1008.
12. M.R. Chowdhury, J. Steffes, B.D. Huey, J.R. McCutcheon, 3D printed polyamide membranes for desalination, *Science*, 361 (2018) 682.
13. M. Zeng, M. Chen, D. Huang, S. Lei, X. Zhang, L. Wang, Z. Cheng, Engineered two-dimensional nanomaterials: An emerging paradigm for water purification and monitoring, *Materials Horizons*, 8 (2021) 758–802.
14. T. Liu, X. Liu, N. Graham, W. Yu, K. Sun, Two-dimensional MXene incorporated graphene oxide composite membrane with enhanced water purification performance, *Journal of Membrane Science*, 593 (2020) 117431.
15. D. Pakulski, W. Czepa, S.D. Buffa, A. Ciesielski, P. Samorì, Atom-thick membranes for water purification and blue energy harvesting, *Advanced Functional Materials*, 30 (2020) 1902394.
16. O. Leenaerts, B. Partoens, F.M. Peeters, Graphene: A perfect nanoballoon, *Applied Physics Letters*, 93 (2008) 193107.
17. J.S. Bunch, S.S. Verbridge, J.S. Alden, A.M. van der Zande, J.M. Parpia, H.G. Craighead, P.L. McEuen, Impermeable atomic membranes from graphene sheets, *Nano Letters*, 8 (2008) 2458–2462.
18. M.E. Suk, N.R. Aluru, Water transport through ultrathin graphene, *The Journal of Physical Chemistry Letters*, 1 (2010) 1590–1594.
19. D. Cohen-Tanugi, J.C. Grossman, Water desalination across nanoporous graphene, *Nano Letters*, 12 (2012) 3602–3608.
20. R.R. Nair, H.A. Wu, P.N. Jayaram, I.V. Grigorieva, A.K. Geim, Unimpeded permeation of water through helium-leak–tight graphene-based membranes, *Science*, 335 (2012) 442.
21. A. Nicolaï, B.G. Sumpter, V. Meunier, Tunable water desalination across graphene oxide framework membranes, *Physical Chemistry Chemical Physics*, 16 (2014) 8646–8654.
22. S. Dervin, D.D. Dionysiou, S.C. Pillai, 2D nanostructures for water purification: Graphene and beyond, *Nanoscale*, 8 (2016) 15115–15131.
23. S. Banerjee, D.D. Dionysiou, S.C. Pillai, Self-cleaning applications of TiO2 by photo-induced hydrophilicity and photocatalysis, *Applied Catalysis B: Environmental*, 176–177 (2015) 396–428.
24. F. Perreault, A. Fonseca de Faria, M. Elimelech, Environmental applications of graphene-based nanomaterials, *Chemical Society Reviews*, 44 (2015) 5861–5896.
25. K.A. Mahmoud, B. Mansoor, A. Mansour, M. Khraisheh, Functional graphene nanosheets: The next generation membranes for water desalination, *Desalination*, 356 (2015) 208–225.

26. L. Huang, M. Zhang, C. Li, G. Shi, Graphene-based membranes for molecular separation, *The Journal of Physical Chemistry Letters*, 6 (2015) 2806–2815.
27. H. Yu, Y. He, G. Xiao, Y. Fan, J. Ma, Y. Gao, R. Hou, J. Chen, Weak-reduction graphene oxide membrane for improving water purification performance, *Journal of Materials Science & Technology*, 39 (2020) 106–112.
28. G.H. Jeong, S.P. Sasikala, T. Yun, G.Y. Lee, W.J. Lee, S.O. Kim, Nanoscale assembly of 2D materials for energy and environmental applications, *Advanced Materials*, 32 (2020) 1907006.
29. M.D. Fischbein, M. Drndić, Electron beam nanosculpting of suspended graphene sheets, *Applied Physics Letters*, 93 (2008) 113107.
30. N. Inui, K. Mochiji, K. Moritani, N. Nakashima, Molecular dynamics simulations of nanopore processing in a graphene sheet by using gas cluster ion beam, *Applied Physics A*, 98 (2010) 787–794.
31. T. Humplik, J. Lee, S.C. O'Hern, B.A. Fellman, M.A. Baig, S.F. Hassan, M.A. Atieh, F. Rahman, T. Laoui, R. Karnik, E.N. Wang, Nanostructured materials for water desalination, *Nanotechnology*, 22 (2011) 292001.
32. S.C. O'Hern, M.S.H. Boutilier, J.-C. Idrobo, Y. Song, J. Kong, T. Laoui, M. Atieh, R. Karnik, Selective ionic transport through tunable subnanometer pores in single-layer graphene membranes, *Nano Letters*, 14 (2014) 1234–1241.
33. G. Liu, W. Jin, N. Xu, Graphene-based membranes, *Chemical Society Reviews*, 44 (2015) 5016–5030.
34. D.-Y. Koh, R.P. Lively, Membranes at the limit, *Nature Nanotechnology*, 10 (2015) 385–386.
35. M. Ikram, R. Tabassum, U. Qumar, S. Ali, A. Ul-Hamid, A. Haider, A. Raza, M. Imran, S. Ali, Promising performance of chemically exfoliated Zr-doped MoS2 nanosheets for catalytic and antibacterial applications, *RSC Advances*, 10 (2020) 20559–20571.
36. F. Pulizzi, Membranes for special water treatment, *Nature Nanotechnology*, 15 (2020) 895–895.
37. Y. Wei, J. Wu, H. Yin, X. Shi, R. Yang, M. Dresselhaus, The nature of strength enhancement and weakening by pentagon–heptagon defects in graphene, *Nature Materials*, 11 (2012) 759–763.
38. K. Celebi, J. Buchheim, R.M. Wyss, A. Droudian, P. Gasser, I. Shorubalko, J.-I. Kye, C. Lee, H.G. Park, Ultimate permeation across atomically thin porous graphene, *Science*, 344 (2014) 289.
39. S.C. O'Hern, D. Jang, S. Bose, J.-C. Idrobo, Y. Song, T. Laoui, J. Kong, R. Karnik, Nanofiltration across defect-sealed nanoporous monolayer graphene, *Nano Letters*, 15 (2015) 3254–3260.
40. Y. Wei, Y. Zhang, X. Gao, Z. Ma, X. Wang, C. Gao, Multilayered graphene oxide membranes for water treatment: A review, *Carbon*, 139 (2018) 964–981.
41. I. Ali, O.M.L. Alharbi, A. Tkachev, E. Galunin, A. Burakov, V.A. Grachev, Water treatment by new-generation graphene materials: Hope for bright future, *Environmental Science and Pollution Research*, 25 (2018) 7315–7329.
42. A. Anand, B. Unnikrishnan, J.-Y. Mao, H.-J. Lin, C.-C. Huang, Graphene-based nanofiltration membranes for improving salt rejection, water flux and antifouling–A review, *Desalination*, 429 (2018) 119–133.
43. M. Ikram, J. Hassan, M. Imran, J. Haider, A. Ul-Hamid, I. Shahzadi, M. Ikram, A. Raza, U. Qumar, S. Ali, 2D chemically exfoliated hexagonal boron nitride (hBN) nanosheets doped with Ni: Synthesis, properties and catalytic application for the treatment of industrial wastewater, *Applied Nanoscience*, 10 (2020) 3525–3528.
44. D. Cohen-Tanugi, L.-C. Lin, J.C. Grossman, Multilayer nanoporous graphene membranes for water desalination, *Nano Letters*, 16 (2016) 1027–1033.

45. D. Cohen-Tanugi, J.C. Grossman, Mechanical strength of nanoporous graphene as a desalination membrane, *Nano Letters*, 14 (2014) 6171–6178.
46. R. Li, L. Zhang, P. Wang, Rational design of nanomaterials for water treatment, *Nanoscale*, 7 (2015) 17167–17194.
47. S.P. Surwade, S.N. Smirnov, I.V. Vlassiouk, R.R. Unocic, G.M. Veith, S. Dai, S.M. Mahurin, Water desalination using nanoporous single-layer graphene, *Nature Nanotechnology*, 10 (2015) 459–464.
48. P.S. Goh, A.F. Ismail, N. Hilal, Nano-enabled membranes technology: Sustainable and revolutionary solutions for membrane desalination? *Desalination*, 380 (2016) 100–104.
49. A. Raza, U. Qumar, J. Hassan, M. Ikram, A. Ul-Hamid, J. Haider, M. Imran, S. Ali, A comparative study of dirac 2D materials, TMDCs and 2D insulators with regard to their structures and photocatalytic/sonophotocatalytic behavior, *Applied Nanoscience*, 10 (2020) 3875–3899.
50. Z. Song, Z. Xu, X. Huang, J.-Y. Kim, Q. Zheng, On the fracture of supported graphene under pressure, *Journal of Applied Mechanics*, 80 (2013) 040911.
51. Y. Liu, X. Chen, Mechanical properties of nanoporous graphene membrane, *Journal of Applied Physics*, 115 (2014) 034303.
52. P.S. Singh, S.V. Joshi, J.J. Trivedi, C.V. Devmurari, A.P. Rao, P.K. Ghosh, Probing the structural variations of thin film composite RO membranes obtained by coating polyamide over polysulfone membranes of different pore dimensions, *Journal of Membrane Science*, 278 (2006) 19–25.
53. F. Kim, L.J. Cote, J. Huang, Graphene oxide: Surface activity and two-dimensional assembly, *Advanced Materials*, 22 (2010) 1954–1958.
54. Q. Shen, Y. Lin, Y. Kawabata, Y. Jia, P. Zhang, N. Akther, K. Guan, T. Yoshioka, H. Shon, H. Matsuyama, Engineering heterostructured thin-film nanocomposite membrane with functionalized graphene oxide quantum dots (GOQD) for highly efficient reverse osmosis, *ACS Applied Materials & Interfaces*, 12 (2020) 38662–38673.
55. M. Zhang, K. Guan, Y. Ji, G. Liu, W. Jin, N. Xu, Controllable ion transport by surface-charged graphene oxide membrane, *Nature Communications*, 10 (2019) 1253.
56. J. Shen, G. Liu, K. Huang, Z. Chu, W. Jin, N. Xu, Subnanometer two-dimensional graphene oxide channels for ultrafast gas sieving, *ACS Nano*, 10 (2016) 3398–3409.
57. Y. Zhang, T.-S. Chung, Graphene oxide membranes for nanofiltration, *Current Opinion in Chemical Engineering*, 16 (2017) 9–15.
58. X. Zhu, Y. Zhou, J. Hao, B. Bao, X. Bian, X. Jiang, J. Pang, H. Zhang, Z. Jiang, L. Jiang, A charge-density-tunable three/two-dimensional polymer/graphene oxide heterogeneous nanoporous membrane for ion transport, *ACS Nano*, 11 (2017) 10816–10824.
59. Y. Ying, W. Ying, Q. Li, D. Meng, G. Ren, R. Yan, X. Peng, Recent advances of nanomaterial-based membrane for water purification, *Applied Materials Today*, 7 (2017) 144–158.
60. Y. Li, W. Zhao, M. Weyland, S. Yuan, Y. Xia, H. Liu, M. Jian, J. Yang, C.D. Easton, C. Selomulya, X. Zhang, Thermally reduced nanoporous graphene oxide membrane for desalination, *Environmental Science & Technology*, 53 (2019) 8314–8323.
61. M. Ikram, E. Umar, A. Raza, A. Haider, S. Naz, A. Ul-Hamid, J. Haider, I. Shahzadi, J. Hassan, S. Ali, Dye degradation performance, bactericidal behavior and molecular docking analysis of Cu-doped TiO_2 nanoparticles, *RSC Advances*, 10 (2020) 24215–24233.
62. J. Chen, B. Yao, C. Li, G. Shi, An improved hummers method for eco-friendly synthesis of graphene oxide, *Carbon*, 64 (2013) 225–229.
63. D.C. Marcano, D.V. Kosynkin, J.M. Berlin, A. Sinitskii, Z. Sun, A. Slesarev, L.B. Alemany, W. Lu, J.M. Tour, Improved synthesis of graphene oxide, *ACS Nano*, 4 (2010) 4806–4814.
64. A.K. Geim, Graphene: Status and prospects, *Science*, 324 (2009) 1530.

65. G. Eda, M. Chhowalla, Chemically derived graphene oxide: Towards large-area thin-film electronics and optoelectronics, *Advanced Materials*, 22 (2010) 2392–2415.
66. D.A. Dikin, S. Stankovich, E.J. Zimney, R.D. Piner, G.H.B. Dommett, G. Evmenenko, S.T. Nguyen, R.S. Ruoff, Preparation and characterization of graphene oxide paper, *Nature*, 448 (2007) 457–460.
67. S. Cerveny, F. Barroso-Bujans, Á. Alegría, J. Colmenero, Dynamics of water intercalated in graphite oxide, *The Journal of Physical Chemistry C*, 114 (2010) 2604–2612.
68. A. Lerf, A. Buchsteiner, J. Pieper, S. Schöttl, I. Dekany, T. Szabo, H.P. Boehm, Hydration behavior and dynamics of water molecules in graphite oxide, *Journal of Physics and Chemistry of Solids*, 67 (2006) 1106–1110.
69. D. Pacilé, J.C. Meyer, A. Fraile Rodríguez, M. Papagno, C. Gómez-Navarro, R.S. Sundaram, M. Burghard, K. Kern, C. Carbone, U. Kaiser, Electronic properties and atomic structure of graphene oxide membranes, *Carbon*, 49 (2011) 966–972.
70. M. Ikram, J. Hassan, A. Raza, A. Haider, S. Naz, A. Ul-Hamid, J. Haider, I. Shahzadi, U. Qamar, S. Ali, Photocatalytic and bactericidal properties and molecular docking analysis of TiO_2 nanoparticles conjugated with Zr for environmental remediation, *RSC Advances*, 10 (2020) 30007–30024.
71. U. Qumar, J. Hassan, S. Naz, A. Haider, A. Raza, A. Ul-Hamid, J. Haider, I. Shahzadi, I. Ahmad, M. Ikram, Silver decorated 2D nanosheets of GO and MoS2 serve as nanocatalyst for water treatment and antimicrobial applications as ascertained with molecular docking evaluation, *Nanotechnology*, 32 (2021) 255704.
72. J. Safaei, P. Xiong, G. Wang, Progress and prospects of two-dimensional materials for membrane-based water desalination, *Materials Today Advances*, 8 (2020) 100108.
73. B. Mi, Graphene oxide membranes for ionic and molecular sieving, *Science*, 343 (2014) 740.
74. W.-S. Hung, Q.-F. An, M. De Guzman, H.-Y. Lin, S.-H. Huang, W.-R. Liu, C.-C. Hu, K.-R. Lee, J.-Y. Lai, Pressure-assisted self-assembly technique for fabricating composite membranes consisting of highly ordered selective laminate layers of amphiphilic graphene oxide, *Carbon*, 68 (2014) 670–677.
75. R.K. Joshi, P. Carbone, F.C. Wang, V.G. Kravets, Y. Su, I.V. Grigorieva, H.A. Wu, A.K. Geim, R.R. Nair, Precise and ultrafast molecular sieving through graphene oxide membranes, *Science*, 343 (2014) 752.
76. M. Sivakumar, D.-K. Liu, Y.-H. Chiao, W.-S. Hung, Synergistic effect of one-dimensional silk nanofiber and two-dimensional graphene oxide composite membrane for enhanced water purification, *Journal of Membrane Science*, 606 (2020) 118142.
77. C.Y. Won, N.R. Aluru, Water permeation through a subnanometer boron nitride nanotube, *Journal of the American Chemical Society*, 129 (2007) 2748–2749.
78. J. Kou, J. Yao, L. Wu, X. Zhou, H. Lu, F. Wu, J. Fan, Nanoporous two-dimensional MoS2 membranes for fast saline solution purification, *Physical Chemistry Chemical Physics*, 18 (2016) 22210–22216.
79. Z. Cao, V. Liu, A. Barati Farimani, Why is single-layer MoS_2 a more energy efficient membrane for water desalination? *ACS Energy Letters*, 5 (2020) 2217–2222.
80. J. Ma, X. Tang, Y. He, Y. Fan, J. Chen, Robust stable MoS_2/GO filtration membrane for effective removal of dyes and salts from water with enhanced permeability, *Desalination*, 480 (2020) 114328.
81. A. Pendse, S. Cetindag, M.-H. Lin, A. Rackovic, R. Debbarma, S. Almassi, B.P. Chaplin, V. Berry, J.W. Shan, S. Kim, Charged layered boron nitride-nanoflake membranes for efficient ion separation and water purification, *Small*, 15 (2019) 1904590.
82. S. Zhou, S. Wang, H. Li, W. Xu, C. Gong, J.C. Grossman, J.H. Warner, Atomic structure and dynamics of defects in 2D MoS_2 bilayers, *ACS Omega*, 2 (2017) 3315–3324.
83. E. Scalise, M. Houssa, G. Pourtois, V.V. Afanas'ev, A. Stesmans, First-principles study of strained 2D MoS2, *Physica E: Low-dimensional Systems and Nanostructures*, 56 (2014) 416–421.

84. M. Chhowalla, H.S. Shin, G. Eda, L.-J. Li, K.P. Loh, H. Zhang, The chemistry of two-dimensional layered transition metal dichalcogenide nanosheets, *Nature Chemistry*, 5 (2013) 263–275.
85. L. Sun, H. Huang, X. Peng, Laminar MoS2 membranes for molecule separation, *Chemical Communications*, 49 (2013) 10718–10720.
86. L. Sun, Y. Ying, H. Huang, Z. Song, Y. Mao, Z. Xu, X. Peng, Ultrafast molecule separation through layered WS2 nanosheet membranes, *ACS Nano*, 8 (2014) 6304–6311.
87. M. Heiranian, A.B. Farimani, N.R. Aluru, Water desalination with a single-layer MoS_2 nanopore, *Nature Communications*, 6 (2015) 8616.
88. M. Macha, S. Marion, V.V.R. Nandigana, A. Radenovic, 2D materials as an emerging platform for nanopore-based power generation, *Nature Reviews Materials*, 4 (2019) 588–605.
89. C.E. Ren, K.B. Hatzell, M. Alhabeb, Z. Ling, K.A. Mahmoud, Y. Gogotsi, Charge- and size-selective ion sieving through $Ti_3C_2T_x$ MXene membranes, *The Journal of Physical Chemistry Letters*, 6 (2015) 4026–4031.
90. M. Khazaei, M. Arai, T. Sasaki, C.-Y. Chung, N.S. Venkataramanan, M. Estili, Y. Sakka, Y. Kawazoe, Novel electronic and magnetic properties of two-dimensional transition metal carbides and nitrides, *Advanced Functional Materials*, 23 (2013) 2185–2192.
91. S. Lin, M.J. Buehler, Mechanics and molecular filtration performance of graphyne nanoweb membranes for selective water purification, *Nanoscale*, 5 (2013) 11801–11807.
92. X. Chen, M. Qiu, H. Ding, K. Fu, Y. Fan, A reduced graphene oxide nanofiltration membrane intercalated by well-dispersed carbon nanotubes for drinking water purification, *Nanoscale*, 8 (2016) 5696–5705.
93. J. Liang, Y. Huang, L. Zhang, Y. Wang, Y. Ma, T. Guo, Y. Chen, Molecular-level dispersion of graphene into poly(vinyl alcohol) and effective reinforcement of their nanocomposites, *Advanced Functional Materials*, 19 (2009) 2297–2302.
94. R. Han, X. Ma, Y. Xie, D. Teng, S. Zhang, Preparation of a new 2D MXene/PES composite membrane with excellent hydrophilicity and high flux, *RSC Advances*, 7 (2017) 56204–56210.
95. R. Han, Y. Xie, X. Ma, Crosslinked P84 copolyimide/MXene mixed matrix membrane with excellent solvent resistance and permselectivity, *Chinese Journal of Chemical Engineering*, 27 (2019) 877–883.
96. Y. Lin, J.W. Connell, Advances in 2D boron nitride nanostructures: Nanosheets, nanoribbons, nanomeshes, and hybrids with graphene, *Nanoscale*, 4 (2012) 6908–6939.
97. W. Lei, D. Portehault, D. Liu, S. Qin, Y. Chen, Porous boron nitride nanosheets for effective water cleaning, *Nature Communications*, 4 (2013) 1777.
98. J. Hassan, M. Ikram, A. Ul-Hamid, M. Imran, M. Aqeel, S. Ali, Application of chemically exfoliated boron nitride nanosheets doped with Co to remove organic pollutants rapidly from textile water, *Nanoscale Research Letters*, 15 (2020) 1–13.
99. M. Ikram, I. Hussain, J. Hassan, A. Haider, M. Imran, M. Aqeel, A. Ul-Hamid, S. Ali, Evaluation of antibacterial and catalytic potential of copper-doped chemically exfoliated boron nitride nanosheets, *Ceramics International*, 46 (2020) 21073–21083.
100. W. Yuan, J. Chen, G. Shi, Nanoporous graphene materials, *Materials Today*, 17 (2014) 77–85.
101. T.S. Sreeprasad, T. Pradeep, Graphene for environmental and biological applications, *International Journal of Modern Physics B*, 26 (2012) 1242001.
102. T. Kuila, S. Bose, P. Khanra, A.K. Mishra, N.H. Kim, J.H. Lee, Recent advances in graphene-based biosensors, *Biosensors and Bioelectronics*, 26 (2011) 4637–4648.
103. J. Liu, J. Tang, J.J. Gooding, Strategies for chemical modification of graphene and applications of chemically modified graphene, *Journal of Materials Chemistry*, 22 (2012) 12435–12452.

104. H. Yue, W. Wei, Z. Yue, B. Wang, N. Luo, Y. Gao, D. Ma, G. Ma, Z. Su, The role of the lateral dimension of graphene oxide in the regulation of cellular responses, *Biomaterials*, 33 (2012) 4013–4021.
105. A. Schinwald, F.A. Murphy, A. Jones, W. MacNee, K. Donaldson, Graphene-based nanoplatelets: A new risk to the respiratory system as a consequence of their unusual aerodynamic properties, *ACS Nano*, 6 (2012) 736–746.
106. J.T. Robinson, S.M. Tabakman, Y. Liang, H. Wang, H. Sanchez Casalongue, D. Vinh, H. Dai, Ultrasmall reduced graphene oxide with high near-infrared absorbance for photothermal therapy, *Journal of the American Chemical Society*, 133 (2011) 6825–6831.
107. Y. Chang, S.-T. Yang, J.-H. Liu, E. Dong, Y. Wang, A. Cao, Y. Liu, H. Wang, In vitro toxicity evaluation of graphene oxide on A549 cells, *Toxicology Letters*, 200 (2011) 201–210.
108. O. Akhavan, E. Ghaderi, A. Akhavan, Size-dependent genotoxicity of graphene nanoplatelets in human stem cells, *Biomaterials*, 33 (2012) 8017–8025.
109. M.-C. Matesanz, M. Vila, M.-J. Feito, J. Linares, G. Gonçalves, M. Vallet-Regi, P.-A.A.P. Marques, M.-T. Portolés, The effects of graphene oxide nanosheets localized on F-actin filaments on cell-cycle alterations, *Biomaterials*, 34 (2013) 1562–1569.
110. A.B. Seabra, A.J. Paula, R. de Lima, O.L. Alves, N. Durán, Nanotoxicity of graphene and graphene oxide, *Chemical Research in Toxicology*, 27 (2014) 159–168.
111. A.C. Ferrari, F. Bonaccorso, V. Fal'ko, K.S. Novoselov, S. Roche, P. Bøggild, S. Borini, F.H.L. Koppens, V. Palermo, N. Pugno, J.A. Garrido, R. Sordan, A. Bianco, L. Ballerini, M. Prato, E. Lidorikis, J. Kivioja, C. Marinelli, T. Ryhänen, A. Morpurgo, J.N. Coleman, V. Nicolosi, L. Colombo, A. Fert, M. Garcia-Hernandez, A. Bachtold, G.F. Schneider, F. Guinea, C. Dekker, M. Barbone, Z. Sun, C. Galiotis, A.N. Grigorenko, G. Konstantatos, A. Kis, M. Katsnelson, L. Vandersypen, A. Loiseau, V. Morandi, D. Neumaier, E. Treossi, V. Pellegrini, M. Polini, A. Tredicucci, G.M. Williams, B. Hee Hong, J.-H. Ahn, J. Min Kim, H. Zirath, B.J. van Wees, H. van der Zant, L. Occhipinti, A. Di Matteo, I.A. Kinloch, T. Seyller, E. Quesnel, X. Feng, K. Teo, N. Rupesinghe, P. Hakonen, S.R.T. Neil, Q. Tannock, T. Löfwander, J. Kinaret, Science and technology roadmap for graphene, related two-dimensional crystals, and hybrid systems, *Nanoscale*, 7 (2015) 4598–4810.
112. C. Bussy, D. Jasim, N. Lozano, D. Terry, K. Kostarelos, The current graphene safety landscape – A literature mining exercise, *Nanoscale*, 7 (2015) 6432–6435.
113. A. Bianco, M. Prato, Safety concerns on graphene and 2D materials: A Flagship perspective, *2D Materials*, 2 (2015) 030201.

5 Investigating Thin-Film Composite Membranes Prepared by Interaction between Trimesoyl Chloride with M-Phenylenediamine and Piperazine on Nylon 66 and Performance in Isopropanol Dehydration

Wan Zulaisa Amira Wan Jusoh and Sunarti Abdul Rahman
Universiti Malaysia Pahang

Abdul Latif Ahmad
Universiti Sains Malaysia

Nadzirah Mohd Mokhtar
Universiti Malaysia Pahang

Hasrinah Hasbullah
Universiti Teknologi Malaysia

DOI: 10.1201/9781003167327-6

CONTENTS

5.1 INTRODUCTION

Isopropanol (IPA) is produced by indirect hydrogenation or fermentation of cellulosic materials [1]. IPA is widely used in modern semiconductor and microelectronic industries as the solvent and cleaning solution [2]. It is also utilized as a chemical intermediate to produce mono-isopropyl amine or isopropyl acetate, weed killer, rubbing alcohol, vitamin B12 and many others. In the aforementioned applications, there are by-products in many other processes where IPA is involved, which generate the IPA and water mixtures. Consequently, IPA waste recycling is extremely significant from both environmental and economic perspectives. Thus, to conserve the environment with low overall cost treatment, many studies have been carried out and proposed throughout the years. Several experiments have taken off and intended over the years in an attempt to protect the ecosystem [3]. Although distillation is more favorable for purifying IPA waste, attention should be focused on the fact that industrial wastes can produce azeotropes, suggesting that pervaporation (PV) is more appropriate [4].

Pristine hydrophilic membranes were mainly used in the IPA dehydration process. However, it is found to be limited in mechanical strength, low thermal stability, change of the surface integrity, low yield as well as swelling. To overcome these limitations, thin-film composite (TFC) membrane can be very effective in enhancing the membrane flux and selectivity performance. The substrate layer of TFC membrane demonstrates mechanical stability and does not interfere with the mass transport. Prior to TFC formation, the support membrane has to act as a platform intended for aqueous and organic monomers for interfacial polymerization (IP) reaction to occur and form a thin layer known as polyamide (PA).

The important remark is to ensure a good interfacial binding between aqueous and organic monomers and then with the subtract layer for improving the polymer function and properties to be applied to processes. Nevertheless, according to Tsai et al. [5], the loose PA layers are associated with the low polymerization rate and the low polymerized layer thickness formation on the final TFC membrane. Decline of crosslinking rate is due to less amount of aqueous or organic monomers molecules to complete the interfacial polymerization reaction. This would lead to TFC membranes with low crosslinking density or known as the loose formation of PA layer on the subtract. Hence, choosing strong compatibility between aqueous amine solutions and subtract is a very important step.

The morphological stability and mechanical strength of the TFC membrane play vital roles in separation. Moreover, the compatibility between aqueous solution, organic solution and support membrane is the most significant decision during preparation. However, the relationship between them is rarely reported. Among the polymers subtracts used, nylon 66 (N66) has attracted much attention in the desalination field because of its relatively high mechanical strength and very high hydrophilic nature for appropriate film formation of PA. In this research, a comprehensive study in preparing TFC membranes by using the aqueous monomer, Trimesoyl Chloride (TMC), amine monomer, M-Phenylenediamine (MPD) and Piperazine (PIP) on the N66was conducted. The effect of the immersion period in amine solution on PA formation was also discussed. The PA formation on N66, the final structure and morphology, chemical composition and roughness were observed. The mechanical strength analysis was also utilized to describe the crosslinking density on the final TFC membranes before applying in PV.

5.2 EXPERIMENTAL

The parameters and steps for interfacial polymerization to prepare the TFC membranes are discussed in this section. TFC membranes were fabricated, characterized for the final morphologies and structure, functional group identification, roughness and lastly, the mechanical strength. Finally, the performance of the fabricated membrane is tested for IPA dehydration in PV system.

5.2.1 Materials

Commercial flat sheet N66 membrane (SKU: NY013001) was purchased from Sterlitech Co. (WA, USA). The N66 membranes with pores size of 0.1 μm are used as substrate for the TFC membrane in this study. All the membranes were dried in a vacuum oven at 60°C for 10 minutes and proceeded to the IP process. The reaction monomer, MPD and PIP were used as the aqueous phase, whereas TMC was used as the organic phase. Both MPD (flakes, 99.0%) and TMC (98.0%) were purchased from Sigma-Aldrich. Hexane supplied by Merck was used to prepare the organic solutions. Analytical grade IPA (≥99.8%) from Merck was employed to conduct PV experiments. All chemicals were used as received without further purification. Distilled water was used to prepare the aqueous amine solutions as well as for the preparation of PV.

5.2.2 Interfacial Polymerization Reaction on Nylon 66 Substrates

TFC membranes were prepared by the IP method provided by Hua et al. [6]. The TFC membranes were produced after exposing substrates membranes in the aqueous phase and subsequently to the TMC in the organic phase. The membranes were contacted with the 2 wt.% aqueous amine solutions for 3 and 5 minutes at 25°C as presented in Table 5.1, where TFC-MPD and TFC-PIP represent the TFC membrane prepared by MPD and PIP, respectively. After removing amine solutions excess using an air blower, the membranes were immersed into a hexane solution containing 0.1 wt.% of TMC to carry out the IP for 0.5–4 minutes. The membranes were then washed using hexane followed by distilled water before drying in a circulation oven at 70°C for 10 minutes to secure the structure of the TFC formed. Next, the fabricated TFC membrane went through post-treatment in the methanol bath for 2 minutes. The membranes were kept in DI water until it is ready to be applied in a precursory study to secure the structure and avoid defects on the membrane surfaces.

5.2.3 Characterizations

The morphologies and elements of the membrane samples were observed by FESEM. The chemical structures of the membrane surface before and after IP were analyzed by using FTIR to confirm the formation of the PA selective layer. To study the surface hydrophilicity of the resultant TFC membranes, the water contact angle was measured by the Test System of JY-82 video contact at room temperature. Meanwhile, the roughness of the TFC membranes was examined by AFM which can also confirm the hydrophilicity of the fabricated membranes. For the absorption properties, a swelling test on the TFC membrane with various IPA-water compositions was carried out. Lastly, the tensile strength was measured by CT3 Texture Analyzer for the mechanical characteristic of the membrane's samples.

TABLE 5.1
Preparation of the TFC Membrane by Varied Immersion Time in Aqueous Solution

Type	Aqueous Solution (wt.%)	Immersion Time (min) in Aqueous Solution	Organic TMC Solution (wt.%)	Reaction Time (min)	IPA Composition in Feed Solution (%)
Pristine, N66	0	0	0	0	90
TFC-MPD3	2.0	3.0	0.1	2.0	90
TFC-MPD5	2.0	5.0	0.1	2.0	90
TFC-PIP3	2.0	3.0	0.1	2.0	90
TFC-PIP5	2.0	5.0	0.1	2.0	90

5.2.4 Pervaporation Separation Tests

The PV system used in the study is shown in Figure 5.1. Five hundred milliliters of feed solution with desired IPA concentration in the mixture was circulated through the membrane, with the flow rate of 18 L h^{-1}. The concentration of the IPA feed solution was prepared using 90 wt.% of IPA solution mix with 10 wt.% of distilled water. The lumen side of the membrane was connected to a vacuum pump, which was the permeate side. The feed temperature was set at 40°C. The permeate side pressure was maintained below 50 mbar (5 kPa). The system was conditioned for 2 hours to ensure that the flux and composition of permeate were stabilized before the collection of the permeate samples as applied by Zuo et al. [2] in his research study. The sample from permeate side was condensed in a cold trap and collected at a time interval of 1 hour. The mass of the collected sample was weighed by a Mettler Toledo balance.

The compositions of the feed and permeate samples were determined by a refractive index using Abbe's refractometer (Atago-3T, Japan) with an accuracy of ±0.001 units by referring to the standard graph of refractive index versus percent composition of the water–isopropanol mixture prepared. At least three permeate samples were collected and their average was reported as the PV performance. Two main parameters should be considered when selecting a membrane for a specific mixture: the permeate flux across the membrane presented as J (kg $m^{-2}h^{-1}$) in Equation 5.1, and the membrane selectivity, α, to measure the quality of separation as Equation 5.2 [7]:

$$J = \frac{Q}{A \cdot \Delta t} \tag{5.1}$$

$$\propto = \frac{P_{water}/P_{IPA}}{F_{water}/F_{IPA}} \tag{5.2}$$

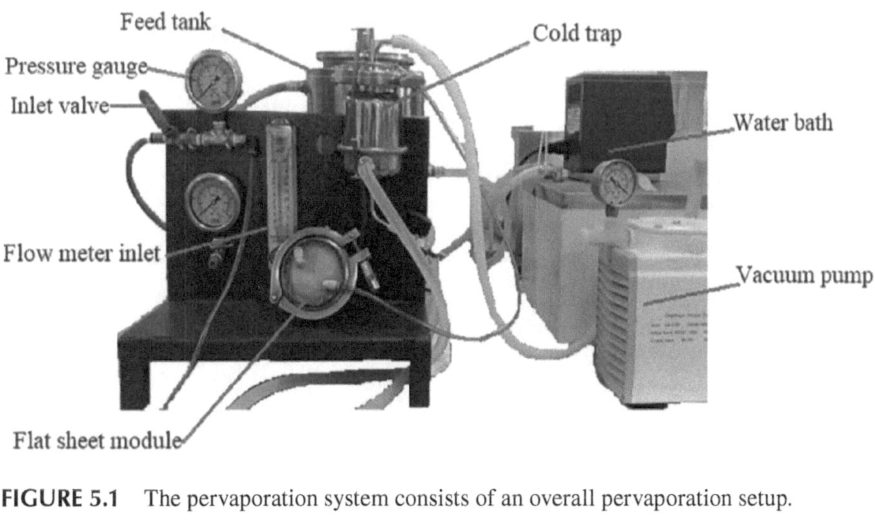

FIGURE 5.1 The pervaporation system consists of an overall pervaporation setup.

5.3 RESULT AND DISCUSSION

5.3.1 Characterization Results of the Nylon 66 and Fabricated TFC Membranes

Both chemical properties and PA separation layer structure of TFC membranes are very significant to the stability and performance. The chemical behavior and reactant concentrations normally impact the membrane's overall performance, while the IP timings concerning preparative stipulations have an effect on the stability of the barrier layer and in a way on the membrane performance. For a pair of reactants taken to shape the PA layer, there are most useful concentrations of reactants to acquire a robust active separation layer with top performance. Thus, pristine N66, TFC-MPD3 and TFC-PIP3 were selected for the analysis of FESEM, FTIR, surface roughness and tensile strength test accordingly.

5.3.1.1 Morphology Structure

Figure 5.2 demonstrates the surface of the FESEM image of the pristine and fabricated TFC membranes. The morphological stability of the microporous support for the TFC membrane plays important role in the fabrication of the TFC membrane. The distribution pores of the pristine membrane were clearly distinguished as a uniform structure.

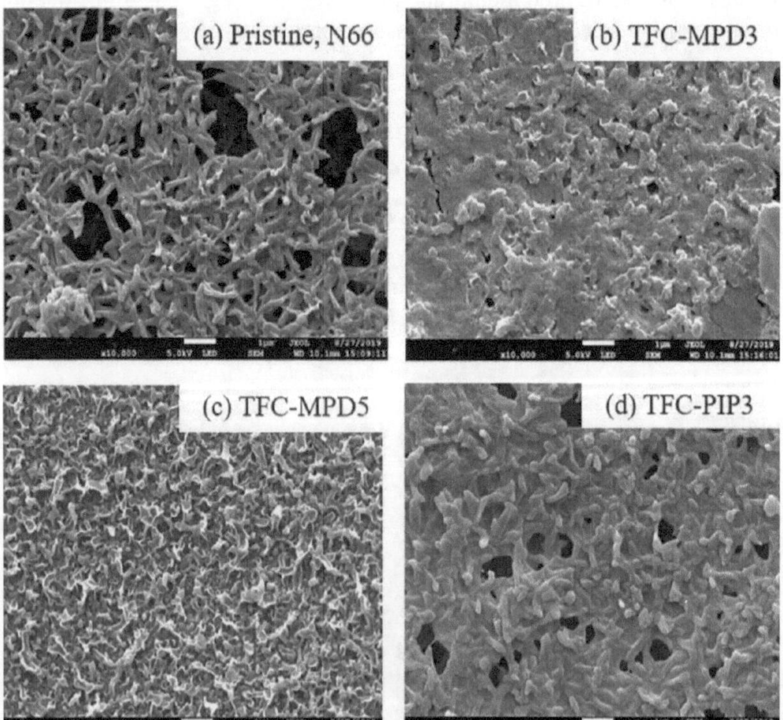

FIGURE 5.2 Top surface FESEM images of (a) Pristine N66, (b) TFC-MPD3, (c) TFC-MPD5 and (d) TFC-PIP3 prepared from 2.0 wt.% MPD and 0.1 wt.% TMC solutions. The scale bar in the lower represents 1 μm (10k magnification).

Pristine N66 shows uniform and average small pores of 1.0 µm. This shows that the distribution of the pores was appropriate for N66 to be utilized as the substrate membrane. These pores' properties provide a high surface area for IP processing as they can accommodate the absorbed MPD in the pores, thus producing smooth TFC membrane surfaces [8–10]. The TFC must not only be very thin but also have a good cross-linked thin layer of PA [11].

Fully aromatic PA formed from MPD and TMC mainly exhibits ridge-and-valley structures [11–13]. The active layers of PA formed on fabricated TFC membranes and demonstrated by a ridge-and-valley structure can be observed in FESEM image, especially for the TFC-MPD5. As shown in Figure 5.2b, the membranes of the TFC-MPD3 present a flatter surface and a lack of ridge-and-valley structures possibly due to lower crosslinking reaction between monomers. The structure increases linearly with increasing contact time in MPD solution. The ridge-and-valley structure can be seen in the second layer on a core layer on the substrate [14]. At first, the IP reaction between organic and aqueous monomers will form an ultrathin film on the N66 pores, then the IP reaction slowly decreased as the MPD started to diffuse out of the pores. Because of strong intermolecular force between MPD and the polar functional group in the N66 substrate pore layers, the MPD is forced out from the pores and forms the second layer of ridge-and-valley structure on the first layer [15]. Higher MPD monomer adsorption can create a denser PA layer [16]. Compared to the contact time of 3 minutes in MPD solution (TFC-MPD3), 5 minutes (TFC-MPD5), however, was a sufficient contact time for crosslinking reaction (both MPD solution and TMC solution) to form the preferred separation layer structure. As for TFC-MPD3, the ridge-and-valley structure could occur as the membrane was not properly wetted in the amine solutions due to the lower contact time in MPD [17].

The TFC layers were deposited on membrane support when the FESEM image shows that all pores on the surfaces were covered. For TFC-PIP3 in Figure 5.2d, however, the pores can still be observed. It should be noted that the visible holes in the TFC-PIP3 FESEM picture refer to the N66 membrane, not the PA skin surface (compared with the FESEM image of the base N66 support in Figure 5.2a). Such holes are noticeable on the support surface due to the establishment of an ultrathin layer of the PA as the first developed layer [18]. This suggests that PIP and TMC concentrations or reaction time needs to be increased for better coverage of the TFC-PIP3. Overall, FESEM images show that at the same concentration and time, the TMC interactions with MPD on the N66 substrate are better than the interactions between TMC and PIP. Because of their chemical structure properties, the interaction between molecules within the TMC-MPD is stronger than TMC-PIP with the same concentration. The final structure of the fabricated membrane is very dependent on the balance rate between TMC and MPD during IP process, the relative factors to travel of the thick lattice and the high-permeable districts within the bi-modal PA active layer [19]. Tsuru et al. [13] have mentioned in their report that the bigger the ridge-and-valley structure, the higher the water permeation area.

5.3.1.2 Chemical Composition Analysis

Figure 5.3 shows a comparison between the FTIR spectra for pristine and TFC membranes (TFC-MPD3, TFC-MPD5 and TFC-PIP3). The absorption wavelengths indicate the existence of a nitro compound that consists of nitro functional groups inside

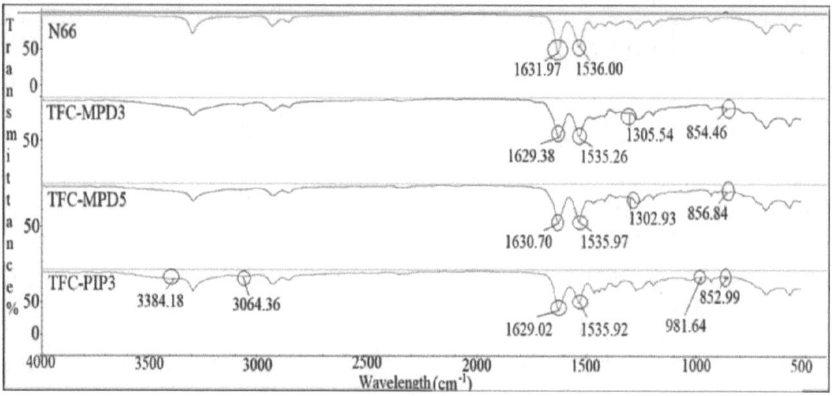

FIGURE 5.3 FTIR spectra of N66 and TFC membranes.

the fabricated TFC membranes. The absorption band of N–O and C=O in stretching at 1,535–1,536 and 1,629–1,631 cm^{-1}, respectively represents the nitro compound and amides that can be spotted in all membrane spectra and explains that pristine and TFC are in the same functional group [20]. Peaks of 852–859 cm^{-1} on these fabricated TFC membranes indicate a strong C–Cl bond from acyl monomer, TMC which were undetected in pristine, N66. The presence of aromatic breathing ring, C_6H_6, was detected at the peak points of 1,302–1,305 cm^{-1} where strong stretching of C–N shows amine functional group of MPD deposited on the membrane surfaces of TFC-MPD3 and TFC-MPD5, and therefore, confirming that the fully aromatic PA was successfully deposited on the pristine membranes [21,22]. The peaks of 3,384.18 cm^{-1} distinguished the aliphatic owned by TFC-PIP3 where N–H stretching shows aliphatic primary amine. The spectra of 3,064 and 981 cm^{-1} representing the C–H stretching in alkene and strong bending in alkene, respectively, give more clue on the formation of aliphatic amine by TMC and PIP. Overall, the chemical composition shown by the FTIR result indicates that the organic and aqueous monomers are successfully deposited on top of TFC membranes and are expected to enhance the separation performance. These results are also consistent with the FESEM results that provide vivid evidence of the good interfacial binding between TMC-aqueous monomer-N66.

5.3.1.3 Surfaces Roughness

Generally, low roughness on the membrane surface is expected to have a relatively better crosslinking area and thus enhances the flux results [23]. The 3 μm × 3 μm three-dimensional AFM images of the fabricated TFC membrane results measured are shown in Figure 5.4. The lowest value (38°–40°) was measured for the smoothest TFC-MPD3 membrane, prepared at room temperature, while the highest value (59°) was measured for the N66 membrane, which had the highest roughness value. The reductions in roughness have confirmed that PA ultrathin layer is formed on TFC-MPD3 and TFC-PIP3 membranes. All the fabricated TFC membranes had comparable surface roughness and were consistent with the surface morphology from FESEM measurements. Since TFC-MPD5 membranes formed ridge-and-valley structures in FESEM analysis, the roughness value is slightly higher than the other membranes as expected [24]. This suggests that TFC-MPD5 membrane production

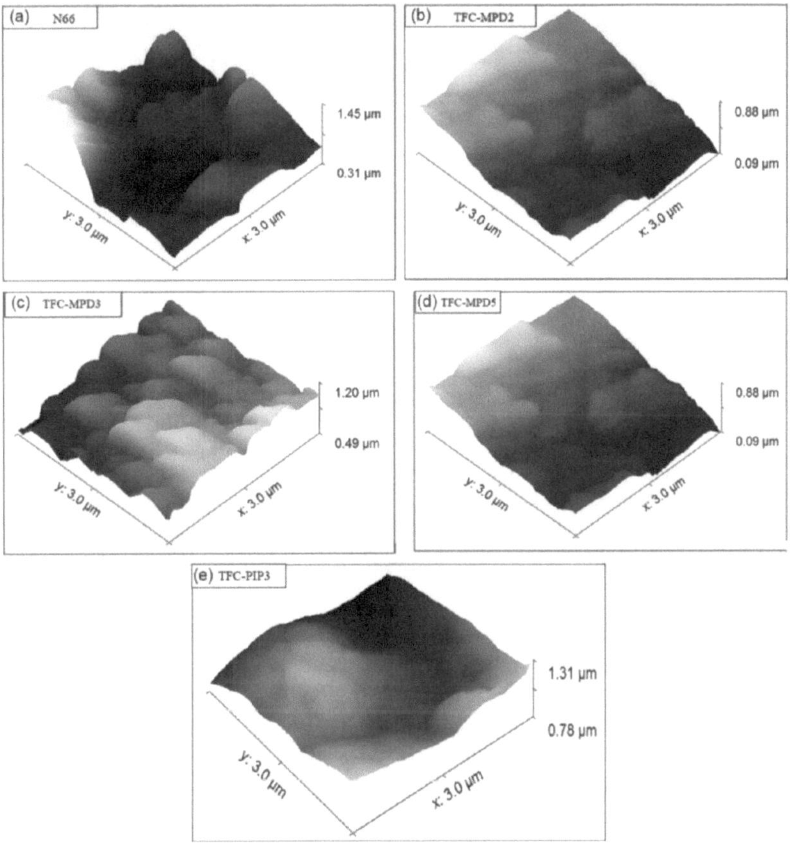

FIGURE 5.4 Surface AFM morphologies of (a) N66, (b) TFC-MPD2, (c) TFC-MPD3, (d) TFC-MPD5 and (e) TFC-PIP3 membranes.

at higher organic solution monomers composition would increase the roughness of the fabricated membrane.

5.3.1.4 Mechanical Strength Test

On the morphological stability side, the mechanical properties of the microporous support for the TFC membrane also play an important role. TFC layers with strong mechanical strength support were more ideal to avoid the pore expansion, the collapse of the porous structure and thickness reduction during separation [25]. There should be an optimal compromise between tensile strength and elongation at break for a strong pervaporation polymer membrane. A higher tensile strength value means the membrane possesses higher crystallinity properties, meanwhile, elongation associates with the glass transition temperature of plastics (higher the elongation value, the lower the glass transition temperature of plastics value) [26]. N66 material is recognized for its profoundly crystalline structure due to the solid building hydrogen bonding between polymer chains [27]. These characteristics allow a sturdy and intense molecular structure of polymeric membrane. The high quality of N66 may make it alluring as a TFC film substrate.

TABLE 5.2
The Tensile Strength of Pristine and Fabricated TFC Membranes

Membrane Type	Tensile Strength/Max Stress ($N\ m^{-2}$ @ Pa)	Young's Modulus, E (MPa)
Pristine	25,075	284.948
TFC-MPD3	29,275	448.487
TFC-MPD5	33,150	593.093
TFC-PIP3	25,875	344.09

Pristine N66, TFC-MPD3, TFC-MPD5 and TFC-PIP3 membranes exhibited tensile strengths of 25,075, 27,400, 29,275, 33,150 and 25,875 Pa, respectively. From Table 5.2, the highest tensile strength increases approximately from 3% to 32% that can be observed in TFC membranes compared to pristine N66 membranes. TFC-MPD5 membrane exhibited the highest mechanical strength followed by the TFC-MPD3 and then pristine, N66. Meanwhile, TFC-PIP shows a lower value compared with other fabricated TFC membranes. This effect can be explained by the fact that the thicker formation of the PA layer on the membrane surfaces leads to higher mechanical strength that resulted in a net increase in structural stiffness to support the result from FESEM and roughness. From this standpoint, it can be considered that the lower mechanical strength of TFC-PIP3 can be attributed to the ultra-active layer formation as shown in FESEM, which seems to downplay the intermolecular interactions between $OCNH_2$ group in the substrate and amine functional group in active layer. These variations in tensile strength of fabricated membranes are consistent with others analysis that increasing the IP crosslinking rate depends on the type of organic and aqueous monomers. Moreover, the crosslinking rate would also increase the mechanical stiffness of the active layer on the TFC membrane.

5.3.2 Effect of Immersion Time in MPD Solution on the TFC Membrane

Among the factors which affected PV performance are the degree of crosslinking, thickness and mechanical strength of the membrane. The overall results are shown in Figure 5.5. TFC membranes are prepared by 0.1 wt.% TMC and 2.0 wt.% MPD solutions in various immersion times. MPD and TMC monomer reacts with each other through the liquid–liquid interface to generate a thin-film polymer layer. Due to the low solubility of TMC molecules in water, the IP process mainly takes place in the organic phase (MPD in the organic phase). This leads to excessive aqueous over organic monomer and influences the MPD to diffuse into the organic phase. Carboxylic acid bulk is enhanced by lowering down the composition of aqueous to organic monomer (w/w) that explained the hydrophilicity enrichment of fabricated TFC membranes. Therefore, the balance between thinner and great cross-linked film structure is important to form TFC membranes with good permeation flux and high selectivity.

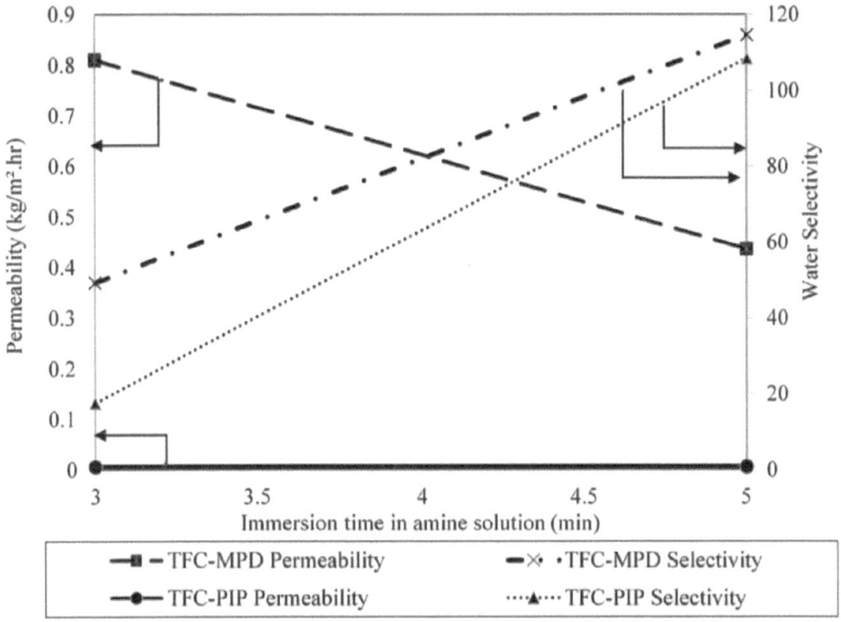

FIGURE 5.5 The permeate flux and water selectivity on the N66 and TFC membranes prepared by various immersion times in MPD and PIP for 90 wt.% IPA concentration.

Immersion time in MPD for 5 minutes shows the highest selectivity of 114.6 but presented the lowest flux of 0.44 kg $m^{-2}h^{-1}$. It is normal for the selectivity and flux to have trade-off relations [15]. Due to the trade-off between permeability and selectivity, the water to IPA selectivity result of TFC-MPD5 is highest compared with other synthesis membranes. On the contrary, it has the lowest water permeability result. This again confirms morphology images are in the same agreement with the PV test result as reported by many researchers [29]. Zuo et al. [2] in their study have mentioned that the separation factor increases with the contacting time. Immersion times are required to be at an adequate ratio of amine–acyl monomer to complete the reaction to form a very thin TFC layer for the best result of selectivity. A long immersion time may escalate the MPD to diffuse into an organic phase which affects the reaction rate, thickness and characteristics of the TFC membrane produced [28]. Longer exposition time led to poorer permeability can be attributed to some unreacted MPD [29]. The presence of the free amine monomers on the membrane surface increases the hydrophilicity properties. The carboxylic acid groups could establish a relatively steady and strong hydrogen bond with water. Excessive carboxylic group on the surface eventually decelerates the water molecules to permeate into the PA matrix. The carboxylic group too has a tendency to decrease the surface roughness and increase the mobility for PA crosslinking [30]. Previous studies suggested that the best crosslinking reaction was obtained with the TFC membranes prepared in similar conditions [31].

TFC membranes produced by the reaction of TMC and PIP (TFC-PIP3 and TFC-PIP5) present similar result in permeability with TFC produced by MPD but lower in selectivity result. The distinction on the overall performance of the fabricated TFC membranes may be accredited to the aromatic and aliphatic nature owned by MPD and PIP, respectively, as shown in FTIR result. The monomers MPD and PIP may propose variance reactivity toward acid chloride which results in the formation of a PA skin layer. Because MPD is a primary aromatic amine, the crosslinking speed between diamine and acid chloride in TMC is higher resulting in thicker and stronger membranes formation which is confirmed in the roughness and tensile strength test. Fast formation of PA structure reduces other side reactions, such as acid chloride hydrolysis [32]. According to Mohan and Kullová [33], and as revealed by TFC-MPD5 in this study, prompter crosslink by aromatic diamide creates a PA skin surface network hence exhibiting higher separations result. The use of aliphatic amines, however, results in a skin layer with a more transparent and open structure, which results in lower water and IPA selectivity and is consistent with the TMC-PIP3 result [34].

5.4 CONCLUSION

FESEM result shows that fabricated TFC membranes employed the dense layer as the active layer. The best water selectivity result from this study was by TFC-MPD5 which has a ridge-and-valley structure, which indicates the best and strong formation (33,150 N m^{-2}) of a complete aromatic PA (in the range of 1,302–1,305 cm^{-1}) on the N66 substrate which is promising for IPA-water separation. Increasing immersion time in amine solution will also increase the crosslinking rate to PA formation which contributed to the increased thickness of final TFC membranes. In general, the membrane produced by TMC-MPD has better properties and performance than TMC-PIP. TFC membranes prepared by MPD show selectivity of 17.6 and 114.621 for preparing in 3 and 5 minutes in amine solutions, respectively. Meanwhile, TFC prepared by PIP amine solutions shows relatively lower the selectivity of 17.6 and 108.5 for TFC-PIP3 and TFC-PIP5, respectively.

ACKNOWLEDGMENTS

The authors would like to thank the sponsor for the financial support for this research through grants PGRS180323 and RDU1803113.

REFERENCES

1. N.R. Rodriguez, M.C. Kroon, Isopropanol dehydration via extractive distillation using low transition temperature mixtures as entrainers, *J. Chem. Thermodyn.* 85 (2015) 216–221. doi:10.1016/j.jct.2015.02.003.
2. J. Zuo, D. Hua, V. Maricar, Y.K. Ong, T.-S.S. Chung, J. Zuo, T.-S.S. Chung, Dehydration of industrial isopropanol (IPA) waste by pervaporation and vapor permeation membranes, *J. Appl. Polym. Sci.* 135 (2017) 1–7. doi:10.1002/app.45086.

3. I. Dahl, C. Kjølseth, H. Fjeld, Ø. Prytz, P. Inge, C. Estournès, R. Haugsrud, T. Norby, Open archive toulouse archive ouverte (OATAO), *Solid State Ionics.* 181 (2010) 268–275. doi:10.1016/j.ssi.2010.01.014.
4. A. Andre, T. Nagy, A.J. Toth, E. Haaz, D. Fozer, J.A. Tarjani, P. Mizsey, Distillation contra pervaporation: Comprehensive investigation of isobutanol-water separation, *J. Clean. Prod.* 187 (2018) 804–818. doi:10.1016/j.jclepro.2018.02.157.
5. H.A. Tsai, T.Y. Wang, S.H. Huang, C.C. Hu, W.S. Hung, K.R. Lee, J.Y. Lai, The preparation of polyamide/polyacrylonitrile thin film composite hollow fiber membranes for dehydration of ethanol mixtures, *Sep. Purif. Technol.* 187 (2017) 221–232. doi:10.1016/j.seppur.2017.06.060.
6. D. Hua, T.-S. Chung, G.M. Shi, C. Fang, Teflon AF2400/Ultem composite hollow fiber membranes for alcohol dehydration by high-temperature vapor permeation, *AIChE J.* 62 (2016) 1747–1757. doi:10.1002/aic.15158.
7. Y.T. Ong, K.F. Yee, Y.K. Cheng, S.H. Tan, A review on the use and stability of supported liquid membranes in the pervaporation process, *Sep. Purif. Rev.* 43 (2012) 62–88. doi:10.1080/15422119.2012.716134.
8. L. Huang, N.N. Bui, M.T. Meyering, T.J. Hamlin, J.R. McCutcheon, Novel hydrophilic nylon 6,6 microfiltration membrane supported thin film composite membranes for engineered osmosis, *J. Memb. Sci.* 437 (2013) 141–149. doi:10.1016/j.memsci.2013.01.046.
9. M.F. Jimenez-Solomon, P. Gorgojo, M. Munoz-Ibanez, A.G. Livingston, Beneath the surface: Influence of supports on thin film composite membranes by interfacial polymerization for organic solvent nanofiltration, *J. Memb. Sci.* 448 (2013) 102–113. doi:10.1016/j.memsci.2013.06.030.
10. X. Zhang, J. Tian, Z. Ren, W. Shi, Z. Zhang, Y. Xu, S. Gao, F. Cui, High performance thin-film composite (TFC) forward osmosis (FO) membrane fabricated on novel hydrophilic disulfonated poly(arylene ether sulfone) multiblock copolymer/polysulfone substrate, *J. Memb. Sci.* 520 (2016) 529–539. doi:10.1016/j.memsci.2016.08.005.
11. D.H.N.N. Perera, Q. Song, H. Qiblawey, E. Sivaniah, Regulating the aqueous phase monomer balance for flux improvement in polyamide thin film composite membranes, *J. Memb. Sci.* 487 (2015) 74–82. doi:10.1016/j.memsci.2015.03.038.
12. F. Pacheco, R. Sougrat, M. Reinhard, J.O. Leckie, I. Pinnau, 3D visualization of the internal nanostructure of polyamide thin films in RO membranes, *J. Memb. Sci.* 501 (2016) 33–44. doi:10.1016/j.memsci.2015.10.061.
13. T. Tsuru, S. Sasaki, T. Kamada, T. Shintani, T. Ohara, H. Nagasawa, K. Nishida, M. Kanezashi, T. Yoshioka, Multilayered polyamide membranes by spray-assisted 2-step interfacial polymerization for increased performance of trimesoyl chloride (TMC)/m-phenylenediamine (MPD)-derived polyamide membranes, *J. Memb. Sci.* 446 (2013) 504–512. doi:10.1016/j.memsci.2013.07.031.
14. B. Khorshidi, T. Thundat, B.A. Fleck, M. Sadrzadeh, Thin film composite polyamide membranes: Parametric study on the influence of synthesis conditions, *RSC Adv.* 5 (2015) 54985–54997. doi:10.1039/C5RA08317F.
15. L. Huang, J.R. McCutcheon, Impact of support layer pore size on performance of thin film composite membranes for forward osmosis, *J. Memb. Sci.* 483 (2015) 25–33. doi:10.1016/j.memsci.2015.01.025.
16. Z.X. Low, Q. Liu, E. Shamsaei, X. Zhang, H. Wang, Preparation and characterization of thin-film composite membrane with nanowire-modified support for forward osmosis process, *Membranes (Basel).* 5 (2015) 136–149. doi:10.3390/membranes5010136.
17. J. Zuo, Y. Wang, S.P. Sun, T.S. Chung, Molecular design of thin film composite (TFC) hollow fiber membranes for isopropanol dehydration via pervaporation, *J. Memb. Sci.* 405–406 (2012) 123–133. doi:10.1016/j.memsci.2012.02.058.

18. B. Khorshidi, T. Thundat, B.A. Fleck, M. Sadrzadeh, A novel approach toward fabrication of high performance thin film composite polyamide membranes, *Sci. Rep.* 6 (2016) 1–10. doi:10.1038/srep22069.
19. B. Khorshidi, T. Thundat, B.A. Fleck, M. Sadrzadeh, A novel approach toward fabrication of high performance thin film composite polyamide membranes, *Sci. Rep.* 6 (2016) 1–10. doi:10.1038/srep22069.
20. F. Esfandian, M. Peyravi, A.A. Ghoreyshi, M. Jahanshahi, A.S. Rad, Fabrication of TFC nanofiltration membranes via co-solvent assisted interfacial polymerization for lactose recovery, *Arab. J. Chem.*, King Saud University, 2016. doi:10.1016/j.arabjc.2017.01.004.
21. S.J. Park, W. Choi, S.E. Nam, S. Hong, J.S. Lee, J.H. Lee, Fabrication of polyamide thin film composite reverse osmosis membranes via support-free interfacial polymerization, *J. Memb. Sci.* 526 (2017) 52–59. doi:10.1016/j.memsci.2016.12.027.
22. J. Yin, Y. Yang, Z. Hu, B. Deng, Attachment of silver nanoparticles (AgNPs) onto thin-fi lm composite (TFC) membranes through covalent bonding to reduce membrane biofouling, *J. Memb. Sci.* 441 (2013) 73–82. doi:10.1016/j.memsci.2013.03.060.
23. H.F. Ridgway, J. Orbell, S. Gray, Molecular simulations of polyamide membrane materials used in desalination and water reuse applications: Recent developments and future prospects, *J. Memb. Sci.* 524 (2017) 436–448. doi:10.1016/j.memsci.2016.11.061.
24. B. Khorshidi, B. Soltannia, T. Thundat, M. Sadrzadeh, Synthesis of thin film composite polyamide membranes: Effect of monohydric and polyhydric alcohol additives in aqueous solution, *J. Memb. Sci.* 523 (2017) 336–345. doi:10.1016/j.memsci.2016.09.062.
25. J.A. Idarraga-Mora, A.S. Childress, P.S. Friedel, D.A. Ladner, A.M. Rao, S.M. Husson, Role of nanocomposite support stiffness on TFC membrane water permeance, *Membranes (Basel).* 8 (2018) 3–5. doi:10.3390/membranes8040111.
26. N.R. Singha, S.B. Kuila, P. Das, S.K. Ray, Separation of toluene-methanol mixtures by pervaporation using crosslink IPN membranes, *Chem. Eng. Process. Process Intensif.* 48 (2009) 1560–1565. doi:10.1016/j.cep.2009.09.002.
27. V. Vatanpour, M. Sheydaei, M. Esmaeili, Box-Behnken design as a systematic approach to inspect correlation between synthesis conditions and desalination performance of TFC RO membranes, *Desalination.* 420 (2017) 1–11. doi:10.1016/j.desal.2017.06.022.
28. M. Shen, S. Keten, R.M. Lueptow, Rejection mechanisms for contaminants in polyamide reverse osmosis membranes, *J. Memb. Sci.* 509 (2016) 36–47. doi:10.1016/j.memsci.2016.02.043.
29. Y. Wang, N. Widjojo, P. Sukitpaneenit, T.S. Chung, Membrane pervaporation (2013). doi:10.1002/9781118493441.ch10.
30. W. Choi, S. Jeon, S.J. Kwon, H. Park, Y.I. Park, S.E. Nam, P.S. Lee, J.S. Lee, J. Choi, S. Hong, E.P. Chan, J.H. Lee, Thin film composite reverse osmosis membranes prepared via layered interfacial polymerization, *J. Memb. Sci.* 527 (2017) 121–128. doi:10.1016/j.memsci.2016.12.066.
31. F. Galiano, F. Falbo, A. Figoli, Polymeric pervaporation membranes: Organic-organic separation, *Nanostruct. Polym. Membr.* 2 (2016) 287–310. doi:10.1002/9781118831823.ch7.
32. D.J. Mohan, L. Kullová, A study on the relationship between preparation condition and properties/performance of polyamide TFC membrane by IR, DSC, TGA, and SEM techniques, Desalin. *Water Treat.* 51 (2013) 586–596. doi:10.1080/19443994.2012.693655.
33. J. Xu, H. Yan, Y. Zhang, G. Pan, Y. Liu, The morphology of fully-aromatic polyamide separation layer and its relationship with separation performance of TFC membranes, *J. Memb. Sci.* 541 (2017) 174–188. doi:10.1016/j.memsci.2017.06.057.
34. W. Choi, S. Jeon, S.J. Kwon, H. Park, Y.I. Park, S.E. Nam, P.S. Lee, J.S. Lee, J. Choi, S. Hong, E.P. Chan, J.H. Lee, Thin film composite reverse osmosis membranes prepared via layered interfacial polymerization, *J. Memb. Sci.* 527 (2017) 121–128. doi:10.1016/j.memsci.2016.12.066.

6 Sustainable Carbonaceous Nanomaterials for Wastewater Treatment

State-of-the-Art and Future Insights

Muhammad Ikram
Government College University Lahore

Ali Raza and Salamat Ali
Department of Physics
University of Sialkot

Anwar Ul-Hamid
King Fahd University of Petroleum & Minerals

CONTENTS

DOI: 10.1201/9781003167327-7

6.1 INTRODUCTION

The twenty-first century has been designated as the age of environmental study [1]. Through increasing global population, an expansion of industrial and agricultural activities, pollution of soils, aquatic, air, and biodiversity, and, ultimately, the modification of global climate, environmental issues have risen to the top of science and political priority lists [2–4]. Presently, there exists a worldwide determination to recognize the impact of human actions on the environment with respect to the development of new technologies to diminish the environmental and health implications. Aside from diverse approaches to reporting on the above urgent environmental issues, recent signs of advancement in the field of nanotechnology have spurred an increased interest in using nanomaterials' exceptional properties for use in environmentally sustainable applications. In this regard, various nanomaterials hold distinctive properties, due to their association with nanoscale dimensions which can be engaged to expand the enactment of prevailing processes or to design novel approaches. Nanomaterials have originated numerous applications for energy production, contaminant sensing, and water treatment, thus growing literature calls on how innovative nanomaterials can be utilized to state main environmental challenges [5–8].

Currently, carbon nanoparticles are reflected as resourceful materials that may be engaged toward enhancement in polluted water treatment. Globally, broad research was carried out that stemmed from the innovation of novel carbon-based materials which can be effectively employed in environmental protection approaches, and for wastewater treatment [9–12]. The latest literature reports the core objectives such as pathogenic microorganisms (micropollutants) that can be extracted from wastewater with the aid of carbon nanoparticles headed for numerous technological methods [13–17]. Their unique physiochemical properties regulate a range of practical applications. Generally, graphene material, its derivative graphene oxide (GO), single-walled carbon nanotubes (SWCNT), and multi-walled carbon nanotubes (MWCNT) are the most frequently inspected carbon-based nanomaterials. These discussed materials can be utilized as complex hybrid materials, or commonly in their pure forms [18–21].

Water supply methods and wastewater treatment are the next generation approaches that depend upon its economic feasibility and ecological approachability [9,22]. These nanomaterials own a high surface-area-to-volume ratio that enables them to boost reactivity with pathogens or environmental pollutants such as viruses, fungi, or bacteria [23–26]. Graphene derivatives correspond to the significant branch of carbon derivative that is designated as graphene oxide. A functional group comprising oxygen is bound to the graphene layer forming stable suspensions in aqueous media because of its composition, which requires a high density attributing to oxygen functional groups namely, carboxyl, carbonyl, and epoxy residing in carbon lattice [2,27–31]. This property permits GO to be utilized in practical applications such as in water treatment. Furthermore, GO is characterized by supplementary unique features that comprise tunable pore size, abundant and selective sites for adsorption, small intra-particle diffusion distance, superior charge carrier mobility, large specific surface area, and exceptional electrical conductivity. High adsorption capacity and mechanical strength permit GO to produce hybrid nanocomposites with many materials. Significantly GO can be characterized by low production cost and simple manufacturing [3,32–35].

6.2 CARBONACEOUS NANOMATERIALS

The distinctive characteristics of hybrid-carbon sp^2 bonds will combine carbon nanomaterials with extraordinary nanoscale physicochemical properties; usually, they exhibit excellent electrical, mechanical, and chemical characteristics. These nanomaterials are classified into many types based on their composition, electrical, and chemical properties (Figure 6.1). In different applications worldwide, they have attracted attention for environmental removal, pollutants remediation, drug delivery, photocatalysis, catalysis, and supercapacitors, respectively. In the coming future, the position of carbonated nanocomposites may be increasingly promising by the booming development of nanotechnologies.

6.2.1 Applications in Wastewater Treatment

6.2.1.1 Adsorption

Adsorption has been suggested as the most useful technique for the removal of contamination from the liquid phase owing to its favorable characteristics e.g., low cost, ease of operation, and no by-products. However, traditional adsorbent efficiency is constantly restricted and hence inhibits the process of removing contaminants. Nanostructured compounds contain a broad specific range and associated adsorption

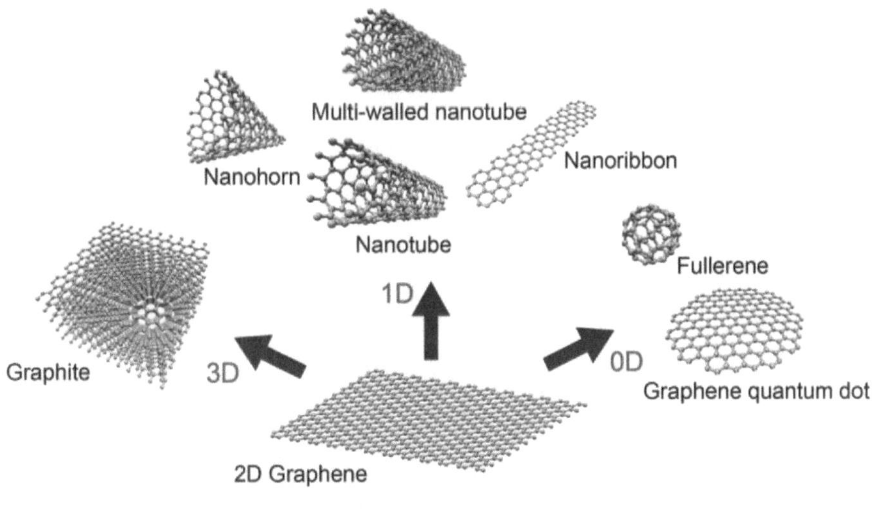

FIGURE 6.1 Various forms of carbonaceous nanomaterials. (Reproduced with permission from Ref. [36]. Copyright 2018 Taylor & Francis Group.)

sites with tuned surface qualities in comparison with standard adsorbents, which offer them some advantages, including high performances and/or quick adsorption speeds covering a wide range of pH [15,37,38].

6.2.1.2 Removal of Organic Contaminants

Numerous carbonaceous compounds, including various dyes [39], variety of antibiotics [40], polycyclic aromatic rings hydrocarbons [41], phenol group compounds [42], and harmful pesticides [43], have shown exceptional adsorption efficiency in graphene-based nanostructured materials. RGO's malathion adsorption concentrations can be 800, 1,100, and 1,200 mg g^{-1}, respectively, in liquid phases, endosulfan, and chlorpyrifos [43]. Many carbonaceous nanomaterials, containing CNTs as well as fullerenes, and nanocomposites, have been widely analyzed and tested for their ability to remove organic compounds from wastewater. Their potential adsorption capability is primarily attributed to their high specific surface area, random pores distribution, and possible surface properties, enabling them to be vigorous reactive organic compounds through hydrophobic effect, hydrogen bonding along with EDA interactions. Furthermore, studies have shown that normal environmental conditions and surface characteristics of adsorbents may influence adsorptive capability, and hence a slight difference between adsorption efficiency. The aggregation of CNTs in the liquid phase creates several interstitial spaces associated with grooves, providing efficient energy adsorption areas, thereby serving to dissolve organic compounds. GO containing various oxygenic groups (such as COOH as well as –OH) on the edges may degrade hydrophobic influence through non-polar organic compounds, thereby preventing adsorption. Large-sized compounds, such as algal toxins, are inaccessible to nanoporous activated carbon (NAC) due to their microporous composition, although it may be useful for comparably small molecules [15,44].

6.2.1.3 Remediation of Toxic Metallic Ions

A group of researchers has conducted extensive literature work on the heavy metal adsorption properties owing to GO as well as oxidizing CNTs, offering collective incorporation of GO with CNTs exhibited higher adsorption efficiency associated with rapid adsorption rate toward metal remediation (Figure 6.2) [45,46]. The ability of organic materials, to adsorb a variety of heavy metal ions, such as Cd^{2+}, Zn^{2+}, Cu^{2+} has been analyzed [47]. Active participation owing to carbonate compounds to

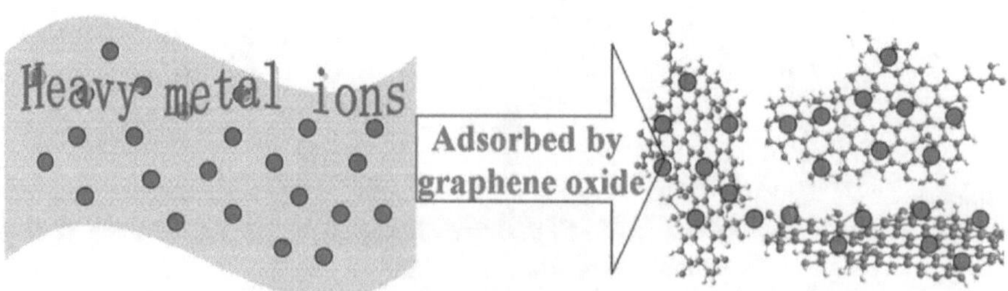

FIGURE 6.2 Removal of heavy ions by graphene oxide layers. (Reproduced with permission from Ref. [48]. Copyright 2011 American Chemical Society.)

adsorb heavy metal ions has been extensively examined and identified with metals such as Cd^{2+}, Cu^{2+}, and Zn^{2+} [48]. Chemical bonding caused by electrostatic attraction will dominate the adsorption technique, containing GO functional groups along with oxidized CNTs that provide the majority of adsorption areas existing in heavy metal ions. The highly accessible adsorption edges and the limited intra-particle diffusion gap may be attributed to fast adsorption kinetics. Also, the removal of heavy metals was provided much consideration by different nanostructured materials and nanocomposites.

6.2.1.4 Photocatalysis

The photocatalytic process is a progressive oxidation method to remove pathogens and trace out contaminants present in wastewater, which is regarded as a favorable prior treatment technique for toxic molecules as well as non-biodegradable containments to upgrade biodegradability [49–51]. At the moment, several studies focused on photocatalysis enhancement through carbonaceous materials are being vigorously investigated. Carbonaceous nanomaterials usage such as CNTs, RGO, and fullerenes chemically interacting with catalysts is a hot topic (TiO_2, etc.). By introducing TiO_2 on separated graphene nanosheets, they may occupy synergetic properties derived by fabrication methods, thus increasing photocatalysis efficiency for desired pollutants. Yang et al. [52] developed TiO_2/graphene composite assembly along with CNTs and fullerenes nanomaterials photocatalysts while converting benzyl alcohol into benzaldehyde, by claiming that the associated nanocomposites could expand adsorption sites toward visible light response. Various hybrid photocatalysts were used for organic pollutants decontamination in addition to TiO_2/carbonaceous nanomaterials such as ZnS/RGO photocatalysts [53] as well as ZnO-graphene. In reality, graphene and carbon nanotubes could have wide surface area advantages and/or particle restriction for photocatalysts, similar to TiO_2. The possessive carbonaceous nanocomposites may upgrade photocatalysis performance in many ways, such as by providing superior quality adsorption edge sites, introducing photo electrons movements, and interfering with electron–hole pair recombination; and also, bandgap tuning or photosensitization[54].

6.2.1.5 Disinfections

Disinfection is an efficient way to remove organic and biological contaminants such as pathogens from wastewater sources. Typical methods such as chlorination and ozonation would generate unsafe DBPs in the decontamination procedure and endanger human health. Nanotechnology advancements shed a youthful insight into reducing the weaknesses of these conventional approaches. Carbonaceous nanocomposites have considerable antimicrobial performance caused by less oxidation potential, maybe greatly inactivated microbes irradiated visible light and/or direct interaction, and possess a lower ability to produce DBPs. For the last few years, graphene-based nanomaterials have been extensively examined as disinfectants [55,56]. Magnetic graphene produced by Ganesh et al. [57] demonstrated disinfected performance (42 ug L^{-1}) for *E. coli* with 98% bacteria destroying toward disinfection applications for minimal toxicity for zebrafish. However, magnetic reduced graphene oxide functionalized with glutaraldehyde nanomaterials and synthesized and presented by Wu et al. [58] has demonstrated a strong disinfection propensity of up to 99% toward *S. aureus* (gram-positive) and *E. coli* bacteria toward infrared irradiated laser light. The antimicrobial activity, high surface area, thereby offering substantial conductivity owing to graphene derivatives nanocomposites make possible use. Additionally, anti-microorganisms mechanism owing to carbonaceous nanocomposites was defined by some important aspects. The first is damaging cell membrane integrity through direct contact, whereas the second is disrupting real microbial progress through oxidative stress. Furthermore, additional studies into optimizing disinfection performance and the environmental impact on carbonaceous nanomaterials must be prioritized.

6.2.1.6 Membrane Process

In many areas of water purification, membrane methodology was considered an efficient and economical technique, and its efficiency is caused heavily by selectivity as well as permeability owing to membrane materials. Further, large energy consumption depending upon membrane fouling has hampered processes' widespread utility. Significant research has indicated that incorporating carbonaceous nanocomposites into membranes can improve selectivity as well as permeability, and/or membrane fouling hindrance. Carbon nanotubes may be associated with some polymers or nanocomposites to create multi-aspect membranes that can be used successfully in membrane technology, as seen in Figure 6.3. Among those who have contributed to this work are Song et al. [59]. CNTs can be injected into double-layer thin-film

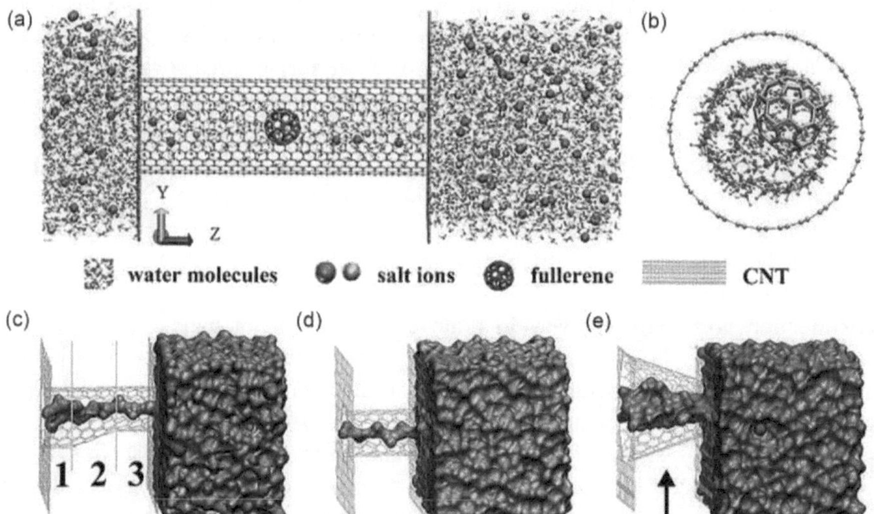

FIGURE 6.3 First membrane structure of CNT (a) side view and (b) front view aspects (Reproduced with permission from Ref. [62] Copyright 2019 Elsevier B.V. Schematics images of various CNTs.); (c) heterostructured CNT, (d) tubular structure (6, 6) CNT, and (e) conic nature CNTs. (Reproduced with permission from Ref. [63]. Copyright 2017 Elsevier B.V.)

membranes to characterize and/or develop morphology, composition, saturation efficiency, and antifouling capability owing to integrated membranes. TFN membranes were thought to have possible water fluxing and extraordinary antifouling capacity due to the incorporation of CNTs. According to Ho et al. [60], the integrated GO and oxidized CNTs had a possible impact on the characterization of membranes containing surface charge porous area, mechanical strength, hardness, pore size and shape, and other indicators. As a result, complex matrix membranes can improve membranes with respect to hydrophilicity as well as antifouling properties by enhancing the rejection rate relevant to organic and/or inorganic contaminants. Another kind of carbonaceous filter is free-standing carbonaceous nanofiber membranes, which have the valuable surface area along with greater degree of uniformity, and a wide number of active sites. Beyond traditional adsorption techniques, they associate attractive efficiency owing to carbonaceous nanofiber and favorable advantage attributed to nanofiber membrane blocking, combined with high permeability as well as antifouling capability. Carbonaceous nanofiber membranes can rapidly strip dyes (MB) with high density (1,580 L $m^{-2}h^{-1}$), which is greater as compared with the most industrial ultrafiltration membranes possessing comparable properties [61]. In addition, other catalysts may also be integrated simultaneously with various mechanisms in carbonaceous membranes for pollutants removal.

6.2.2 Fullerenes

Since their discovery in 1985, fullerenes have been regarded as the third allotropic form of carbon material type. They are surrounded by cage nature structures made up of twelve pentagon rings and are typically seen in the form of six-member rings, unified by locations of carbon atoms. As an example, the buckyball (C_{60}) contains 32 sides, out of which 12 are pentagons whereas 20 are hexagons, and the total carbon atoms are chained in a truncated icosahedron with 60 vertices suggesting it to be greatly symmetric [64,65]. Structures containing decreased hexagons have a higher sp3 bonding behavior associated with large strain energies and incorporated with a greater supply of reactive sites identified with carbon. Furthermore, fullerenes show such chemical properties that are numerous and diverse, allowing for differences in manipulating and associated environmental applications. To achieve the goal of functional applications, various structural measuring strategies such as covalent bonding, endohedral, and/or super-molecular transformations were employed. Moreover, a variety of approaches including OH as well as COOH groups will improve the dispersion mechanism, particularly toward pristine fullerenes associated with polymers, thereby changing the mechanical and electrical properties [15].

Brunet et al. [66] demonstrated that C_{60} hydrophilic functionalized fullerenes can be used to kill the microorganism's pathogens in water by utilizing a photocatalytic activity, the mechanism of which is shown in Figure 6.4. Fullerenes are also ideal clean green materials for hydrogen storage due to their low carbon and hydrogen bond energies, which enable fullerene molecules to be easily transformed from C–C to C–H bonds. Because of their chemistry and cage-like molecular configuration, fullerenes can store up to 6.1% hydrogen, and their structure can be easily reversed due to higher C–C bond energies [29].

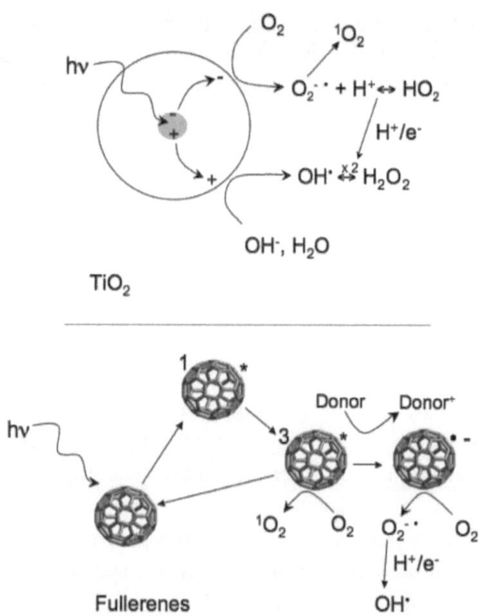

FIGURE 6.4 Mechanisms of ROS production by fullerenes. (Reproduced with permission from Ref. [66]. Copyright 2009 American Chemical Society.)

Pickering et al. [67] successfully created water-soluble fullerene materials and used them as a sensitizer to produce reactive oxygen species (ROS) in aqueous media when exposed to ultraviolet and visible light. ROS are photodegraded organic contaminants in water, and water-soluble fullerenes (fullerols) also act as antioxidants. After the photodegradation process, fullerols, in particular, can be easily extracted via water. Fullerenes are thought to adsorb species by adsorbates penetrating into spaces or defects within carbon nanoclusters, and their limited aggregation potential and large surface area, in addition to defects, make them useful nanomaterials for heavy metal ion adsorption from water [68]. Alekseeva et al. [69] performed a comparative study of fullerenes and composite polystyrene film to extract Cu^{+2} ions and discovered that fullerenes were more effective. They also discovered that fullerenes use the Langmuir adsorption model for Cu^{+2} ions. In the first case, they discovered that fullerenes have a high Cu^{+2} removal efficiency, and they developed an isothermal equilibrium of Cu^{+2} adsorption on fullerenes using the Langmuir model. Fullerenes, on the other hand, responded strongly to water adsorption approaches, but their cost is significantly higher, reducing their widespread usage. A trace amount of fullerenes can be used to improve the adsorption efficiency of materials such as activated carbon, lignin, and zeolite [70]. The hydrophobic character of fullerenes can be improved, resulting in materials that are ideally suited for adsorption and that can be recycled more easily [71]. Antimicrobial materials with the potential to be used in water disinfection have been identified as being developed by C_{60} grafting fullerenes with polyvinylpyrrolidone (PVP). Membrane technology has gained importance in the purification of organic materials, salts, particles, and gases through water. Membrane efficiency is determined by the material composition, which determines selectivity, reactivity, and mechanical strength. Because of their ease of functionalization, higher electron affinity, high strength, and ability to customize the form, fullerenes have considerable potential for use in membrane technology. Fullerenes may be useful in grafting nano-adsorbents to improve their adsorption efficiency [29].

6.2.3 Carbon Nanotubes

In 1999, carbon nanotubes (CNTs) were found by Lijima and might be SWCNTs and MWCNTs for instance layered rolled-up graphene and multilayered rolled-up graphene, respectively [72–74]. CNTs have sparked a lot of interest in the field of nanotechnology after they were discovered because of their unique physicochemical properties. These cylindrical carbon allotropes nanomaterials are used in field emission, semiconductors, energy storage systems, biomedical and industrial, water and air filters, and other applications. Their diameters range from 1 nm to several nanometers, they have a high specific surface area (150–1,500 m^2g^{-1}), and they contain mesopores, making them an ideal candidate for heavy metal ion removal through adsorption and advanced oxidation processes (Figure 6.5). Furthermore, CNTs can be easily functionalized with a variety of organic molecules, making them ideal for adsorbate selection and enhancing their adsorption capacity. The sorption of heavy metal ions across carbon nanotubes is affected by surface morphology, ion exchange strategies, and electrochemical ability [29,75–77].

6.2.3.1 Adsorption Mechanism

Thc adsorption process is based on the physicochemical properties of the adsorbent, which include unique surface area, hydrophobicity, surface functional groups, and chemical bonds such as hydrogen bonds, electrostatic, and π–π bonds [20,79]. Figure 6.6 shows schematic display of an adsorption phenomenon of a vast number of adsorbents. The surface wettability of CNTs is altered by functional groups, resulting

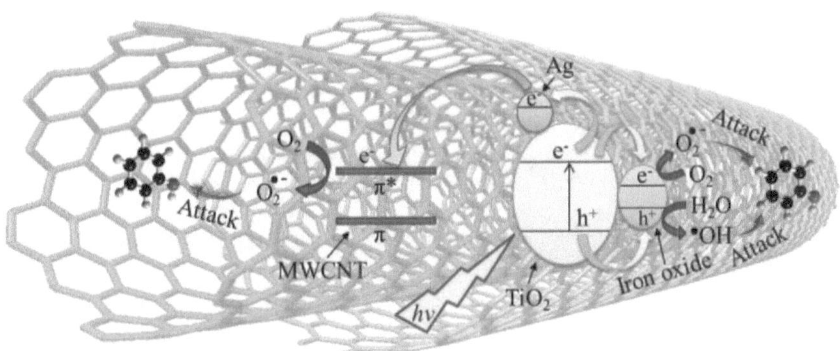

FIGURE 6.5 Schematics of photocatalytic activation mechanism and phenol degradation by CNTs. (Reproduced with permission from Ref. [78]. Copyright 2017 Brazilian Chemical Society.)

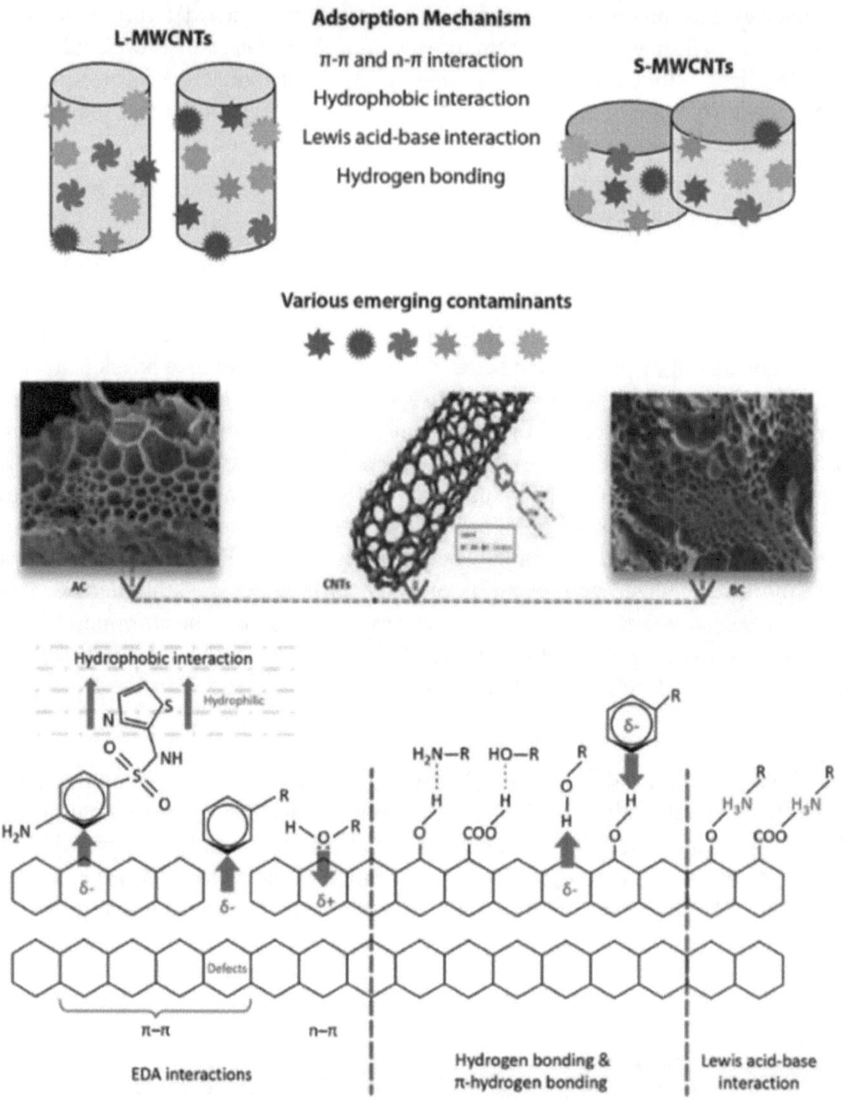

FIGURE 6.6 Illustration of adsorption mechanism by CNTs. (Reproduced with permission from Refs. [82,83]. Copyright 2016 and 2015, respectively, Elsevier B.V.)

in hydrophilicity and improved adsorption of polar compounds with low molecular weight.

Researchers investigated the adsorptive acetaminophen elimination of MWCNTs modified chemically with NaOH, HNO_3 or H_2SO_4, ozone, and chitosan [80]. The addition of carboxylic groups, –OH, and –COO groups, to the surface of MWCNTs changes the pore size and shape, resulting in increased acetaminophen adsorption without allowing the crystal structure to rupture. The adsorption abilities of chemically adapted MWCNTs were in order of (250 mg g^{-1}) for ozone-treated

MWCNTs > chitosan-coated MWCNTs (205 mg g^{-1}) > acidic MWCNTs (160 mg g^{-1}) > NaOH treated (130 mg g^{-1}) > (90 mg g^{-1}) for MWCNTs. Similarly, MWCNTs that had been modified or oxidized with hypochlorite were tested for ciprofloxacin elimination. According to scientists, the rate of increase in adsorption potential slowed during the π–π electron donor–acceptor interaction [81]. Therefore, the evolution of hydrophilicity and dispersibility showed a significant role for ciprofloxacin adsorption into oxidized MWCNTs. Several studies have demonstrated CNTs adsorption interaction and an organic compound.

Adsorption phenomena can be dominated by energy decomposition analysis (EDA) interactions, such as π–π and n–π EDA interactions. During the adsorption process, π-electron depleted and π-electron-rich rings or CNTs aromatic zones, as well as goal adsorbents, are the main causes of interactions. CNTs can act as both π-donors and π-acceptors. Aromatic structures with high electron density can produce larger quadrupoles resulting in rich electron adsorbates as a π-donor and CNTs surface as a π-acceptor [84]. Conversely, lowering aromatic structure electron density opposites the quadrupoles moment and form adsorbate electron-deficient (π-acceptor). But CNTs, for example, will behave like electron-rich (π-donors) graphene sheets with a polarized edge, and adsorbate donor electron may bind to a high-electric density defective area [85,86]. A different relationship is likely when electron unshared pairs of amino or hydroxyl groups serve as electron donors and associate with electron-deficient edges of CNTs that act as electron acceptors [87]. Aromatic compounds enclosing hydroxyl group and CNTs can demonstrate improved adsorption due to the heavy EDA interaction between electrons donated hydroxyl group and CNTs π-electron-deficient areas. Ionized amino and hydroxyl groups strengthen n–π electron EDA interaction at high pH due to their strong electron-donating groups [82]. Hydrogen bonding is aided by carbon molecules or CNTs with functional groups such as –COOH, –OH, and NH_2. In this case, the adsorbates benzene rings serve as hydrogen-bonding acceptors whereas these functional groups can act as hydrogen-bonding donors and form hydrogen bonds with CNTs graphite surfaces. The functional groups on the surface of CNTs that serve as hydrogen-bonding donors, such as –COOH and –OH, could also form bonds with organic elements. Instead of adsorbate chemicals, strong H-bonding interactions between CNT functional groups and water can cause competitive sorption between CNTs and organic compounds. As a result, minor hydrogen-bonding effects on solute adsorption without a hydrogen-bonding tendency will occur [82,87]. The Lewis acid–base concept, which has been elucidated as an additional touch for target molecule adsorption using amino groups as a significant method described above, can be interpreted as a bonding contact between adsorbates and adsorbents. The electronic properties of adsorbate materials may have a significant impact on the adsorption process [88].

6.2.3.2 CNTs as Photocatalyst

Photocatalytic activity is one of the progressive technologies being employed for the purification of water with uses in semiconductors [89,90]. Numerous semiconducting materials are utilized such as Fe_3O_4, ZnO, and TiO_2; thus, these materials have low quantum efficiency, and further, their ultraviolet response is not high [91–93].

CNTs are exciting prospect materials for catalysis due to their improved quantum performance, nanosize, increased chemical stability, empty tube shape, and stretched light adsorption region due to their high specific surface area [94]. Gao et al. [95] successfully used ultrathin photocatalyst-based SWCNTs-TiO_2 for oil–water purification. Park et al. [96] successfully removed MB from water by incorporating titania on SWCNTs aerogel. Xu et al. [97] successfully removed pyridine from water using photocatalysts made from MWCNTs hydroxyl and PbO3 nanostructured anodes (Figure 6.7a and b) and used it successfully for pyridine removal from water. Lee et al. [98] generated MWCNTs-TiO_2 for photodegradation of MB.

6.2.4 Graphene

6.2.4.1 Nanoporous Graphene for Water Filtration and Desalination

This procedure has been regarded as a revolutionary form of water desalination treatment at MIT, where it could be used as a sustainable RO membrane. Water permeability is poor in traditional RO membranes, necessitating high operating

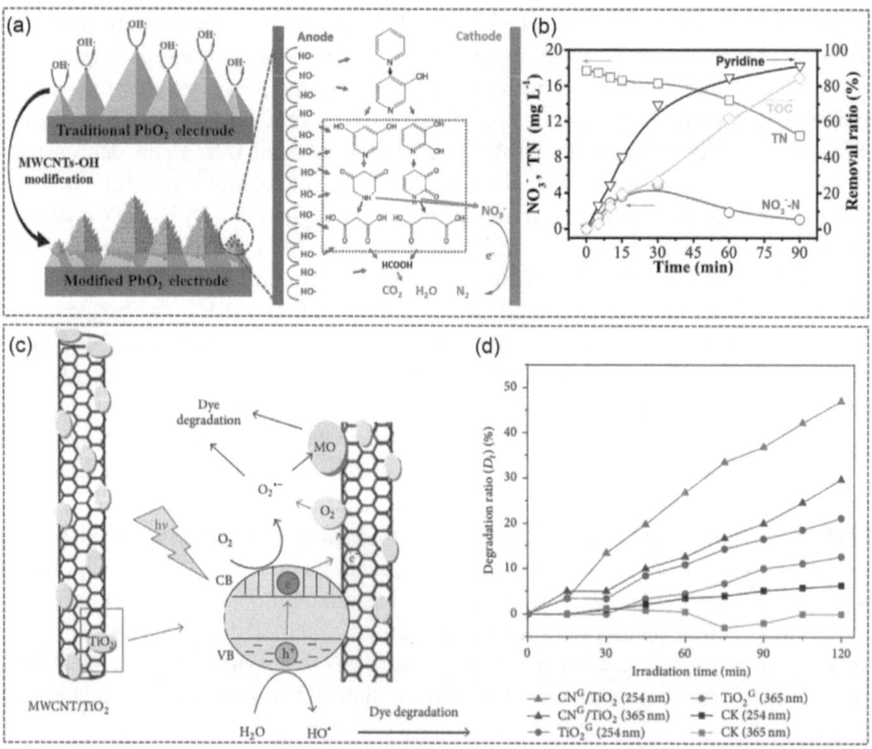

FIGURE 6.7 Mechanistic view and output results of photocatalysis using CNTs: (a) MWCNTs-OH-PbO_2 electrode forming procedure and reaction scheme, and (b) as a function of electrolysis time, the concentrations of NO_3^-N and total nitrogen (TN) change corresponding with the removal rates of pyridine and total organic carbon (TOC). (Reproduced with permission from Ref. [97]. Copyright 2017 Elsevier B.V.) (c) Schematic depiction revealing mechanism towards photocatalytic degradation for MWCNTs/TiO_2 irradiated with UV light, (d) Methyl orange photodegradation, Reproduced with permission from ref. [98] Copyright 2016 Hindawi.

power for the desalination strategy [99,100]. This new membrane includes the modification of graphene via generating nanopores above the surface of graphene. The use of well-structured pore channels will simplify the flow of water and thus achieve a quick flow using this process. Furthermore, not only salt can be removed, but some molecules, based on their molecular size, can also be filtered by adjusting the size of the pores. To remove ions and other solute compounds, the pores employ physical concepts such as hydrophobicity and charge transformation [101–104]. The entire desalination process is depicted in Figure 6.8. Graphene has a number of pros over other traditional membranes used in water desalination research. Since graphene is just one atom thick, its hardness can be overlooked, and this is its most important advantage. Graphene, on the other hand, has high mechanical strength and low-pressure requirements, allowing for easier water transport [100,101,105]. As a result, the issue of how to perforate something harder than a diamond arises. The nanopores that are needed for water desalination on graphene can be created using two methods. Furthermore, since the scale of the pores must be precise, the perforating route must be carried out with special care and precision. Desalination would not be possible due to the small pore sizes that prevent water molecules from passing through [101].

Nowadays, scientists have established a proficient approach for the production of essential pores above graphene sheets. In this step, gallium ion cannon is used to examine the graphene sheet from top to bottom and left to right, while an unfocused laser is fired at the graphene sheet, allowing gallium ions to scatter throughout the graphene sheet, causing disruptions in the material's lattice and hitting off carbon atoms. As a consequence, the etchable defects are formed, and graphene is used in an acidic potassium permanganate solution, which is commonly used to open carbon nanotubes. The amount of time a graphene sheet is immersed in an oxidizing solution changes the pores that form above it. More time to immerse graphene in the solution caused to enlarge the size of the pores. The outcome of using such a technique was single-layer graphene that confined five trillion 0.4 nm holes/cm [100]. For desalination initiatives, the full pore diameter must be approximately 7 A° through experimental research [99,101]. More research on the declared process for use in large graphene filters should be conducted. After pores were made, some hydrophilic bonds were created in order to increase water flow through the pores. The higher number of hydrogen bonds present inside the pores is the reason for this [100,101]. Thanks to its higher water permeability than the best RO membranes currently in use, Tanugi and Grossman [101] determined that the new membrane for water treatment was superior to traditional RO membranes. Furthermore, two considerations influenced salt removal: strain and pore size. It was discovered that aggregating the pore size and strain reduces the membrane's salt removal capacity. Graphene is another carbon allotrope that can be used in water desalination. This membrane has not yet been manufactured, but it is based on simulation results; its promise is comparable to that of nanoporous graphene. The primary challenge for the NG membrane is the direct proportionality of the amount of pores to water flow, and increasing the number of pores allows the membrane's mechanical stability to deteriorate. Pore scattering adds to the complexity of this technique, and mass fabrication of graphene sheets must be lectured through thorough study [101,106].

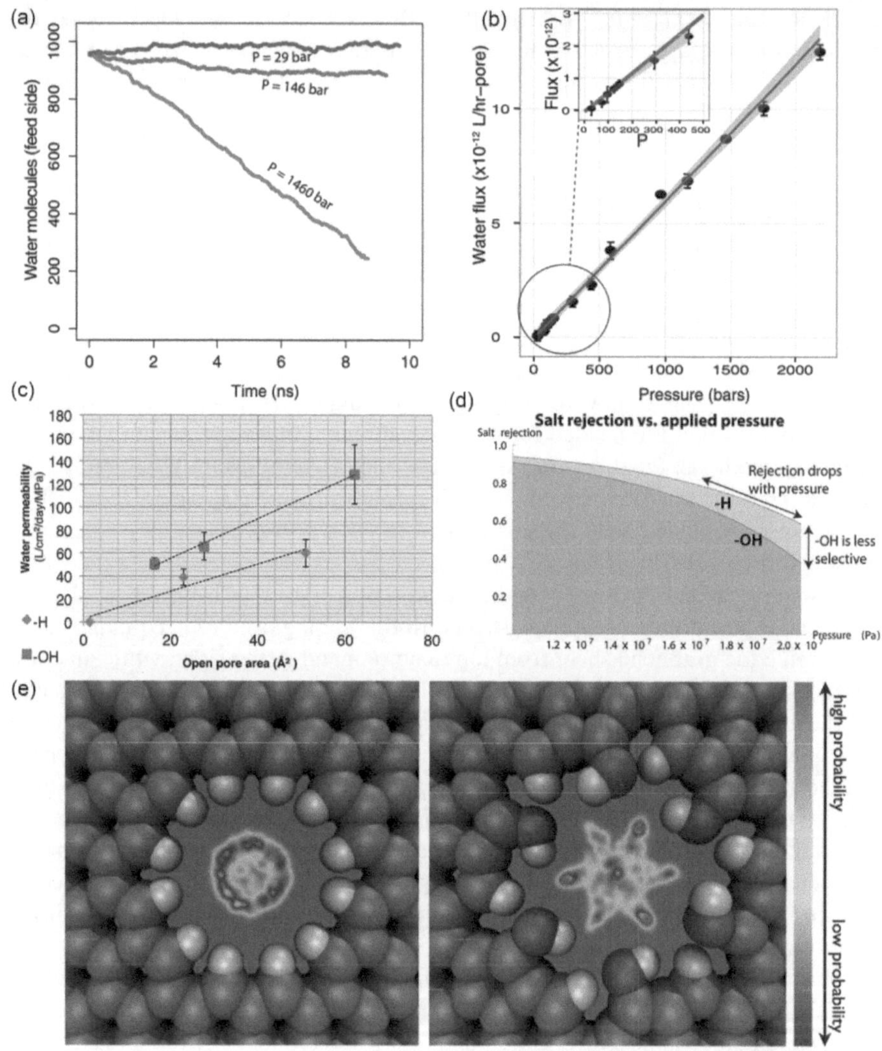

FIGURE 6.8 (a) For varying pressures, the number of water molecules on the feed side vs. simulation time, (b) water flux vs. applied pressure (Reproduced with permission from Ref. [99] Copyright 2014 AIP Publishing.), (c) for different pore sizes, water permeability was calculated for nanoporous graphene functionalized with hydrogen and hydroxyl groups, (d) for hydrogenated (upper line) and hydroxylated (lower line) pore salt rejection vs. applied pressure (bottom line), and (e) within a hydrogenated (left) and hydroxylated (right) pore, with open-pore areas of 23 and 28 Å^2, respectively, oxygen density maps are plotted. (Reproduced with permission from Ref. [101]. Copyright 2012 American Chemical Society.)

6.2.4.2 Advanced Oxidation Process

In recent years, many metal oxide nanoparticles are combined with carbonaceous electron scavenger materials to grip over the shortcomings and growth of photocatalytic response, as presented in Figure 6.9. For illustration, in the recent 20 years, numerous research articles have been conveyed on the photocatalytic response of graphene-based materials. One characteristic of the growth and employment of graphene-based systems is because of their unique properties such as economical, harmless nature, easily available and processable, obstructing corrosion and leaching, and dropping the combination of nanoparticles. Furthermore, it can be altered with numerous surfactants to boost the adsorption efficacy [4].

IUPAC is one of the standards that defines a catalyst as a material that increases the pace of chemical reactions without changing the full standards connected to Gibbs energy variation in that reaction; hence, the process is known as catalysis [38]. Furthermore, photocatalysis refers to catalyst reactions that are thought to be dependent on, particularly, light energy and are carried out using a special setup. Photocatalyst types depend on the catalyst phase, and catalysts in the same reactant phase are defined as homogeneous, while heterogeneous phase performance is known. Special steps with heterogeneous photocatalysis are used to develop a series of reactions that mainly eliminate organic aqueous impurities [107]. Graphene showing zero bandgap acts as a semimetal, presenting light absorbance ranging large wavelengths that become efficiently helpful toward photocatalysis [108,109]. Graphene containing a few layers may play an active role as semiconductors that are considered better photocatalysts. Heterogeneous photocatalysis mechanism involves electrons excitation associated with photon energy from valence band (VB) toward conduction band (CB) along with holes formation in valence bands caused by incident light energy that becomes greater than bandgap of semiconducting materials [110]. Resultantly, electrons, as well as holes, carry on redox reactions become responsible for the photocatalytic mechanism. These redox reactions result in the formation of various types of radicals/intermediates, such as hydroxyl radicals, which serve as effective oxidizers on the surface of the photocatalyst and are hence commonly involved in the removal of wastewater organic contaminants containing dyes as well as other removable organic entities [34]. It is regarded necessary for the provision of enough reactants for which charge carriers are considered to be utilized rather than interfering with the aquatic ecosystem because radical generation produced by them is observed efficiently active in nature.

As photocatalysis requires adsorbent presence for the provision of active sites of chemical reactions, graphene is considered suitable for achieving this goal, being a highly efficient adsorbent, resulting in enhanced photocatalytic activity [3]. Therefore, TiO_2 proved to be a good photocatalyst possessing promising photocatalytic properties, the first time demonstrated and published by Akira Fujishima et al. in 1972 [111]. TiO_2/graphene photocatalytic nanocomposites exhibited versatile photocatalytic property that was widely researched as well as cataloged successfully [112–114]. UV irradiation along with visible irradiation presented enhanced photocatalytic activity related to TiO_2/graphene nanocomposites [115]. Efficiently enhanced catalytic activity was caused by conduction band electrons transfer of

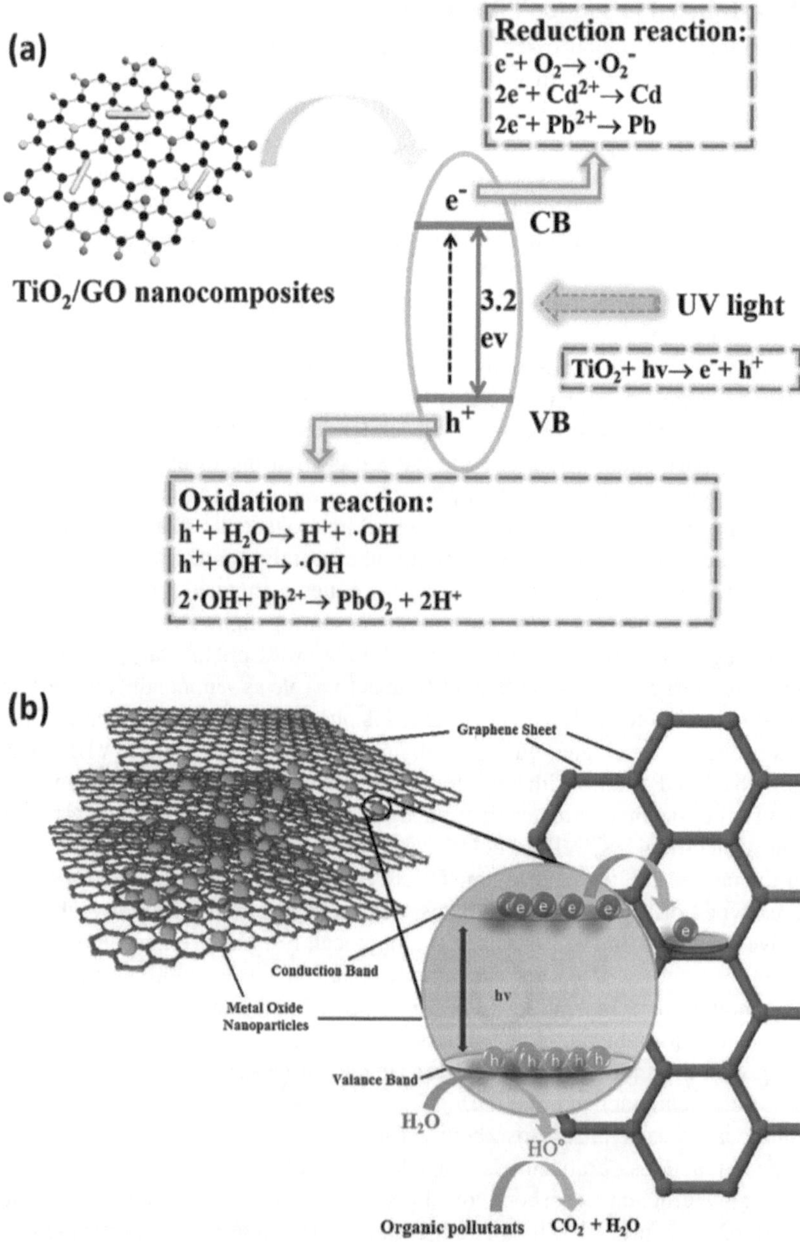

FIGURE 6.9 (a) Charge transfer and separation in TiO_2/GO nanocomposites, as well as the different steps involved in heavy metal ion reduction, are depicted in this schematic diagram. (Reproduced with permission from Ref. [123] Copyright 2018 Royal Society of Chemistry.) (b) Graphene's ability to scavenge electrons from metal oxide conduction bands. (Reproduced with permission from Ref. [56]. Copyright 2014 Royal Society of Chemistry.)

TiO_2 toward graphene, thereby enlarging bandgap whereas decreasing electron–hole pairs recombination that may occur in TiO_2 [116]. Graphene was found to be unique and effective in the eradication of RhB due to its thermodynamic properties. Furthermore, graphene acted as an electron mediator in composite forming associated with SnO_2, while in TiO_2 nanocomposites, electrons trend in the form of freely motion into graphene sheets or may readily interact with oxygen functional groups [117]. Further, similar behavior has been observed in gold/graphene (Au/graphene) photocatalytic composite. However, 2,4-Dicholorophenoxyacetic acid associated with TiO_2/graphene proved more efficient in degradation mechanism that was considered four times efficient, whereas dodecylbenzenesulfonate (DBS) showed three times improved performance as compared with Degussa P25(commercial TiO_2) where graphene was reasonably doped by platinum to form graphene/TiO_2 nanocomposites [118,119]. Enhanced organic entities degradation may have been attributed to effective adsorbing of graphene where TiO_2 charge transfer occurred toward graphene, whereas titanium bridged bond toward carbon associated with oxygen resulted with decrement bandgap value was observed [120]. P25-graphene was found to be more effective than single Degussa P25 in eliminating Reactive Black 5 when exposed to UV light. Above 90% pollutant oxidation was successfully tested for a 100-minute irradiation period with UV belonging to a high surface region with a narrow bandgap [121]. Flowering G/titania sheets (G/TNS) were created using a hydrothermal process that efficiently produced Ti crystalline phases containing anatase as well as sodium titanate. Graphene, acting as a basis for Ti growth and reserving TiO_2 growth alongside degraded 4-chlorophenol, demonstrated performance of greater than 99.2% over a 120-minute degradation time [122]. Graphene use enhanced titanium (Ti) photocatalytic performance owing to two logical reasons: firstly, enhancing transfer rate attributed to photoexcited electrons by virtue of hindrance of electron–hole pairs recombination; and secondly, narrowing bandgap values that result in efficient enhancement toward visible light absorption spectrum as demonstrated in Figure 6.9. Hydroxyl radicals formed were observed, such as reactive oxygen species that clearly exhibited 4-chlorophenol degradation verifying via DFT measurements and retained their effective role rather with passage of further five cycles [38,122].

6.3 CHALLENGES AND FUTURE PROSPECTS

Carbonaceous nanocomposites hold extraordinary specific surface areas, chemical stability, as well as exceptional thermal, electrical, and optical properties; therefore, it has been reflected as most potential aspirants for the elimination of biological and chemical pollutants from wastewater. The use of these nanomaterials for wastewater treatment at the lab scale was the subject of previous studies. Future advances and full-scale use of these nanomaterials, however, would inevitably pose several problems, necessitating more intensive testing.

1. The present production approaches for carbonaceous nanomaterials are quite difficult and unproductive, even several studies have been directed to challenge them. Additionally, vigorous and competent fabrication techniques are immediately required. Although commercial large-scale fabrication of

carbonaceous nanomaterials is intriguing, it needs to be discussed because of the wide variety of applications.

2 The agglomeration of graphene-related nanomaterials and carbon nanotubes in the aqueous section is a new flaw in water refining. Grouped nanomaterials will decrease the surface area and active sites available to them, leading to lower efficacy for pollutant removal. To strengthen nanomaterials, many metal oxides and functional groups have been added. Furthermore, researchers can focus more on targeted modification in order to improve removal efficiency, selectivity, and affinity for definite pollutants.

3 The effectiveness of carbonaceous nanomaterials in handling physical natural and wastewater from various factories must be confirmed; almost all of the experiments were performed in lab settings. Furthermore, research into the effectiveness and application of various carbonaceous nanomaterials in wastewater treatment can be performed under more precise conditions. In addition, the long-term growth of carbonaceous nanomaterials must be taken into account.

4 The cytotoxicity of carbonaceous nanomaterials for humans and organisms is yet to be fully investigated; as a result, the disparity between potential environmental properties and carbonaceous nanomaterials characteristics should be resolved as soon as possible. With the development of carbonaceous nanomaterials, hazard evaluation for these nanomaterials is urgently needed to negotiate environmental remediation. Furthermore, carbonaceous nanomaterials have a variety of advantages and disadvantages in wastewater management; nanomaterials will likely be used to solve a variety of environmental problems.

6.4 CONCLUSION

Environmental and distinctive refinement of heavy metals from respective wastewater contaminants via carbon nanomaterials has been discussed. With considerable success, these carbon nanomaterials have been used in the decontamination of heavy metals from wastewater. The inspiration for successful use derives from their appealing properties such as plenty of recycling, patience to desorb the adsorbed metal ions, and high surface area. Furthermore, carbon nanomaterials are readily manufactured and functionalized with other nanomaterials, resulting in a multifunctional nano-adsorbent. Carbon-based fabrics are highly biocompatible for both the atmosphere and living organisms. Furthermore, various parameters such as contact time, pH value, and adsorbent form are essential. Based on this analysis of the literature, it is possible to infer that carbon nanomaterials have fascinating physicochemical properties and have a high response when used in wastewater treatment and environmental remediation.

REFERENCES

1. J. Lubchenco, Entering the Century of the Environment: A New Social Contract for Science, *Science*, 279 (1998) 491.

2. F. Perreault, A. Fonseca de Faria, M. Elimelech, Environmental Applications of Graphene-Based Nanomaterials, *Chemical Society Reviews*, 44 (2015) 5861–5896.
3. X. Li, J. Yu, S. Wageh, A.A. Al-Ghamdi, J. Xie, Graphene in Photocatalysis: A Review, *Small*, 12 (2016) 6640–6696.
4. P. Singh, P. Shandilya, P. Raizada, A. Sudhaik, A. Rahmani-Sani, A. Hosseini-Bandegharaei, Review on Various Strategies for Enhancing Photocatalytic Activity of Graphene Based Nanocomposites for Water Purification, *Arabian Journal of Chemistry*, 13 (2020) 3498–3520.
5. M. Mitra, S.T. Ahamed, A. Ghosh, A. Mondal, K. Kargupta, S. Ganguly, D. Banerjee, Polyaniline/Reduced Graphene Oxide Composite-Enhanced Visible-Light-Driven Photocatalytic Activity for the Degradation of Organic Dyes, *ACS Omega*, 4 (2019) 1623–1635.
6. X. Qu, P.J.J. Alvarez, Q. Li, Applications of Nanotechnology in Water and Wastewater Treatment, *Water Research*, 47 (2013) 3931–3946.
7. P.V. Kamat, Graphene-Based Nanoarchitectures. Anchoring Semiconductor and Metal Nanoparticles on a Two-Dimensional Carbon Support, *The Journal of Physical Chemistry Letters*, 1 (2010) 520–527.
8. X. Wei, M.U. Akbar, A. Raza, G. Li, A Review on Bismuth Oxyhalide Based Materials for Photocatalysis, *Nanoscale Advances*, 3 (2021) 3353–3372.
9. K.K. Chenab, B. Sohrabi, A. Jafari, S. Ramakrishna, Water Treatment: Functional Nanomaterials and Applications from Adsorption to Photodegradation, *Materials Today Chemistry*, 16 (2020) 100262.
10. N. Madima, S.B. Mishra, I. Inamuddin, A.K. Mishra, Carbon-Based Nanomaterials for Remediation of Organic and Inorganic Pollutants from Wastewater. *A review, Environmental Chemistry Letters*, 18 (2020) 1169–1191.
11. A. Lee, J.W. Elam, S.B. Darling, Membrane Materials for Water Purification: Design, Development, and Application, *Environmental Science: Water Research & Technology*, 2 (2016) 17–42.
12. A. Raza, U. Qumar, A. Haider, S. Naz, J. Haider, A. Ul-Hamid, M. Ikram, S. Ali, S. Goumri-Said, M.B. Kanoun, Liquid-Phase Exfoliated MoS_2 Nanosheets Doped with p-type Transition Metals: A Comparative Analysis of Photocatalytic and Antimicrobial Potential Combined with Density Functional Theory, *Dalton Transactions*, 50 (2021) 6598–6619.
13. O. Zaytseva, G. Neumann, Carbon Nanomaterials: Production, Impact on Plant Development, Agricultural and Environmental Applications, *Chemical and Biological Technologies in Agriculture*, 3 (2016) 17.
14. M.M. Messina, M.E. Coustet, J. Ubogui, R. Ruiz, F.D. Saccone, P.C. dos Santos Claro, F.J. Ibañez, Simultaneous Detection and Photocatalysis Performed on a 3D Graphene/ZnO Hybrid Platform, *Langmuir*, 36 (2020) 2231–2239.
15. S.J. Shan, Y. Zhao, H. Tang, F.Y. Cui, A Mini-review of Carbonaceous Nanomaterials for Removal of Contaminants from Wastewater, *IOP Conference Series: Earth and Environmental Science*, 68 (2017) 012003.
16. A. Raza, M. Ikram, M. Aqeel, M. Imran, A. Ul-Hamid, K.N. Riaz, S. Ali, Enhanced Industrial Dye Degradation Using Co Doped in Chemically Exfoliated MoS_2 Nanosheets, *Applied Nanoscience*, 10 (2020) 1535–1544.
17. S. Mureed, S. Naz, A. Haider, A. Raza, A. Ul-Hamid, J. Haider, M. Ikram, R. Ghaffar, M. Irshad, A. Ghaffar, A. Saeed, Development of Multi-concentration Cu: Ag Bimetallic Nanoparticles as a Promising Bactericidal for Antibiotic-Resistant Bacteria as Evaluated with Molecular Docking Study, *Nanoscale Research Letters*, 16 (2021) 91.
18. M. Nasrollahzadeh, M. Sajjadi, S. Iravani, R.S. Varma, Carbon-based Sustainable Nanomaterials for Water Treatment: State-of-art and Future Perspectives, *Chemosphere*, 263 (2021) 128005.

19. S.C. Smith, D.F. Rodrigues, Carbon-based Nanomaterials for Removal of Chemical and Biological Contaminants from Water: A Review of Mechanisms and Applications, *Carbon*, 91 (2015) 122–143.
20. S. Kurwadkar, T.V. Hoang, K. Malwade, S.R. Kanel, W.F. Harper, G. Struckhoff, Application of Carbon Nanotubes for Removal of Emerging Contaminants of Concern in Engineered Water and Wastewater Treatment Systems, *Nanotechnology for Environmental Engineering*, 4 (2019) 12.
21. K. Piaskowski, P.K. Zarzycki, Carbon-Based Nanomaterials as Promising Material for Wastewater Treatment Processes, *International Journal of Environmental Research and Public Health*, 17 (2020) 1–14.
22. R.K. Thines, N.M. Mubarak, S. Nizamuddin, J.N. Sahu, E.C. Abdullah, P. Ganesan, Application Potential of Carbon Nanomaterials in Water and Wastewater Treatment: A Review, *Journal of the Taiwan Institute of Chemical Engineers*, 72 (2017) 116–133.
23. J. Zhu, J. Hou, Y. Zhang, M. Tian, T. He, J. Liu, V. Chen, Polymeric Antimicrobial Membranes Enabled by Nanomaterials for Water Treatment, *Journal of Membrane Science*, 550 (2018) 173–197.
24. M.T. Amin, A.A. Alazba, U. Manzoor, A Review of Removal of Pollutants from Water/Wastewater Using Different Types of Nanomaterials, *Advances in Materials Science and Engineering*, 2014 (2014) 825910.
25. C.H. Lee, B. Tiwari, D. Zhang, Y.K. Yap, Water Purification: Oil–Water Separation by Nanotechnology and Environmental Concerns, *Environmental Science: Nano*, 4 (2017) 514–525.
26. M. Ikram, M.I. Khan, A. Raza, M. Imran, A. Ul-Hamid, S. Ali, Outstanding Performance of Silver-Decorated MoS_2 Nanopetals Used as Nanocatalyst for Synthetic Dye Degradation, *Physica E: Low-dimensional Systems and Nanostructures*, 124 (2020) 114246.
27. F. Yu, Y. Li, S. Han, J. Ma, Adsorptive Removal of Antibiotics from Aqueous Solution Using Carbon Materials, *Chemosphere*, 153 (2016) 365–385.
28. C. Xu, P. Ravi Anusuyadevi, C. Aymonier, R. Luque, S. Marre, Nanostructured Materials for Photocatalysis, *Chemical Society Reviews*, 48 (2019) 3868–3902.
29. R. Baby, B. Saifullah, M.Z. Hussein, Carbon Nanomaterials for the Treatment of Heavy Metal-Contaminated Water and Environmental Remediation, *Nanoscale Research Letters*, 14 (2019) 341.
30. M. Ikram, R. Tabassum, U. Qumar, S. Ali, A. Ul-Hamid, A. Haider, A. Raza, M. Imran, S. Ali, Promising Performance of Chemically Exfoliated Zr-doped MoS_2 Nanosheets for Catalytic and Antibacterial Applications, *RSC Advances*, 10 (2020) 20559–20571.
31. M. Ikram, E. Umar, A. Raza, A. Haider, S. Naz, A. Ul-Hamid, J. Haider, I. Shahzadi, J. Hassan, S. Ali, Dye Degradation Performance, Bactericidal Behavior and Molecular Docking Analysis of Cu-doped TiO_2 Nanoparticles, *RSC Advances*, 10 (2020) 24215–24233.
32. Y. Wang, C. Pan, W. Chu, A.K. Vipin, L. Sun, Environmental Remediation Applications of Carbon Nanotubes and Graphene Oxide: Adsorption and Catalysis, *Nanomaterials*, 9 (2019).
33. Q. Xiang, J. Yu, M. Jaroniec, Graphene-Based Semiconductor Photocatalysts, *Chemical Society Reviews*, 41 (2012) 782–796.
34. X. An, J.C. Yu, Graphene-Based Photocatalytic Composites, *RSC Advances*, 1 (2011) 1426–1434.
35. M. Faraldos, A. Bahamonde, Environmental Applications of Titania-Graphene Photocatalysts, *Catalysis Today*, 285 (2017) 13–28.
36. H. Nakano, H. Tetsuka, M.J.S. Spencer, T. Morishita, Chemical Modification of Group IV Graphene Analogs, *Science and Technology of Advanced Materials*, 19 (2018) 76–100.

37. F.R. Bagsican, A. Winchester, S. Ghosh, X. Zhang, L. Ma, M. Wang, H. Murakami, S. Talapatra, R. Vajtai, P.M. Ajayan, J. Kono, M. Tonouchi, I. Kawayama, Adsorption Energy of Oxygen Molecules on Graphene and Two-Dimensional Tungsten Disulfide, *Scientific Reports*, 7 (2017) 1774.
38. K. Thakur, B. Kandasubramanian, Graphene and Graphene Oxide-Based Composites for Removal of Organic Pollutants: A Review, *Journal of Chemical & Engineering Data*, 64 (2019) 833–867.
39. L. Ai, C. Zhang, Z. Chen, Removal of Methylene Blue from Aqueous Solution by a Solvothermal-Synthesized Graphene/Magnetite Composite, *Journal of Hazardous Materials*, 192 (2011) 1515–1524.
40. Y. Gao, Y. Li, L. Zhang, H. Huang, J. Hu, S.M. Shah, X. Su, Adsorption and Removal of Tetracycline Antibiotics from Aqueous Solution by Graphene Oxide, *Journal of Colloid and Interface Science*, 368 (2012) 540–546.
41. J. Wang, Z. Chen, B. Chen, Adsorption of Polycyclic Aromatic Hydrocarbons by Graphene and Graphene Oxide Nanosheets, *Environmental Science & Technology*, 48 (2014) 4817–4825.
42. J. Xu, L. Wang, Y. Zhu, Decontamination of Bisphenol A from Aqueous Solution by Graphene Adsorption, *Langmuir*, 28 (2012) 8418–8425.
43. S.M. Maliyekkal, T.S. Sreeprasad, D. Krishnan, S. Kouser, A.K. Mishra, U.V. Waghmare, T. Pradeep, Graphene: A Reusable Substrate for Unprecedented Adsorption of Pesticides, *Small*, 9 (2013) 273–283.
44. M. Ikram, J. Hassan, A. Raza, A. Haider, S. Naz, A. Ul-Hamid, J. Haider, I. Shahzadi, U. Qamar, S. Ali, Photocatalytic and Bactericidal Properties and Molecular Docking Analysis of TiO_2 Nanoparticles Conjugated with Zr for Environmental Remediation, *RSC Advances*, 10 (2020) 30007–30024.
45. R. Sitko, E. Turek, B. Zawisza, E. Malicka, E. Talik, J. Heimann, A. Gagor, B. Feist, R. Wrzalik, Adsorption of Divalent Metal Ions from Aqueous Solutions Using Graphene Oxide, *Dalton Transactions*, 42 (2013) 5682–5689.
46. M.A. Tofighy, T. Mohammadi, Adsorption of Divalent Heavy Metal Ions from Water Using Carbon Nanotube Sheets, *Journal of Hazardous Materials*, 185 (2011) 140–147.
47. N.M. Mubarak, J.N. Sahu, E.C. Abdullah, N.S. Jayakumar, Removal of Heavy Metals from Wastewater Using Carbon Nanotubes, *Separation & Purification Reviews*, 43 (2014) 311–338.
48. G. Zhao, J. Li, X. Ren, C. Chen, X. Wang, Few-Layered Graphene Oxide Nanosheets as Superior Sorbents for Heavy Metal Ion Pollution Management, *Environmental Science & Technology*, 45 (2011) 10454–10462.
49. R. Ameta, M.S. Solanki, S. Benjamin, S.C. Ameta, Chapter 6 – Photocatalysis, in: S.C. Ameta, R. Ameta (Eds.) *Advanced Oxidation Processes for Waste Water Treatment*, Academic Press, 2018, pp. 135–175.
50. J. Zhang, B. Tian, L. Wang, M. Xing, J. Lei, Mechanism of Photocatalysis, in: J. Zhang, B. Tian, L. Wang, M. Xing, J. Lei (Eds.) *Photocatalysis: Fundamentals, Materials and Applications*, Springer Singapore, Singapore, 2018, pp. 1–15.
51. M. Ikram, J. Hassan, M. Imran, J. Haider, A. Ul-Hamid, I. Shahzadi, M. Ikram, A. Raza, U. Qumar, S. Ali, 2D Chemically Exfoliated Hexagonal Boron Nitride (hBN) Nanosheets Doped with Ni: Synthesis, Properties and Catalytic Application for the Treatment of Industrial Wastewater, *Applied Nanoscience*, 10 (2020) 3525–3528.
52. S.C. Panchangam, C.S. Yellatur, J.-S. Yang, S.S. Loka, A.Y.C. Lin, V. Vemula, Facile Fabrication of TiO_2-Graphene Nanocomposites (TGNCs) for the Efficient Photocatalytic Oxidation of Perfluorooctanoic Acid (PFOA), *Journal of Environmental Chemical Engineering*, 6 (2018) 6359–6369.

53. S. Thangavel, K. Krishnamoorthy, S.-J. Kim, G. Venugopal, Designing ZnS Decorated Reduced Graphene-Oxide Nanohybrid Via Microwave Route and their Application in Photocatalysis, *Journal of Alloys and Compounds*, 683 (2016) 456–462.
54. Y.-H. Chiu, T.-F. Chang, C.-Y. Chen, M. Sone, Y.-J. Hsu, Mechanistic Insights into Photodegradation of Organic Dyes Using Heterostructure Photocatalysts, *Catalysts*, 9 (2019).
55. S. Talebian, G.G. Wallace, A. Schroeder, F. Stellacci, J. Conde, Nanotechnology-based Disinfectants and Sensors for SARS-CoV-2, *Nature Nanotechnology*, 15 (2020) 618–621.
56. R.K. Upadhyay, N. Soin, S.S. Roy, Role of Graphene/Metal Oxide Composites as Photocatalysts, Adsorbents and Disinfectants in Water Treatment: A Review, *RSC Advances*, 4 (2014) 3823–3851.
57. G. Gollavelli, C.-C. Chang, Y.-C. Ling, Facile Synthesis of Smart Magnetic Graphene for Safe Drinking Water: Heavy Metal Removal and Disinfection Control, *ACS Sustainable Chemistry & Engineering*, 1 (2013) 462–472.
58. M.-C. Wu, A.R. Deokar, J.-H. Liao, P.-Y. Shih, Y.-C. Ling, Graphene-Based Photothermal Agent for Rapid and Effective Killing of Bacteria, *ACS Nano*, 7 (2013) 1281–1290.
59. X. Song, L. Wang, C.Y. Tang, Z. Wang, C. Gao, Fabrication of Carbon Nanotubes Incorporated Double-Skinned Thin Film Nanocomposite Membranes for Enhanced Separation Performance and Antifouling Capability in Forward Osmosis Process, *Desalination*, 369 (2015) 1–9.
60. K.C. Ho, Y.H. Teow, W.L. Ang, A.W. Mohammad, Novel GO/OMWCNTs mixed-Matrix Membrane with Enhanced Antifouling Property for Palm Oil Mill Effluent Treatment, *Separation and Purification Technology*, 177 (2017) 337–349.
61. H.-W. Liang, X. Cao, W.-J. Zhang, H.-T. Lin, F. Zhou, L.-F. Chen, S.-H. Yu, Robust and Highly Efficient Free-Standing Carbonaceous Nanofiber Membranes for Water Purification, *Advanced Functional Materials*, 21 (2011) 3851–3858.
62. M. Foroutan, V.F. Naeini, M. Ebrahimi, Carbon Nanotubes Encapsulating Fullerene as Water Nano-Channels with Distinctive Selectivity: Molecular Dynamics Simulation, *Applied Surface Science*, 489 (2019) 198–209.
63. M. Razmkhah, A. Ahmadpour, M.T.H. Mosavian, F. Moosavi, What is the Effect of Carbon Nanotube Shape on Desalination Process? A Simulation Approach, *Desalination*, 407 (2017) 103–115.
64. M. Anafcheh, F. Ektefa, Cyclosulfurization of C_{60} and C_{70} Fullerenes: A DFT Study, *Structural Chemistry*, 26 (2015) 1115–1124.
65. Y. Pan, X. Liu, W. Zhang, Z. Liu, G. Zeng, B. Shao, Q. Liang, Q. He, X. Yuan, D. Huang, M. Chen, Advances in Photocatalysis Based on Fullerene C_{60} and Its Derivatives: Properties, Mechanism, Synthesis, and Applications, *Applied Catalysis B: Environmental*, 265 (2020) 118579.
66. L. Brunet, D.Y. Lyon, E.M. Hotze, P.J.J. Alvarez, M.R. Wiesner, Comparative Photoactivity and Antibacterial Properties of C_{60} Fullerenes and Titanium Dioxide Nanoparticles, *Environmental Science & Technology*, 43 (2009) 4355–4360.
67. K.D. Pickering, Photochemistry and Environmental Applications of Water-Soluble Fullerene Compounds, in: Civil and Environmental Engineering, Rice University, Houstan, 2005.
68. R. Lucena, B.M. Simonet, S. Cárdenas, M. Valcárcel, Potential of Nanoparticles in Sample Preparation, *Journal of Chromatography A*, 1218 (2011) 620–637.
69. O.V. Alekseeva, N.A. Bagrovskaya, A.V. Noskov, Sorption of Heavy Metal Ions by Fullerene and Polystyrene/Fullerene Film Compositions, *Protection of Metals and Physical Chemistry of Surfaces*, 52 (2016) 443–447.

70. V.V. Samonin, V.Y. Nikonova, M.L. Podvyaznikov, Carbon Adsorbents on the Basis of the Hydrolytic Lignin Modified with Fullerenes in Producing, *Russian Journal of Applied Chemistry*, 87 (2014) 190–193.
71. V.V. Samonin, V.Y. Nikonova, M.L. Podvyaznikov, Sorption Properties of Fullerene-Modified Activated Carbon with Respect to Metal Ions, *Protection of Metals*, 44 (2008) 190–192.
72. A. Aqel, K.M.M.A. El-Nour, R.A.A. Ammar, A. Al-Warthan, Carbon Nanotubes, Science and Technology Part (I) Structure, Synthesis and Characterisation, *Arabian Journal of Chemistry*, 5 (2012) 1–23.
73. A. Eatemadi, H. Daraee, H. Karimkhanloo, M. Kouhi, N. Zarghami, A. Akbarzadeh, M. Abasi, Y. Hanifehpour, S.W. Joo, Carbon Nanotubes: Properties, Synthesis, Purification, and Medical Applications, *Nanoscale Research Letters*, 9 (2014) 393.
74. N. Saifuddin, A.Z. Raziah, A. R. Junizah, Carbon Nanotubes: A Review on Structure and Their Interaction with Proteins, *Journal of Chemistry*, 2013 (2013) 676815.
75. Z. Shi, Y. Lian, F. Liao, X. Zhou, Z. Gu, Y. Zhang, S. Iijima, Purification of Single-Wall Carbon Nanotubes, *Solid State Communications*, 112 (1999) 35–37.
76. J.H. Lehman, M. Terrones, E. Mansfield, K.E. Hurst, V. Meunier, Evaluating the Characteristics of Multiwall Carbon Nanotubes, *Carbon*, 49 (2011) 2581–2602.
77. A. Raza, U. Qumar, J. Hassan, M. Ikram, A. Ul-Hamid, J. Haider, M. Imran, S. Ali, A Comparative Study of Dirac 2D Materials, TMDCs and 2D Insulators with Regard to their Structures and Photocatalytic/Sonophotocatalytic Behavior, *Applied Nanoscience*, 10 (2020) 3875–3899.
78. J.O. Marques Neto, C.R. Bellato, C.H.F.d. Souza, R.C.d. Silva, P.A. Rocha, Synthesis, Characterization and Enhanced Photocatalytic Activity of Iron Oxide/Carbon Nanotube/Ag-doped TiO_2 Nanocomposites, *Journal of the Brazilian Chemical Society*, 28 (2017) 2301–2312.
79. D. Zhang, B. Pan, M. Wu, B. Wang, H. Zhang, H. Peng, D. Wu, P. Ning, Adsorption of Sulfamethoxazole on Functionalized Carbon Nanotubes as Affected by Cations and Anions, *Environmental Pollution*, 159 (2011) 2616–2621.
80. L. Yanyan, T.A. Kurniawan, A.B. Albadarin, G. Walker, Enhanced Removal of Acetaminophen from Synthetic Wastewater Using Multi-Walled Carbon Nanotubes (MWCNTs) Chemically Modified with NaOH, HNO_3/H_2SO_4, Ozone, and/or Chitosan, *Journal of Molecular Liquids*, 251 (2018) 369–377.
81. F. Yu, S. Sun, S. Han, J. Zheng, J. Ma, Adsorption Removal of Ciprofloxacin by Multi-Walled Carbon Nanotubes with Different Oxygen Contents from Aqueous Solutions, *Chemical Engineering Journal*, 285 (2016) 588–595.
82. H. Zhao, X. Liu, Z. Cao, Y. Zhan, X. Shi, Y. Yang, J. Zhou, J. Xu, Adsorption Behavior and Mechanism of Chloramphenicols, Sulfonamides, and Non-antibiotic Pharmaceuticals on Multi-walled Carbon Nanotubes, *Journal of Hazardous Materials*, 310 (2016) 235–245.
83. M.B. Ahmed, J.L. Zhou, H.H. Ngo, W. Guo, Adsorptive Removal of Antibiotics from Water and Wastewater: Progress and Challenges, *Science of The Total Environment*, 532 (2015) 112–126.
84. M. Keiluweit, M. Kleber, Molecular-Level Interactions in Soils and Sediments: The Role of Aromatic π-Systems, *Environmental Science & Technology*, 43 (2009) 3421–3429.
85. N. Chakrapani, Y.M. Zhang, S.K. Nayak, J.A. Moore, D.L. Carroll, Y.Y. Choi, P.M. Ajayan, Chemisorption of Acetone on Carbon Nanotubes, *The Journal of Physical Chemistry B*, 107 (2003) 9308–9311.
86. S.B. Fagan, A.G. Souza Filho, J.O.G. Lima, J.M. Filho, O.P. Ferreira, I.O. Mazali, O.L. Alves, M.S. Dresselhaus, 1,2-Dichlorobenzene Interacting with Carbon Nanotubes, *Nano Letters*, 4 (2004) 1285–1288.

87. W. Chen, L. Duan, L. Wang, D. Zhu, Adsorption of Hydroxyl- and Amino-Substituted Aromatics to Carbon Nanotubes, *Environmental Science & Technology*, 42 (2008) 6862–6868.
88. D. Zhu, J.J. Pignatello, Characterization of Aromatic Compound Sorptive Interactions with Black Carbon (Charcoal) Assisted by Graphite as a Model, *Environmental Science & Technology*, 39 (2005) 2033–2041.
89. H. Wang, L. Zhang, Z. Chen, J. Hu, S. Li, Z. Wang, J. Liu, X. Wang, Semiconductor Heterojunction Photocatalysts: Design, Construction, and Photocatalytic Performances, *Chemical Society Reviews*, 43 (2014) 5234–5244.
90. N. Serpone, A.V. Emeline, Semiconductor Photocatalysis – Past, Present, and Future Outlook, *The Journal of Physical Chemistry Letters*, 3 (2012) 673–677.
91. C.W. Tan, K.H. Tan, Y.T. Ong, A.R. Mohamed, S.H.S. Zein, S.H. Tan, Energy and Environmental Applications of Carbon Nanotubes, *Environmental Chemistry Letters*, 10 (2012) 265–273.
92. M. Shaban, A.M. Ashraf, M.R. Abukhadra, TiO_2 Nanoribbons/Carbon Nanotubes Composite with Enhanced Photocatalytic Activity; Fabrication, Characterization, and Application, *Scientific Reports*, 8 (2018) 781.
93. S.K. Sharma, R. Gupta, G. Sharma, K. Vemula, A.R. Koirala, N.K. Kaushik, E.H. Choi, D.Y. Kim, L.P. Purohit, B.P. Singh, Photocatalytic Performance of Yttrium-Doped CNT-ZnO Nanoflowers Synthesized from Hydrothermal Method, *Materials Today Chemistry*, 20 (2021) 100452.
94. M.S. Mauter, M. Elimelech, Environmental Applications of Carbon-Based Nanomaterials, *Environmental Science & Technology*, 42 (2008) 5843–5859.
95. R.K. Gupta, G.J. Dunderdale, M.W. England, A. Hozumi, Oil/Water Separation Techniques: A Review of Recent Progresses and Future Directions, *Journal of Materials Chemistry A*, 5 (2017) 16025–16058.
96. H.-A. Park, S. Liu, P.A. Salvador, G.S. Rohrer, M.F. Islam, High Visible-Light Photochemical Activity of Titania Decorated on Single-Wall Carbon Nanotube Aerogels, *RSC Advances*, 6 (2016) 22285–22294.
97. Z. Xu, H. Liu, J. Niu, Y. Zhou, C. Wang, Y. Wang, Hydroxyl Multi-Walled Carbon Nanotube-Modified Nanocrystalline PbO_2 Anode for Removal of Pyridine from Wastewater, *Journal of Hazardous Materials*, 327 (2017) 144–152.
98. Q. Duan, J. Lee, Y. Liu, H. Qi, Preparation and Photocatalytic Performance of MWCNTs/TiO_2 Nanocomposites for Degradation of Aqueous Substrate, *Journal of Chemistry*, 2016 (2016) 1262017.
99. D. Cohen-Tanugi, J.C. Grossman, Water Permeability of Nanoporous Graphene at Realistic Pressures for Reverse Osmosis Desalination, *The Journal of Chemical Physics*, 141 (2014) 074704.
100. A. Aghigh, V. Alizadeh, H.Y. Wong, M.S. Islam, N. Amin, M. Zaman, Recent Advances in Utilization of Graphene for Filtration and Desalination of Water: A Review, *Desalination*, 365 (2015) 389–397.
101. D. Cohen-Tanugi, J.C. Grossman, Water Desalination Across Nanoporous Graphene, *Nano Letters*, 12 (2012) 3602–3608.
102. M. Ikram, A. Raza, M. Imran, A. Ul-Hamid, A. Shahbaz, S. Ali, Hydrothermal Synthesis of Silver Decorated Reduced Graphene Oxide (rGO) Nanoflakes with Effective Photocatalytic Activity for Wastewater Treatment, *Nanoscale Research Letters*, 15 (2020) 95.
103. U. Qumar, J. Hassan, S. Naz, A. Haider, A. Raza, A. Ul-Hamid, J. Haider, I. Shahzadi, I. Ahmad, M. Ikram, Silver Decorated 2D nanosheets of GO and MoS_2 Serve as Nanocatalyst for Water Treatment and Antimicrobial Applications as Ascertained with Molecular Docking Evaluation, *Nanotechnology*, 32 (2021) 255704.

104. A. Raza, U. Qumar, A. Haider, S. Naz, J. Haider, A. Ul-Hamid, M. Ikram, S. Ali, S. Goumri-Said, M.B. Kanoun, Liquid-Phase Exfoliated MoS_2 Nanosheets Doped with p-type Transition Metals: A Comparative Analysis of Photocatalytic and Antimicrobial Potential Combined with Density Functional Theory, *Dalton Transactions*, 50 (2021) 6598–6619.
105. A.K. Geim, K.S. Novoselov, The Rise of Graphene, *Nat Mater*, 6 (2007) 183–191.
106. Y. Liu, X. Chen, Mechanical Properties of Nanoporous Graphene Membrane, *Journal of Applied Physics*, 115 (2014) 034303.
107. L.K. Putri, B.-J. Ng, W.-J. Ong, H.W. Lee, W.S. Chang, S.-P. Chai, Heteroatom Nitrogen- and Boron-Doping as a Facile Strategy to Improve Photocatalytic Activity of Standalone Reduced Graphene Oxide in Hydrogen Evolution, *ACS Applied Materials & Interfaces*, 9 (2017) 4558–4569.
108. C. Byrne, G. Subramanian, S.C. Pillai, Recent Advances in Photocatalysis for Environmental Applications, *Journal of Environmental Chemical Engineering*, 6 (2018) 3531–3555.
109. G. Wei, X. Quan, S. Chen, H. Yu, Superpermeable Atomic-Thin Graphene Membranes with High Selectivity, *ACS Nano*, 11 (2017) 1920–1926.
110. S. Chowdhury, R. Balasubramanian, Graphene/Semiconductor Nanocomposites (GSNs) for Heterogeneous Photocatalytic Decolorization of Wastewaters Contaminated with Synthetic Dyes: A Review, *Applied Catalysis B: Environmental*, 160–161 (2014) 307–324.
111. A. Fujishima, K. Honda, Electrochemical Photolysis of Water at a Semiconductor Electrode, *Nature*, 238 (1972) 37–38.
112. V. Štengl, S. Bakardjieva, T.M. Grygar, J. Bludská, M. Kormunda, TiO_2-Graphene Oxide Nanocomposite as Advanced Photocatalytic Materials, *Chemistry Central Journal*, 7 (2013) 41.
113. G. Lui, J.-Y. Liao, A. Duan, Z. Zhang, M. Fowler, A. Yu, Graphene-Wrapped Hierarchical TiO_2 Nanoflower Composites with Enhanced Photocatalytic Performance, *Journal of Materials Chemistry A*, 1 (2013) 12255–12262.
114. X. Yin, H. Zhang, P. Xu, J. Han, J. Li, M. He, Simultaneous N-doping of Reduced Graphene Oxide and TiO_2 in the Composite for Visible Light Photodegradation of Methylene Blue with Enhanced Performance, *RSC Advances*, 3 (2013) 18474–18481.
115. Y. Xu, Z. Liu, X. Zhang, Y. Wang, J. Tian, Y. Huang, Y. Ma, X. Zhang, Y. Chen, A Graphene Hybrid Material Covalently Functionalized with Porphyrin: Synthesis and Optical Limiting Property, *Advanced Materials*, 21 (2009) 1275–1279.
116. P. Wang, Y. Zhai, D. Wang, S. Dong, Synthesis of Reduced Graphene Oxide-Anatase TiO_2 Nanocomposite and Its Improved Photo-Induced Charge Transfer Properties, *Nanoscale*, 3 (2011) 1640–1645.
117. J. Zhang, Z. Xiong, X.S. Zhao, Graphene–Metal–Oxide Composites for the Degradation of Dyes Under Visible Light Irradiation, *Journal of Materials Chemistry*, 21 (2011) 3634–3640.
118. Y.H. Ng, I.V. Lightcap, K. Goodwin, M. Matsumura, P.V. Kamat, To What Extent Do Graphene Scaffolds Improve the Photovoltaic and Photocatalytic Response of TiO_2 Nanostructured Films? *The Journal of Physical Chemistry Letters*, 1 (2010) 2222–2227.
119. B. Neppolian, A. Bruno, C.L. Bianchi, M. Ashokkumar, Graphene Oxide Based Pt–TiO_2 Photocatalyst: Ultrasound Assisted Synthesis, Characterization and Catalytic Efficiency, *Ultrasonics Sonochemistry*, 19 (2012) 9–15.
120. K.K. Manga, Y. Zhou, Y. Yan, K.P. Loh, Multilayer Hybrid Films Consisting of Alternating Graphene and Titania Nanosheets with Ultrafast Electron Transfer and Photoconversion Properties, *Advanced Functional Materials*, 19 (2009) 3638–3643.

121. J. Li, S.l. Zhou, G.-B. Hong, C.-T. Chang, Hydrothermal Preparation of P25–Graphene Composite with Enhanced Adsorption and Photocatalytic Degradation of Dyes, *Chemical Engineering Journal*, 219 (2013) 486–491.
122. F. Li, P. Du, W. Liu, X. Li, H. Ji, J. Duan, D. Zhao, Hydrothermal Synthesis of Graphene Grafted Titania/Titanate Nanosheets for Photocatalytic Degradation of 4-Chlorophenol: Solar-Light-Driven Photocatalytic Activity and Computational Chemistry Analysis, *Chemical Engineering Journal*, 331 (2018) 685–694.
123. H. Zhang, X. Wang, N. Li, J. Xia, Q. Meng, J. Ding, J. Lu, Synthesis and Characterization of TiO_2/Graphene Oxide Nanocomposites for Photoreduction of Heavy Metal Ions in Reverse Osmosis Concentrate, *RSC Advances*, 8 (2018) 34241–34251.

7 Magnetic Materials and Their Application in Water Treatment

Muhammad Ikram, Syed Ossama Ali Ahmad, and Atif Ashfaq
Government College University Lahore

Anwar Ul-Hamid
King Fahd University of Petroleum & Minerals

CONTENTS

DOI: 10.1201/9781003167327-8

7.1 INTRODUCTION

In an interview telecast by BBC TV (Fun to image, 1983), Richard P. Feynman (American Physicist) stated that explaining the "magnetic phenomenon" in a simple way is very difficult [1]. Nevertheless, magnetism has been a key phenomenon in the advancement of human technology ever since the advent of civilization and has played a central role in the survival of all life forms, as the Earth's magnetic field protects us from the hazardous solar wind coming from outer space. Especially for the past few decades, numerous works have been reported by researchers from all over the world on magnetic materials being used for the fabrication of smart novel devices for technological advancement in various fields like catalysis, robotics, biomedicine, spintronics, data storage and engineering to name a few [2–8]. This signifies the importance of comprehending and exploiting the complex phenomenon of magnetism for a better and sustainable future.

Abrupt population growth and rapid industrialization have caused the introduction of toxic compounds into the environment in large quantities. This wild discharge of hazardous chemicals into the water, air and soils has posed a serious threat to all lifeforms and environment of our planet. Moreover, continuously depleting fresh water resources due to excessive use of water for the last century is of great concern nowadays [9]. Additionally, the use of conventional water treatment methods has not proven to be much effective and yielding anymore, due to the addition of emerging pollutants that are very difficult to remove from water. Such pollutants widely include dyes, paints, cosmetics, coatings, medicines and pharmaceuticals that have proven to be extremely life-threatening due to their carcinogenic nature and other physicochemical properties [10–12]. Hence, novel and innovative procedures involving magnetic and magnet-sensitive materials are being adopted with environmental and economic benefits as compared to other conventional methods [13,14].

This chapter deals with the study of magnetic materials, synthesis and their application in the treatment of polluted water. A compact description of various types of magnetic materials, numerous synthesis approaches for their preparation and useful characterization techniques to study their magnetic properties has been provided for a better understanding. Lastly, a promising application of these magnetic materials for the treatment of water, future perspective and challenges have been discussed in detail.

7.2 MAGNETISM

7.2.1 Basic Concepts and Definition

On a basic level, magnetism is defined as a physical phenomenon governed by magnetic fields (produced via electric current), which regulate the orientation and attraction/repulsion effects on iron and other magnetizable materials. Moving charges in materials produce both electric and magnetic fields; hence, these two phenomena are often correlated with each other (electromagnetism). Analogous to the electric field, the magnetic field also has direction and magnitude represented by magnetic field lines generating from one end to the other. This implies that magnetic materials are also polarized and have two opposite ends (North and South Poles) just

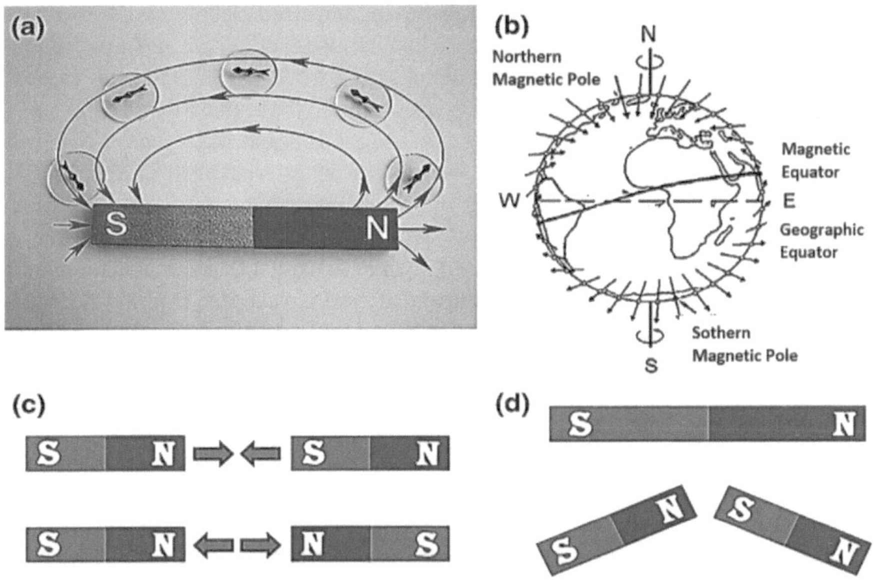

FIGURE 7.1 (a) Schematic representation of magnetic field lines and a compass pointing toward the direction of the field. (b) The magnetic field of the Earth. Magnetic field lines, magnetic equator and poles are marked. (c) Schematic representation of both attraction and repulsion phenomena between magnetic poles. (d) Schematic representation of magnetic poles in a broken magnet. (Reproduced from Ref. [15].)

like electric dipoles. So, by analogy, the like poles repel (N–N poles will repel) and the unlike poles attract (N–S poles will attract) each other. However, unlike electric charges, it is impossible to separate magnetic poles, i.e., magnetic monopole does not exist (Figure 7.1).

7.2.2 Types of Magnetism

The "Modern Theory of Magnetism" states that the motion of electrons is the basic reason for the magnetic properties of materials. Technically, all materials in nature are magnetic but the interaction between magnetic moments of electrons (weak or strong) decides the magnetic behavior of materials. In some materials, this interaction is very strong, and hence, they show pronounced magnetic behavior, while in some cases, this interaction is very weak, and hence the materials show weak magnetic response. Based on the type of interaction present, six different types of magnetic behavior have been studied up till today which are discussed below.

Diamagnetism is a weak form of magnetism that arises due to the non-uniform arrangement of magnetic moment (unpaired electrons are not present) under an external magnetic field. When placed under a magnetic field, diamagnetic materials show negative susceptibility and magnetization hence repelling the applied external field [16].

Paramagnetism is shown by materials having unpaired electrons and give a net magnetic moment in external magnetic fields. Paramagnetic materials give slightly positive values of susceptibility and magnetization when placed under an external magnetic field hence aligning themselves according to the applied field [17].

Ferromagnetism is an important form of magnetism and is present in materials that show spontaneous magnetization even when no external field is present. The spontaneous alignment is ascribed to the Exchange Energy phenomenon where electrons with similar spin states are aligned together to lower electrostatic energy between unpaired electrons. When placed in an external field, ferromagnetic materials align themselves with the applied field and after removal of the external field, they stay aligned (permanently magnetized). Also, above a certain temperature (Curie temperature) ferromagnetic materials lose their ferromagnetism and become paramagnetic [17].

Antiferromagnetism is shown by materials in which there is an equal anti-alignment of neighboring electrons' spins (magnetic moments are canceled out) hence producing zero net magnetization. Just like ferromagnetic materials, these materials also show paramagnetic behavior after a certain temperature i.e., Neel temperature [17].

Ferrimagnetism is another type of magnetism that is analogous to antiferromagnetism such that ferromagnetic materials also show anti-alignment of electrons' spins but it is not equal to the aligned spins. As a result, magnetic moments do not completely cancel out each other resulting in a net magnetization. These materials also possess spontaneous magnetization and are converted into paramagnetic materials at the Curie point [18].

Superparamagnetism is a typical form of magnetization shown by the single domain (nanoparticles) ferromagnetic materials below Curie temperature. Such

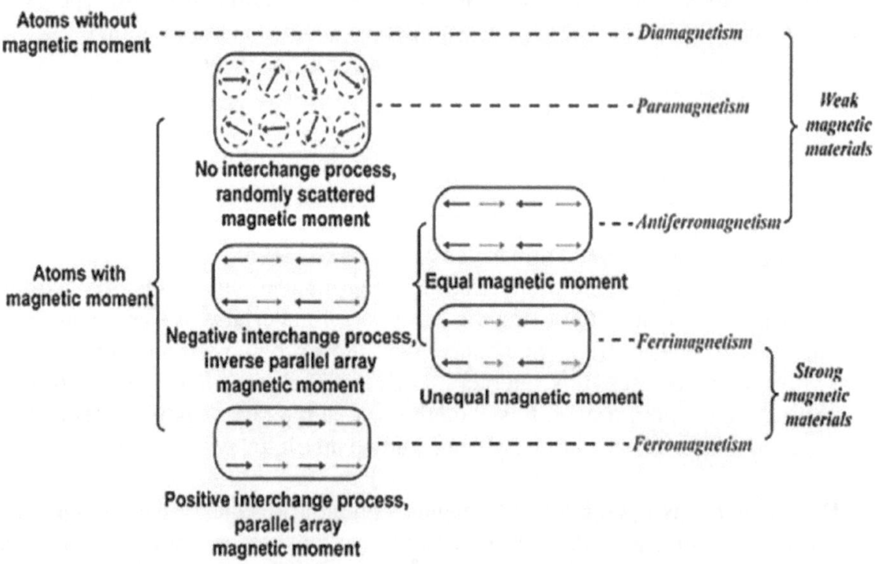

FIGURE 7.2 Schematic representation of different types of magnetism. (Reproduced from Ref. [20].)

materials have temperature-fluctuated magnetism and show zero net magnetization when no external magnetic field is available. In the presence of an external field, these materials show exceptional susceptibility as compared to paramagnetic materials and hence they are named "superparamagnetic" [19].

7.3 DETERMINATION OF MAGNETIC PROPERTIES

Since some knowledge of the magnetism phenomenon has been developed, the characterization techniques to study the magnetic properties of materials are discussed in this section.

7.3.1 Magnetic Force Microscopy (MFM)

MFM is a typical scanning probe microscopy used for magnetic domain analysis and magnetic field imaging on the surface of magnetic (generally ferromagnetic) materials [21]. A sharp magnetic tip (made up of highly coercive materials) with constant magnetization is used to scan the surface of the sample. This spectroscopy involves two major steps (Figure 7.3) where firstly, AFM (Atomic Force Microscopy) is conducted to obtain topographic information of the surface, while in the second step, MFM is executed from a fixed height (lift height: 10–200 nm range) away from sample's surface [22]. This lift height eliminates the electrostatic interactions and only magnetostatic interactions are recorded via long-range tip-surface interaction. In this way, the morphological properties of magnetic domains present on the surface of the sample can be easily determined [23].

7.3.2 Magnetization Hysteresis (M–H or B–H Curves)

Magnetization hysteresis is the most significant and widely used characterization technique for the evaluation of the magnetic properties of materials. The tool used for such analysis is called a magnetometer that measures the magnetic moment of materials and their dependency on various controllable parameters. Graphs are plotted for total magnetization induced within the sample (M) and magnetic field intensity or magnetic induction (B), which reveal the magnetic behavior of the sample (Figure 7.4). Remnant magnetization, saturation magnetization and coercivity are the key factors that determine the type of materials (soft or hard materials) and their magnetic response [24]. The determination and analysis of these magnetization curves are of utmost importance for the absolute characterization of magnet-sensitive materials.

The following section comprises various synthesis routes that have been frequently adopted so far for the preparation of magnetic materials (mostly iron-based magnetic materials), their advantages, disadvantages and feasibility.

7.4 SYNTHESIS ROUTES

In this section, some of the most widely used fabrication routes for the synthesis of magnetic NPs have been discussed briefly. The advantages and drawbacks of each method have also been provided for comparative analysis.

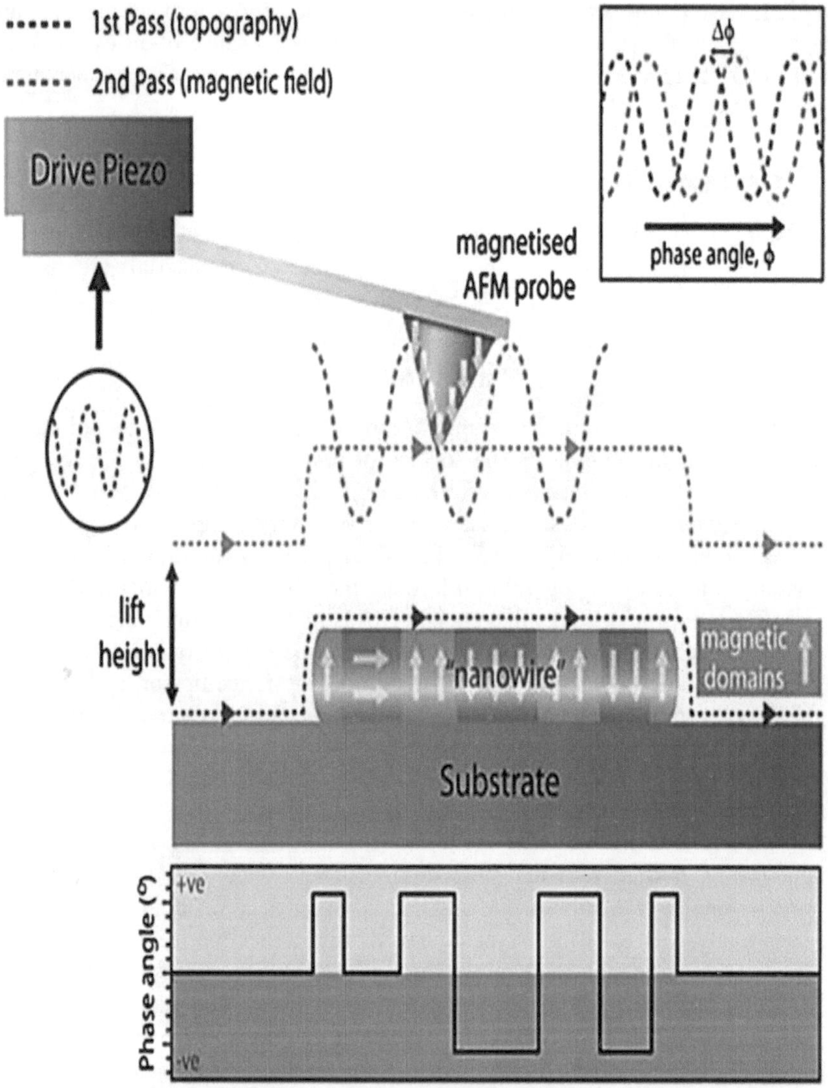

FIGURE 7.3 Schematic representation of MFM principles applied to a magnetic nanowire. Magnetic interactions between the AFM probe and the sample are monitored through recording changes in the phase lag of the probe relative to the drive voltage signal. (Reproduced from Ref. [22].)

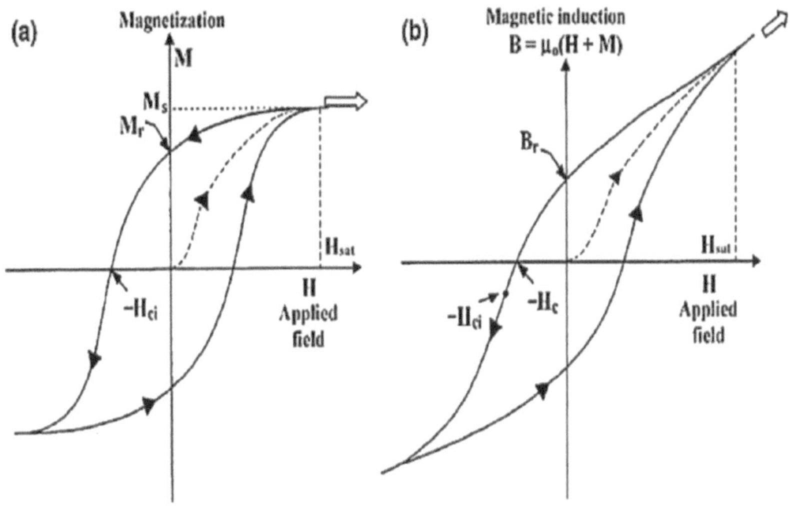

FIGURE 7.4 Example of hysteresis curves for a ferromagnetic material. (a) M versus H curve, where Ms is the saturation magnetization, Mr is the remnant magnetization at H = 0, and Hci is the intrinsic coercivity (i.e., the reverse field that reduces M to 0). (b) B versus H curve, where Br is the remnant induction (or remanence) at H = 0, and Hc is the coercivity (i.e., the reverse field required to reduce B to 0). (Note that Hci ≠ Hc for hard magnets, while Hci ≈ Hc for soft magnets. Reproduced from Ref. [24].)

7.4.1 Co-Precipitation

This method involves intimate mixing of both ferrous and ferric ions at room temperature or elevated temperatures with adequate molar concentration of both species dissolved in basic medium. The reaction mechanism involved in the process is as follows:

$$Fe^{2+} + 2Fe^{3+} + 8OH^- \leftrightarrow Fe(OH)_2 + 2Fe(OH)_3 \rightarrow Fe_3O_4 + 4H_2O.$$

Apart from molar concentration, the size of nanoparticles (NPs) produced via this method depends upon pH of the medium, reaction temperature, types of precursors, stirring rate, ionic strength of medium and dropping rate for maintaining pH of the solution [25–27]. Co-precipitation is widely used due to its unrivaled advantage of high conversion yield but there are a few drawbacks, such as high pH value and poor control over particle's size and morphology [28].

7.4.2 Thermal Decomposition

This route involves the thermal treatment of iron-containing salts in non-aqueous medium at elevated temperatures in an open or closed vessel to prepare magnetic NPs. Salt precursors frequently used in thermal decomposition include $Fe_3(CO)_{12}$, $Fe(CO)_5$, $Fe((C_5H_5)_2)$, $Fe_4[Fe(CN)_6.14H_2O]$ (Prussian Blue) and $Fe(acac)_3$ (acac = acetylacetonate). Organic stabilizers are also added sometimes to slow down the nucleation to control the size and growth of NPs. This procedure is significantly adopted to synthesize NPs of different shapes like core–shell composites or nanotubes [28]. The thermal decomposition technique provides better control of the size and morphology of NPs with improved crystallinity due to heat treatment. However, this procedure suffers from drawbacks of the hydrophobic surface of particles and high-temperature values.

7.4.3 Hydrothermal and Solvothermal Synthesis

Hydrothermal (Solvent is water) or Solvothermal (Solvent other than solvent) is a wet chemical route that involves crystallization from an aqueous solution at high temperature (130–250 °C) in a sealed container at high pressure (0.3–4.0 MPa) [29]. Advantages of this method are the high crystallinity of magnetic products and good control over the shape of the particle by varying both concentration and reaction time [30].

7.4.4 Microemulsion

This method involves the addition of iron oxide precursors in a thermodynamically stable mixture of two immiscible liquids (water and oil generally) along with the addition of various surfactants. The amphiphilic nature of the surfactants derives them to self-organize on the water/oil interface, forming different kinds of supramolecular orders (lamellar phases, cylindrical, spherical, worm and onion-like structures and reverse micelles, etc.) [31,32]. This method gives better control over the morphology of the prepared magnetic NPs, however, a high degree of aggregation is a severe drawback that demands other post-synthetic procedures to overcome.

7.4.5 Sol–Gel Synthesis

A wet chemical route where the precursors are added into a colloidal solution (generally referred to as "sol") and a polycondensation phenomenon results in the formation of an integrated structure (generally referred to as "gel") of NPs or bulk materials. Initially, the precursors (iron alkoxides, nitrates, acetates, chloride salts, etc.) are dissolved into the sol at room temperature and products are generated after hydrolysis and condensation reactions. The fabricated products are later calcined at high temperatures to achieve desired crystallinity and purity [33]. Highly hydrophilic iron NPs can be fabricated via this approach, which makes them easily dissolvable in many solvents; however, iron alkoxides are very costly and many alcoholic byproducts form during the synthesis, which are main disadvantages of this approach [34].

7.4.6 Microwave-Assisted Synthesis

This method involves the fabrication of magnetic NPs via both aqueous and non-aqueous mediums with the help of microwaves. Upon irradiating with microwaves, the molecules get excited and the alignment of dipoles along the external field generates an intense heating effect [35]. In contrast to other methods, this technique requires a very less amount of treatment time and energy consumption and allows the preparation of magnetic NPs through organic solvents. However, the produced NPs through this method have been reported to be relatively less reactive than other synthesis routes.

These were some significant approaches that are frequently adopted by researchers to produce magnetic NPs worldwide. Other than these methods, spray pyrolysis, electrochemical synthesis and bacterial synthesis are also used for this purpose.

7.5 MAGNETIC MATERIALS AND WASTEWATER TREATMENT

In this section, the application of magnetic materials in the treatment of wastewater is discussed. The removal of organic pollutants in water via photocatalysis and adsorption techniques deploying magnetic materials are focused. Furthermore, the magnetic materials have been categorized into two major groups including iron-based magnetic materials and titania-based magnetic materials. Some significant examples regarding the application of these magnet-sensitive materials for wastewater treatment have been discussed in detail.

7.5.1 Iron-Based Magnetic Materials

Fenton reactions (a particular class of advanced oxidation processes) have been widely explored by researchers for the treatment of polluted water. Fenton reactions involve the degradation of various organic pollutants via different oxidizing species (especially OH radicals) in water or other solvents [36]. In an acidic environment, highly reactive •OH radical species are generated after the interaction of ferrous ions with H_2O_2 as shown in the below equation:

$$\mathrm{Fe(II)} + \mathrm{H_2O_2} \rightarrow \mathrm{Fe(III)} + \mathrm{OH^-} + \mathrm{OH}.$$

The highly reactive radical species generated at acidic pH interacts with the surrounding organic pollutants to degrade them into non-toxic products, while Fe(II) is regenerated after the reduction of Fe(III) atoms. However, this phenomenon generally occurs in the presence of light that is the basic essence of the photocatalytic degradation of organic pollutants.

Magnetite has been successfully used for the degradation of phenol in water, following the photo-Fenton process. Numerous samples of magnetite (S_1, S_2, S_3 and S_4) with varying Fe(II)/Fe(III) contents of 0.34, 0.42, 0.43 and 0.30, respectively were fabricated as iron sources to effectively oxidize the phenol [37]. No activity was shown by any sample in the dark while a complete degradation of phenol was observed upon light irradiation for almost all samples. The highest activity was shown by S_2 and S_3 due to their higher Fe(II) content in contrast to other samples. Moreover, the addition of hydroquinone (HQ) boosted the activity of samples, both in light and in dark as well, due to the accelerated reduction of Fe(III) into Fe(II). The possibility of using external magnets to remove the magnet-sensitive photocatalyst opened promising opportunities for recycling the photocatalyst. The reaction kinetics of all samples in the degradation of phenol are given in Figure 7.5.

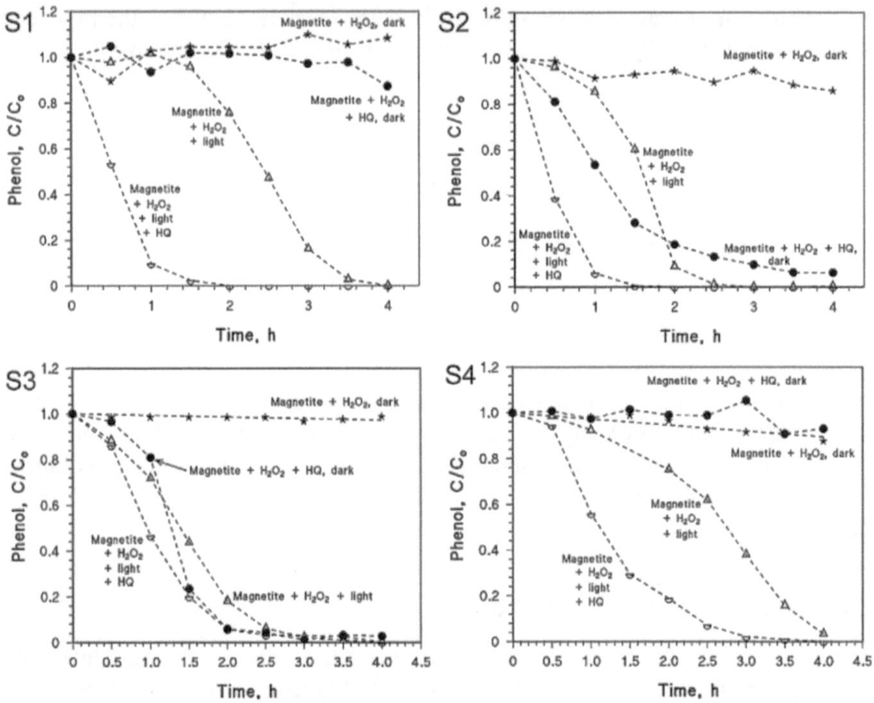

FIGURE 7.5 Time trend of phenol (initial concentration 0.1 mM, pH ~ 3 by $HClO_4$) under photo-Fenton conditions and in blank runs. (Reproduced from Ref. [37].)

Stabilizing the magnetic NPs is a very important factor for the effective treatment of polluted water and many studies have been reported in this regard. For instance, humic-like structures (BBS) have been reported as a stabilizing agent in the photo-Fenton reaction to degrade caffeine in slightly basic media (pH = 6). The performance of the hybrid NPs was compared with the bare magnetite sample and the results depicted an improved activity of hybrid samples. Moreover, the activity was reported to be dependent on the BBS concentration in the host material and the extent of Fe-leaching [38]. However, the approach suffers from the limitation of lower reusability due to the depletion of Fe reservoir. A shortcut route has been proposed to overcome this limitation, where Fe(II) ions are added in between the procedure to boost the capability of Fe-leaching [39].

Recently, the degradation of diphenhydramine with iron-based materials produced under different conditions was studied by Pastarana-Martinez and colleagues [40]. The varying synthesis conditions resulted in the formation of pristine magnetite, magnetite/maghemite and hematite structures. The highest degradation performance was shown by the pristine magnetite sample in an acidic medium, owing to higher Fe(II) content and lower iron-leaching. Doped magnetic materials have also been reported for the effective photocatalytic degradation of organic pollutants as well. For instance, Li et al. fabricated Zn-doped Fe_2O_4 NPs for the degradation of rhodamine B dye under UV–Vis light source. Complete degradation along with 23%

mineralization was reported for the prepared magnetic NPs within 5 hours of operation [41].

Recently, magnetic materials have received much attention to be utilized as adsorbates in water treatment, offering easier recovery due to magnet-sensitive nature [42]. A study was reported on the use of magnet-sensitive materials coated with carbon as adsorbent for the degradation of PAHs in water [3]. The removal efficiency of a fixed amount of carbon-coated magnetite adsorbent was studied against different PAHs like acenaphthene, phenanthrene and anthracene, etc. The removal efficiency was reported to be dependent on PAHs structure, hydrophobicity and the number of aromatic rings present. A microporous nanocomposite of carbon and iron (Fe/C) with superhydrophobic nature has also been reported for the effective removal of oil from the water via adsorption [43]. Another study has been reported for the effective removal of heavy atoms (Ni^{2+}, Pb^{2+} and Cd^{2+}) with the help of cyclodextrin/magnetite NPs. The carboxylic and hydroxyl functional groups offered by the cyclodextrin, significantly interact with the heavy metals resulting in better adsorption. Moreover, the competitive effect of various metals on the adsorption efficiency was also evaluated by utilizing binary and ternary metal solutions. A notable decrease in adsorption capacity for each metal was observed in the binary and ternary solutions in contrast to single metal solution. This decrement in the adsorption was accredited to the competition among the metals to acquire the active sites available on the adsorbent [44]. The percentage adsorption efficiency of each heavy metal in various solutions is given in Figure 7.6.

A new class of magnetic materials named zero-valent iron (ZVI) having unique magnetic properties and relatively higher Curie point are being extensively explored nowadays. They have proven to be a potential replacement for conventional magnetic

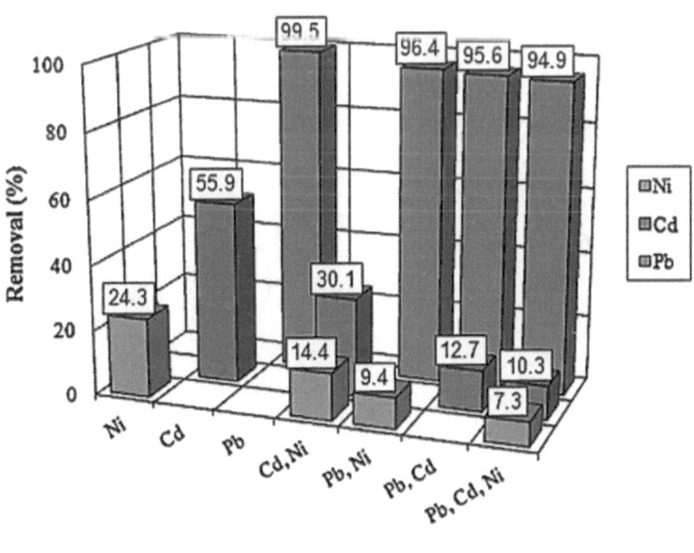

FIGURE 7.6 Percentage removal of heavy metals from single, binary and ternary mixtures by using cyclodextrin-magnetite NPs (each metal concentration: 100 mg L^{-1}, adsorbents: 120 mg, temperature: 25 C, pH: 5.5 and contact time: 2 hours). (Reproduced from Ref. [44].)

materials which suffers from the disadvantage of corrosion over time [45]. The nano-sized ZVI particles offer larger surface area and better reactivity that makes them highly desirable for pollutant removal in water [46]. The conventional ball milling or thermal reduction of Fe oxides in the presence of hydrogen flux is the synthetic routes to fabricate these ZVI NPs [47,48]. Higher adsorbing and reducing efficiency of these ZVI NPs make them a potential candidate for polluted water treatment [49]. A versatile and simple technique that has been adopted by many researchers is to hydraulically inject these NPs into treatment site, which gives opportunities for both surface and depth contaminated water treatment. Furthermore, this technique can be deployed to both mobile and immobile pollutant sites as well (Figure 7.7). For immobile sites, the upstream injected ZVI NPs degrade and immobilize the pollutants in water, while in mobile sites, the ZVI NPs are adsorbed onto an aquifer to form a filtration system [50]

The ZVI NPs have been extensively reported as potential adsorbents and photocatalysts to remove organic pollutants in the past few years. For instance, Jing et al. fabricated ZVI NPs for the photocatalytic reduction of organic dyes and reported degradation of

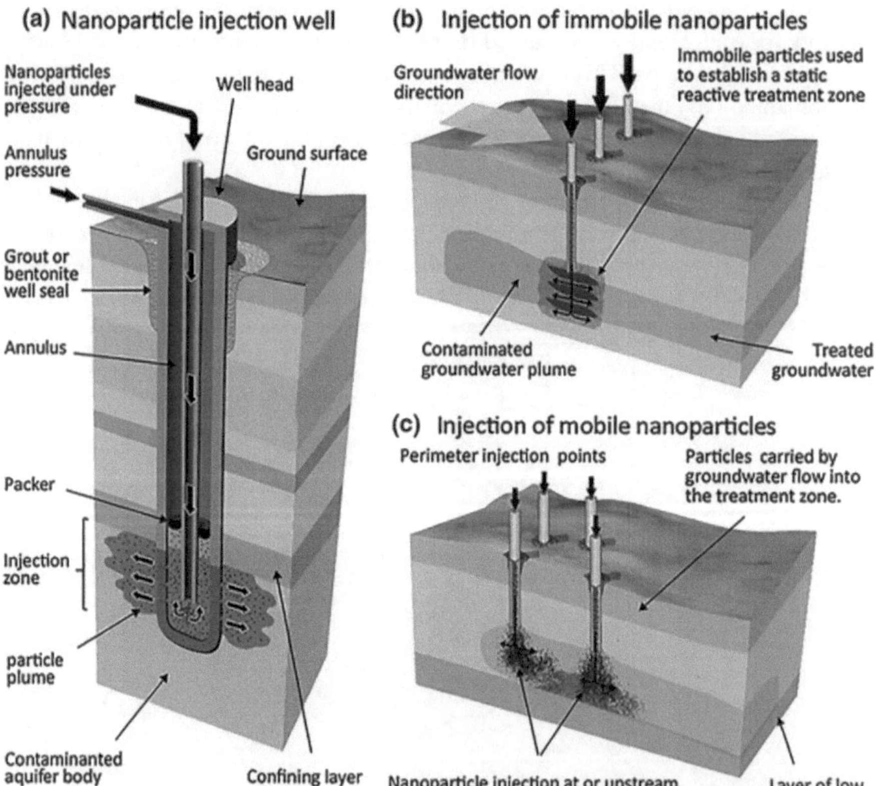

FIGURE 7.7 Schematic representation showing the layout of a typical NPs injection well (a), and the use of NPs injection technology for treating contaminant bodies that are either mobile (b) or immobile (c). (Reproduced from Ref. [50].)

90% within 1 hour of operation [51]. Chen and colleagues also reported dye degradation of 99.75% for the ZVI NPs in just 6 hours at basic pH [52]. Both studies revealed the independency of ZVI NPs degradation efficiency on H^+ atoms (acidic medium).

Ferrites are another class of magnetic materials that have been used for water treatment applications. They generally belong to the ceramic family with a general formula MFe_2O_4, where M denotes a metal other than Fe. Ferrites are further categorized as soft and hard ferrites depending on the extent of coercivity present in them. Soft ferrites with relatively lower coercivity include Ni–Zn and Mn–Zn, while hard ferrites having larger coercivity include Sr, Ba and Ni [53]. Usually, these materials are fabricated by thermally decomposing a mixture of magnetite and metal carbonate. For improved performance, ferrites are usually integrated with silica or polymers matrix that lessens the dissolution probability of metal components in the composites [54]. Ferrites and their magnetic composites are generally used as adsorbents in pollutant removal from water. Recently, many studies were reported on the adsorption of heavy metals (As, Hg, Pb, Cr, etc.) in polluted water with the help of ferrites [53,55]. Spherical Co-ferrites with higher surface area of 230 $m^2 g^{-1}$ have been recently reported for the adsorption of U(VI) ions from water. Outstanding results with an adsorption capacity of 170 mg g^{-1} were observed for the fabricated Co-ferrites [56]. In another study, MN-ferrites modified with chitosan were investigated for their adsorption abilities to remove Cu(II) ions from wastewater and excellent results were achieved [57]. A sorption capacity of 34.2 mg g^{-1} within 10 minutes of operation was reported for Zn-ferrites for the removal of Cr(VI) by Jia et al. The high porosity of the prepared adsorbents resulted in superior removal efficiency of the material [58].

7.5.2 TiO_2-Based Magnetic Materials

There are many magnetic NPs reported to date for the decomposition of organic pollutants and most of them are TiO_2 based, since it is chemically stable, abundantly available and has better oxidation efficiency. TiO_2 mainly possesses three natural forms named rutile, brookite and anatase. Anatase is mostly used for a photocatalytic activity that absorbs only 5%–8% of the light spectra coming from the Sun. Hence, numerous efforts are being made to extend the solar spectrum absorbance to increase the photocatalytic activity. In this section, three different configurations of titania-based magnetic materials (mostly photocatalysts) have been discussed including: the coating of magnet-sensitive core by TiO_2, which was initially used as a magnetic photocatalyst. Secondly, the core of magnetic material was coated with an interlayer element and TiO_2 as an outer-shell having configuration as Core@Coating@TiO_2. This effort significantly increases the efficacy of photocatalyst. Here, the purpose of utilizing interlayer element was to avoid recombination as well as dissolution of magnetic core. The third configuration is related to the surface decoration of the aforementioned photocatalyst with suitable metals and nonmetals for speedy reactions.

7.5.2.1 Core@TiO_2 Catalysts

As shown in Table 7.1, several structural configurations of TiO_2-based magnetic NPs applicability in wastewater treatment have been formulated. Multiple composites of

TABLE 7.1
Some Magnetic Materials Reported for the Degradation of Organic Pollutants in Water

Material	Synthesis Route	Characterization	Pollutant	Operational Conditions	Performance	Cycles	Ref.
$Fe_3O_4@TiO_2$	Fe_3O_4: Fe(II) and Fe(III) chlorides co-precipitation TiO_2: sonochemical method, titanium(IV) isopropoxide (TTIP) as Ti precursor	Size: 40 nm, Sat. mag.: 14 emu·g^{-1}	RhB	[RhB]=28.7 mg·L^{-1}, [MNPC]=0.455 g·L^{-1}, UV light	98.3% (75 minutes); Fe in solution not analyzed	6	[59]
$Fe_3O_4@TiO_2$	$Fe_3O_4@TiO_2$: sonochemical method using $FeSO_4·7H_2O$ and NaOH, TTIP as Ti precursor	Size: 50 nm, Sat. mag.: 57.0 emu·g^{-1}	MB	[MB]=1600 mg·L^{-1}, [MNPC]=50 g·L^{-1}, vis light	81% (6 minutes); Fe in solution not analyzed	6	[60]
$Fe_3O_4@SiO_2@TiO_2$	Fe_3O_4: high-temperature hydrolysis reaction using $FeCl_3$, NaOH, poly(acrylic acid) and diethylene glycol SiO_2: modified Stöber method, TEOS as Si source TiO_2: TBT as Ti precursor	Size: 338.7 nm, Surf. area: 35.5 m^2·g^{-1}	RhB	[RhB]=4.8 mg·L^{-1}, [MNPC]=0.5 g·L^{-1}, UV light	100% (40 minutes); Fe_3O_4 in solution = 0.0083% after 18 cycles	18	[66]
$Fe_3O_4@SiO_2@TiO_2$	Fe_3O_4: solvothermal reaction using $FeCl_3·6H_2O$, NaAc, tri-sodium citrate and ethylene glycol SiO_2: modified sol–gel process, TEOS as Si source TiO_2: $TiOSO_4$ as Ti precursor	Size: 325 nm, Surf. area: 211.37 m^2·g^{-1}, Sat. mag.: 38.7 emu·g^{-1}	RhB	[RhB]=9.6 mg·L^{-1}, [MNPC]=0.40 g·L^{-1}, UV light	100% (30 minutes); Fe in solution not analyzed	10	[67]
$Fe_3O_4@SiO_2@TiO_2$	Fe_3O_4: thermal decomposition of iron carboxylate salts SiO_2: water-in-oil reverse microemulsion, TEOS as Si source TiO_2: solvothermal method, TBT as Ti precursor	Size: 200 nm, Surf. area: 157.2 m^2·g^{-1}	MB	[MB]=10 mg·L^{-1}, [MNPC]=1 g·L^{-1}, UV light	100% (50 minutes); Fe in solution not analyzed	10	[68]

(Continued)

TABLE 7.1 (*Continued*)

Some Magnetic Materials Reported for the Degradation of Organic Pollutants in Water

Material	Synthesis Route	Characterization	Pollutant	Operational Conditions	Performance	Cycles	Ref.
Fe_3O_4@ TiO_2Carboxymethyl ß-cyclodextrin	Carboxymethyl-cyclodextrin-Fe_3O_4: solvothermal method using FeOOH powder, oleic acid, 1-octadecene and carboxymethyl ß-cyclodextrin (CMCD)	Size: (10-15) nm, Sat. mag.: 28 $emu{\cdot}g^{-1}$	BPA, DBP	[BPA]=[DBP]=20 $mg{\cdot}L^{-1}$, [MNPC]=1 $g{\cdot}L^{-1}$, UV light	100% (90 minutes); Fe in solution not analyzed	10	[61]
$Fe_{2.25}W_{0.75}O_4$	Hydrothermal method using $FeSO_4{\cdot}7H_2O$ as Fe source, $Na_2WO_4{\cdot}4H_2O$ as W source and sodium dodecyl benzene sulfonate (SDBS)	Size: 30–50 nm, Sat. mag.: 5.1 $emu{\cdot}g^{-1}$, Bandgap: 3.55 eV	MO	[MO]=10 mg L^{-1}, [MNPC]=1 g L^{-1}, UV light	91% (50 minutes)	5	[80]
Fe_3O_4@α-Fe_2O_3	Fe_3O_4: $FeSO_4$ as Fe source α-Fe_2O_4: surface oxidation method using $NaNO_3$ and KNO_3	Size: 50 nm, Sat. mag.: 48.7 $emu{\cdot}g^{-1}$	MO	[MO]=20 mg L^{-1}, [MNPC]=0.2 g L^{-1}, UV light	90.1% (120 minutes)	5	[81]
Fe_3O_4@C@ BiOCl	Fe_3O_4: co-precipitation method using $FeCl_3{\cdot}$ $6H_2O$ and $FeSO_4{\cdot}6H_2O$ C: glucose hydrothermal method BiOCl: solvothermal method using $Bi(NO_3)_3{\cdot}5H_2O$ as Bi source and ethylene glycol	Size: 4000 nm, Surf. area: 17 $m^2{\cdot}g^{-1}$, Bandgap: 3.25 eV	RhB	[RhB]=4.8 mg L^{-1}, [MNPC]=0.5 g L^{-1}, vis light	98% (30 minutes)	5	[82]

iron oxide with TiO_2 are also shown, where magnetic materials are utilized as the magnetic core of TiO_2. These reported magnetic NPs were used for dye degradation under UV light, approaching high efficacy in most cases [59–61]. Jing et al. used the magnetite-TiO_2 catalyst for degradation of carcinogenic compound and degraded it around ~90% within 2 hours [62]. Moreover, TiO_2 as a core has been tested for removing organic pollutants under UV light. For example, Cheng et al. used zinc-ferrite-TiO_2 for the removal of methyl orange (MO). This prepared substance absorbed the light toward the visible region and increased the efficacy to 90% within 3h, while bare TiO_2 degraded dye just up to 13% [63].

7.5.2.2 Core@Coating@TiO_2 Catalysts

All above-mentioned materials have shown an excellent degradation efficacy, but there could be a risk of core photodissolution that is caused by contact between titania and magnetic core. This process can generate recombination centers and also deteriorate the catalyst performance, as reported in [64,65]. So, there is a need for an insulating material that could avoid the core dissolution. An intermediate coating of SiO_2 has been reported to evade the interfacial connection between both materials. The conduction band of SiO_2 is significantly higher than TiO_2 while a lower valence band. This semiconductor restricts the electron injection from TiO_2 to iron oxide and improves the degradation efficacy of the photocatalyst [66–68].

To further increase the photocatalytic activity, mesoporous TiO_2 can be utilized, as reported in [67], Fe_3O_4@SiO_2@TiO_2 MNPC degraded rhodamine B (RhB) completely within 30 min. This catalyst improved its performance after using mesoporous TiO_2 due to the high surface area provided by titania. Furthermore, the removal of organic pollutants is also possible under visible light utilizing nanocomposites. Xue et al. used the nanocomposites of Fe_3O_4@SiO_2@TiO_2 as a magnetic material and degraded the MB under visible light [69]. The degradation percentage calculated for nanocomposites was ~88%, which was only 5.3% earlier without the nanocomposite. Some authors found that the degradation of pollutants using TiO_2-based magnetic NPs is also related to the thickness of the titania layer [70,71].

7.5.2.3 Core@(Coating)@TiO_2-doped Catalysts

The photocatalytic performance was extended further with metal-doped (Ag, Au, Pt) TiO_2-based magnetic NPs. Mostly used metal for this purpose is Ag. Ag-doped materials showed better performance than undoped materials, as published in [72–74]. Chi et al. synthesized Ag-doped Fe_3O_4@SiO_2@TiO_2 that completely decomposed RhB under UV light within 10 minutes, while undoped material could only remove it 93% during same time [74]. Au-doped materials also showed remarkable performance for photocatalytic activity. As Fe_3O_4@TiO_2-Au eliminated MB up to 80% whereas undoped specimen could only decompose it 35%, under the same conditions [75]. On the other side, rare earth metals (like cerium and neodymium) have also been employed as dopants for increasing the efficacy of TiO_2-based magnetic NPs [76–79].

The use of non–titania-based photocatalysts have also extended, as reported in Table 7.1, the ternary oxides [80], used as bare photocatalyst against MO and showed remarkable performance up to ~91% degradation. Moreover, the extended configuration having a core of magnetic material, covered with photocatalytic layers has also reported [81,82]. Both catalysts performed against MO and RhB and showed ~90% and ~98% efficacy, respectively.

7.6 RECOVERY AND REUSE OF MAGNETIC MATERIALS

Slurry batch reactors are conventionally used for photocatalytic processes that include nano photocatalysts suspended in a solvent. But, the removal of magnetic substances from the solution after treatment is extremely important and the only difficulty in their recovery is caused by their small size. A promising solution has been proposed in this regard, where the magnetic NPs are fixed onto a substrate that is inert in nature. However, some drawbacks of this methodology are the reduction of available active sites on the catalyst surface, expansion of the mass transfer limitations and increase in the operational difficulty which consequently reduces the photocatalytic performance [83–85]. Nevertheless, slurry reactors have been extensively used for photocatalytic applications.

Magnetic separation has various benefits in contrast to conventional techniques used for the recovery of catalysts including centrifugation, sedimentation and membrane filtration. This method is easier to operate, faster than conventional recovery techniques and more importantly, it is energy efficient [86]. Moreover, there are no possibilities of contamination in this technique because the magnetic field can easily penetrate different materials and no physical interaction occurs with magnetic materials, making it a non-invasive technique. In addition to that, the magnetic separation technique is totally independent of various factors like temperature, ionic composition and pH of the medium, which allows for its extensive range of operational conditions [87].

In many reported studies, batch reactors have been used for the experimental degradation of pollutants present in water. The magnetic materials used in the process are finally separated from the reaction medium by placing a permanent magnet near the reactor. The schematic representation of the magnetic recovery process is illustrated in Figure 7.8. This technique can be easily implied and magnet-sensitive materials can be reused many times. Ye et al. reported a study where the fabricated $TiO_2@Fe_3O_4@SiO_2$ magnetic materials were successfully recycled and reused for 18 cycles by providing an external magnetic field by a magnet (NdFeB), and no significant decrease in photocatalytic activity was observed. Many researchers have also reported significant results as shown in Table 7.1, where the fabricated magnetic NPs have been utilized for up to ten cycles with no significant effects on the material's quantity and photocatalytic efficiency [67].

Generally, using the same material, almost the same degradation rate can be achieved. But after certain cycles as mentioned in Table 7.1, a slight decrease in photocatalytic activity has been observed. Two factors might explain this loss. Firstly, the material is not completely recovered [59] that affects its performance in the subsequent cycle. Secondly, there might be changes that occurred in magnetic NPs properties (aggregation, fouling, etc.) during the recycling process [89–91]. The technique demonstrated in Figure 7.8 might cause severe aggregation of NPs which decreases the magnetic NPs surface area and as a result decreases the number of active sites

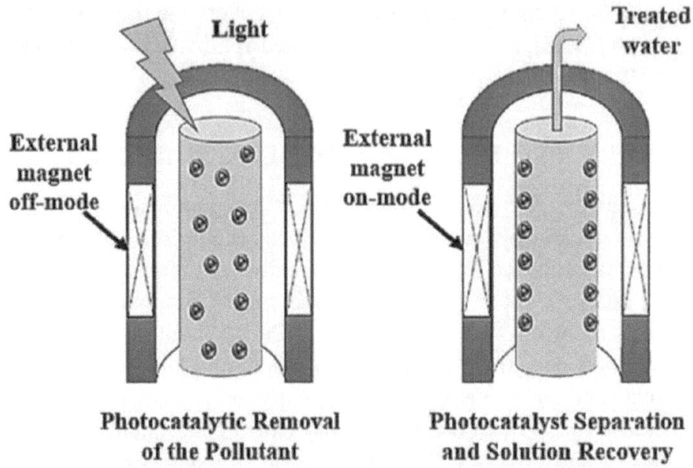

FIGURE 7.8 Magnetic NPs recovery methodology after the photocatalytic treatment of organic pollutants. (Reproduced from Ref. [88].)

for photocatalytic activity. It implies that the mentioned technique suffers from some significant flaws which make it inappropriate for efficient catalyst recovery and several advancements are needed to improve the reusability of magnetic NPs. Moreover, rare earth magnets have a low magnetic field, so this technique is limited to lab-scale applications. For the recovery of magnetic NPs on large scale, novel approaches need to be devised in this regard.

The following equation is used to understand the mechanism involved in the separation of particles caused by a magnetic force (F_B) [92–94]:

$$F_B = \mu_o V_P (M_P n) H_a$$

μ_0 = Permeability of free space; M_P = Magnetization of particle; V_P = Volume of particle; H_a = Applied magnetic field intensity.

As given in the above equation, F_B is directly proportional to the volume of the particle. Therefore, the greater the size of the particle, the greater would be the force that enables material separation from the treated solution. However, a reduction in surface-to-volume ratio can be induced by increasing particle diameter, thus reducing the photocatalytic efficacy of the catalyst. All these factors must be optimized to obtain the desired results for improved magnetic recovery and photocatalysis kinetics. Magnetic force could be enhanced by using alloys or ferromagnetic materials (Co, Fe, FeNi, CoFe, etc.) that facilitate more responsive magnetic cores while keeping particle size small. These materials get fully magnetized with magnetization values greater than typically reported iron oxides cores, thus assisting in separation. Iron cores have depicted higher magnetization values than their oxides (approximately 3–10 times), so they are a better choice for fabricating magnetic photocatalysts [95]. Apart from that, some other factors should also be optimized to generate proper magnetic fields in the system which would lead to effective magnetic NPs separation.

Following Equation (A), high field gradients and high magnetic fields are necessary for efficient separation of particles, mainly for particles of very small size.

As far as the methodology based on permanent magnets is concerned, the inadequate gradients and fields generated by such magnets limit their application in large-scale separation of small suspended NPs [96]. Consequently, proper magnet-based separators should be developed for effective recovery of magnetic NPs in large-scale water treatment processes. Many reports have been published for the separation of small magnetic solids on a large scale. Most commonly used filter is HGMS (High Gradient Magnetic Separator) for the separation of magnetic solids at the industrial level, its configuration allows gradients and high fields inside the separator column [86]. With the use of this filter, suspension of particles is promoted through a column (having ferromagnetic filaments which are magnetized by applying field externally) to particle capture around matrix as well as giving clean solution at the outlet. There are also some disadvantages that this filter presents for magnetic NPs separation. The first one is that there is a possibility of mechanically trapped solid substances due to an inhomogeneous matrix present in the filter, so magnetic and non-magnetic substances are retained in filters [97]. This is also an unenviable effect in the rinsing process because it is more difficult to flush the filter at the end. As there is a possibility of particle agglomeration which can restrict further material recovery and makes impossible further use of magnetic NPs [86].

There are other configurations that might be successful for magnetic separation. Herein OGMS (Open-Gradient Magnetic Separator) is presented as an appropriate alternative for the recovery of magnetic NPs. Figure 7.9 illustrates a view of photocatalysis when OGMS is applied. This type of filter does not utilize a ferromagnetic material for the generation of large gradients in the device, rather it uses special types of magnets in a suitable arrangement around the walls of the separator [98]. When an external field is applied, particles are forced around the walls and the treated solution can be secured from the outlet.

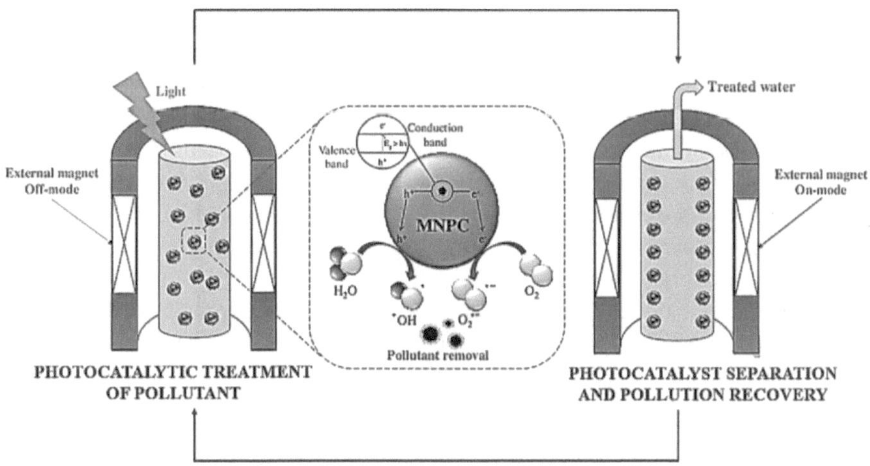

FIGURE 7.9 Proposed open-gradient magnet separator system for the magnetic NPs separation. (Reproduced from Ref. [88].)

By using OGMS, the blocking issue of HGMS has been resolved which was a technical fault for large-scale HGMS. Additionally, possibility for the generation of switchable fields in column permits the magnetic separation step in the reactor used for photocatalysis. The magnetic gradients employed in column can be calculated through computational method and optimization as well as scale-up of the system is compared with the HGMS column [99]. By using superconducting magnets in OGMS, magnet gradients can be increased which also improves the filter efficacy.

7.7 CONCLUSION AND FUTURE TRENDS/CHALLENGES

Despite the great potential of magnetic materials for water treatment, there are major questions regarding the characteristics, design of the process, application as well as implementation that need to be covered before their establishment. In the last few years, studies have been developed on a lab scale and focused on its further applications on a commercial level. Although this method is suitable to check the formulation methods of new catalysts, additional research is needed to develop the treatment of polluted water on a large scale, utilizing magnetic NPs. The examined works regarding these materials have characterized the properties of magnetic NPs by multiple methods prior to their use, but the additional characterization was not carried out after their use. So, rigorous characterization is required with time, providing much attention to potential modifications of those characteristics as a result of magnetic NPs photocatalytic application and later after magnetic separation. As described earlier, the particle agglomeration might happen after some cycles which reduces the performance of magnetic NPs. This can be verified by minimizing or avoiding some negative parameters discussed earlier or by improving the properties of materials during preparation which as a result can optimize the lifetime of the catalysts.

Furthermore, we examined important properties with time that should be estimated related to magnetic properties (magnetic susceptibility), the structure of the material (are of surface, distribution of size) and photocatalysis (bandgap of materials as well as kinetic rates). With the surface area and distribution of size, the agglomeration of material can be controlled. Additionally, by the magnetic susceptibility estimation and magnetization of the particles, the magnetic core's state that responds to magnetic materials should not alter during multiple cycles and can be inspected. In the end, when utilizing magnetic NPs having iron, an important factor that must be checked is the leaching of iron from catalyst toward reaction sites. Iron ions can increase reactive species by other reactions instead of photocatalysis (Fenton-like reaction and photolysis) which could remove the target compound.

A new type of photocatalysts is needed to obtain novel magnetic NPs for the degradation of organic pollutants having lower bandgap. In this work, the most reported photocatalysts are TiO_2-based. However, the greater bandgap of TiO_2 does not make it an efficient photocatalyst for light-driven technology. Alternative photocatalysts have been proposed such as Ag- or Au-doped TiO_2 to get better results. On the other side, new separators should be developed for better recovery of photocatalysts. In this chapter, OGMS system has been purposed which does not utilize ferromagnetic materials, dissimilar to the conventional HGMS system. Using this system, an efficient field, as well as gradients, is produced through magnetic arrangement near walls that optimize its performance. Regarding the core of material that is responsible for the separation of pollutants, it has been suggested to have more responsive magnetic cores like ferromagnetic metals. Once a suitable separator is designed, one can proceed toward the efficient working of both photocatalysis and magnetic separation. One possibility could be the use of the same materials for both purposes. For this, electromagnets or powerful superconducting magnets should be used which can provide switchable magnetic sources.

REFERENCES

1. Richard Feynman: Physics is fun to imagine | TED Talk, https://www.ted.com/talks/richard_feynman_physics_is_fun_to_imagine.
2. A.H. Lu, E.L. Salabas, F. Schüth, Magnetic nanoparticles: Synthesis, protection, functionalization, and application, *Angew. Chemie - Int. Ed.* 46 (2007) 1222–1244. doi:10.1002/anie.200602866.
3. R. Nisticò, F. Franzoso, F. Cesano, D. Scarano, G. Magnacca, M.E. Parolo, L. Carlos, Chitosan-derived iron oxide systems for magnetically guided and efficient water purification processes from polycyclic aromatic hydrocarbons, *ACS Sustain. Chem. Eng.* 5 (2017) 793–801. doi:10.1021/acssuschemeng.6b02126.
4. A.K. Gupta, M. Gupta, Synthesis and surface engineering of iron oxide nanoparticles for biomedical applications, *Biomaterials.* 26 (2005) 3995–4021. doi:10.1016/j.biomaterials.2004.10.012.
5. R.A. Duine, K.J. Lee, S.S.P. Parkin, M.D. Stiles, Synthetic antiferromagnetic spintronics, *Nat. Phys.* 14 (2018) 217–219. doi:10.1038/s41567-018-0050-y.
6. H.W. Lee, K.C. Kim, J. Lee, Review of Maglev train technologies, *IEEE Trans. Magn.* 42 (2006) 1917–1925. doi:10.1109/TMAG.2006.875842.
7. A.-H. Lu, W. Schmidt, N. Matoussevitch, H. Bönnemann, B. Spliethoff, B. Tesche, E. Bill, W. Kiefer, F. Schüth, Nanoengineering of a magnetically separable hydrogenation catalyst, *Angew. Chemie.* 116 (2004) 4403–4406. doi:10.1002/ange.200454222.
8. Y.W. Jun, J.H. Lee, J. Cheon, Chemical design of nanoparticle probes for high-performance magnetic resonance imaging, *Angew. Chemie - Int. Ed.* 47 (2008) 5122–5135. doi:10.1002/anie.200701674.
9. S. Jeevanantham, A. Saravanan, R. V. Hemavathy, P.S. Kumar, P.R. Yaashikaa, D. Yuvaraj, Removal of toxic pollutants from water environment by phytoremediation: A survey on application and future prospects, *Environ. Technol. Innov.* 13 (2019) 264–276. doi:10.1016/j.eti.2018.12.007.
10. W. Kleemann, Magnetoelectric spintronics, J. Appl. Phys., American Institute of PhysicsAIP (2013). 27013. doi:10.1063/1.4811823.
11. B. Wu, Y. Zhang, X. Zhang, S. Cheng, Health risk from exposure of organic pollutants through drinking water consumption in Nanjing, China, *Bull. Environ. Contam. Toxicol.* 84 (2010) 46–50. doi:10.1007/s00128-009-9900-8.
12. B. Petrie, R. Barden, B. Kasprzyk-Hordern, A review on emerging contaminants in wastewaters and the environment: Current knowledge, understudied areas and recommendations for future monitoring, *Water Res.* 72 (2015) 3–27. doi:10.1016/j.watres.2014.08.053.
13. P.R. Gogate, A.B. Pandit, A review of imperative technologies for wastewater treatment I: Oxidation technologies at ambient conditions, *Adv. Environ. Res.* 8 (2004) 501–551. doi:10.1016/S1093-0191(03)00032-7.
14. P.R. Gogate, A.B. Pandit, A review of imperative technologies for wastewater treatment II: Hybrid methods, *Adv. Environ. Res.* 8 (2004) 553–597. doi:10.1016/S1093-0191(03)00031-5.

15. R. Wiltschko, W. Wiltschko, Magnetoreception, *Adv. Exp. Med. Biol.* 739 (2012) 126–141. doi:10.1007/978-1-4614-1704-0_8.
16. L. Solymar, D. Walsh, Lectures on the electrical properties of materials (1993). http://inis.iaea.org/Search/search.aspx?orig_q=RN:26065836.
17. C. Tannous, J. Gieraltowski, The Stoner-Wohlfarth model of ferromagnetism, *Eur. J. Phys.* 29 (2008) 475–487. doi:10.1088/0143-0807/29/3/008.
18. E. Paterson, The iron oxides: Structure, properties, reactions, occurrences and uses, *Clay Miner.* 34 (1999) 209–210. doi:10.1180/claymin.1999.034.1.20.
19. Q.A. Pankhurst, J. Connolly, S.K. Jones, J. Dobson, Applications of magnetic nanoparticles in biomedicine, *J. Phys. D. Appl. Phys.* 36 (2003) R167. doi:10.1088/0022-3727/36/13/201.
20. J. Liu, Z. Wu, Q. Tian, W. Wu, X. Xiao, Shape-controlled iron oxide nanocrystals: Synthesis, magnetic properties and energy conversion applications, *CrystEngComm.* 18 (2016) 6303–6326. doi:10.1039/c6ce01307d.
21. A. Schwarz, R. Wiesendanger, Magnetic sensitive force microscopy, *Nano Today.* 3 (2008) 28–39. doi:10.1016/S1748-0132(08)70013-6.
22. H.D.A. Mohamed, S.M.D. Watson, B.R. Horrocks, A. Houlton, Magnetic and conductive magnetite nanowires by DNA-templating, *Nanoscale.* 4 (2012) 5936–5945. doi:10.1039/c2nr31559a.
23. C. Schönenberger, S.F. Alvarado, Understanding magnetic force microscopy, *Zeitschrift Für Phys. B Condens. Matter.* 80 (1990) 373–383. doi:10.1007/BF01323519.
24. H.W.F. Sung, C. Rudowicz, Physics behind the magnetic hysteresis loop – A survey of misconceptions in magnetism literature, *J. Magn. Magn. Mater.* 260 (2003) 250–260. doi:10.1016/S0304-8853(02)01339-2.
25. S. Wu, A. Sun, F. Zhai, J. Wang, W. Xu, Q. Zhang, A.A. Volinsky, Fe_3O_4 magnetic nanoparticles synthesis from tailings by ultrasonic chemical co-precipitation, *Mater. Lett.* 65 (2011) 1882–1884. doi:10.1016/j.matlet.2011.03.065.
26. F. Yazdani, M. Seddigh, Magnetite nanoparticles synthesized by co-precipitation method: The effects of various iron anions on specifications, *Mater. Chem. Phys.* 184 (2016) 318–323. doi:10.1016/j.matchemphys.2016.09.058.
27. R. Nisticò, G. Magnacca, M. Antonietti, N. Fechler, Highly porous silica glasses and aerogels made easy: The hypersaline route, *Adv. Porous Mater.* 2 (2014) 37–41. doi:10.1166/apm.2014.1049.
28. W. Wu, Z. Wu, T. Yu, C. Jiang, W.S. Kim, Recent progress on magnetic iron oxide nanoparticles: Synthesis, surface functional strategies and biomedical applications, *Sci. Technol. Adv. Mater.* 16 (2015). doi:10.1088/1468-6996/16/2/023501.
29. W. Wu, Q. He, C. Jiang, Magnetic iron oxide nanoparticles: Synthesis and surface functionalization strategies, *Nanoscale Res. Lett.* 3 (2008) 397–415. doi:10.1007/s11671-008-9174-9.
30. W. Wu, X. Xiao, S. Zhang, J. Zhou, L. Fan, F. Ren, C. Jiang, Large-scale and controlled synthesis of iron oxide magnetic short nanotubes: Shape evolution, growth mechanism, and magnetic properties, *J. Phys. Chem. C.* 114 (2010) 16092–16103. doi:10.1021/jp1010154.
31. C. Solans, P. Izquierdo, J. Nolla, N. Azemar, M.J. Garcia-Celma, Nano-emulsions, *Curr. Opin. Colloid Interface Sci.* 10 (2005) 102–110. doi:10.1016/j.cocis.2005.06.004.
32. J.A. Lopez Perez, M.A. Lopez Quintela, J. Mira, J. Rivas, S.W. Charles, Advances in the preparation of magnetic nanoparticles by the microemulsion method, *J. Phys. Chem. B.* 101 (1997) 8045–8047. doi:10.1021/jp972046t.
33. R. Ciriminna, A. Fidalgo, V. Pandarus, F. Béland, L.M. Ilharco, M. Pagliaro, The sol-gel route to advanced silica-based materials and recent applications, *Chem. Rev.* 113 (2013) 6592–6620. doi:10.1021/cr300399c.

34. E.M. Modan, A.G. Plăiașu, Advantages and disadvantages of chemical methods in the elaboration of nanomaterials, *Ann. "Dunarea Jos" Univ. Galati. Fascicle IX, Metall. Mater. Sci.* 43 (2020) 53–60. doi:10.35219/mms.2020.1.08.
35. J.P. Tierney, P. Lidström, *Microwave Assisted Organic Synthesis*, Blackwell Publishing Ltd., Oxford, UK, 2005. doi:10.1002/9781444305548.
36. G. Ruppert, R. Bauer, G. Heisler, The photo-fenton reaction – An effective photochemical wastewater treatment process, *J. Photochem. Photobiol. A Chem.* 73 (1993) 75–78. doi:10.1016/1010-6030(93)80035-8.
37. M. Minella, G. Marchetti, E. De Laurentiis, M. Malandrino, V. Maurino, C. Minero, D. Vione, K. Hanna, Photo-Fenton oxidation of phenol with magnetite as iron source, *Appl. Catal. B Environ.* 154–155 (2014) 102–109. doi:10.1016/j.apcatb.2014.02.006.
38. F. Franzoso, R. Nisticò, F. Cesano, I. Corazzari, F. Turci, D. Scarano, A. Bianco Prevot, G. Magnacca, L. Carlos, D.O. Mártire, Biowaste-derived substances as a tool for obtaining magnet-sensitive materials for environmental applications in wastewater treatments, *Chem. Eng. J.* 310 (2017) 307–316. doi:10.1016/j.cej.2016.10.120.
39. A. Bianco Prevot, F. Baino, D. Fabbri, F. Franzoso, G. Magnacca, R. Nisticò, A. Arques, Urban biowaste-derived sensitizing materials for caffeine photodegradation, *Environ. Sci. Pollut. Res.* 24 (2017) 12599–12607. doi:10.1007/s11356-016-7763-1.
40. L.M. Pastrana-Martínez, N. Pereira, R. Lima, J.L. Faria, H.T. Gomes, A.M.T. Silva, Degradation of diphenhydramine by photo-Fenton using magnetically recoverable iron oxide nanoparticles as catalyst, *Chem. Eng. J.* 261 (2015) 45–52. doi:10.1016/j.cej.2014.04.117.
41. X. Li, Y. Hou, Q. Zhao, L. Wang, A general, one-step and template-free synthesis of sphere-like zinc ferrite nanostructures with enhanced photocatalytic activity for dye degradation, *J. Colloid Interface Sci.* 358 (2011) 102–108. doi:10.1016/j.jcis.2011.02.052.
42. O.V. Kharissova, H.V.R. Dias, B.I. Kharisov, Magnetic adsorbents based on micro- and nano-structured materials, *RSC Adv.* 5 (2015) 6695–6719. doi:10.1039/c4ra11423j.
43. Y. Chu, Q. Pan, Three-dimensionally macroporous Fe/C nanocomposites as highly selective oil-absorption materials, *ACS Appl. Mater. Interfaces.* 4 (2012) 2420–2425. doi:10.1021/am3000825.
44. A.Z.M. Badruddoza, Z.B.Z. Shawon, W.J.D. Tay, K. Hidajat, M.S. Uddin, Fe_3O_4/cyclodextrin polymer nanocomposites for selective heavy metals removal from industrial wastewater, *Carbohydr. Polym.* 91 (2013) 322–332. doi:10.1016/j.carbpol.2012.08.030.
45. S.M. Ponder, J.G. Darab, J. Bucher, D. Caulder, I. Craig, L. Davis, N. Edelstein, W. Lukens, H. Nitsche, L. Rao, D.K. Shuh, T.E. Mallouk, Surface chemistry and electrochemistry of supported zerovalent iron nanoparticles in the remediation of aqueous metal contaminants, *Chem. Mater.* 13 (2001) 479–486. doi:10.1021/cm000288r.
46. A.F. Ngomsik, A. Bee, M. Draye, G. Cote, V. Cabuil, Magnetic nano- and microparticles for metal removal and environmental applications: A review, *Comptes Rendus Chim.* 8 (2005) 963–970. doi:10.1016/j.crci.2005.01.001.
47. X. Zhao, W. Liu, Z. Cai, B. Han, T. Qian, D. Zhao, An overview of preparation and applications of stabilized zero-valent iron nanoparticles for soil and groundwater remediation, *Water Res.* 100 (2016) 245–266. doi:10.1016/j.watres.2016.05.019.
48. S. Li, W. Yan, W.X. Zhang, Solvent-free production of nanoscale zero-valent iron (nZEVI) with precision milling, *Green Chem.* 11 (2009) 1618–1626. doi:10.1039/b913056j.
49. F. Fu, D.D. Dionysiou, H. Liu, The use of zero-valent iron for groundwater remediation and wastewater treatment: A review, *J. Hazard. Mater.* 267 (2014) 194–205. doi:10.1016/j.jhazmat.2013.12.062.
50. R.A. Crane, T.B. Scott, Nanoscale zero-valent iron: Future prospects for an emerging water treatment technology, *J. Hazard. Mater.* 211–212 (2012) 112–125. doi:10.1016/j.jhazmat.2011.11.073.

51. J. Fan, Y. Guo, J. Wang, M. Fan, Rapid decolorization of azo dye methyl orange in aqueous solution by nanoscale zerovalent iron particles, *J. Hazard. Mater.* 166 (2009) 904–910. https://doi.org/10.1016/j.jhazmat.2008.11.091.
52. H.Y. Shu, M.C. Chang, C.C. Chen, P.E. Chen, Using resin supported nano zero-valent iron particles for decoloration of acid blue 113 azo dye solution, *J. Hazard. Mater.* 184 (2010) 499–505. doi:10.1016/j.jhazmat.2010.08.064.
53. D.H.K. Reddy, Y.S. Yun, Spinel ferrite magnetic adsorbents: Alternative future materials for water purification? *Coord. Chem. Rev.* 315 (2016) 90–111. doi:10.1016/j.ccr.2016.01.012.
54. C.M.B. Henderson, J.M. Charnock, D.A. Plant, Cation occupancies in Mg, Co, Ni, Zn, Al ferrite spinels: A multi-element EXAFS study, *J. Phys. Condens. Matter.* 19 (2007) 76214. doi:10.1088/0953-8984/19/7/076214.
55. Y. Xiao, H. Liang, W. Chen, Z. Wang, Synthesis and adsorption behavior of chitosan-coated $MnFe_2O_4$ nanoparticles for trace heavy metal ions removal, *Appl. Surf. Sci.* 285 (2013) 498–504. doi:10.1016/j.apsusc.2013.08.083.
56. J. Wei, X. Zhang, Q. Liu, Z. Li, L. Liu, J. Wang, Magnetic separation of uranium by CoFe2O4 hollow spheres, *Chem. Eng. J.* 241 (2014) 228–234. doi:10.1016/j.cej.2013.12.035.
57. Y. Meng, D. Chen, Y. Sun, D. Jiao, D. Zeng, Z. Liu, Adsorption of Cu 2+ ions using chitosan-modified magnetic Mn ferrite nanoparticles synthesized by microwave-assisted hydrothermal method, *Appl. Surf. Sci.* 324 (2015) 745–750. doi:10.1016/j.apsusc.2014.11.028.
58. Z. Jia, Q. Qin, J. Liu, H. Shi, X. Zhang, R. Hu, S. Li, R. Zhu, The synthesis of hierarchical $ZnFe_2O_4$ architecture and their application for Cr(VI) adsorption removal from aqueous solution, *Superlattices Microstruct.* 82 (2015) 174–187. doi:10.1016/j.spmi.2015.01.028.
59. W. Jiang, X. Zhang, X. Gong, F. Yan, Z. Zhang, Sonochemical synthesis and characterization of magnetic separable $Fe_3O_4 - TiO_2$ nanocomposites and their catalytic properties, *Int. J. Smart Nano Mater.* 1 (2010) 278–287. doi:10.1080/19475411.2010.528873.
60. R. Chalasani, S. Vasudevan, Cyclodextrin-functionalized $Fe_3O_4@TiO_2$: Reusable, magnetic nanoparticles for photocatalytic degradation of endocrine-disrupting chemicals in water supplies, *ACS Nano.* 7 (2013) 4093–4104. doi:10.1021/nn400287k.
61. M. Abbas, B. Parvatheeswara Rao, V. Reddy, C. Kim, Fe_3O_4/TiO_2 core/shell nanocubes: Single-batch surfactantless synthesis, characterization and efficient catalysts for methylene blue degradation, *Ceram. Int.* 40 (2014) 11177–11186. doi:10.1016/j.ceramint.2014.03.148.
62. J. Jing, J. Li, J. Feng, W. Li, W.W. Yu, Photodegradation of quinoline in water over magnetically separable Fe_3O_4/TiO_2 composite photocatalysts, *Chem. Eng. J.* 219 (2013) 355–360. doi:10.1016/j.cej.2012.12.058.
63. P. Cheng, W. Li, T. Zhou, Y. Jin, M. Gu, Physical and photocatalytic properties of zinc ferrite doped titania under visible light irradiation, *J. Photochem. Photobiol. A Chem.* 168 (2004) 97–101. doi:10.1016/j.jphotochem.2004.05.018.
64. Y. Ao, J. Xu, D. Fu, L. Ba, C. Yuan, Deposition of anatase titania onto carbon encapsulated magnetite nanoparticles, *Nanotechnology.* 19 (2008) 405604. doi:10.1088/0957-4484/19/40/405604.
65. D. Beydoun, R. Amal, G. Low, S. McEvoy, Occurrence and prevention of photodissolution at the phase junction of magnetite and titanium dioxide, *J. Mol. Catal. A Chem.* 180 (2002) 193–200. doi:10.1016/S1381-1169(01)00429-0.
66. J. Joo, Y. Ye, D. Kim, J. Lee, S. Jeon, Magnetically recoverable hybrid TiO_2 nanocrystal clusters with enhanced photocatalytic activity, *Mater. Lett.* 93 (2013) 141–144. doi:10.1016/j.matlet.2012.10.067.

67. Q. Yuan, N. Li, W. Geng, Y. Chi, X. Li, Preparation of magnetically recoverable Fe_3O_4@SiO_2@meso-TiO_2 nanocomposites with enhanced photocatalytic ability, *Mater. Res. Bull.* 47 (2012) 2396–2402. doi:10.1016/j.materresbull.2012.05.031.
68. M. Ye, Q. Zhang, Y. Hu, J. Ge, Z. Lu, L. He, Z. Chen, Y. Yin, Magnetically recoverable core-shell nanocomposites with enhanced photocatalytic activity, *Chem. – A Eur. J.* 16 (2010) 6243–6250. doi:10.1002/chem.200903516.
69. C. Xue, Q. Zhang, J. Li, X. Chou, W. Zhang, H. Ye, Z. Cui, P.J. Dobson, High photocatalytic activity of Fe_3O_4-SiO_2-TiO_2 functional particles with core-shell structure, *J. Nanomater.* 2013 (2013). doi:10.1155/2013/762423.
70. K. Kang, M. Jang, M. Cui, P. Qiu, B. Park, S.A. Snyder, J. Khim, Preparation and characterization of magnetic-core titanium dioxide: Implications for photocatalytic removal of ibuprofen, *J. Mol. Catal. A Chem.* 390 (2014) 178–186. doi:10.1016/j.molcata.2014.03.023.
71. W. Su, T. Zhang, L. Li, J. Xing, M. He, Y. Zhong, Z. Li, Synthesis of small yolk-shell Fe_3O_4@TiO_2 nanoparticles with controllable thickness as recyclable photocatalysts, *RSC Adv.* 4 (2014) 8901–8906. doi:10.1039/c3ra47461e.
72. M.W. Xu, S.J. Bao, X.G. Zhang, Enhanced photocatalytic activity of magnetic TiO_2 photocatalyst by silver deposition, *Mater. Lett.* 59 (2005) 2194–2198. doi:10.1016/j.matlet.2005.02.065.
73. Y. Zhang, X. Yu, Y. Jia, Z. Jin, J. Liu, X. Huang, A facile approach for the synthesis of Ag-coated Fe_3O_4@TiO_2 core/shell microspheres as highly efficient and recyclable photocatalysts, *Eur. J. Inorg. Chem.* 2011 (2011) 5096–5104. doi:10.1002/ejic.201100707.
74. Y. Chi, Q. Yuan, Y. Li, L. Zhao, N. Li, X. Li, W. Yan, Magnetically separable Fe_3O_4@SiO_2@TiO_2-Ag microspheres with well-designed nanostructure and enhanced photocatalytic activity, *J. Hazard. Mater.* 262 (2013) 404–411. doi:10.1016/j.jhazmat.2013.08.077.
75. C. Li, R. Younesi, Y. Cai, Y. Zhu, M. Ma, J. Zhu, Photocatalytic and antibacterial properties of Au-decorated Fe_3O_4@mTiO_2 core-shell microspheres, *Appl. Catal. B Environ.* 156–157 (2014) 314 322. doi:10.1016/j.apcatb.2014.03.031.
76. B.K. Sunkara, R.D.K. Misra, Enhanced antibactericidal function of W4+-doped titania-coated nickel ferrite composite nanoparticles: A biomaterial system, *Acta Biomater.* 4 (2008) 273–283. doi:10.1016/j.actbio.2007.07.002.
77. Z.L. Shi, C. Du, S.H. Yao, Preparation and photocatalytic activity of cerium doped anatase titanium dioxide coated magnetite composite, *J. Taiwan Inst. Chem. Eng.* 42 (2011) 652–657. doi:10.1016/j.jtice.2010.10.001.
78. Z. Shi, H. Lai, S. Yao, Preparation and photocatalytic activity of magnetic samarium-doped mesoporous titanium dioxide at the decomposition of methylene blue under visible light, *Russ. J. Phys. Chem. A.* 86 (2012) 1326–1331. doi:10.1134/S0036024412060337.
79. S. Rana, J. Rawat, M.M. Sorensson, R.D.K. Misra, Antimicrobial function of Nd3+-doped anatase titania-coated nickel ferrite composite nanoparticles: A biomaterial system, *Acta Biomater.* 2 (2006) 421–432. doi:10.1016/j.actbio.2006.03.005.
80. J. Guo, X. Zhou, L. Chen, Y. Lu, X. Zhang, W. Hou, One-pot synthesis of ferromagnetic Fe2.25W0.75O 4 nanoparticles as a magnetically recyclable photocatalyst, *J. Nanoparticle Res.* 14 (2012) 1–6. doi:10.1007/s11051-012-0992-4.
81. Y.R. Yao, W.Z. Huang, H. Zhou, Y.F. Zheng, X.C. Song, Self-assembly of dandelion-like Fe_3O_4@C@BiOCl magnetic nanocomposites with excellent solar-driven photocatalytic properties, *J. Nanoparticle Res.* 16 (2014) 1–9. doi:10.1007/s11051-014-2451-x.
82. Y. Tian, D. Wu, X. Jia, B. Yu, S. Zhan, Core-shell nanostructure of α-Fe_2O_3/Fe_3O_4: Synthesis and photocatalysis for methyl orange, *J. Nanomater.* 2011 (2011). doi:10.1155/2011/837123.
83. U.I. Gaya, A.H. Abdullah, Heterogeneous photocatalytic degradation of organic contaminants over titanium dioxide: A review of fundamentals, progress and

problems, *J. Photochem. Photobiol. C Photochem. Rev.* 9 (2008) 1–12. doi:10.1016/j.jphotochemrev.2007.12.003.
84. M.N. Chong, B. Jin, C.W.K. Chow, C. Saint, Recent developments in photocatalytic water treatment technology: A review, *Water Res.* 44 (2010) 2997–3027. doi:10.1016/j.watres.2010.02.039.
85. I. Gehrke, A. Geiser, A. Somborn-Schulz, Innovations in nanotechnology for water treatment, *Nanotechnol. Sci. Appl.* 8 (2015) 1. doi:10.2147/NSA.S43773.
86. J. Gómez-Pastora, E. Bringas, I. Ortiz, Recent progress and future challenges on the use of high performance magnetic nano-adsorbents in environmental applications, *Chem. Eng. J.* 256 (2014) 187–204. doi:10.1016/j.cej.2014.06.119.
87. N. Pamme, Magnetism and microfluidics, *Lab Chip.* 6 (2006) 24–38. doi:10.1039/b513005k.
88. J. Gómez-Pastora, S. Dominguez, E. Bringas, M.J. Rivero, I. Ortiz, D.D. Dionysiou, Review and perspectives on the use of magnetic nanophotocatalysts (MNPCs) in water treatment, *Chem. Eng. J.* 310 (2017) 407–427. doi:10.1016/j.cej.2016.04.140.
89. P.P. Hankare, R.P. Patil, A. V Jadhav, K.M. Garadkar, R. Sasikala, Enhanced photocatalytic degradation of methyl red and thymol blue using titania-alumina-zinc ferrite nanocomposite, *Appl. Catal. B Environ.* 107 (2011) 333–339. doi:10.1016/j.apcatb.2011.07.033.
90. J. Xia, A. Wang, X. Liu, Z. Su, Preparation and characterization of bifunctional, Fe_3O_4/ZnO nanocomposites and their use as photocatalysts, *Appl. Surf. Sci.* 257 (2011) 9724–9732. doi:10.1016/j.apsusc.2011.05.114.
91. Z.P. Yang, X.Y. Gong, C.J. Zhang, Recyclable Fe_3O_4/hydroxyapatite composite nanoparticles for photocatalytic applications, *Chem. Eng. J.* 165 (2010) 117–121. doi:10.1016/j.cej.2010.09.001.
92. S.A. Khashan, E.P. Furlani, Scalability analysis of magnetic bead separation in a microchannel with an array of soft magnetic elements in a uniform magnetic field, *Sep. Purif. Technol.* 125 (2014) 311–318. doi:10.1016/j.seppur.2014.02.007.
93. E.P. Furlani, Magnetic biotransport: Analysis and applications, *Materials (Basel).* 3 (2010) 2412–2446. doi:10.3390/ma3042412.
94. E.P. Furlani, Y. Sahoo, K.C. Ng, J.C. Wortman, T.E. Monk, A model for predicting magnetic particle capture in a microfluidic bioseparator, *Biomed. Microdevices.* 9 (2007) 451–463. doi:10.1007/s10544-007-9050-x.
95. I.K. Herrmann, R.N. Grass, W.J. Stark, High-strength metal nanomagnets for diagnostics and medicine: Carbon shells allow long-term stability and reliable linker chemistry, *Nanomedicine.* 4 (2009) 787–798. doi:10.2217/nnm.09.55.
96. G. Mariani, M. Fabbri, F. Negrini, P.L. Ribani, High-gradient magnetic separation of pollutant from wastewaters using permanent magnets, *Sep. Purif. Technol.* 72 (2010) 147–155. doi:10.1016/j.seppur.2010.01.017.
97. G.D. Moeser, K.A. Roach, W.H. Green, T. Alan Hatton, P.E. Laibinis, High-gradient magnetic separation of coated magnetic nanoparticles, *AIChE J.* 50 (2004) 2835–2848. doi:10.1002/aic.10270.
98. M. Ahoranta, J. Lehtonen, R. Mikkonen, Magnet design for superconducting open gradient magnetic separator, in: *Physica C (Superconductivity and its Applications)*, North-Holland, 2003: pp. 398–402. doi:10.1016/S0921-4534(02)02213-X.
99. S. Fukui, H. Nakajima, A. Ozone, M. Hayatsu, M. Yamaguchi, T. Sato, H. Imaizumi, S. Nishijima, T. Watanabe, Study on open gradient magnetic separation using multiple magnetic field sources, *IEEE Trans. Appl. Supercond.* 12 (2002) 959–962. doi:10.1109/TASC.2002.1018559.

Section 2

Applications

8 Direct Membrane Filtration for Wastewater Treatment

Elorm Obotey Ezugbe, Emmanuel K. Tetteh, Sudesh Rathilal, Edward K. Armah, and Gloria Amo-Duodu
Durban University of Technology

Dennis Asante-Sackey
Kumasi Technical University

CONTENTS

DOI: 10.1201/9781003167327-10

8.1 INTRODUCTION

The term 'wastewater' can simply be defined as 'used water'. This includes contaminated water from residential sources (laundry, toilet, bathroom, sinks, etc.), technically known as domestic wastewater, effluent from manufacturing facilities, known as industrial wastewater, and water from offices, hotels, stores, known as urban/commercial wastewater [1,2]. Typically, wastewater contains all manner of contaminants such as pathogens, organics (COD, turbidity, and color), salts, and emerging contaminants, and therefore, requires extensive treatment before being discharged or reused.

Effective wastewater treatment has become a necessity as it has a bearing on the availability of potable water for domestic, agricultural, and industrial use. In the World Economic Forum 2019, freshwater scarcity was identified as one of the most pressing global challenges posing threats to future socioeconomic growth [3,4]. Herein, reclaimed wastewater has been considered an alternate water supply for non-potable (indirect) or potable use, as well as a viable route to water sustainability [5,6].

Wastewater treatment has evolved over the years, accommodating the ever-increasing human population and its effects on industrial and agricultural activities. Through this evolution, direct membrane filtration processes have been in the limelight of research, providing many options for effective wastewater treatment and resource recovery. The ease of design and maintenance, low capital cost, high water quality, less sludge production, the flexibility of use, and the ability for remote use, among others, are the advantages of direct membrane filtration processes [7,8].

Generally, to meet stringent wastewater discharge specifications, conventional anaerobic digestion (AD) processes are used to remove organics followed by a membrane process to produce high-quality permeate water [9]. Besides the inefficiency of the biological-based wastewater treatment process due to the diversity and complexity of wastewater compositions, they are also energy-driven (aeration system) and mostly affected by environmental factors such as temperature, feed composition variation, and hydraulic retention time [10]. Other types of pre-treatment methods employed before membrane filtration include coagulation, adsorption with powdered or granular activated carbon (PAC), ozonation, dissolved air flotation clarifiers, sand filtration, and other solid-phase settleable systems [11,12]. These processes contribute significantly to the total cost of membrane treatment facilities [13,14].

One of the major setbacks in direct membrane processes is fouling. Fouling hinders the cost-effectiveness of membrane processes especially for treating high-strength organic matter [15,16]. Fouling, which results from the deposit of particles (foulants) on membrane surfaces, such as hydrophilic dissolved organic matter, salts, and colloids, has gained massive global attention. On this note, interventions such as membrane surface modulation with sensitive nanomaterials (metal or metal oxide nanoparticles [17], zeolites [18], clay nanoparticles [19], carbon nanotubes [20], and metal-organic frameworks (MOFs) [21]) have emerged as a promising technique for circumventing existing membrane limitations, improving the functional properties of the membrane [22,23].

This chapter presents an overview of the various direct membrane filtration processes in wastewater treatment, highlighting the various driving forces, their advantages and disadvantages, membrane modules, and factors affecting the application of direct membrane filtration processes. Additionally, methods to improve the applications of direct membrane filtration processes as well as mechanisms to mitigate membrane fouling are proposed. Finally, the prospects of developing direct membrane filtration systems as a viable technology to address the challenges of the global water crisis are recommended.

8.2 MEMBRANES, TYPES OF MEMBRANES, AND MEMBRANE PROCESSES

A membrane is defined as a thin barrier (Figure 8.1) that moderates the movements or permeation of various components across it in a selective way. The membrane acts as a permselective barrier between two phases where the separation of the desired species occurs in the presence of a driving force [24].

Membrane development dates to the 18th century. Since then, many milestones have been achieved, improving membranes toward a variety of applications including wastewater treatment [24,26]. Membranes can be classified as follows:

i. Symmetric (isotropic) and asymmetric (anisotropic) membranes: Symmetric membranes have a uniform composition and nature across the entire membrane structure. These are mostly applied in microfiltration (MF). Asymmetric membranes, however, are non-uniform with structured layers and varied porosity across the entire membrane structure. Asymmetric membranes are applied in MF, ultrafiltration (UF), nanofiltration (NF), and reverse osmosis (RO) [27,28].
ii. Dense and porous: Dense/microporous membranes are made of a dense film structure, presenting non-detectable pores across the membrane structure. Even though these membranes may have pores, these pores may only be

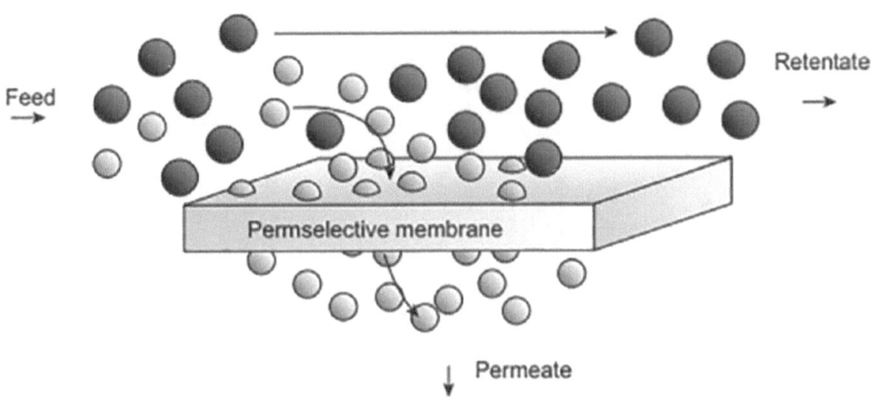

FIGURE 8.1 Schematic diagram of membrane separation [25].

detected using electron microscopy. Mass transfer across dense membranes takes place by diffusion under the driving force of pressure or chemical potential or electrical potential gradient. Porous membranes, however, have rigidly interconnected pores which are randomly distributed across the void structure. Separation in a porous membrane is mainly controlled by size exclusion and the hydrodynamic conditions of the flow [24,29].

iii. Organic (polymeric) and inorganic membranes: Membrane material makeup may be organic or inorganic. Polymeric membranes are made from materials such as polysulfone, polyethersulfone, cellulose acetate, polymethylpentene, polyimide, polyetherimide, etc. Even though these membranes are most widely used due to their low cost, their application is affected by mechanical instability and severe fouling. Inorganic membranes are those that are made from materials such as nanoporous carbon, ceramics, carbon molecular sieves, amorphous silica, etc. These are more chemically resistant, thermally, and mechanically stable compared to polymeric membranes [30,31].

In terms of application, direct membrane processes may be classified according to the driving force. This includes pressure-driven processes [reverse osmosis (RO), nanofiltration (NF), ultrafiltration (UF), and microfiltration (MF)] and osmotically driven processes [forward osmosis (FO), pressure-retarded osmosis (PRO), and pressure-assisted osmosis (PAO)]. Other driving forces include electrical [electrodialysis (ED)] and thermal [membrane distillation (MD)] [8]. Retentate (residue) from these processes can be post-digested for other products [renewable energy (H_2, CH_4) or fertilizers] as shown in Figure 8.2. These processes are briefly described in the following sections.

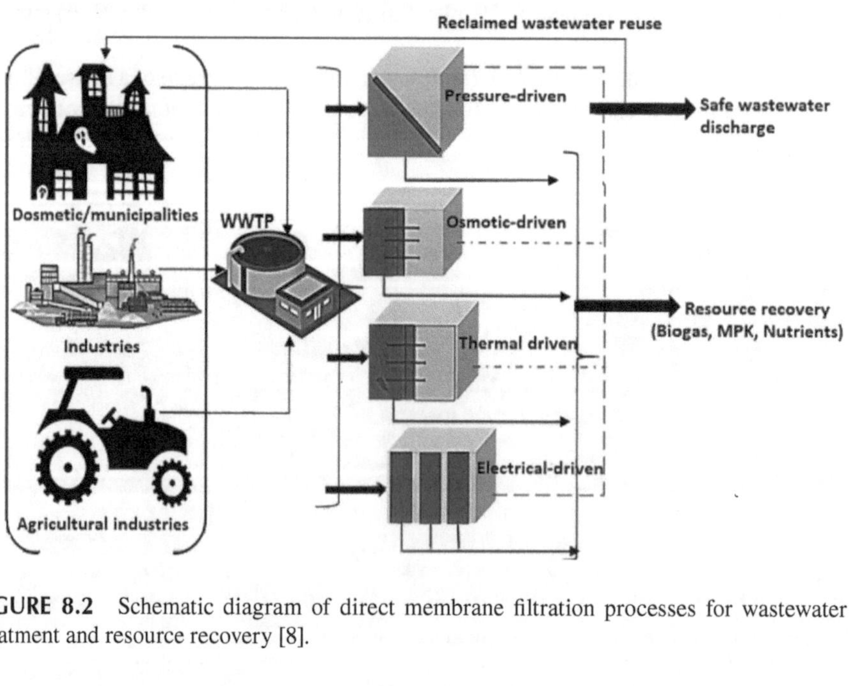

FIGURE 8.2 Schematic diagram of direct membrane filtration processes for wastewater treatment and resource recovery [8].

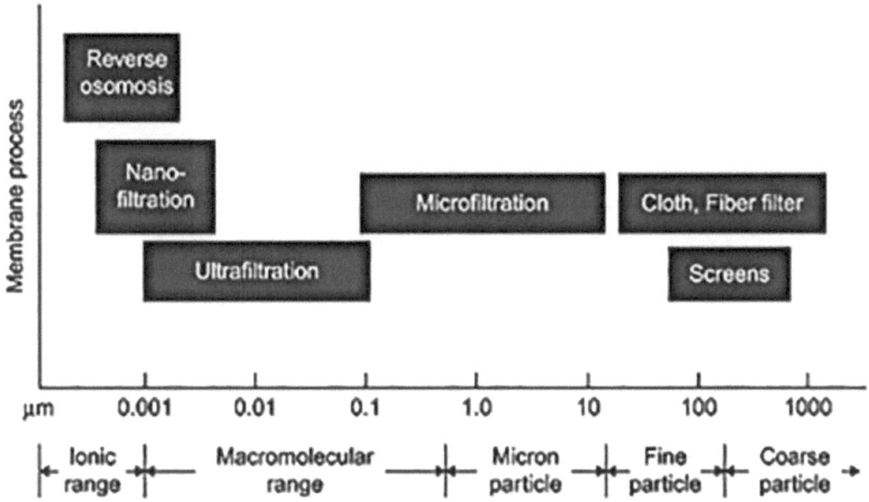

FIGURE 8.3 Pore size distribution in pressure-driven membrane processes [32].

8.2.1 Pressure-Driven Membrane Processes

These processes depend on external hydraulic pressure to force permeate water across a membrane, leaving behind a concentrated stream, referred to as brine. Pressure-driven membrane processes are the most widely applied membrane process in water and wastewater treatment.

During their application, the pressure exerted on the solution at one side of the membrane acts as a driving force to separate the feed into a permeate and a retentate. The permeate water is normally pure water, while the retentate is a concentrated solution that must be discarded or treated in some way. Filtration techniques range from dense (RO and NF) to porous membranes (UF and MF) and membranes may be polymeric, organo-mineral, ceramic, or metallic. Contaminants removed from feed solutions include salts, small organic molecules, macromolecules, or particles. It is, however, worthy to note that the removal efficiency of contaminants from wastewater using membranes depends on the type of technique employed. Figure 8.3 summarizes the range of pore sizes for pressure-driven processes.

From Figure 8.3, RO and NF have a rejection range of <0.001 µm, which makes them suitable for application as post-treatment processes, while UF and MF with larger pore sizes may be used as pre-treatment processes. These are known as low-pressure membrane processes. Pressure-driven membrane processes have been applied in the treatment of wastewaters having varied compositions. Table 8.1 shows some applications of pressure-driven membrane processes in wastewater treatment.

8.3 OSMOTIC-DRIVEN MEMBRANE PROCESSES (ODMPs)

Osmotic-driven membrane processes depend on osmotic pressure differences to enhance the movement of water across a semipermeable membrane. These membrane

TABLE 8.1
Some Applications of Pressure-Driven Membrane Processes in Wastewater Treatment

Pressure-Driven Process	Type of Feed Treated	Contaminant Removal Efficiencies	Reference
MF	Oily wastewater	TOC = 92.4%	[33]
UF	Vegetable oil factory wastewater	COD = 91%, TOC = 87%, TSS = 100%, PO_4^{3-} = 85% and Cl^- = 40%	[34]
NF	Effluent of biologically treated municipal wastewater	Conductivity = 89%, salinity = 90.3%, color = 90%, COD = 86.6%, TOC = 88%	[35]
RO	Distillery wastewater treatment (pre-treatment by NF)	TDS = 99.80%, COD = 99.90%, K^+ = 99.99%	[36]
RO	Landfill leachate effluent (pre-treated by up-flow anaerobic sludge blanket)	COD = 99%–99.5%, N-NH_4^+ = 99%–99.8%	[37]

processes are innovative technologies with extensive applications in water treatment and desalination (forward osmosis, FO), power generation (pressure-retarded osmosis, PRO), and dewatering of aqueous solutions (direct osmotic concentration, DOC, and osmotic dilution, ODN) [38].

In the area of wastewater treatment, FO has a wide application, being utilized extensively for different kinds of feed streams. FO is strictly guided by the osmotic pressure gradient between the feed solution (FS) and the draw solution (DS). This osmotic pressure gradient represents a difference in concentrations between the FS (the one with the lower concentration) and the DS (having the higher concentration). Water molecules move from the region of higher concentration (FS) to the region of lower concentration (DS) until an equilibrium is reached [39]. In a typical FO system, a recovery process is required to reclaim pure water as well as reconstitute the draw solution to maintain the osmotic pressure gradient to drive the process. Figure 8.4 depicts the principle of FO.

FO membranes are highly hydrophilic and asymmetric in nature, having a dense active layer for solute rejection and a thin support layer to enhance water permeation. The most common membranes used in FO include cellulose triacetate (CTA) membrane and thin-film composite (TFC) membrane [41,42]

8.3.1 Draw Solutes and Recovery Processes for FO

Draw solutes play a significant role in the efficiency of FO. These are osmotic agents that generate the required osmotic pressure to drive the FO process. Examples of draw solutes include NaCl, $MgCl_2$, Na_2SO_4, KCl, KNO_3, sucrose, fructose, EDTA, SO_2, and NH_3/CO_2. The recovery process adopted depends on the type of draw solute used. In most cases, salt-based draw solutes are recovered using pressure-driven processes such as RO and NF. Gases and volatile compounds such as SO_2 and NH_3/CO_2 are

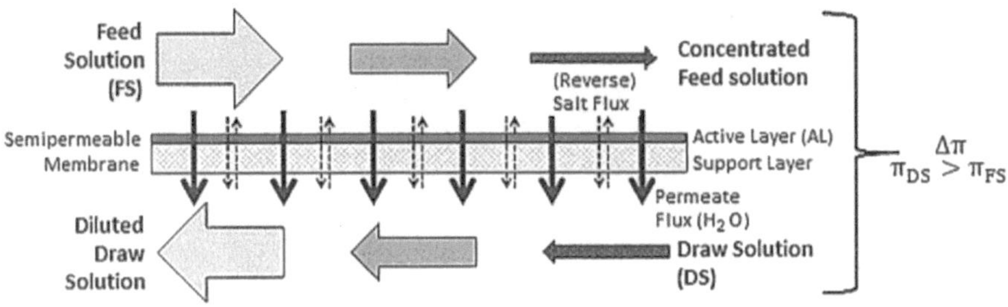

FIGURE 8.4 Principle of FO in counter-current flow mode, FS and DS flow in opposite directions to each other [40].

TABLE 8.2
Some Applications of FO in Wastewater Treatment

Type of Feed Treated	DS Type	Results	Reference
RO concentrate from dairy wastewater treatment plant	1 M NaCl	Average permeate flux of 15.1 L $m^{-2}h^{-1}$	[46]
Wastewater from grain processing	brine from CO_2 sequestration site	Flux = 5–15 L $m^{-2}h^{-1}$	[47]
Secondary effluent from industrial wastewater treatment plant	1 M NaCl, Na_2SO_4, $MgCl_2$	67% water recovery, flux = 13 L $m^{-2}h^{-1}$	[48]
Municipal wastewater	0.2–4.0 M NaCl	100% rejection of SO_4^{2-}, PO_4^{3-}; Mg^{2+} = 90% rejection, K^+ = 60% rejection	[49]
Municipal wastewater	2 M $MgCl_2$	NH_4^+ < 20 mg L^{-1}; TOC < 20 mg L^{-1}, PO_4^{3-} > 70% rejection, polysaccharide > 70% rejection	[50]

recovered by thermal separation. For sulfate-based draw solutes like $Al_2(SO_4)_3$, Mg(SO_4), and Cu(SO_4), recovery by precipitation is usually adopted, while stimuli-based recovery processes are employed for hydrogels and magnetite nanoparticles [43].

Increased demand for water and promising results from FO studies in the recent years have sped up the development of FO technologies, including many membrane development works aimed at improving and commercializing FO. Many studies looked at how new membranes developed performed in a variety of applications such as different feed and draw solutions, testing modes, and operating conditions, like flow rate, temperatures, and concentrations [44,45]. Some applications of FO are shown in Table 8.2.

8.4 THERMALLY DRIVEN MEMBRANE PROCESSES – MEMBRANE DISTILLATION

Membrane distillation (MD) is a process in which substances are separated based on their vapor pressure differences. In MD, heat applied causes the more volatile

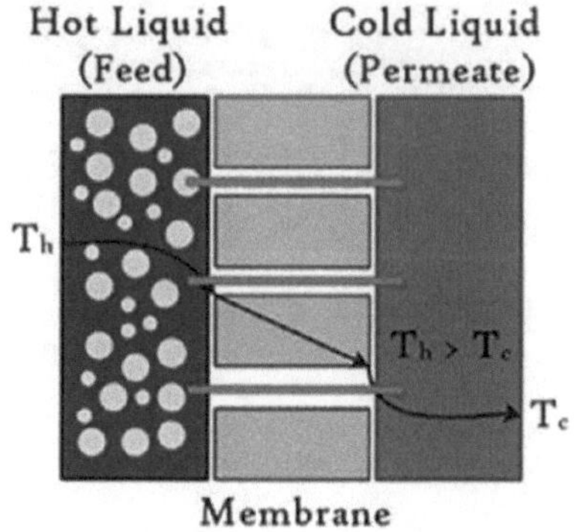

FIGURE 8.5 Principle of MD. (Adapted from Ref. [51].)

substances to move across the microporous hydrophilic membrane, and thus the driving force of this process is the thermal gradient across the microporous membrane [51]. Figure 8.5 shows the principle of MD. The hydrophilic nature of the membrane prevents the passing of other materials through the membrane and only allows for the volatile vapor molecules to pass through.

Due to the nature of the process, membranes used for MD are characterized by low resistance to mass transfer (to enhance free flow of mass), low thermal conductivity (to enhance maintenance of heat in the system), and low wetting ability. In addition, membranes for this process must be thermally stable at high temperatures. They should also be chemically stable and have high permeabilities. Most important of all, membranes for this process must be highly hydrophobic, hence fabrication of these membranes must be done with hydrophobic polymers with low surface energies [52].

Conventionally, there are four main configurations adopted in MD. These configurations play an important role in the separation efficiency and cost of running the process. These are direct contact membrane distillation (DCMD), air gap membrane distillation (AGMG), sweeping gas membrane distillation (SGMD), and vacuum membrane distillation (VMD). These are shown in Figure 8.6. The main difference between these configurations is their operational modes. This operational difference is in the way the vaporized permeate is collected/condensed and recovered. Of all these configurations, the DCMD is the most widely applied [53]. Other types of MD configuration have been developed to improve conventional configurations in terms of energy utilization and permeate flow. These include thermostatic sweeping gas membrane distillation (TSGMD), multi-effect membrane distillation (MEMD), vacuum multi-effect membrane distillation (V-MEMD), material gap membrane distillation (MGMD), and permeate gap membrane distillation (PGMD) [51].

The application of MD has a number of advantages. The process can be operated using low-grade heat such as low-temperature industrial streams and heat generated from renewable sources (wind, solar, tidal, etc.). Again, compared to pressure-driven processes such as RO, MD requires low hydrostatic pressure. In terms of contaminant rejection, MD is very efficient, showing up to 100% rejection efficiency for colloids, salts, macromolecules, and other non-volatile compounds. All these notwithstanding,

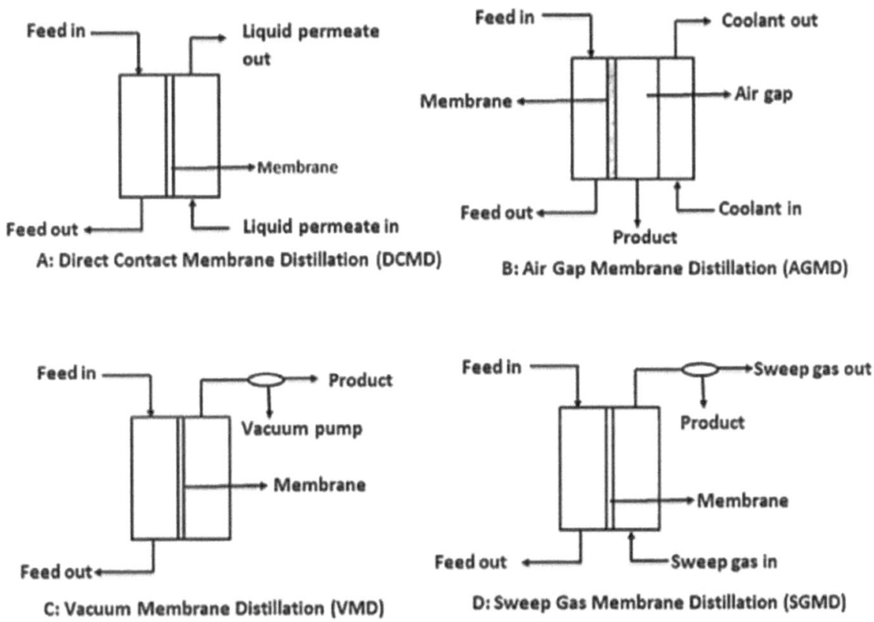

FIGURE 8.6 Conventional MD configurations [54].

TABLE 8.3
Some Applications of Thermally Driven Membrane Processes in Wastewater Treatment

Type of Feed Treated	Removal/Rejection Efficiency	Reference
Textile wastewater	Color = 95.3%; COD = 90.8%; TDS = 93.7%	[56]
Dairy wastewater	> 99% rejection of total organic carbons	[57]
Olive mill wastewater	< 2 g L^{-1} of TOC in permeate after treatment	[58]
Industrial dyeing wastewater	COD = 96%, color = 100%	[59]
Fermentation wastewater	COD = 95%, TOC = 95%	[60]

MD has some drawbacks. The process is highly susceptible to temperature polarization that leads to low permeate fluxes. The lack of commercially available membrane modules makes large-scale application of the process expensive and limited. In addition, the economic viability of the process cannot be fully ascertained as the process lacks history [55]. Table 8.3 shows some applications of MD in wastewater treatment.

8.5 ELECTRICALLY DRIVEN MEMBRANE PROCESSES

Electrically driven membrane processes are mainly adapted for the removal of charged fractions from a solution or suspension. These processes depend on electrical

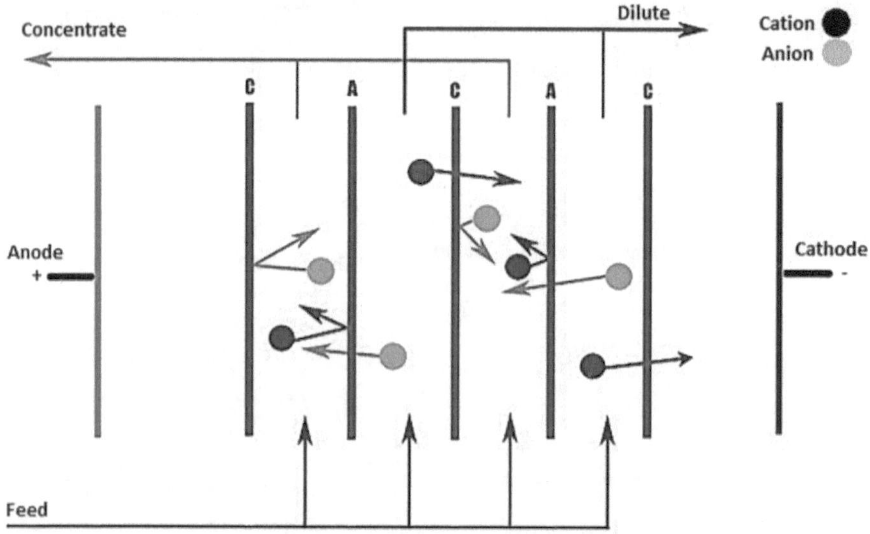

FIGURE 8.7 Schematic diagram showing the principle of ED. A = anion exchange membrane; C = cation exchange membrane, arranged alternately between two electrodes – the anode and the cathode [61].

voltage differences provided by an electric field to remove ions from wastewater streams [61].

The most widely used electrically driven membrane process is electrodialysis (ED)/electrodialysis reversal (EDR). ED/EDR utilizes ion-selective membranes to separate ions from a wastewater stream under the influence of an electric field. The result of this process is the production of two streams: the dilute stream, which is depleted of ions, and the concentrate stream, which is rich in ions [62,63]. In principle, during ED/EDR, a train of ion-exchange membranes known as cation exchange membranes (CEM) and anion exchange membranes (AEM) are arranged alternately, between which are the flow channels for the feed and the product streams. This arrangement, as shown in Figure 8.7, forms the ED stack [64].

As the feed solution flows through the system, cations migrate to the cathode, passing through the CEM, while anions migrate toward the anode through the AEM. It is worthy to note that as the migration goes on, CEM blocks anions from passing through them while allowing cations. Similarly, AEM blocks cations from passing through them, while allowing anions to pass through [64]. After a long while of operation of ED, the positions of the electrodes are reversed while keeping the ED stack intact. This is known as EDR. This is useful in controlling membrane fouling [65]. Many applications of ED/EDR in wastewater treatment are documented in the literature. A typical case is an EDR plant in Moody Gardens, Galveston, Texas, which processes a 1,000 $m^3 day^{-1}$ of municipal wastewater for reuse, supplying fishponds and plants with water [66]. Some other applications of this process are listed in Table 8.4 below.

ED/EDR has proved to be advantageous and competitive in the wastewater treatment terrain, showing more tolerance for feed with low quality. In addition, ED/EDR

TABLE 8.4
Applications of Electrically Driven Membrane Processes in Wastewater Treatment

Type of Feed Treated	Results	Reference
Electroplating wastewater	95%–99% removal of $NiCl_2$, $NiSO_4$, and H_3BO_3	[67]
Spent electroless nickel plating wastewater	>50% removal of phosphite, sulfate, and sodium	[68]
Wastewaters from cyanide-free plating process	99.7% removal of Cu	[69]
Wastewater from tanning industries	~90% removal of Cl^-, ~50% removal of SO_4^{2-}, ~100% removal of Cr(III)	[70]
Municipal wastewater (which already underwent secondary treatment)	100% effectiveness in meeting discharge limits	[71]
Secondary-treated municipal wastewater	Removal efficiencies: TDS = 70.3%, conductivity = 72%, Calcium = 84%, Chloride = 76%, Fluoride = 64%, alkalinity = 60%	[72]
Wastewater from China Steel corporation	Removal efficiencies: Cl^- = 98%, SO_4^{2-} = 80% and COD = 51%	[65]

membranes have shown more resistance to membrane fouling, especially for feed with TDS < 10,000 mg L^{-1}. In terms of pressure requirements, ED/EDR operates at low pressure yet achieves high removal of contaminants [73,74].

All these notwithstanding, ED/EDR has some drawbacks. The process does not remove non-ionized particles like bacteria and viruses, which are very harmful. Again, at salinities above 10,000 mg L^{-1}, ED/EDR will be very expensive to operate, as more energy will be required to separate the highly concentrated feed [64,75].

8.6 MODULE TYPES AND CONFIGURATION

Large-scale applications of membranes require large surface areas within the range of hundreds to thousands of square meters. To achieve this, membranes are packaged strategically in packings known as modules. A membrane module is the smallest practical unit that contains a set of membrane areas and their supporting structures [76]. Membrane modules and configuration have a great bearing on the membrane process in terms of cost, membrane fouling, and reliability. For example, one of the most basic requirements of a membrane module is for it to be easily replaceable [76]. An overview of the most common membrane modules is given in the next subtopics.

8.6.1 Plate-and-Frame Module

This is one of the first membrane modules to be designed. In this module, the membrane, feed spacers, and product spacers are layered together between two end plates. The feed mixture is forced across the membrane's surface, some of which

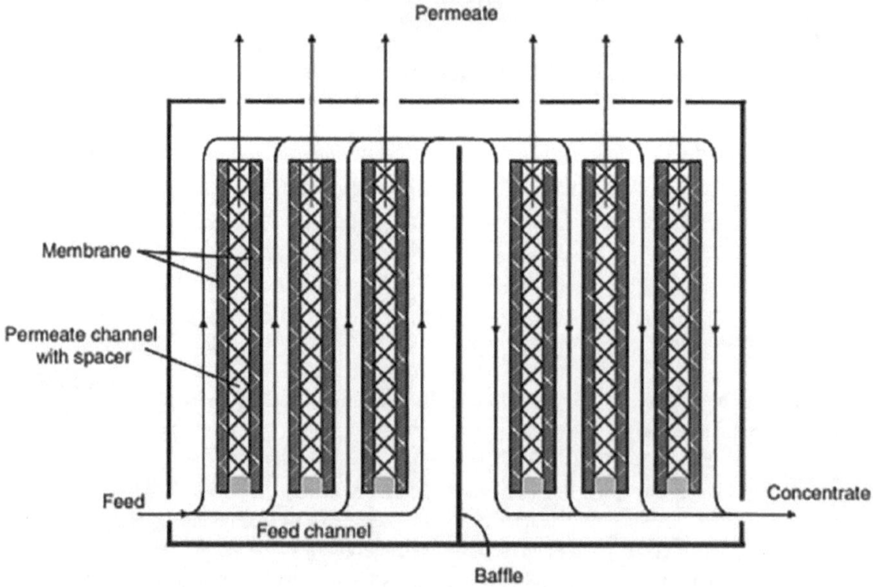

FIGURE 8.8 Schematic diagram of a plate-and-frame module. (Adapted from Ref. [78].)

flows through the membrane into the collection channel. The permeate water enters the membrane envelope and is collected through the central collection channel. Plate-and-frame module provides better flow control over the membrane penetration and feed sides, but this requires a great number of spacer plates and seals that can lead to higher production costs [77]. Figure 8.8 shows a schematic diagram of a plate-and-frame membrane module.

This membrane module is hardly used these days. Many of the applications of the plate-and-frame membrane module are found in ultrafiltration and pervaporation. For a typical plate-and-frame module, the packing density is within the range of 148–492 m^2/m^3 [79].

8.6.2 Tubular Membrane Module

This module is made up of an outer casing, referred to as a shell, within which the tubular membrane is built. The membrane is usually made of porous paper or fiberglass. This module has the lowest surface area (< 300 m^2m^{-3}) and is mostly used for ultrafiltration. The tubular membrane module is adapted to the treatment of feed solutions with high contents of suspended solids and can easily be cleaned physically using foam balls [80]. One of the notable advantages of tubular membrane modules is their ability to resist severe fouling due to their operation in turbulent conditions (>3,000 Re).

8.6.3 Spiral Wound

The spiral wound module is by far the commonest module applied especially in reverse osmosis. The module is made up of the membrane and permeates spacers wound around a porous collection tube usually in the center of the module as shown in Figure 8.9. Feed

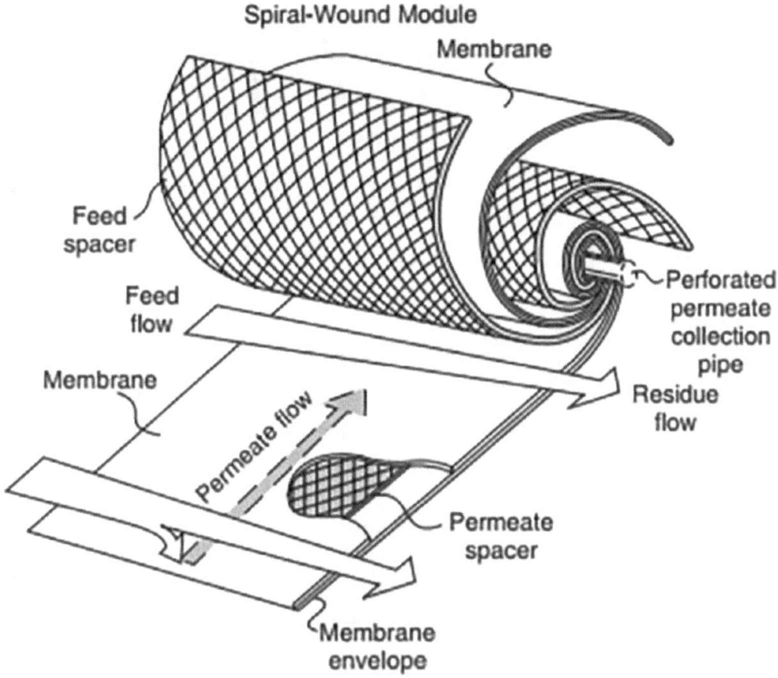

FIGURE 8.9 Spiral wound membrane module. (Adapted from Ref. [82].)

to be treated is pumped across the membrane surface, where permeate passes through the membrane element and enters the membrane envelope, spirals inward to the central perforated collection pipe, and then channeled into a permeate tank. A typical spiral wound membrane module has a packing density in the range of 300–1,000 m^2m^{-3} [81].

8.6.4 Hollow Fiber

This membrane module is known for its high packing density (500–5,000 m^2m^{-3}), resulting in high rates of permeate production. In this module, a bundle of hollow fiber membranes is housed in a pressure vessel. Typically, the membranes within the module are 10–20 cm in diameter and 1.0–1.6 m in length [83]. The feed to be treated is supplied either through the inside of the hollow fibers (known as inside or bore side feed) as shown in Figure 8.10 or from the outside of the hollow fibers (known as outside or shell side feed). Permeate water passes through the walls of the fiber to the other side for collection [84].

8.7 FACTORS AFFECTING DIRECT MEMBRANE FILTRATION

8.7.1 Flow Models in Membranes

There are basically two flow models in membrane processes. These are the cross flow and the dead-end flow. During the cross flow, the bulk solution flowing under pressure is parallel to the surface of the membrane through which water

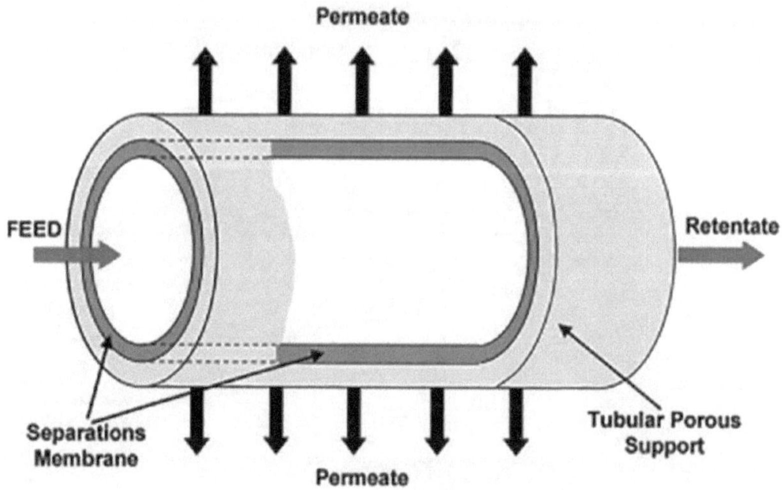

FIGURE 8.10 Hollow fiber membrane module [85].

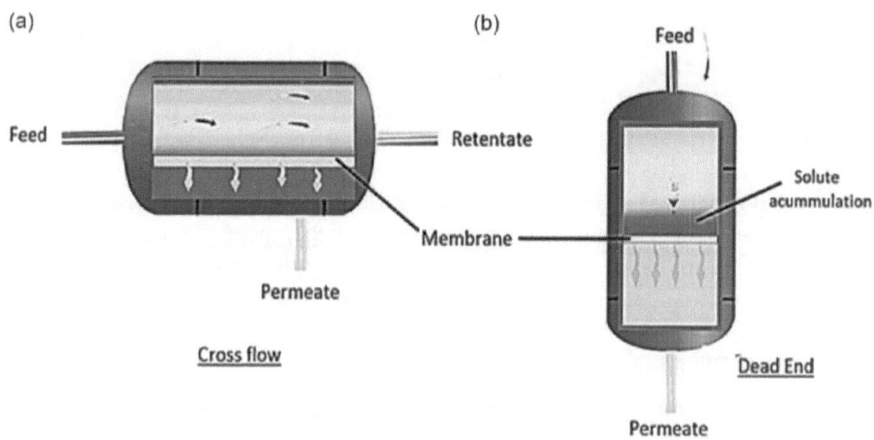

FIGURE 8.11 Cut section of filtration vessels: (a) cross-flow mode and (b) dead-end mode.

permeates perpendicularly to the flow direction of the bulk solution (Figure 8.11a). The energy-intensive cross-flow mode increases the membrane life span and reduces permeate drop. While the cross-flow model is used in batch, semibatch, and continuous processes, the second model, the dead-end flow is used mostly in batch processes. In dead-end flow, the flow direction of the bulk solution is perpendicular to the membrane surface without a retentate stream (Figure 8.11b). Although simple in design, the dead end is not simple to operate and high maintenance is involved due to the accumulation of solutes and decrease in permeability.

There are two primary causes of flux decline which are of major concern to the membrane treatment industry, namely, concentration polarization and fouling.

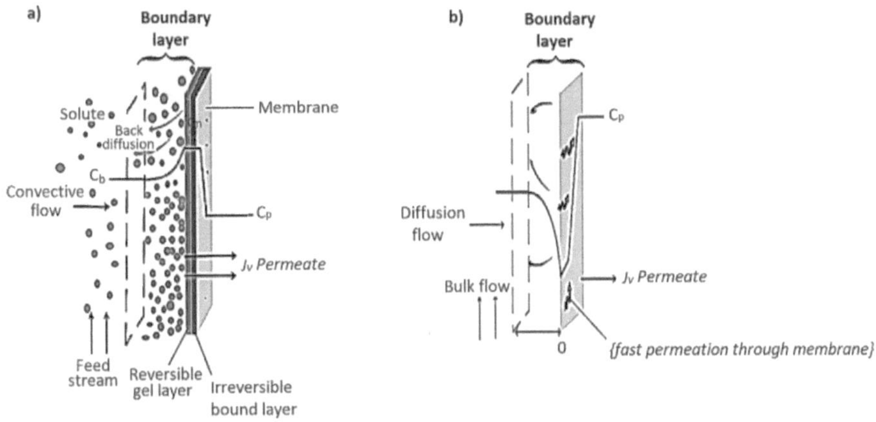

FIGURE 8.12 A representation of concentration profiles: (a) mass transfer limited by membrane and (b) mass transfer limited by boundary layer.

8.7.2 Concentration Polarization

The inability of solutes or particles transferred by convection to permeate through the membrane pore sizes results in the accumulation of the solutes at the boundary layer adjoined to the surface of the membrane. Due to the difference in particle concentration differences at the membrane surface and the permeate side, the solute diffuses back into the bulk solution until an equilibrium is established, leading to the formation of a steady-state boundary layer [86,87]. This phenomenon is referred to as concentration polarization. Figure 8.12a shows the concentration profile for pressure-driven membranes where the mass transfer is limited by the membrane and occurring fouling (discussed in detail in Section 8.8.3). Figure 8.12b depicts the concentration profile for solute transport limited by diffusion such as membrane crystallization, membrane distillation, dialysis, electrodialysis, etc.

The negative effect of CP on membrane filtration units include a reduction in trans-membrane flux, a decrease in membrane rejection capacity, and enhanced membrane fouling. Also, precipitation would occur when concentration at the membrane surface exceeds the solubility limit. Different models such as film theory, gel layer, osmotic pressure model, resistance in-series model, resistance-osmotic pressure model have been used to describe concentration polarization. The application and limitations of some of these models are presented in Table 8.5.

8.7.3 Fouling

Fouling is a phenomenon that occurs through the adsorption and accumulative deposition of bulk solution solutes on the surface of the membrane (reversible fouling) or in the membrane pores (irreversible fouling). The tendency and behavior of fouling are complex and vary due to the mode of operation, properties of the membrane, and the nature of foulants. Classification of fouling is therefore based on the foulants, broadly as colloidal or particulate, organic, inorganic or scaling and biofouling as

TABLE 8.5
Application Areas and Limitations of Different CP Models [88]

Model	Application	Limitation
Film theory	Evaluates permeate flux based on chemical potential gradient	Assumption of a constant mass transfer coefficient and the neglect of localized CP concept close to the surface of the membrane
Gel layer	Evaluates permeate flux based on constant macromolecules gel layer resistance and membrane resistance	Assumes a constant gel layer concentration, mass transfer coefficient from theories of convective heat transfer to the impermeable surface is adapted, incorrect conclusion that correlates higher flux to increasing solute concentration in gel
Osmotic pressure	Determines the osmotic pressure adjacent to the surface of the membrane	Not applicable to MF, UF, and loose-end NF
Resistance-in-series	Estimation of permeate flux at different fouling stages	Specific to behavior of foulants such as mono-dispersed colloids, CaSO4
Resistance-osmotic pressure	Incorporates the osmotic pressure, resistance, and CP	Unknown performance for multicomponent systems

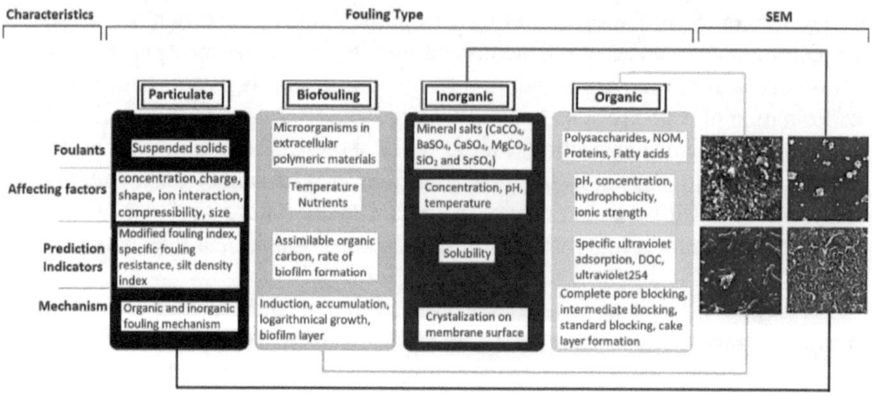

FIGURE 8.13 Characteristics of the various fouling types. (Adapted and modified from [90,91].)

depicted in Figure 8.13. The relative proportion of the sources of fouling has been estimated as 30% particulate fouling, 50% organic fouling, and 20% associated with inorganic fouling. The SEM images depict various characteristics that includes a conglomerate of small particles for the particulate, slimy deposits for biofouling, crystalline structures for inorganic, and thin smooth covering for organic [89].

One other fouling mechanism aside from adsorption and precipitation is pore blocking. Pore blocking is a function of the size and shape of the solute in connection with the pore size distribution of the membrane. There are four types, viz complete pore blocking, partial or intermediate blocking of membrane opening, constriction

TABLE 8.6
Summary of Pore-Blocking Mechanism

Pore-Blocking Type	Schematic	Equation	Assumptions
Complete $n = 2$	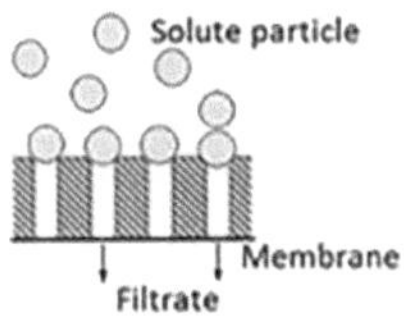	KcbV = Q0 (1-e-kbt)	Without superimposition of particles occurring, each particle blocks a pore. Blocked area is proportional to the permeate volume.
Intermediate $n = 1$	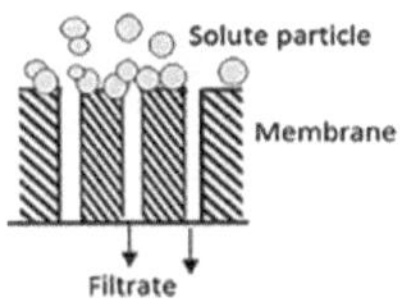	KibV = In (1+KibQ0t)	Overlapping of particles occurs while others block the membrane pores. Not every particle blocks the pores.
Standard $n = 1.5$	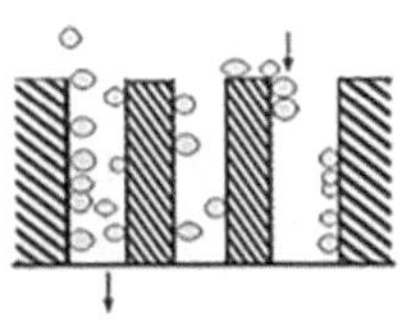	Ksbt2 = tv-1Q0	Size of particle allows for deposit into entrance and deposit within the walls of the pores; Decrease of the pore volume is proportional to volume of particle deposited.
Cake $n = 0$	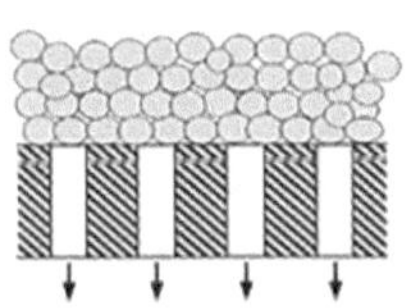	Kcf = 2tv-2Q0	Size of the particles is large enough to block the membrane pore at the surface leading to a cake formation on the surface.

n, blocking index; K, constant specific to each blocking mechanism; Q_0, initial permeate flow; t, filtration time, and V, total filtered volume.

in the membrane pore walls (standard blocking), and cake formation. The schematic view, mathematical expression, and assumptions for the various pore-blocking types are expounded in Table 8.6 [92,93].

8.8 FACTORS OF CONCERN

To deal with flux decline in membrane processes, it is very crucial to identify the contributing factors. Looking into the complete membrane treatment system, factors of concern can be categorized into membrane type and model, feed stream composition, and operating parameters in relation to the membrane module.

8.8.1 Membrane Type

Membranes with larger pore size, greater porosity, and higher MWCO have lower inherent filtration resistance and higher flux. However, this will affect treatment efficiency resulting in more solute particles permeating through the membrane pore while CP and membrane surface fouling would have been alleviated [94]. For a membrane to possess a better anti-fouling characteristic and increased flux, membrane pore size must be uniformly distributed and interconnected. Typical of the pore size is their dissimilarity with foulant size. Also, the interaction between the membrane and the foulants cannot be neglected as the surface of membranes with the same electric charges as foulants lead to electrostatic repulsion and reduced adsorption of foulants onto the surface. Membrane technologies are geared toward developing hydrophilic membranes (reduced hydrophobicity) that increase permeation and selectivity through the incorporation of attractive interactions between membrane and water and low affinity between membrane and organic matter [95]. Hydrophobic microporous membranes are adopted in pervaporation, membrane gas absorption, and membrane distillation preferably for the removal of small amounts of solvent from the feed stream consisting of mostly water. Furthermore, the roughness affects the liquid contact angle. Smooth surfaces improve the anti-fouling ability of the membranes as roughness facilitates the accumulation [96].

8.8.2 Feed Stream Composition

The feed streams requiring treatment are from different sources. This implies that they are made up of different components that affect membrane performance. As discussed earlier, fouling types are associated with the type of foulants that directly make up the composition of the feed streams. Irrespective of the composition, the pH and ionic strength affect the interaction of the membrane and foulants to the extent of changing the membrane structure and characteristics.

Due to protonation and de-protonation of the functional groups, the sign and amount of charge of foulants may differ at different pH conditions. The higher the pH, the more negative the zeta potential and completion of de-protonation. Consequently, when the pH is farther away from the isoelectric point of the membrane, the foulants possess a high surface charge density, thereby the electrostatic repulsion tendency is strong [97]. A lower pH value induces intermolecular adhesion where the membrane surface charge and foulant charge are opposite. Therefore, the foulants with decreased molecular size are easily adsorbed onto the surface of the membrane [98].

The surface charge of the membrane or the dominant functional group of the foulants may vary with ionic strength and most dependently on the pH, ionic strength, and foulant size combinative behavior on flux. While fouling decreases with increasing ionic strength, increasing ionic strength results in decreasing flux [99].

8.8.3 Operating Parameters

Operating parameters such as cross-flow velocity, trans-membrane pressure/permeate flux, and temperature affects the CP and fouling on the membrane surface. This has a direct effect on the rate of mass transfer and a hydrodynamic effect between the feed stream and membrane surface. Optimizing these process conditions serve as a CP and fouling control measure.

8.8.3.1 Cross-Flow Velocity

The existence of shear forces evolves from the relative tangential motion of the feed stream over the membrane surface. Shear force dominates in cross-flow membrane modules spiral wound and hollow fiber where increasing the cross-flow velocity improves the shear stress, increases the membrane flux (at steady-state conditions), and reduces the membrane filtering area [100]. However, when the cross-flow velocity exceeds a threshold, it leads to flow mal-distribution, higher pressure drop, dehydration of the membrane, and increased specific energy demand. The resulting effect is increased operational cost [101]. Very high shear ranging $1\text{–}3\times10^5\,s^{-1}$ is beneficial in causing an increase in both permeate flux and membrane selectively [102].

8.8.3.2 Trans-Membrane Pressure and Permeate Flux

An increase and decrease in trans-membrane pressure and permeate flux affect the CP and fouling. When the TMP is high, particle concentration on the surface of the membrane increases, and permeation flux also ascends slowly. Due to the accumulation of solute particle deposition, the flux becomes independent of the pressure at increasing pressures. Operating at a high initial permeate flux above the critical flux creates a large hydrodynamic force at the solution–membrane interface resulting in severe CP and increased foulant adsorption on the membrane surface. Hence, the TMP should not be too low, as such a condition affects the flux, CP, and fouling [100,103].

8.8.3.3 Temperature

Viscosity decreases with an increase in temperature. This increases the solubility of suspended particles, resulting in increased mass transfer diffusion coefficient and bulk transport of solute to the surface of the membrane. This improves the cross-flow velocity by reducing the thickness of the CP layer, hence increasing the membrane flux [16,104]. However, the higher temperatures might affect the membrane life span. It will also induce pore blocking depending on the effect of increasing temperature on the pore size of the membrane and accessibility of foulants into the pores.

8.9 METHODS OF FOULING CONTROL IN MEMBRANE PROCESSES

Membrane separation is primarily a mechanism of size exclusion. In theory, the particles that are rejected end up fouling the membrane. Some of the ways of mitigating membrane fouling are briefly described in the next subtopics.

8.9.1 Pre-Treatment

Pre-treatment refers to the initial treatment of wastewater prior to the use of membrane separation processes. Feed pre-treatment plays a critical role in the success of the membrane process. Pre-treatment is typically used to remove foulants, improve recovery and system productivity, and extend membrane life. Pre-treatment may also be used to protect the membranes from physical damage [105].

In principle, pre-treatment alters the physical, chemical, and/or biological properties of wastewater to improve the efficiency of membrane separation. The pre-treatment method used is determined by the source of the feed water. Some of these methods are discussed below.

Coagulation/flocculation helps in the reduction of turbidity, color, COD, and total suspended solids (TSS) of wastewater. During this process, chemicals known as coagulants or flocculants such as poly-aluminum chloride, ferric chloride, aluminum hydroxide, or aluminum sulfate are added to the wastewater to agglomerate smaller particles into bigger shapes to enhance easy settlement [105]. After settlement, the supernatant is fed to the membrane process for treatment. In other forms of coagulation, such as inline coagulation, coagulated water is sent directly to the membrane process without settlement. This process was found to be equally effective in mitigating membrane fouling and maintaining permeation flux [106].

Pre-filtration involves the use of low-pressure filtration processes, screening, or coarse filtration to remove suspended organic materials, micro and macro pollutants, suspended solids, inorganic particles, biological debris, etc. from wastewater before further treatment by membrane processes. Pre-filtration processes may be used before each membrane unit or in the entire membrane system [105]. Examples of pre-filtration processes include packed bed filters, strainers, filter cloths, cartridge filters, MF and UF, bag filters, etc. The pore size specifications of pre-filters range from as low as 100 to 3,000 μm depending on the nature of the effluent and the expected raw water quality. Pre-filtration has been applied in several instances to pre-treat different forms of wastewater [107].

Other forms of pre-treatment procedures include biological pre-treatment processes (use of bacteria, algae, or fungi), chemical conditioning (pH adjustment, disinfection, scale inhibition), and ion exchange [108,109].

8.9.2 Membrane Cleaning

Membrane cleaning is used to restore the permeation flux of a membrane that has been lost due to fouling. This entails removing accumulated materials from the membrane to allow permeating water to pass through.

Membrane cleaning is achieved in mainly two ways – physical and chemical cleaning. Other forms of cleaning procedures may encompass the combination of physical and chemical cleaning processes. Mostly, the modes of these cleaning procedures are carried out as in-situ or ex-situ.

During in-situ cleaning procedures, membrane elements are cleaned in place, without dismantling the membrane unit. This is known as cleaning in place (CIP) [110]. Cleaning chemicals are passed through the system multiple times depending on the nature of the soiling of the membrane. Examples of chemicals used include detergents (sodium hydroxide, potassium hydroxide, sodium carbonate, hydrochloric acid, nitric acid, phosphoric acid, etc.), sanitizers (hypochlorite, hydrogen peroxide, ozone, peracetic acid, etc.), or disinfectants [110,111]. During ex-situ cleaning, the membrane element or module is removed from the membrane tank before cleaning procedures are applied to it. This is known as cleaning out of place (COP) [112,113]. Usually, the membrane element will be submerged in a cleaning tank and soaked

TABLE 8.7
Advantages and Disadvantages of CIP and COP [115,116]

Cleaning Procedure	Advantages	Disadvantage(s)
CIP	Much faster than manual cleaning	High initial investment
	Less labor intensive, no disassembly or reassembly	
	Highly repeatable results	
	Documentation	Highly technical and requires constant training of workers
	Safer for workers, less chemical exposure	
	Aids in controlling the cost of water and chemicals	
COP	Usually less expensive than CIP systems	More labor intensive due to dismantling of membrane housing units
	Delivers consistent results	Risk of damage of parts of machines during dismantling and re-assembling
	Saves money when compared with manual cleanings by reducing the amount of time, chemicals, and water used	
	Reduces the amount of time that operators are exposed to elevated temperatures and high chemical concentrations	

with chemicals. This is referred to as submerged chemical cleaning [114]. CIP and COP have their respective advantages and disadvantages as shown in Table 8.7.

Chemical cleaning and physical cleaning procedures are applied based on the nature of membrane fouling. For reversible fouling, where foulants are merely deposited on the membrane surface, such as in osmotically driven membranes, physical cleaning may be used [117]. In instances of irreversible fouling, mainly in pressure-driven membrane processes, chemical cleaning is employed. Chemical cleaning agents like the alkaline solutions (sodium hydroxide, potassium hydroxide, and sodium carbonate) or acid-based solutions (hydrochloric acid, nitric acid, phosphoric acid, etc.) work based on membrane-foulant and foulant-chemical chemistries to remove foulants from the membrane elements. Alkaline-based cleaning solutions are used in the removal of organic stains whereas acid-based cleaning solutions are applied in the removal of inorganic foulants [118]. Examples of physical cleaning methods include ultrasonic cleaning, water washing, and high-frequency vibration [119,120].

8.10 CONCLUSIONS AND RECOMMENDATIONS

The chapter emphasizes the beneficial functionality of direct membrane filtration techniques for wastewater treatment and resource recovery. Evidently, the wastewater treatment setting is now dominated by one type of membrane filtration technique or the other. The wide range of options available, the flexibility of application, the superior permeate quality, as well as the high rate of water recovery make their applications for water reclamation a viable option.

Furthermore, the different direct membrane filtration techniques, in terms of their driving forces, operational mechanism, and suitability for a variety of wastewater compositions are highlighted. Pressure-driven membrane processes (RO, NF, UF, and MF) are the most widely applied, yet are very susceptible to fouling. ODMPs, even though they require less energy for separation, mostly require a recovery process for permeating water and drawing solute, while ED/EDR provides the option for the treatment of saline streams based on principles of ion exchange. Being able to utilize low heat for quality separation, MD provides the option of exploring wastewater treatment using renewable energy and waste heat from industrial processes. To mitigate membrane fouling and maximize throughput, some operational parameters were discussed, and fouling control measures were recommended. In light of all the above, it is recommended that further studies be conducted on fouling-resistant membranes. In addition, to explore the full potential of direct membrane filtration processes, economic analysis, and life cycle assessment also require needful attention.

REFERENCES

1. USGS, *Wastewater treatment Water Reuse.*
2. Lofrano, G. and J. Brown, Wastewater management through the ages: A history of mankind. *Science of The Total Environment*, 2010. **408**(22): p. 5254–5264.
3. Klaus, S., The Global Competitiveness Report: World Economic Forum. 2019.
4. WEF, *Global Risk Report* 2019, World Economic Forum.
5. Bhatt, A., P. Arora, and S.K. Prajapati, *Occurrence, fates and potential treatment approaches for removal of viruses from wastewater: A review with emphasis on SARS-CoV-2.* Journal of environmental chemical engineering, 2020: p. 104429.
6. Zhang, H., H. Sun, and Y. Liu, Water reclamation and reuse. *Water Environment Research*, 2020. **92**(10): p. 1701–1710.
7. Perera, M.K., J.D. Englehardt, and A.C. Dvorak, Technologies for recovering nutrients from wastewater: A critical review. *Environmental Engineering Science*, 2019. **36**(5): p. 511–529.
8. Hube, S., et al., Direct membrane filtration for wastewater treatment and resource recovery: A review. *Science of The Total Environment*, 2020. **710**: p. 136375.
9. Ravi, J., et al., Polymeric membranes for desalination using membrane distillation: A review. *Desalination*, 2020. **490**: p. 114530.
10. Kweinor Tetteh, E. and S. Rathilal, Evaluating pre-and post-coagulation configuration of dissolved air flotation using response surface methodology. *Processes*, 2020. **8**(4): p. 383.
11. Asante-Sackey, D., et al., Effect of ion exchange dialysis process variables on aluminium permeation using response surface methodology. *Environmental Engineering Research*, 2019. **25**(5): p. 714–721.
12. Kweinor Tetteh, E. and S. Rathilal, Kinetics and nanoparticle catalytic enhancement of biogas production from wastewater using a magnetized biochemical methane potential (MBMP) system. *Catalysts*, 2020. **10**(10): p. 1200.
13. Iqbal, A., et al., *Anaerobic digestion.* Water environment research: A research publication of the water environment federation, 2019. **91**(10): p. 1253–1271.
14. Van, D.P., et al., A review of anaerobic digestion systems for biodegradable waste: Configurations, operating parameters, and current trends. *Environmental Engineering Research*, 2020. **25**(1): p. 1–17.

15. Mohamadi, S., H. Hazrati, and J. Shayegan, Influence of a new method of applying adsorbents on membrane fouling in MBR systems. *Water and Environment Journal*, 2020. **34**: p. 355–366.
16. Du, X., et al., A review on the mechanism, impacts and control methods of membrane fouling in MBR system. *Membranes*, 2020. **10**(2): p. 24.
17. Hosseini, S., et al., Preparation and surface modification of PVC/SBR heterogeneous cation exchange membrane with silver nanoparticles by plasma treatment. *Journal of Membrane Science*, 2010. **365**(1–2): p. 438–446.
18. Sun, H., B. Tang, and P. Wu, Hydrophilic hollow zeolitic imidazolate framework-8 modified ultrafiltration membranes with significantly enhanced water separation properties. *Journal of Membrane Science*, 2018. **551**: p. 283–293.
19. Morihama, A. and J. Mierzwa, Clay nanoparticles effects on performance and morphology of poly (vinylidene fluoride) membranes. *Brazilian Journal of Chemical Engineering*, 2014. **31**: p. 79–93.
20. Thamaraiselvan, C., et al., Characterization of a support-free carbon nanotube-microporous membrane for water and wastewater filtration. *Separation and Purification Technology*, 2018. **202**: p. 1–8.
21. Li, J., et al., Metal-organic framework membranes for wastewater treatment and water regeneration. *Coordination Chemistry Reviews*, 2020. **404**: p. 213116.
22. Palmarin, M.J., S. Young, and J. Chan, Recovery of a hybrid and conventional membrane bioreactor following long-term starvation. *Journal of Water Process Engineering*, 2020. **34**: p. 101027.
23. Erkan, H.S., et al., Performance evaluation of conventional membrane bioreactor and moving bed membrane bioreactor for synthetic textile wastewater treatment. *Journal of Water Process Engineering*, 2020. **38**: p. 101631.
24. Abdullah, N., et al., Chapter 2- membranes and membrane processes: Fundamentals, in *Current Trends and Future Developments on (Bio-) Membranes*, A. Basile, S. Mozia, and R. Molinari, Editors. 2018, Elsevier. p. 45–70.
25. Lu, G., et al., Inorganic membranes for hydrogen production and purification: a critical review and perspective. *Journal of Colloid and Interface Science*, 2007. **314**(2): p. 589–603.
26. Cot, L., et al., Inorganic membranes and solid state sciences. *Solid State Sciences*, 2000. **2**(3): p. 313–334.
27. Prapulla, S.G. and N.G. Karanth, *Fermentation (Industrial) | Recovery of Metabolites*, in *Encyclopedia of Food Microbiology (Second Edition)*, C.A. Batt and M.L. Tortorello, Editors. 2014, Academic Press: Oxford. p. 822–833.
28. Sagle, A. and B. Freeman, Fundamentals of membranes for water treatment. *The Future of Desalination in Texas*, 2004. **2**(363): p. 137.
29. Bazzarelli, F., L. Giorno, and E. Piacentini, *Dense Membranes*, in *Encyclopedia of Membranes*, E. Drioli and L. Giorno, Editors. 2016, Springer Berlin Heidelberg: Berlin, Heidelberg. p. 530–532.
30. Mohshim, D.F., et al., Latest Development on Membrane Fabrication for Natural Gas Purification: A Review. *Journal of Engineering*, 2013. **2013**: p. 101746.
31. Kayvani Fard, A., et al., Inorganic membranes: Preparation and application for water treatment and desalination. *Materials*, 2018. **11**(1): p. 74.
32. Cui, Z.F., Y. Jiang, and R.W. Field, Chapter 1- Fundamentals of pressure-driven membrane separation processes, in *Membrane Technology*, Z.F. Cui and H.S. Muralidhara, Editors. 2010, Butterworth-Heinemann: Oxford. p. 1–18.
33. Hua, F.L., et al., Performance study of ceramic microfiltration membrane for oily wastewater treatment. *Chemical Engineering Journal*, 2007. **128**(2): p. 169–175.
34. Mohammadi, T. and A. Esmaeelifar, Wastewater treatment of a vegetable oil factory by a hybrid ultrafiltration-activated carbon process. *Journal of Membrane Science*, 2005. **254**(1): p. 129–137.

35. Bunani, S., et al., Application of nanofiltration for reuse of municipal wastewater and quality analysis of product water. *Desalination*, 2013. **315**: p. 33–36.
36. Nataraj, S.K., K.M. Hosamani, and T.M. Aminabhavi, Distillery wastewater treatment by the membrane-based nanofiltration and reverse osmosis processes. *Water Research*, 2006. **40**(12): p. 2349–2356.
37. Tałałaj, I.A., P. Biedka, and I. Bartkowska, Treatment of landfill leachates with biological pretreatments and reverse osmosis. *Environmental Chemistry Letters*, 2019. **17**(3): p. 1177–1193.
38. Cath, T.Y., et al. Standard methodology for evaluating membrane performance in osmotically driven membrane processes. *Desalination*, 2013. **312**, 31–38. doi:10.1016/j.desal.2012.07.005.
39. Wei, J., Q. She, and X. Liu, Insights into the influence of membrane permeability and structure on osmotically-driven membrane processes. *Membranes*, 2021. **11**(2): p. 153.
40. Haupt, A. and A. Lerch, Forward osmosis application in manufacturing industries: A short review. *Membranes*, 2018. **8**: p. 47. doi:10.3390/membranes8030047.
41. Suwaileh, W., et al. Forward osmosis membranes and processes: A comprehensive review of research trends and future outlook. *Desalination*, 2020. **485**: p. 114455. doi:10.1016/j.desal.2020.114455.
42. Suwaileh, W.A., et al. Advances in forward osmosis membranes: Altering the sub-layer structure via recent fabrication and chemical modification approaches. *Desalination*, 2018. **436**, 176–201. doi:10.1016/j.desal.2018.01.035.
43. Luo, H., et al., A review on the recovery methods of draw solutes in forward osmosis. *Journal of Water Process Engineering*, 2014. **4**: p. 212–223.
44. Salehi, H., M. Rastgar, and A. Shakeri, Anti-fouling and high water permeable forward osmosis membrane fabricated via layer by layer assembly of chitosan/graphene oxide. *Applied Surface Science*, 2017. **413**: p. 99–108.
45. Yang, H., et al., Highly water-permeable and stable hybrid membrane with asymmetric covalent organic framework distribution. *Journal of Membrane Science*, 2016. **520**: p. 583–595.
46. Haupt, A. and A. Lerch, Forward osmosis treatment of effluents from dairy and automobile industry–results from short-term experiments to show general applicability. *Water Science and Technology*, 2018. **78**(3): p. 467–475.
47. Salih, H.H. and S.A. Dastgheib, Treatment of a hypersaline brine, extracted from a potential CO2 sequestration site, and an industrial wastewater by membrane distillation and forward osmosis. *Chemical Engineering Journal*, 2017. **325**: p. 415–423.
48. Wünsch, R., et al., Water recovery by forward osmosis from challenging industrial effluents towards zero liquid discharge: selection of a suitable draw solution. *Proceedings of the 12. Aachener Tagung Wassertechnologie*, Aachen, Germany, 2017: p. 24–25.
49. Gao, Y., et al., Direct concentration of municipal sewage by forward osmosis and membrane fouling behavior. *Bioresource Technology*, 2018. **247**: p. 730–735.
50. Sun, Y., et al., Membrane fouling of forward osmosis (FO) membrane for municipal wastewater treatment: A comparison between direct FO and OMBR. *Water Research*, 2016. **104**: p. 330–339.
51. Biniaz, P., et al., Water and wastewater treatment systems by novel integrated membrane distillation (MD). *ChemEngineering*, 2019. **3**(1): p. 8.
52. Khayet, M. and T. Matsuura, *Membrane Distillation: Principles and Applications*. 2011: Elsevier.
53. Shirazi, M.M. and A. Kargari, A review on applications of membrane distillation (MD) process for wastewater treatment. *Journal of Membrane Science and Research*, 2015. **1**(3): p. 101–112.
54. Ezugbe, E.O. and S. Rathilal, Membrane technologies in wastewater treatment: A review. *Membranes*. 2020. **10**, 89.
55. Sanmartino, J.A., et al., *Desalination by Membrane Distillation*. 2016. p. 77–109.

56. Mokhtar, N.M., et al., The potential of direct contact membrane distillation for industrial textile wastewater treatment using PVDF-Cloisite 15A nanocomposite membrane. *Chemical Engineering Research and Design*, 2016. **111**: p. 284–293.
57. Hausmann, A., et al. Direct contact membrane distillation of dairy process streams. *Membranes*. 2011. **1**, 48–58.
58. Carnevale, M.C., et al., Direct contact and vacuum membrane distillation application for the olive mill wastewater treatment. *Separation and Purification Technology*, 2016. **169**: p. 121–127.
59. Li, F., et al., Direct contact membrane distillation for the treatment of industrial dyeing wastewater and characteristic pollutants. *Separation and Purification Technology*, 2018. **195**: p. 83–91.
60. Wu, Y., et al., Performance and fouling mechanism of direct contact membrane distillation (DCMD) treating fermentation wastewater with high organic concentrations. *Journal of Environmental Sciences*, 2018. **65**: p. 253–261.
61. Beier, S.P., *Electrically Driven Membrane Processes*. 2014.
62. Strathmann, H., *Ion-exchange Membrane Separation Processes*. 2004: Elsevier.
63. Tanaka, Y., *Ion Exchange Membranes: Fundamentals and Applications, Series 12*. Membrane Science and Technology: Elsevier, Netherlands, 2007.
64. Gurreri, L., et al., *Chapter 6- Electrodialysis for wastewater treatment—Part I: Fundamentals and municipal effluents, in Current Trends and Future Developments on (Bio-) Membranes*, A. Basile and A. Comite, Editors. 2020, Elsevier. p. 141–192.
65. Chao, Y.-M. and T.M. Liang *A feasibility study of industrial wastewater recovery using electrodialysis reversal*. Desalination, 2008. **221**, 433–439 DOI: https://doi.org/10.1016/j.desal.2007.04.065.
66. Allison, R.P., Electrodialysis reversal in water reuse applications. *Desalination*, 1995. **103**(1): p. 11–18.
67. Benvenuti, T., et al., Recovery of nickel and water from nickel electroplating wastewater by electrodialysis. *Separation and Purification Technology*, 2014. **129**: p. 106–112.
68. Li, C.L., et al., Recovery of spent electroless nickel plating bath by electrodialysis. *Journal of Membrane Science*, 1999. **157**(2): p. 241–249.
69. Scarazzato, T., et al., Treatment of wastewaters from cyanide-free plating process by electrodialysis. *Journal of Cleaner Production*, 2015. **91**: p. 241–250.
70. Rao, J.R., et al., Electrodialysis in the recovery and reuse of chromium from industrial effluents. *Journal of Membrane Science*, 1989. **46**(2): p. 215–224.
71. Gally, C.R., et al., Electrodialysis for the tertiary treatment of municipal wastewater: Efficiency of ion removal and ageing of ion exchange membranes. *Journal of Environmental Chemical Engineering*, 2018. **6**(5): p. 5855–5869.
72. Goodman, N.B., et al., A feasibility study of municipal wastewater desalination using electrodialysis reversal to provide recycled water for horticultural irrigation. *Desalination*, 2013. **317**: p. 77–83.
73. Reahl, E.R., *San Diego Uses Ionics EDR Technology to Produce Low Salinity Irrigation Water From Reclaimed Municipal Wastewater*. 2005, American Membrane Technology Association.</
74. Strathmann, H. *Electrodialysis, a mature technology with a multitude of new applications*. Desalination, 2010. **264**, 268–288 DOI: https://doi.org/10.1016/j.desal.2010.04.069.
75. Galama, A.H., et al. *Seawater predesalination with electrodialysis*. Desalination, 2014. **342**, 61–69 DOI: https://doi.org/10.1016/j.desal.2013.07.012.
76. Noble, R. and S.A. Stern, *Membrane Separations Technology: Principles and Applications*. 1995, Denver, Colorado, USA: Elsevier Science & Technology. 731.
77. Günther, R., et al., Engineering for high pressure reverse osmosis. *Journal of Membrane Science*, 1996. **121**(1): p. 95–107.

78. Balster, J., >>*Plate and Frame Membrane Module*>>, in *Encyclopedia of Membranes*, E. Drioli and L. Giorno, Editors. 2015, Springer Berlin Heidelberg: Berlin, Heidelberg. p. 1–3.
79. Kucera, J., *Reverse Osmosis: Industrial Processes and Applications*. 2015: John Wiley & Sons.
80. Cheryan, M., *Ultrafiltration and microfiltration handbook*. 1998: CRC press.
81. Mulder, M. and J. Mulder, *Basic principles of membrane technology*. 1996: Springer Science & Business Media.
82. Baker, R.W., *Membrane technology and applications*. 2012: Wiley Online Library.
83. McKeen, L.W., *Permeability properties of plastics and elastomers*. 2016: William Andrew.
84. Ismail, A.F., K.C. Khulbe, and T. Matsuura, *Chapter 5- RO Membrane Module, in Reverse Osmosis*, A.F. Ismail, K.C. Khulbe, and T. Matsuura, Editors. 2019, Elsevier. p. 117–141.
85. McKeen, L., *4-markets and applications for films, containers, and membranes*. Permeability properties of plastics and elastomers, 3rd edn. William Andrew Publishing, Oxford, 2012: p. 59–75.
86. Hadi, S., et al., *Experimental and theoretical analysis of lead Pb2+ and Cd2+ retention from a single salt using a hollow fiber PES membrane*. Membranes, 2020. **10**(7): p. 136.
87. May, P., S. Laghmari, and M. Ulbricht, Concentration Polarization Enabled Reactive Coating of Nanofiltration Membranes with Zwitterionic Hydrogel. *Membranes*, 2021. **11**(3): p. 187.
88. Shirazi, S., C.-J. Lin, and D. Chen, Inorganic fouling of pressure-driven membrane processes—A critical review. *Desalination*, 2010. **250**(1): p. 236–248.
89. Peña, N., et al., Evaluating impact of fouling on reverse osmosis membranes performance. *Desalination and Water Treatment*, 2013. **51**(4–6): p. 958–968.
90. Jiang, S., Y. Li, and B.P. Ladewig, A review of reverse osmosis membrane fouling and control strategies. *Science of the Total Environment*, 2017. **595**: p. 567–583.
91. Sadr, S.M. and D.P. Saroj, *Membrane technologies for municipal wastewater treatment*, in *Advances in membrane technologies for water treatment*. 2015, Elsevier. p. 443–463.
92. Wang, F. and V.V. Tarabara *Pore blocking mechanisms during early stages of membrane fouling by colloids*. Journal of Colloid and Interface Science, 2008. **328**, 464–469 DOI: https://doi.org/10.1016/j.jcis.2008.09.028.
93. Iritani, E. and N. Katagiri, Developments of blocking filtration model in membrane filtration. *KONA Powder and Particle Journal*, 2016. **33**: p. 179–202.
94. Zheng, Y., et al., Membrane fouling mechanism of biofilm-membrane bioreactor (BF-MBR): Pore blocking model and membrane cleaning. *Bioresource technology*, 2018. **250**: p. 398–405.
95. Ahmad, N., et al., Membranes with great hydrophobicity: a review on preparation and characterization. *Separation & Purification Reviews*, 2015. **44**(2): p. 109–134.
96. Barambu, N.U., et al., Membrane surface patterning as a fouling mitigation strategy in liquid filtration: A Review. *Polymers*, 2019. **11**(10): p. 1687.
97. Xu, H., et al., Outlining the roles of membrane-foulant and foulant-foulant interactions in organic fouling during microfiltration and ultrafiltration: A mini-review. *Frontiers in Chemistry*, 2020. **8**.
98. Hoang, T., G. Stevens, and S. Kentish, The influence of feed pH on the performance of a reverse osmosis membrane during alginate fouling. *Desalination and Water Treatment*, 2012. **50**(1–3): p. 220–225.
99. Lim, Y.P. and A.W. Mohammad, Influence of pH and ionic strength during food protein ultrafiltration: Elucidation of permeate flux behavior, fouling resistance, and mechanism. *Separation Science and Technology*, 2012. **47**(3): p. 446–454.
100. Ren, L., et al., Pilot study on the effects of operating parameters on membrane fouling during ultrafiltration of alkali/surfactant/polymer flooding wastewater: optimization and modeling. *RSC advances*, 2019. **9**(20): p. 11111–11122.

101. Suresh, P. and S. Jayanti, Peclet number analysis of cross-flow in porous gas diffusion layer of polymer electrolyte membrane fuel cell (PEMFC). *Environmental Science and Pollution Research*, 2016. **23**(20): p. 20120–20130.
102. Jaffrin, M.Y., Dynamic shear-enhanced membrane filtration: a review of rotating disks, rotating membranes and vibrating systems. *Journal of membrane science*, 2008. **324**(-1–2): p. 7–25.
103. Meng, S., et al., Membrane fouling and performance of flat ceramic membranes in the application of drinking water purification. *Water*, 2019. **11**(12): p. 2606.
104. Kallioinen, M., et al., Effect of high filtration temperature on regenerated cellulose ultrafiltration membranes. *Separation science and technology*, 2007. **42**(13): p. 2863–2879.
105. Koyuncu, I., et al., *3- Advances in water treatment by microfiltration, ultrafiltration, and nanofiltration*, >>in *Advances in Membrane Technologies for Water Treatment*>>, A. Basile, A. Cassano, and N.K. Rastogi, Editors. 2015, Woodhead Publishing: Oxford. p. 83–128.
106. Guigui, C., et al., Impact of coagulation conditions on the in-line coagulation/UF process for drinking water production. *Desalination*, 2002. **147**(1): p. 95–100.
107. Huang, H., K. Schwab, and J.G. Jacangelo Pretreatment for Low Pressure Membranes in Water Treatment: A Review. *Environmental Science & Technology*, 2009. **43**, 3011–3019 DOI: 10.1021/es802473r.
108. Bustillo-Lecompte, C.F. and M. Mehrvar *Treatment of an actual slaughterhouse wastewater by integration of biological and advanced oxidation processes: Modeling, optimization, and cost-effectiveness analysis*. Journal of Environmental Management, 2016. **182**, 651–666 DOI: https://doi.org/10.1016/j.jenvman.2016.07.044.
109. Tiravanti, G., D. Petruzzelli, and R. Passino, Pretreatment of tannery wastewaters by an ion exchange process for Cr(III) removal and recovery. *Water Science and Technology*, 1997. **36**(2): p. 197 207.
110. Thomas, A. and C. Sathian, Cleaning-In-Place (CIP) System in Dairy Plant- Review. IOSR Journal of Environmental Science, *Toxicology and Food Technology*, 2014. **8**: p. 41–44.
111. Tamime, A.Y., *Cleaning-in-place: dairy, food and beverage operations*. Vol. 13. 2009: John Wiley & Sons.
112. Wang, Z., et al. *Membrane cleaning in membrane bioreactors: A review*. Journal of Membrane Science, 2014. **468**, 276–307 DOI: https://doi.org/10.1016/j.memsci.2014.05.060.
113. Shi, X., et al., Fouling and cleaning of ultrafiltration membranes: A review. *Journal of Water Process Engineering*, 2014. **1**: p. 121–138.
114. Ramos, C., et al., Chemical cleaning of membranes from an anaerobic membrane bioreactor treating food industry wastewater. *Journal of Membrane Science*, 2014. **458**: p. 179–188.
115. Crane site, *CIP AND COP SYSTEMS: THE SIMPLE DEFINITIONS*. 2014.
116. Wei, C.-H., et al., Critical flux and chemical cleaning-in-place during the long-term operation of a pilot-scale submerged membrane bioreactor for municipal wastewater treatment. *Water Research*, 2011. **45**(2): p. 863–871.
117. Huyskens, C., et al., Study of (ir)reversible fouling in MBRs under various operating conditions using new on-line fouling sensor. *Separation and Purification Technology*, 2011. **81**(2): p. 208–215.
118. Ang, W.S., S. Lee, and M. Elimelech, Chemical and physical aspects of cleaning of organic-fouled reverse osmosis membranes. *Journal of Membrane Science*, 2006. **272**(-1): p. 198–210.
119. Di Bella, G., D. Di Trapani, and S. Judd, Fouling mechanism elucidation in membrane bioreactors by bespoke physical cleaning. *Separation and Purification Technology*, 2018. **199**: p. 124–133.
120. Liang, H., et al., Cleaning of fouled ultrafiltration (UF) membrane by algae during reservoir water treatment. *Desalination*, 2008. 220(1): p. 267–272.

9 3D Printing Technology for the Next Generation of Greener Membranes towards Sustainable Water Treatment

Mohammadreza Kafi
University of Kashan

Mohammad Mahdi A. Shirazi
University of Kashan
Membrane Industry Development Institute
Green-MEMTECH Team, Incubation Center, Chemistry and Chemical Engineering Research Center of Iran (CCERCI)
Islamic Azad University

Hamidreza Sanaeepur
Arak University

Saeed Bazgir and Paria Banino
Islamic Azad University

Nur Hidayati Othman and Ahmad Fauzi Ismail
Universiti Teknologi MARA

Ahmad Fauzi Ismail
Universiti Teknologi Malaysia

CONTENTS

DOI: 10.1201/9781003167327-11

9.1 INTRODUCTION

Besides food, environmental security, and energy, fresh water is a crucial key element for the sustainable and promising development of societies. This can be explained by the flourishing population growth, and as a consequence, it has led to developing industrialization. As a result, the fresh water demand increased dramatically in many developing and developed countries. Furthermore, climate change, which is tied up with rapid industrialization growth, has further complicated and exacerbated the balance between water supply and its demand. Therefore, severe water shortage and the consequence of water stress can be expected. Recent predictions highlighted the fact that this issue will be even worse in the near future. All these, hence, spotlight the importance and emergence of searching for new and sustainable water sources (Shirazi et al., 2012; Goh et al., 2016; Frappa et al., 2020).

Conventionally, fresh water has been supplied from rivers, deep wells, and seasons' rainfall. These sources cannot meet the worldwide increasing fresh water demand. During the last decades, seawater has been introduced as a promising and infinitive intake source for fresh water production in desalination plants. The thermally driven desalination processes such as multi-effect distillation (MED) and multi-stage flash (MSF) were utilized as the first generation of desalination technologies (Ghaffour et al., 2013). However, these technologies are not promising anymore due to the high energy consumption, which is completely in contrast with the sustainability approach in the circular economy. The next generations of water treatment and desalination technologies have been employing a (semi-)permeable barrier which is known as a "Membrane" (Matsuura, 2001; Ravanchi et al., 2009). In order to supply the fresh and potable water from saline sources (i.e., brackish and seawater), reverse osmosis (RO), which is among the most well-developed and utilized membrane processes, has been used (Goh et al., 2016). Nanofiltration (NF) is also an attractive and promising membrane-based desalination and water treatment technique, which can be operated under lower pressures than that of RO (Zhou et al., 2015). Together with brackish water desalination, NF membranes can also be utilized for wastewater treatment purposes (Mozaffarikhah et al., 2017; Cao et al., 2020). Membrane-based desalination can be further improved through RO/NF integration with other membrane processes. For instance, other pressure-driven membrane processes including microfiltration

(MF) and ultrafiltration (UF) can be used as the pre-treatment step for reducing the fouling/biofouling in RO desalination, which can lead to enhanced overall efficiency and membrane lifetime increase (Kavitha et al., 2019; Choi et al., 2019; Ali et al., 2021).

In all the above-mentioned processes, the membrane plays a crucial role, and the overall process can enjoy far lower energy consumption, higher selectivity, lower carbon footprint, more compact systems, ease of scalability, and more importantly the potential for replacement of toxic chemicals and harsh treatment operations. As a result, the utilized membrane should provide specific requirements for each case. According to the water treatment processes, various membrane features are provided in Table 9.1. Nevertheless, it should be noted that not in all proposed processes, the selectivity of the membrane is already sufficient enough to outperform conventional separation techniques or to provide alternatives for challenging separations (Nunes et al., 2020; Fuzil et al., 2021).

9.1.1 Membrane Materials/Fabrication for Water and Wastewater Treatment

Typically, a few materials, including organic (polymers) and inorganic materials (carbon, metals, and metal oxides), have been utilized for membrane fabrication. Additionally, a number of fabrication techniques have been developed for industrial-scale fabrication of the membranes. However, it should be noted that developing these materials and techniques returns to the time when environmental pollution via toxic chemicals, solvents, and polymers was not taken as seriously as these days. In better words, the term "sustainability in membrane production" was much less investigated in the past few decades (Feng et al., 2015; Otitoju et al., 2016).

Phase inversion is the most investigated fabrication technique for the commercially fabrication of microporous membranes. More specifically, this technique can be categorized as non-solvent-induced phase inversion (NIPS), vapor-induced phase inversion (VIPS), or temperature-induced phase inversion (TIPS). The phase inversion technique has been used for the fabrication of a wide range of polymer membranes, such as polyethersulfone (PES), polysulfone (PSf), polyvinylidene fluoride (PVDF), etc. for MF to RO processes (Table 9.1) (Figoli et al., 2014; Dehban et al., 2020). Another fabrication method used in a large scale is the mechanical stretching, which is a post-synthesis technique for pore formation in the pre-extruded polymer films, such as polytetrafluoroethylene (PTFE). Thoroughly, a porous membrane can be prepared for MF or MD applications owing to the pore sizes (Table 9.1) (Shirazi et al., 2015a). Track-etching technique is another technique that uses high-energy particle radiation exposing the pre-extruded polymeric film to create tracks. The film is consequently immersed in an acid or alkaline solution to etch the damaged polymer materials and form the pores. Track-etching is usually applied for the preparation of porous membranes from polymeric films with outstanding mechanical properties such as polycarbonate. One of the most outstanding advantages of this technique in comparison with the phase inversion is the capability of providing cylindrical pores with uniform dimensions (Tan and Rodrigue, 2019). Although it can provide high size-based selectivity, the resultant membrane suffers from low porosity and pore density. In addition,

TABLE 9.1
Summary of Industrial Membrane Applications for Water Treatment

Process	Membrane Material	Pore Size	Separation Mechanism	Module Configuration	Application	Ref.
MF	• PSf • PES • PVDF • CA • PP • PTFE	> 100 nm	Size exclusion	• FS • Tubular • HF	• Water treatment • Pre-treatment for RO/NF • Colloid removal • Bacteria/virus removal	Nesan et al. (2021)
UF	• PES • PSf • PAN • PVDF	2–100 nm	Size exclusion	• FS • Tubular • HF	• Colloid removal • Bacteria/virus removal • Wastewater treatment	Aryanti et al. (2021)
NF	• PA • PSf • CA	1–2 nm	Size exclusion & solution-diffusion	• FS • HF • Spiral-wound	• Divalent ions removal • Water treatment • Virus removal • Wastewater treatment	Figoli et al. (2020)
RO	• PA • PSf • CA	< 1 nm	Solution-diffusion	• FS • Spiral-wound	• Monovalent ion removal • Desalination • Heavy metal removal	Rahmah et al. (2020)
MD	• PTFE • PVDF • PP • ABS • SAN	> 100 nm	Phase separation	• FS • Tubular • HF • Spiral-wound	• Water treatment and purification • Heavy metal removal • Desalination • Wastewater treatment • Brine concentration	Chin et al. (2020)

ABS, acrylonitrile butadiene styrene; CA, cellulose acetate; FS, flat sheet; HF, hollow fiber; MF, microfiltration; NF, nanofiltration; PA, polyamide; PAN, polyacrylonitrile; PES, polyethersulfone; PP, polypropylene; PSf, polysulfone; PTFE, polytetrafluoroethylene; PVDF, polyvinylidene fluoride; RO, reverse osmosis; SAN, styrene-acrylonitrile; UF, ultrafiltration

TABLE 9.2
Overview of the Conventional Techniques and the Electrospinning for Membrane Fabrication

	Parameters			Properties		
Technique	Ease of Technique	Reproduction	Cost	Pore size Distribution	Roughness	Charge
Stretching	♦♦	♦♦	♦♦♦	♦♦	♦	♦
Track-etching	♦♦	♦♦♦	♦♦♦	♦	♦	♦
Phase inversion	♦♦♦	♦♦♦	♦	♦♦♦	♦♦	♦♦♦
Interfacial polymerization	♦♦	♦♦	♦	♦♦	♦♦	♦♦♦
Electrospinning	♦♦♦	♦♦	♦♦♦	♦♦	♦♦♦	♦♦

♦♦♦: High
♦♦: Moderate
♦: Low

as the pore structure goes through the membrane bulk (i.e., the entire membrane thickness), the track-etched membranes can cause higher overall transport resistance compared to the thin-film composite (TFC) membranes. Microporous membranes with the TFC structure can be prepared via the phase inversion and interfacial polymerization techniques, and directly be used for seawater desalination and water treatment purposes (Hailemariam et al., 2020; Ng et al., 2021b).

Inorganic membranes can be made of carbon and carbon-derived materials, ceramics, or even metals. Typically, ceramic membranes can be prepared by two main techniques, including sol–gel and particle dispersion casting. For both techniques, the fabrication process should be followed by sintering at a high temperature (Samaei et al., 2018). For metallic membranes innovative techniques such as wire arc spraying have also been investigated (Madaeni et al., 2008). Carbon and graphene/graphene oxide-based membranes have shed some light on the new generation of membrane-based desalination alternatives (Homaeigohar and Elbahri, 2017; Li et al., 2019).

Recently, electrospinning technology has been used for fabricating highly porous membranes for a wide range of water treatment purposes (Shirazi et al., 2017a). In this technique, a high-voltage electric field is utilized for fabricating a solution jet from a polymer-solvent/(additive) solution. A solution jet is elongated from a needle tip (or a spinneret), travels with a whipping movement in the space between the spinneret and the collector, and accumulated on the collector as a highly porous nonwoven mat with random (nano) fibers (Tabe, 2017; Sanaeepur et al., 2021). Table 9.2 presents an overview of conventional techniques as well as the electrospinning method for membrane fabrication.

9.1.2 Driving Forces for Developing Next Generation of Membranes

Severe environmental concerns such as the production of highly polluted wastewater and the emission of significant greenhouse gases (GHGs) are caused by two factors:

on one hand, a dramatic increase in water and energy demand due to population growth and industrialization; on the other hand, increasing the fresh water demand for agro-food sector (Elsaid et al., 2020; Del Borghi et al., 2020). All of these can be considered as the main driving force for developing the membranes and membrane processes as the main supporting technology for cleaner production. Membrane separation processes have had a great impact on urban wastewater sanitation and clean water production (Warsinger et al., 2018). Recently, the water and agro-food sectors have been important end users for membrane technology, particularly for desalination, milk/cheese processing, fruit juice clarification, wastewater treatment, etc. (Bhattacharjee et al., 2017; Csatro-Munoz et al., 2020; Cassano et al., 2020). Unquestionably, membranes play a prominent role in a wider range of applications, such as health and medical treatments, the pulp and paper industry, clean energy, and even climate change mitigation (Shuit et al., 2012; Frappa et al., 2020; Yazdi et al., 2020).

Despite these widespread applications, the contribution of membranes to the global perspective of water and wastewater treatment is still small. In better words, much more practical and feasible efforts should be done for the development of further membrane processes and fabrication of specific membranes. In this regard, academic researchers can play a dominant role in conducting innovative projects for developing new membranes. Research and development (R&D) centers can also work on optimizing the membrane processes and industrializing the membrane fabrication methods. However, there are some serious challenges that should be taken into consideration in this way, including energy consumption, zero waste generation, improved cleanability of the membrane, low fouling/biofouling, and increasing the water recovery (Ahmed et al., 2020; Saleem and Zaidi, 2020; Firouzjaei et al., 2020). Furthermore, the permeability/selectivity trade-off of the membranes is still a challenging issue, which needs a more pragmatic solution (Werber et al., 2016). All these issues increase the driving force for thinking of the next generation of membranes.

On the one hand, membranes are proposed as a clean and environmentally friendly separation equipment with lower energy consumption than the conventional techniques, such as the liquid-liquid extraction and distillation, but on the other hand, owing to the highly toxic solvents, chemicals, and non-degradable polymers employment (i.e., hydrocarbon-based polymers), current membranes and membrane fabrication techniques are far from sustainable and environmentally friendly approaches. Hence, this so-called cleanliness of these polluting membranes may be seriously questioned.

Moreover, there are some serious questions about membrane production impacts and the utilized polymers/chemicals on human health, contamination of water sources, global warming, marine eco-toxicity, etc. (Yadav et al., 2021). Therefore, it is crucial to give some thought to the next generation of membranes and their fabrication methods, which can enjoy much cleaner and more sustainable routes. Some alternative strategies have been explored and developed so far, such as techniques that can use much fewer chemical solvents (e.g., electrospinning) (Shirazi et al., 2020), using green solvents in the fabrication steps (e.g., dimethyl sulfoxide (DMSO)) (Wang et al., 2019a), and the use of biodegradable and biocompatible polymers created from natural or synthetic resources (Jiang and Ladewig, 2020; Bandehali et al., 2021).

Recently, a new fabrication technique, which is known as the three-dimensional (3D) printing technology (or additive manufacturing) has been introduced to the membrane technology (Ngo et al., 2018). This is a solvent-free fabrication technique, which enjoys numerous advantages (Yanar et al., 2020a). Generally, for membrane technology, 3D printing can be utilized in three main sections, including fabrication of channel spacers, membrane fabrication, and module design. Over the past few years, the number of publications related to 3D printing technology for desalination, water, and wastewater treatment has considerably increased (Koo et al., 2021). This technology has attracted a lot of attention in this area, probably due to the numerous significant advantages in printing water-related items, such as plastic packing, spacers in spiral-wound membrane modules, and so on. One of its most prominent benefits is the ability to control both the geometry and architecture of the spacer, absorber, packing, filter, etc. (Dommati et al, 2019). A number of review papers have been recently published, which covered various aspects of 3D printing for water treatment-related issues. Table 9.3 summarizes some of these recent publications in the open literature.

9.1.3 Outlines of This Work

This chapter provides an overview of the application of 3D printing used in membrane engineering for sustainable water treatment. First, the 3D printing technology, materials, and techniques are presented. The mechanical properties of printed materials are also studied (Section 9.2). Next, the applications of 3D printers in membrane engineering are comprehensively reviewed in Section 9.3. This section is focused on three main items, including channel spacers, membrane fabrication, and module design. Additionally, potential challenges in terms of materials and process limitations, scale-up issues, and environmental concerns, as well as costs, are introduced and discussed in Section 9.4. Finally, the future prospects, including the hybrid techniques, are discussed.

9.2 Overview of 3D Printing Technology

3D printing technology was first introduced by the Japanese inventor, Hideo Kodama, in 1981. In this innovative project, UV light was used to solidify the polymer jet. It was further developed by another inventor, Charles Hull, who is known as a pioneer researcher in the development of stereolithography. In 3D printing technology, a layer-by-layer fabrication strategy can be used to create a 3D model. The fabricated object can then be washed with a suitable solvent and solidified using UV light. Using this strategy, a smaller model of the target object can be designed and fabricated before spending time, energy, and money to fabricate the actual size (Dommati et al., 2019).

Over the past decades, especially in the last 40 years, significant advances in materials science and manufacturing technologies have led to the advancement of innovative strategies to address environmental concerns, particularly water pollution. Among them, 3D printing technology, also known as additive manufacturing (AM), is a promising and emerging technology that has been successfully developed on a

TABLE 9.3
Overview of the Published Review Papers in 3D Printing and Water-Related Issues

Year	Title	Highlights	Journal	Ref.
2021	3D printing for membrane desalination: Challenges and future prospects	• Applications for 3D printing throughout the desalination plant process were reviewed. • 3D printing costs are projected to fall by approximately 50%–75% over the next decade. • 3D printing expands membranes, spacers, modules, factory design and optimization. • 3D Printing will lead to lower operating, research, and engineering and procurement costs. • Spacer manufacturing is at the forefront of efforts to commercialize 3D printing in RO membrane desalination. • 3D printing can potentially accelerate the commercial viability of emerging desalination technologies.	*Desalination*	Soo et al. (2021)
2021	The emerging role of 3D printing in water desalination	• 3D printing enables new materials and methods for water desalination. • For thermal desalination, 3D printed solar absorbers are promising. • For membrane desalination, 3D printing offers customized membranes and modules. • 3D printing enables the use of renewable energy for water desalination. • Industry is adopting 3D printing to improve water desalination.	*Science of the Total Environment*	Khalil et al. (2021)
2021	A review on spacers and membranes: Conventional or hybrid additive manufacturing?	• Hybrid additive manufacturing for membrane water treatment. • Comparison of hybrid and conventional additive manufacturing. • Review of 3D printing spacer and membrane. • Future trends in 3D printing for membranes.	*Water Research*	Koo et al. (2021)
2020	A new era of water treatment technologies: 3D printing for membranes	• 3D printing of the polymer membrane support. • 3D printing-based interfacial polymerization. • Key aspects of 3D printing technology. • Reviewing critical developments to date.	*Journal of Industrial and Engineering Chemistry*	Yanar et al. (2020a)

(Continued)

TABLE 9.3 (*Continued*)
Overview of the Published Review Papers in 3D Printing and Water-Related Issues

Year	Title	Highlights	Journal	Ref.
2020	3D printing for membrane separation, desalination, and water treatment	• The potential of 3D printing for water-related applications. • Challenges and opportunities facing the process/material of 3D printing for water treatment. • Challenges and potentials of industrial scale-up.	*Applied Materials Today*	Tijing et al. (2020)
2019	A comprehensive review of the recent development in the 3D printing technique for ceramic membrane fabrication for water purification	• Review of the advantages and limitations of 3D printing processes for ceramic membranes. • A brief background of 3D printing processes and their future prospects. • The potential benefits for fabrication and flexibility with different materials.	*RSC Advances*	Dommati et al. (2019)
2019	3D printing and surface imprinting technologies for water treatment: A review	• Different 3D printing techniques for spacers and membranes. • Recent developments in surface imprinting for membrane design. • Enhancement of 3D printing technologies in areas of materials, resolution, and speed.	*Journal of Water Process Engineering*	Balogun et al. (2019)
2017	Perspective on 3D printing of separation membranes and comparison to related unconventional fabrication techniques	• 3D printing can provide more control over membrane design • A discussion of the potential and limitations of 3D printing for membrane fabrication	*Journal of Membrane Science*	Low et al. (2017)

commercial scale for a wide range of applications. 3D printing can fabricate complex structures much faster than conventional manufacturing processes. Therefore, more customizable products with a very lightweight design can be provided. Moreover, 3D printing technology enjoys some other outstanding benefits, including flexibility in design and prototyping, accessibility, risk reduction, waste minimization, and most importantly, sustainability, because it is a solvent-free production strategy. Some practical applications of 3D printing technology include medical and healthcare issues, the food industry, aerospace, polymer composite, and catalyst preparation. Other sectors such as automotive, fashion, accessories, jewelry and cosmetics, home appliances, construction, foundry, computer, and education industries, as well as the chemical industry, are other end users of this technology (Joshi and Sheikh, 2015; Zhou et al., 2017; Singh et al., 2017; Mantihal et al., 2020; Oladapo et al., 2021). However, there are still some challenging issues that need to be addressed to further commercialize the technology, such as material usability, printing resolution, scalability, and costs (Chen et al., 2019; Tejo-Otero et al., 2020; Mikula et al., 2020). These issues will be discussed in the following sections.

3D printing technology can be performed with different configurations. For all of these options, the target structure must first be designed through computer-aided design (CAD). At this stage, you can use various software such as AutoCAD, SolidWorks, Rhino, etc. to draw 3D sketches. The CAD file must then be converted to STL (stereolithography) format, which allows the printer to create the desired object (Parandoush and Lin, 2017; Zhou et al., 2020). Using some mechanical devices such as laser beams and step motors, the position of the printing spinneret can be completely controlled to create a well-designed object. Finally, the completion of the object surface should be done by removing the supports and polishing (Singh et al., 2017; Koo et al, 2021). Figure 9.1 shows an overview of these key steps in 3D printing technology.

Material selection for 3D printing is an important issue. Depending on the target object and its application, different types of materials can be used, including living cells for medical and biological purposes, concrete for construction, ceramics for high-temperature objects, and polymers and composites for fabricating membranes. In terms of membranes, a wide range of polymers such as polycarbonate (PC), acrylonitrile butadiene styrene (ABS), nylon, acrylonitrile styrene acrylate (ASA), and

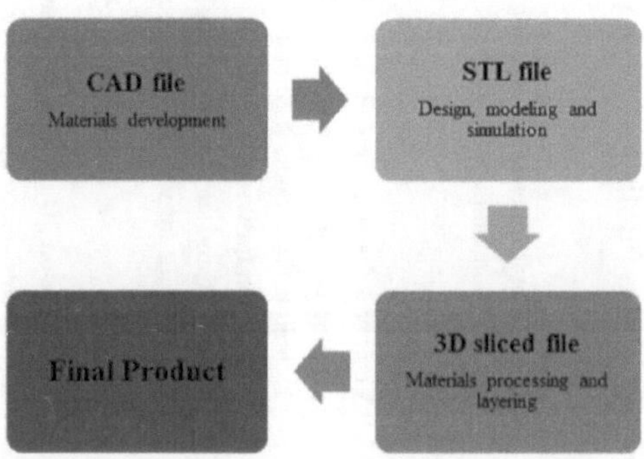

FIGURE 9.1 An overview of the key steps in 3D printing technology.

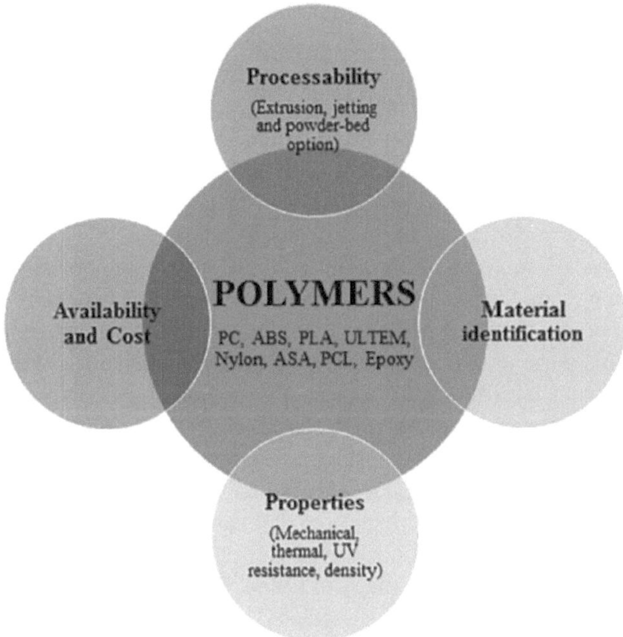

FIGURE 9.2 Material selection for 3D printing in membrane technology.

polycaprolactone (PCL) have been studied to fabricate spacers and membranes. Low et al. (2017) provided a comprehensive list of common polymers for 3D printing with detailed specifications. It is worth noting that some important issues should be considered before the material selection. In addition to availability and cost, which outweigh final production costs, processability (extrusion, jetting, and powder bed option) is a crucial item. Moreover, the mechanical durability of the final printed object, which in this case can be a spacer, a membrane, or a module, is of great importance (Figure 9.2).

About 50 different 3D printing techniques have been registered so far. However, according to ASTM 52900:2015, they can be classified into seven main groups (Balogun et al., 2019). The main classifications of these techniques include binder jetting, powder-bed fusion, material extrusion, laminated object manufacturing, vat polymerization, material jetting, and directed energy deposition. Each technique has its own advantages and limitations. For instance, a wide range of materials can be utilized in the material extrusion technique. Moreover, large size objects can be built using this technique. Hence, the production cost can be lower compared to other techniques. However, poor surface finishing and the need for additional supports are some of its drawbacks. In the powder bed fusion technique, highly durable objects with proper mechanical properties are achievable. Nevertheless, building tall structures is challenging. The vat polymerization technique is applicable to making small objects with high resolution. But, the final product needs to be supported. Binder jetting can use a wide range of materials without the need for support for the final product. However, the production of a rough, highly porous object with poor mechanical strength is its main drawback. Although in the case of jetting material, multicomponent materials can be used to fabricate homogenous objects with high mechanical and thermal strength, the system is expensive and also requires post polishing. Sheet lamination can provide a variety of objects economically, while the fabricated items suffer from poor mechanical strength and durability. The directed energy deposition system can process composite materials with high mechanical properties. However, it is expensive and does not succeed in fabricating small features. Table 9.4 summarizes the overview and different aspects of these techniques.

Before practical use, some important features of the printed object should be considered. Among them, the mechanical strength of 3D-printed materials plays a dominant role. Inevitably different printing techniques can present different mechanical anisotropy. This parameter is proportional to the layer orientation. Since almost all objects are printed layer-on-layer, poor interlayer bonding and consequently poor tensile properties are expected for 3D-printed samples. In addition, each printing technique has its own effect on the mechanical strength of the printed object. In this regard, some parameters such as layer thickness, air gap, printing direction, temperature, printing speed, layer orientation, etc. can affect the mechanical strength of the printed object. The mechanical properties of the printed objects also strongly depend on the direction in which they are tested. That is why manufacturer usually does not provide information about 3D-printed products such as material composition and mechanical properties. Therefore, evaluating material data and comparing properties is quite complex due to the variety of analysis techniques. Tijing et al. (2020) tabulated the desired mechanical features of different 3D printing techniques which have been used in membrane technology. However, this issue needs further and comprehensive investigation.

9.3 3D PRINTING FOR MEMBRANE ENGINEERING

Recently, 3D printing technology for objects related to membrane engineering in water and wastewater treatment has experienced significant advances. This is due to the unique and outstanding advantages of this new fabrication technique over conventional methods such as phase inversion. One of the most important advantages of 3D printing technology is its freedom in material selection and prototype design. This has been playing an important role in the design and fabrication of innovative membranes, modules, and spacers. In better words, more control can be achieved in the design of spacers, modules, and membranes to fabricate these elements in different shapes, designs, and types that are impossible to fabricate using existing techniques (Issac and Kandasubramanian, 2020; Yanar et al., 2020b; Dang et al., 2021). However, some important features should be considered to determine the suitability of 3D printing for membrane technology, including mass production of the target object (i.e., spacer, membrane or module), high accuracy (design capability) and (micro and nanoscale) resolution of the printed object, proper mechanical strength (in the high operating pressures), proper thermal, physical, and chemical properties, low production time (minimized fabrication time), the usability of a wide range of materials (including conventional materials), and low production cost (should be comparable to phase inversion) (Ng et al., 2021a; Khalil et al., 2021).

TABLE 9.4
An Overview of 3D Printing Techniques, Configurations and Classifications

3D Printing	Configuration	General Scheme	Description	Advantages	Disadvantages
Material Extrusion (ME)	• FFF • FDM • PEx	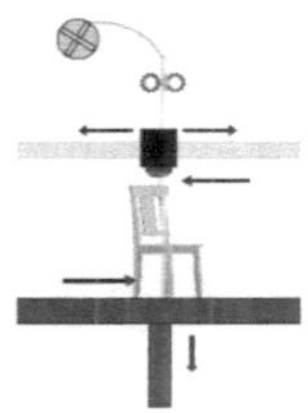	A molten thermoplastic material is formed via the layered-by-layered, until the final object is fabricated.	• Using a wide range of materials with reasonable prices • Able to form objects with large volumes • The most affordable on an industrial scale	• Additional supports required • Lower durability • Post processing is required to clean the support structures • Poor surface finishing
Powder Bed Fusion (PBF)	• SLS • SLM • DMLS • EBM	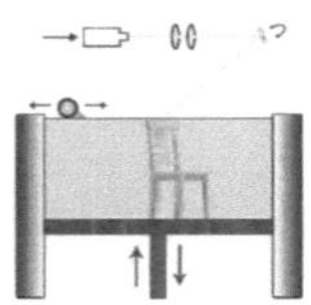	Powdered materials are fused using a high-power laser beam. The final object is formed through a layer-by-layer pattern.	• High durability of the printed object • High mechanical strength • Using a wide range of materials • Able to form composite structures	• Disable to form tall structures • The non-sintered material should be drained for hollow cavities perforation • Post-treatment is required
Vat Polymerization (VP)	• SLA • DLP • TPP • CLIP	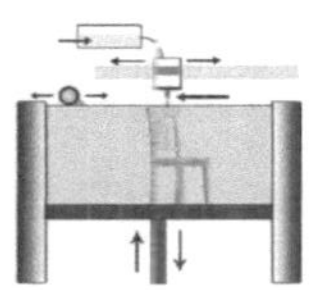	The photopolymer resin used is selectively hardened using a UV laser beam. The final object is formed by the solidification of layers through layer-by-layer formation.	• Feature size of at least 0.2–20 mm can be formed with high resolution • SLA is relatively faster than other printers	• Ceramic structures can not be formed in large volumes • Support structures are required for complex overhangs

(*Continued*)

TABLE 9.4 (*Continued*)
An Overview of 3D Printing Techniques, Configurations and Classifications

3D Printing	Configuration	General Scheme	Description	Advantages	Disadvantages
Binder Jetting (BJ)		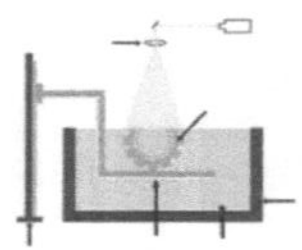	To form the final object, a proper liquid binder is selectively dropped into the powder bed.	• Complex overhangs can be formed without any support • Polymer materials can be printed in different colors • Different types of materials can be used • Able to print large objects • Highly porous, if the printed object is the support layer in membrane engineering	• Post surface finishing is required to provide a smoother surface • Brittle structure with insufficient mechanical strengths • Highly porous, if a smaller pore size is required for the selective layer of the membrane
Material Jetting (MJ)	• DOD • Polyjet	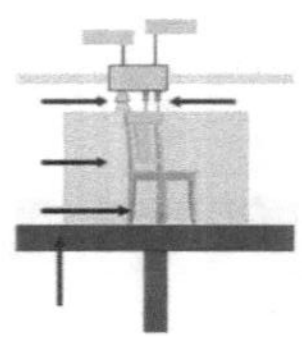	Polyjet printing and Binder jetting are almost similar. The only deference goes back to the binding agents. In this technique, the photopolymer liquid is sprayed and the curing step is performed instantly using UV light.	• Multi-materials can be printed together • Proper thermal and mechanical features can be achieved homogenously • High dimensional accuracy • Colorful 3D printing	• High investment price and operational cost make the final object more expensive • Post processing is required • Insufficient mechanical strength • Improper durability for long-term operation
Sheet Lamination (SL) or Laminated object manufacturing (LOM)		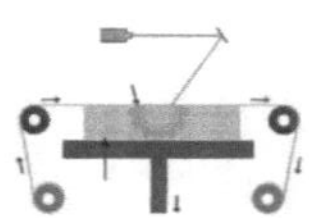	The adhesive coated layers must first be glued together in succession. The final object is then cut in a proper shape using a laser.	• Able to print objects in the entire color spectrum • Economically feasible	• Degradation of the binding material can lead to very poor durability • Disable to print overhang structures • Improper mechanical strength

(Continued)

TABLE 9.4 (*Continued*)
An Overview of 3D Printing Techniques, Configurations and Classifications

3D Printing	Configuration	General Scheme	Description	Advantages	Disadvantages
Directed Energy Deposition (DED)	• LENS	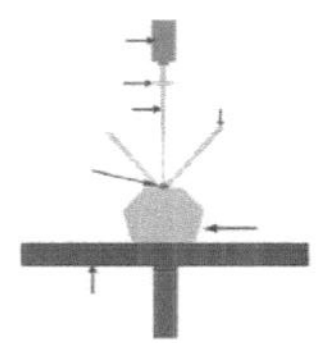	In this technique, the materials used are mixed with a suitable gas agent, then exposed to thermal energy, and finally deposited on the platform. This can provide a molten pool before the solidification step to form the desired pattern.	• Very high mechanical strength of the printed object • Complex objects can be formed without the use of supports due to the use of a multi-axial platform • Able to print composite materials • Very large objects can be printed	• Post-treatment is required to achieve high-grade surface finishing • Expensive printing technique • Insufficient performance in printing small features

FFF, Fused filament fabrication; FDM, Fused deposition modeling; PEx, Paste extraction; SLS, Selective laser sintering; SLM, Selective laser melting; DMLS, Direct metal laser sintering; EBM, Electron beam melting; SLA, Stereolithography; DLP, Digital light processing; TPP, Two-photon polymerization; CLIP, Continuous liquid interface production; DOD, Drop-on-demand; LENS, Laser-engineering net shaping

In this section, the applications of the 3D printing technique for the three main parts of membrane engineering are comprehensively discussed. First, the published results on the 3D-printed channel spacers are reviewed. This is the most studied case for 3D printing opportunities in membrane engineering. Then, membrane fabrication using 3D printers is discussed. Next, the role of this technology in module design is studied. In each section, the latest published results, advantages, and disadvantages are discussed.

9.3.1 Channel Spacers

Feed channel spacers are net-like structures that can be found in various membrane modules for water and wastewater treatment, such as RO, NF, MF, UF, MD, FO, etc. The spacer can affect the hydrodynamic conditions inside the membrane module, mainly in the feed channel, to reduce the effect of fouling and concentration polarization. As a result, it can improve the overall mass transfer. The channel spacer can also separate membrane sheets in the spiral-wound modules, create flow channels, create a micro vortex inside the module, and thus increase turbulency (Abid et al., 2017; Lin et al., 2021). Nevertheless, there are some challenges associated with commercially available spacers including pressure drop inside the feed channel, dead zones of biofilm growth, and improper stiffness to support the membrane without damaging its surface. Therefore, optimization of channel spacers, with emphasis on their geometry, is much needed to save energy and control biofouling (Shirazi et al., 2016; Bucs et al., 2018).

Much research has been done to study new spacer designs to overcome the above-mentioned barriers. Because of the difficulty of fabricating spacers with complex design, many of these attempts rely on computational fluid dynamics (CFD) (Shirazi et al., 2016). However, 3D printing has provided a promising option for designing new channel spacers with complex geometries. The printed spacer prototype can be integrated into the membrane module for the accurate evaluation of performance.

One of the first research results to design a new channel spacer was published by Li et al. in 2005 (Li et al., 2005). In this work, the role of newly designed spacers (twisted tapes, modified filaments, and multilayer structures) to increase mass transfer was experimentally investigated. The nonwoven net spacer was used as the control. The obtained results showed that the best performance was achieved using the multilayer structure spacer and the average Sherwood number increased by about 30%. In another work, Balster et al. (2006) used a multilayer spacer in electrodialysis desalination processes. The spacer was used to reduce the effect of concentration polarization inside the module and thus increase the mass transfer. The multilayer spacer structure consists of a standard middle spacer with two thin outer spacers. The results showed that reducing the diameter of the middle spacer filament can increase mass transfer by 20% more than a commercially available nonwoven spacer, with 30 times less cross-flow power consumption. In 2008, Shrivastava et al. (2008) designed and printed three new, nonwoven spacers for feed channels in the UF and RO processes. The proposed spacers had ladder-like, herringbones, and helical structures and were used to reduce the concentration polarization. The results showed that a significant increase in mass transfer occurs using channel spacers for ladder-like, herringbone, and helical, respectively.

In 2013, Liu et al. (2013) fabricated a new spacer for the feed channel in a UF module inspired by static mixers. Spacer samples were produced by stereolithography apparatus (SLA). In this new design, the fluid flow is diverged from top to bottom thorough the static element objects (Figure 9.3). This flow pattern can provide proper mixing. The effect of the newly designed feed spacer was predicted using a theoretical model. The results showed that static mixing spacers have a better mass transfer at lower flow rates and comparable mass transfer at high flow rates despite having a higher pressure drop than conventional spacers under similar power inputs.

In another work, Fritzmann et al. (2013) used the Polyjet technique to print a dual-layered spacer for submerged membrane filtration. In the new design, each of the proposed spacer layers consisted of parallel helical filaments. The top layer was completely fitted onto the bottom layer. Using this twisted direction structure, the filaments were perpendicular to the flow. The obtained results showed an almost 100% increase in flux. The authors discussed that fouling mitigation is further influenced by the newly designed spacer and the cross-flow velocity as well as the aeration rate. Hence, the authors concluded that the new spacers can reduce energy consumption in an air-sparged submerged membrane system. The same team continued research to study the effect of the proposed spacer on mass transfer and separation performance of UF membranes. The obtained results showed a significant effect of spacer geometry on the selectivity and performance of the UF process in a positive way when used for fractional separation (Fritzmann et al., 2014).

Spiral-wound membrane modules have been widely used for water treatment and desalination purposes through the NF and RO processes. However, membrane cleaning is still a major challenge. Siddiqui et al. (2016) designed and fabricated different

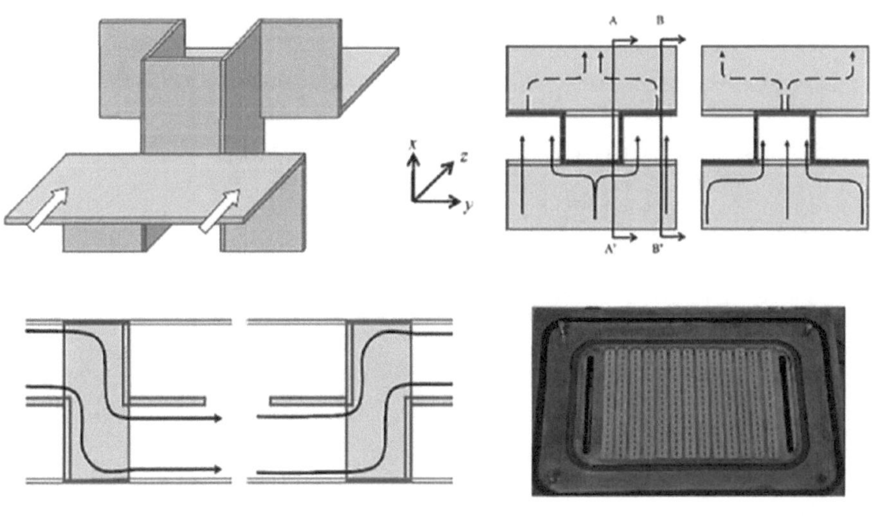

FIGURE 9.3 The geometry of a new static mixer/spacer for a UF module and its flow arrangement (Liu et al., 2013). (Copyright 2013. Reproduced with permission from Elsevier Science Ltd.)

channel spacers for spiral-wound modules using various 3D printing techniques. Figure 9.4 shows the spacers fabricated and their geometric shapes. The results of this research showed that the use of the newly designed spacer can lead to the creation of NF and RO spiral-wound modules with increased cleaning capability. While FDM printing could not provide the proper resolution for thin spacer filaments, the highest mass transfer was achieved using the FDM-fabricated spacers. Moreover, the fabricated spacers were very brittle to be employed properly in long-term operation. The use of the SLA technique resulted in the printing of improper and inflexible spacers that are also soluble in long-term performance in the water. In contrast, the Polyjet technique could provide cheaper and high-resolution spacers.

Recently, 3D-printed spacers for feed channels have been extensively studied for MD-based seawater desalination. The obtained results showed a significant increase in permeate flux by reducing the effects of temperature and concentration polarization. This can be achieved by developing turbulent flow in the feed channel. Moreover,

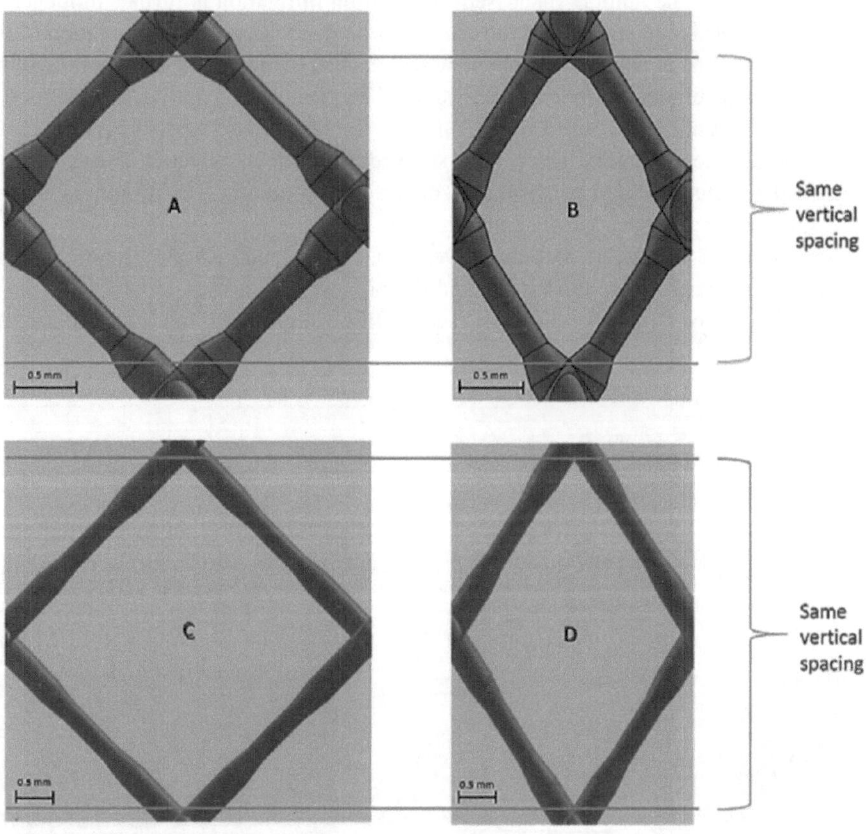

FIGURE 9.4 Spacers with different structures, including (a) standard feed spacer, (b) spacer with modified filament angle, (c) spacer with modified mesh size, and (d) spacer with modified filament angle and mesh size (Siddiqui et al., 2016). (Copyright 2016. Reproduced with permission from Elsevier Science Ltd.)

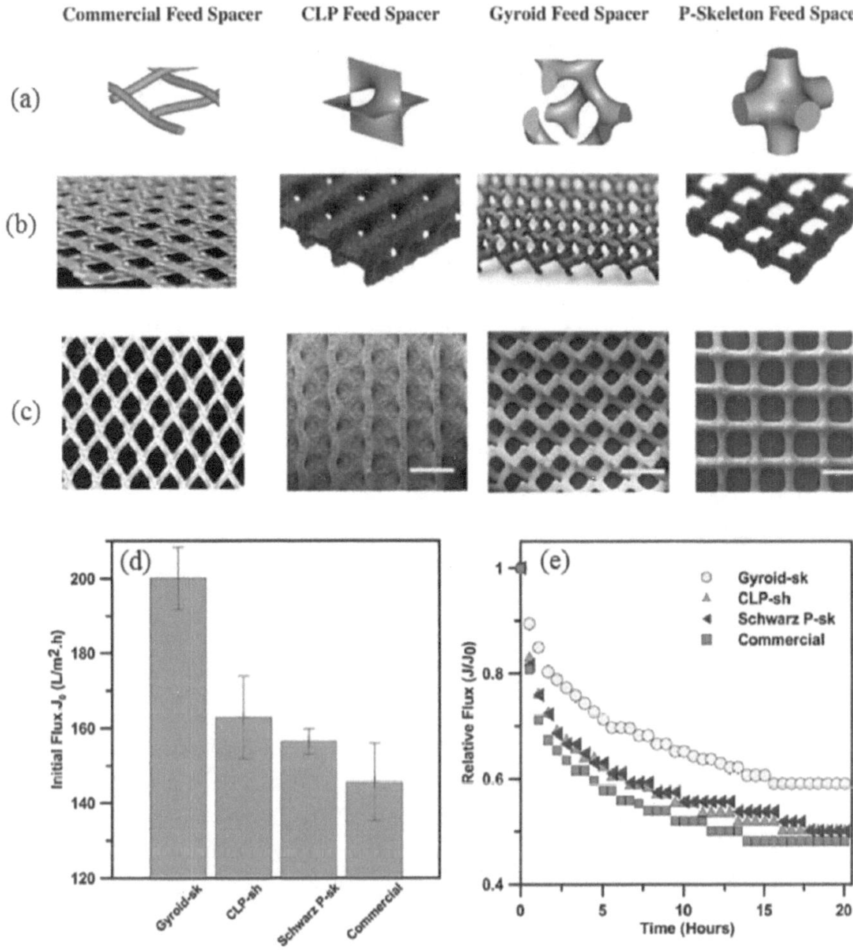

FIGURE 9.5 Images of commercial and printed spacers (a–c) and the effect of proposed spacers on the (d) initial and (e) relative fluxes in the UF process (Sreedhar et al., 2018a). (Copyright 2018. Reproduced with permission from Elsevier Science Ltd.)

fouling mitigation and facilitation can be achieved by using 3D-printed feed spacers with complex geometries. For example, Sreedhar et al. (2018a) printed triply periodic minimal surface (TPMS) spacers with complex geometries to increase the permeate flux in the UF process. The performance of the printed spacer was compared with that of a commercial feed spacer. The results showed a lower pressure drop in the feed channel, lower biofouling on the membrane surface, and an increase in the flux of ~38% when sodium alginate solution was used as the feed sample using TPMS spacers (Figure 9.5). In another work, five different TPMS feed spacers were used to increase heat and mass transfer in the MD process (Thomas et al., 2018). The results showed that the permeate flux and overall heat transfer could be increased by ~60% and ~63%, respectively, using the proposed TPMS spacers compared to commercial

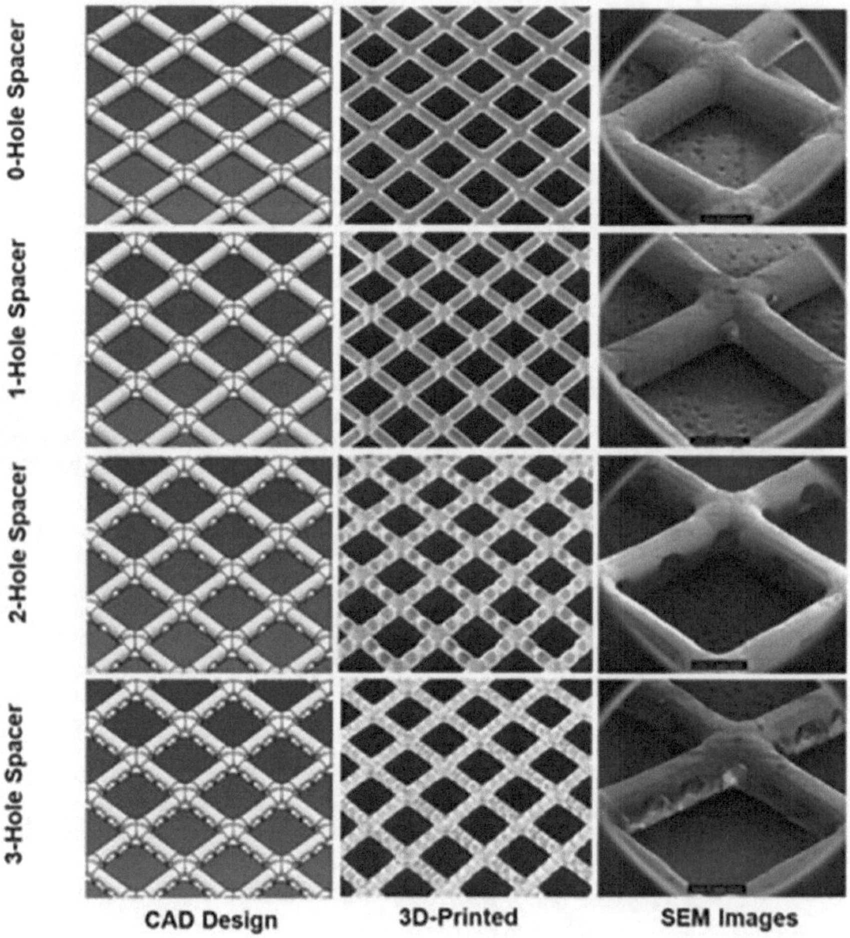

FIGURE 9.6 CAD models, real images, and SEM images of perforated, 3D-printed feed spacers (Kerdi et al., 2018). (Copyright 2018. Reproduced with permission from Elsevier Science Ltd.)

spacers. It is argued that printed TPMS spacers can significantly maintain the permeate flux by providing higher turbulency in the feed channel by disrupting the feed flow. It can also increase the concentration of TDS from 75,000 to 100,000 ppm. In addition to the described results, other teams have worked on using 3D printing techniques to develop advanced spacers for the MD module (Castillo et al., 2019; Thomas et al., 2019; Thomas et al., 2021).

Kerdi et al. (2018) studied the possibilities of 3D-printed perforated channel spacers applied in the UF process. The new design includes 1-hole, 2-hole, and 3-hole cylindrical filaments as feed spacers (Figure 9.6). The authors evaluated the effect of the proposed printed feed spacer on filtration efficiency and pressure drop in the feed channel. CFD modeling was also performed for further study of the experimental results obtained. Using a 1-hole perforated spacer, the pressure drop in the feed channel and the permeate flux can be reduced by ~15% and increased by ~75%, respectively. A further reduction in pressure drop (up to 54%) can be achieved by using a 3-hole perforated spacer, while the permeate flux is reduced by up to ~17%. The final results performed by CFD simulation showed that the formation of fouling on the membrane surface increases with the increasing number of holes in the spacer body. It is claimed that the 1-hole feed spacer can provide the best performance in terms of lower pressure drop in the feed channel, higher permeate flux, and less fouling on the membrane surface (Figure 9.7).

Recently, new approaches to the use of the 3D printing technique in the fabrication of feed channel spacers have been considered. For instance, Park et al. (2021) studied the use of printed spacers to reduce fouling in the NF process. They proposed a new channel spacer with honeycomb-shaped geometry. They discussed that hexagonal geometry is really economically promising and is the most stable structure in nature. That's why they proposed this structure for the 3D printing of spacers. The results showed that a higher fouling reduction was observed for the newly developed feed spacer. In terms of permeate flux, honeycomb-shaped spacers could provide approximately 16% more than conventional spacers. Moreover, hydraulic cleaning was tested for honeycomb-shaped spacers. The results showed that the higher potential of the concentration polarization layer and the cake layer could lead to greater permeate production for the new spacer (Figure 9.8). In another creative work, Sreedhar et al. (2022) developed a new photocatalytic feed spacer using a 3D printing technique for water and wastewater treatment using the UF process. This new spacer was not only used to support the membrane, but also had two other functions, including cleaning the membrane from humic acid, sodium alginate, and bovine serum albumin (BSA), and decomposing permeated pollutants (methylene blue and 4-nitrophenol) (Figure 9.9). It was found that the permeate flux recovery for sodium alginate, humic acid, and BSA was 92%, 60%, and 54%, respectively. The authors concluded that this research opens a new horizon in spacer-centered photocatalytic membrane systems for water and wastewater treatment.

It is worth noting that indirect 3D printing techniques have not yet been used to fabricate the spacers. Lack of study on the function of channel spacers can be the main reason for this. Some features such as mechanical strength, chemical stability, flexibility, etc. also need to be improved. Moreover, there is no information or data on the use of printed spacers in real conditions, such as the processing of seawater or actual wastewater samples, as well as different pH values and operating temperatures. Hybrid 3D printing techniques have not been studied for similar reasons as indirect 3D printing for feed channel spacers. It is worth noting that most research on the fabrication of spacers focuses more on optimizing the design steps than on the overall performance. In addition, more studies are needed to increase the resolution and search for materials with better mechanical strength and long-term performance. Table 9.5 summarizes some of the published results on the use of 3D printed feed spacers for membrane-based water and wastewater treatment.

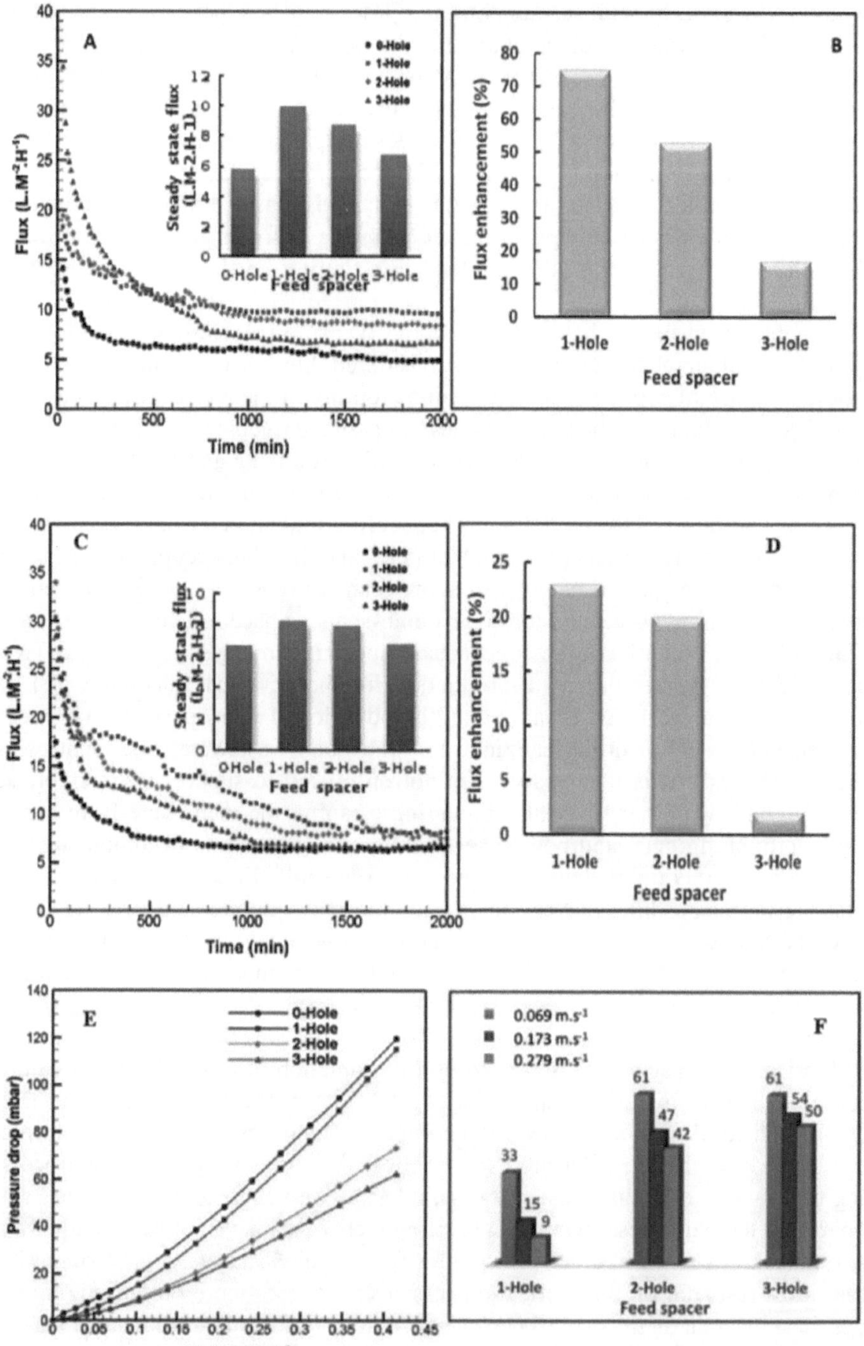

FIGURE 9.7 The effect of the proposed perforated 3D printed feed spacers on the permeate flux and pressure drop in the feed channel (Kerdi et al., 2018). (Copyright 2018. Reproduced with permission from Elsevier Science Ltd.)

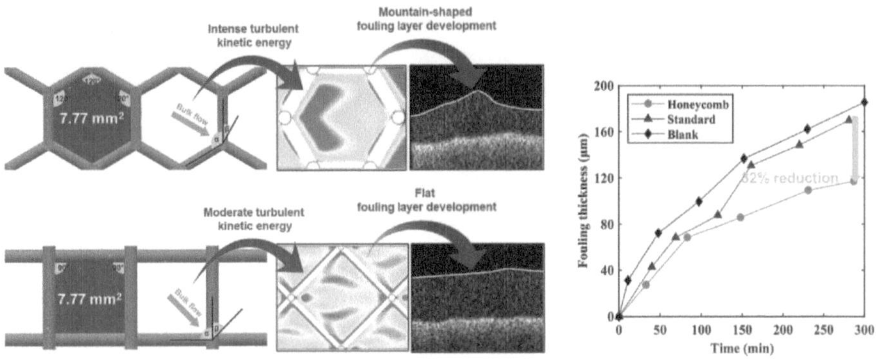

FIGURE 9.8 Channel spacers with 3D printed honeycomb structure to reduce membrane fouling in the NF process (Park et al., 2021). (Copyright 2021. Reproduced with permission from Elsevier Science Ltd.)

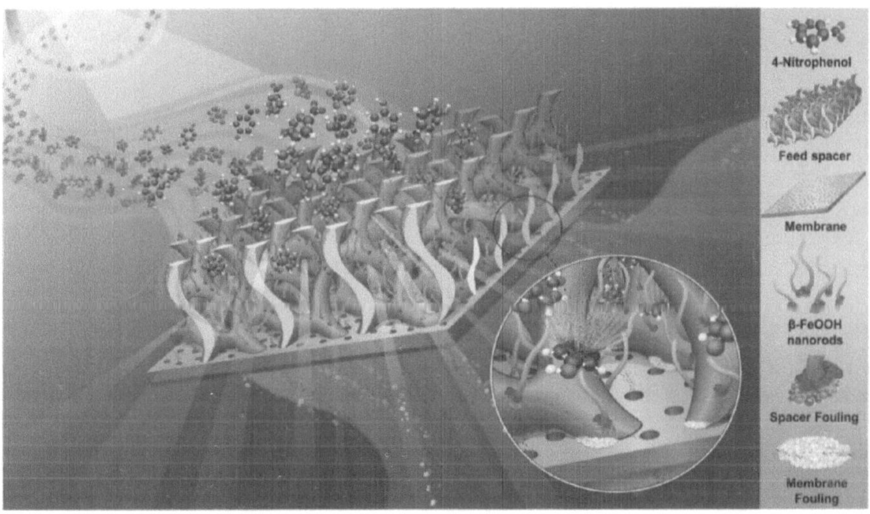

FIGURE 9.9 3D printed photocatalytic feed spacers for degradation of water pollutants and membrane cleaning (Sreedhar et al., 2022). (Copyright 2022. Reproduced with permission from Elsevier Science Ltd.)

9.3.2 Membrane Fabrication

In water treatment, many conventional techniques have been replaced by membranes due to their excellent separation performance and cost-effectiveness (Azizi et al., 2021). It has already been mentioned that water treatment membranes can be classified based on various items, including materials (organic, inorganic, and organic-inorganic hybrids), pore size (MF, UF, NF, RO, MD, FO, ED, etc.), the module configuration (plate and frame, hollow fiber, spiral wound), and driving force for mass transfer (pressure, temperature, concentration, or electrical potential gradient)

TABLE 9.5
A Summary of the Results Published in the Literature for 3D-Printed Channel Spacers in Various Membrane Technologies

3D Printing Technique	Material and System	Configuration and Geometry of Spacers	Process and Application	Ref.
SLS	• Thermoplastic pigment-filled polyamide powder PA 2202 black • Formiga P 110 machine	• Triply periodic minimal surfaces (TPMS) • Schoen Gyroid and Schwarz transverse Crossed Layer of Parallels (tCLP) • Dubbed Gyr-tCLP	MD	Thomas et al. (2020)
MJM	• Visijet M3 Crystal and wax • ProJet® 3500 HD Max	• Two herringbone geometries • Gyroid structure	MF and UF	Van Dang et al. (2020)
SLS	• Nylon powder • sPro60, 3D Systems, USA	• Standard diamond-structured • Honeycomb-structured	Fouling mitigation in NF	Park et al. (2020)
PolyJet	• PP • ProJet 3510 SD model	• Full-contact honeycomb type support (Bio-mimetic honeycomb structure inspired hexagonal type shear distributing support)	FO	Yanar et al. (2020)
SLS	• Polyamide (PA2200) • EOSINT P395 machine	• Hill-like pattern • Wave-like pattern • Wave-like pattern with perforated holes	Submerged MF	Tan et al. (2019)
DLP	• Acrylate monomer • BV-007 (MiiCraft 125, Rays Optics Inc., Hsinchu, Taiwan)	• Symmetric spacer • Column spacer	Membrane filtration	Ali et al. (2019)
SLS	• N.A.	• Triply periodic minimal surfaces spacers (Gyroid and tCLP)	MD	Castilloa et al. (2019)
SLS	• Polyamide (PA2200) • EOSINT P395 machine	• Vibration of 3-D spacers (sinusoidal geometry)	Submerged filtration	Wu et al. (2019)
SLS	• Thermoplastic (PA 2202) • Formiga P 110 machine	• Triply periodic minimal surfaces (TPMS) • tCLP • Gyroid	MD	Thomas et al. (2019)

(Continued)

TABLE 9.5 (*Continued*)
A Summary of the Results Published in the Literature for 3D-Printed Channel Spacers in Various Membrane Technologies

3D Printing Technique	Material and System	Configuration and Geometry of Spacers	Process and Application	Ref.
PolyJet FDM N.A	• ABS, PP and Natural PLA • ProJet 3510 SD model • Acrylate monomer, • BV- 007 (MiiCraft 125, Version 3.4.5, MiiCraft Inc.)	• Diamond-shaped feed spacer • Standard mesh spacers • Perforated spacers with symmetric structure	FO UF	Yanar et al. (2018) Kerdi et al. (2018)
PolyJet	• Photosensitive acrylate • Stratasys, RGD810 (Stratasys, Objet Eden 260V)	• Twisted tape geometry • Kenics static mixer	UF	Armbruster et al. (2018)
N.A	• ABS • Replicator 2X, MarkerBot	• Porous spacer with a hollow cubic assembled structure	Electrodialysis	Bai et al. (2018)
SLS	• Polyamide 2202 • Plastic laser sintering system	• TPMS	RO/UF	Sreedhar et al. (2018a)
SLS	• Polyamide 2202 • Formiga P 110 machine	• TPMS	RO/UF/MD	Thomas et al. (2018)
SLS	• Polyamide 2202 • Formiga P110 by EOS GmbH	• TPMS	UF	Sreedhar et al. (2018b)
SLS FDM PolyJet	• Polyamide 2200 • EOS GmbH Electro Optical Systems	• Net-typed structure with periodic unit cells in square, rhombus, or parallelogram shapes	N.A.	Tan et al. (2017)
SLA FDM PolyJet	• PP	• Mesh spacers with modified geometry	RO/NF	Siddiqui et al. (2016)
SLA	• Harvest Technologies, Belton	• Static mixer	UF/NF/RO	Liu et al. (2013)

(*Continued*)

TABLE 9.5 (*Continued*)

A Summary of the Results Published in the Literature for 3D-Printed Channel Spacers in Various Membrane Technologies

3D Printing Technique	Material and System	Configuration and Geometry of Spacers	Process and Application	Ref.
FDM	• ABS • Stratasys, Eden Prairie, MN, model Prodigy Plus	• Ladder spacer • Helical spacer • Herringbone spacer	UF/RO	Shrivastava et al. (2008)
SLS	N.A.	• Multilayer spacers • Twisted taped spacers	Electrodialysis	Balster et al. (2006)
SLS	N.A.	• Multilayer spacers • Twisted taped spacers	Membrane filtration	Li et al. (2005)

ABS, Acrylonitrile butadiene styrene; DLP, Digital light processing; FDM, Fused dispersion modeling; MJM, Multi-jet manufacturing; N.A., Not available; PLA, Polylactic acid; PP, Polypropylene; SLA, Stereolithography apparatus; SLS, Selective laser sintering; TPMS, Triply periodic minimal surfaces.

(see Table 9.1) (Naseri Rad et al., 2016). Table 9.2 summarizes conventional membrane fabrication techniques, among which phase inversion is the most studied. A positive trend in research works has also been observed in the use of electrospinning to fabricate membranes (Niknejad et al., 2021a; Sanaeepur et al., 2021). Recently, 3D printing technology has attracted considerable attention in the development of more sustainable membranes in water treatment. This new approach to membrane fabrication has advantages such as desirable properties, solvent-free fabrication, and the production of membranes with complex geometries and topographies. In addition, conventional membrane fabrication techniques can be complemented with the help of 3D printing. All of these can provide a better performance, higher mass transfer, fewer polarization effects, increased selectivity, and a lower tendency to fouling (Dommati et al., 2019).

For printing membranes with unique features and 3D structure along with high printing speed, high resolution and accuracy are required to produce membranes with different pore sizes on an industrial scale. In addition, to obtain the desired pore size and its distribution, it is required post-treatment for printed membranes. The resulting membranes can have a special shape, surface morphology, and topography to reduce surface fouling and reduce the polarization effect. However, the 3D printing technique has not yet been widely used to direct membrane fabrication due to some prominent drawbacks. In other words, most of the advanced 3D printing techniques cannot print the desired object effectively with submicron resolution, which is really necessary for water treatment membranes (Femmer et al., 2015b; Low et al., 2017).

Femmer et al. (2015b) worked on an innovative idea to fabricate four rapidly prototyped TPMS heat exchanger geometries. Later, the performance of the fabricated items was evaluated as 3D membrane geometry. According to the obtained results, Schwarz-D showed better heat transfer and less pressure drop. The heat transfer analogy was used to model the mass transfer performance of printed objects. Subsequently, these printed membranes were compared with flat-sheet and hollow fiber membranes (Figure 9.10). The obtained results showed that TPMS modules perform an order of magnitude better than conventional membranes in terms of heat transfer. Therefore, it is concluded that the newly developed 3D geometry can provide increased mass transport when used in membrane applications. In another study, new membrane geometries were evaluated by Femmer et al. (2015a) (Figure 9.11). These works were among the first attempts to use 3D printing technology to fabricate membranes.

Seo et al. (2016) printed an anion-exchange membrane with a micropatterned surface. For the fabrication step, the photo-initiated free radical polymerization and quaternization technique was used. Membrane samples were evaluated in terms of selectivity, water uptake, and ionic resistance to determine their suitability as ion-exchange membranes. Due to the higher quaternized VBC content, water uptake of the polymers increased as a result of increasing ion-exchange capacity (IEC). Samples with IEC values between 0.98 and 1.63 meq g^{-1} were fabricated by changing the VBC content in the range of 15–25 wt.%. The water uptake was affected by the PEGDA content in the polymer. This resulted in water uptake values of 85–410 wt.% by changing PEGDA contents in the range of 0–60 wt.%. Selectivity

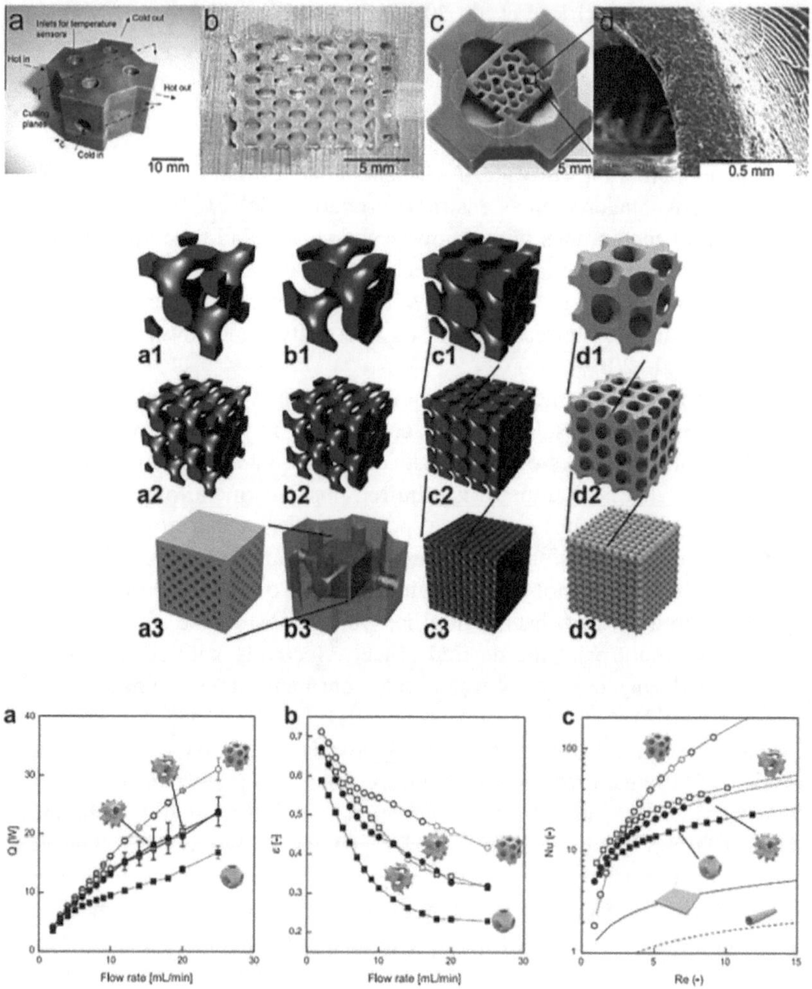

(A) (B) (C)

FIGURE 9.10 (A) The 3D printed membranes, (B) with different geometries, and (C) heat transfer analogy results based on the Nusselt number (Femmer et al., 2015b). (Copyright 2015. Reproduced with permission from Elsevier Science Ltd.)

of the anion-exchange membrane (AEM) samples decreased from 0.91 (168 wt.%, 1.63 meq g^{-1}) to 0.85 (410 wt.%, 1.63 meq g^{-1}) with water uptake increase to 0.88 (162 wt%, 0.98 meq g^{-1}) with IEC decrease. The authors discussed that no significant change was observed in the selectivity results. This provided a general understanding of the correlation between selectivity and water uptake and ion content of the fabricated membranes. Moreover, the obtained results showed that the ionic resistance of patterned membranes is lower compared to flat membranes with the same material volume. The effect of the pattern on the overall ionic resistance measured was explained using a parallel resistance model. The authors concluded that this developed model can help maximize the performance of ion-exchange membranes.

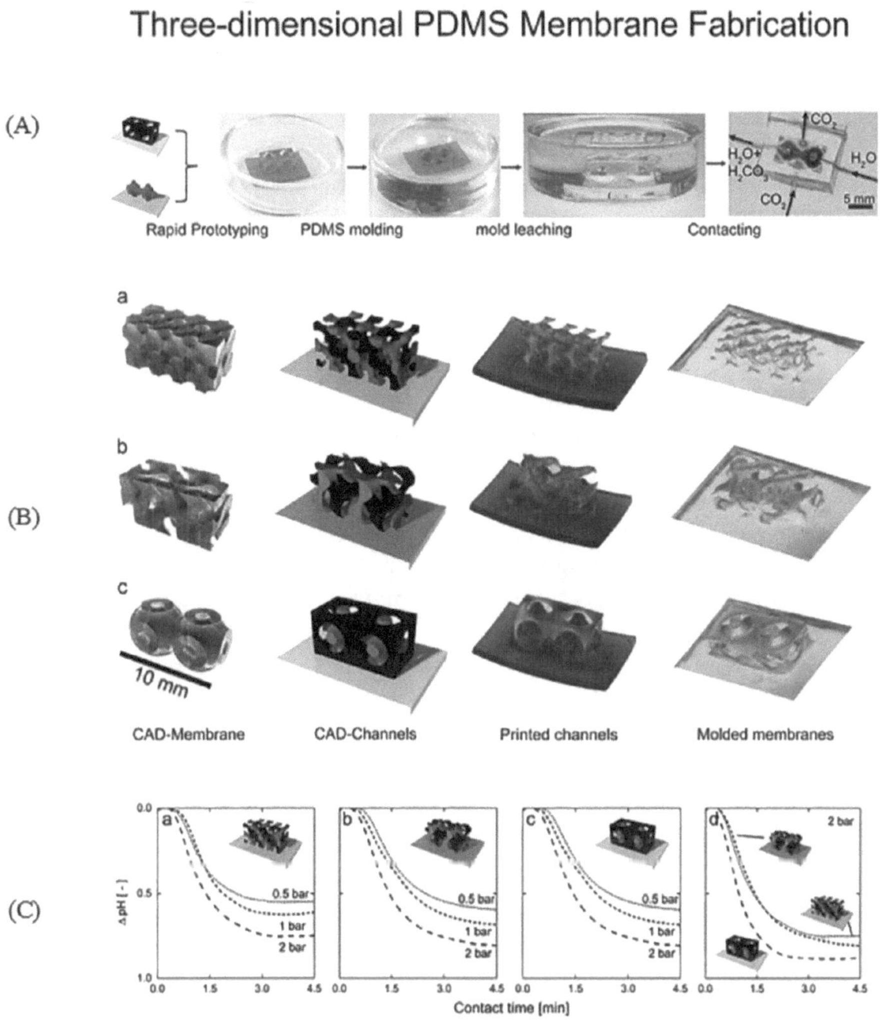

FIGURE 9.11 (A) Printing method, (B) geometry, and (C) pressure drop performances of 3D PDMS membranes (Femmer et al., 2015a). (Copyright 2015. Reproduced with permission from Elsevier Science Ltd.)

This was also achieved by optimizing surface patterns without chemical modification in the membrane.

Gao et al. (2016b) studied a versatile and rapid technique based on a combination of injection printing and template synthesis for the fabrication of nanoscale polymeric materials. This new technique for fabricating nanotubes, nanowires, and thin films was examined. In the case of thin films, this technique was used to fabricate layer-by-layer a thin film made of polyallylamine hydrochloride and polystyrene sulfonate. The thin layer was printed on a polycarbonate track-etched (PCTE) membrane. The resulting membrane could potentially be used as an NF membrane for ion removal from the aqueous medium. The same team continued their research on this technique to fabricate positively and negatively charged mosaic membranes. The authors used PVA-based composite inks (containing poly diallyl dimethyl ammonium chloride or poly sodium-4-styrene sulfonate) to fabricate a thin layer with 30 nm pores on the surface of a polycarbonate track-etched membrane. Mosaic membrane samples with the same coverage of positive and negative surface charges can enrich potassium chloride (KCl) in the rejected solution of the membrane separation. It was concluded that the newly developed membrane can be considered a promising selective membrane for the separation of ionic solutes from solutions (Gao et al., 2016a).

Lv et al. (2017) developed a 3D printing technique for fabricating highly hydrophobic membranes for oil-water separation. This new approach can be an easy and environmentally friendly route to prepare an ordered porous structure incorporated with nanosilica-filled polydimethylsiloxane (PDMS) ink. The authors discussed that the addition of nanosilica can increase mechanical strength. As a result, it can reduce the risk of collapse during printing. With adjustable pore sizes for printed membranes in the range of 0.37 mm, the authors reported a maximum oil-in-water separation of 99.6% and a permeate flux of ~23,700 L $m^{-2}h^{-1}$.

The structure of the thin-film membranes consists of a thin selective layer on a support layer with larger pore sizes that is laminated on a thicker nonwoven substrate. In this structure, the top selective layer plays a dominant role in the separation process. In creative work, Chowdhury et al. (2018) used the electrospraying technique to fabricate a nanometric selective polyamide layer to develop a 3D printed RO membrane. To achieve the desired structure, the monomers (m-phenylene diamine (MPD) and trimesoyl chloride (TMC)) were directly electrosprayed and deposited on an ultrafiltration sublayer to form a polyamide layer. The authors discussed that the thickness and roughness of the thin layer were precisely controlled by the electrospraying of droplets on the sublayer surface in an incremental manner. The thickness of the top layer was measured at about 15 nm. Moreover, the thickness and roughness were measured at about 4 and 2 nm, respectively. It was also concluded that the permeate flux and selectivity of the new membrane are comparable to commercial thin-film membranes.

In order to fabricate new membrane architectures, a new systematic method was introduced by Luelf et al. (2018). To do this, spinnerets with multiple bore channels and different geometries were used. In addition, the authors used a mechanical system to rotate the spinneret, which enabled them to fabricate membranes with complex geometries, including twisted channels inside a monolithic porous membrane and twisted helical outer geometry. The authors concluded that the twisted spiral-type bore channel geometries can provide better mass transfer with less polarization effect on the membrane surface. In another study, Al-Shimmery et al. (2019) fabricated a 3D-printed membrane with a composite structure by depositing a thin layer of PES as the selective layer on ABS support with flat and wavy structures. Vacuum filtration was applied to deposit the thin selective layers on the 3D printed supports. Membrane samples were then characterized and used for oil-in-water separation by cross-flow (Re = 100, 500, and 1,000) UF process (under operating pressure of 1 bar). The authors reported a 30% higher permeate flux for the 3D printed membrane compared to the flat membrane sample. Despite having 96% oil rejection for both membrane samples, the 3D printed membrane had a 52% higher permeate recovery ratio compared to the flat sample. Mazinani et al. (2019) investigated the same strategy for fabricating 3D printed fouling-resistant composite membranes. A thin layer of PES was deposited onto a 3D printed flat and double sinusoidal (wavy) support. BSA solution was used to study the antifouling properties of new membranes during cross-flow UF process (under operating pressure of 1 bar and variable Re numbers between 400 and 1,000). The 3D printed membrane showed approximately 87% recovery in pure water permeates after ten filtration cycles. The authors concluded that the significant fouling resistance of the 3D printed membrane with a wavy surface can be attributed to localized fluid turbulency. All of these can lead to lower operational costs, fewer negative environmental impacts, and longer operational lifetime.

Recently, Liang et al. (2021) introduced a new 3D printed membrane with a fibrous structure to separate particulates from the aqueous stream. The authors used a mixture of PVDF polymer, SiO_2, and DMF as a dope solution to fabricate membrane samples with ordered geometric pore structures using the 3D printing near-field electrospinning (NFES) technique. A series of regular geometric pore structures including triangular, rectangular, hexagonal, and square as well as cylindrical pores were electrospun. The obtained results showed that high permeate flux was up to 1,020.7 L $m^{-2}h^{-1}$ and the particle rejection was 96.7%. The authors discussed that membrane performance, in terms of permeate flux and solute rejection can be efficiently adjusted by designing the membrane pore geometry. For example, approximately 75% improvement in permeate flux with stable particle rejection of >95% was achieved when the length-to-width ratio increased from 1:1 to 2:1.

Summarizing the above studies on 3D printed membranes, the technological lack of integration between 3D printing techniques and membrane engineering is clear. In other words, most published results show the use of material instead of printer development. As a result, high-resolution custom printing of membranes, especially in terms of membrane pore size, requires further development and emergence. In other words, the 3D printing technique should be able to fabricate membranes that are used not only in a wide range of geometries but also with much smaller pore sizes, for water treatment purposes. Table 9.6 summarizes the most advanced membranes associated with 3D printing technology.

9.3.3 Module Fabrication

The term "membrane module" describes a complete package consisting of inlet and outlet channels, membrane, spacers, a pressure support structure, and a general support structure (the outer layer). Membrane modules used in water treatment applications can have configurations, including flat sheets (mostly for laboratory-scale experiments), capillary, hollow fiber, and spiral wounds (Naseri Rad et al., 2016).

As mentioned earlier, 3D printing technology has enabled researchers to fabricate new membranes with complex geometries. Therefore, these new membrane configurations also require new membrane modules. In other words, not only can 3D printing technology be used to fabricate individual parts of a membrane module, such as channel spacers and membrane, but it can also be used to fabricate the entire module

TABLE 9.6
Summary of Recent Studies on 3D Printed Membranes in Water and Wastewater Treatment Applications

3D Printing	Material	Membrane	Accomplishment	Performance	Reference
3D printing+NFES	• PVDF (Solef 6020)	Nanofibrous membrane	• A combination of 3D printing and NFES technique was used to fabricate an ordered porous structure • SiO_2 particles of different sizes can be effectively sieved using the developed membrane • Selective permeability can be determined by adjusted pore size	• High permeate flux of 1,020.7 L $m^{-2}h^{-1}$ was achieved • Particle rejection was measured at about 96.7% • Adjusting the length-to-width ratio to 2:1 can increase the permeate flux to 1,791.4 L $m^{-2}h^{-1}$. However, it maintained a stable rejection ratio of over 95%	Liang et al. (2021)
DLP	• Alumina powder • A proper solvent • Photopolymer binding agent • Dispersant • Color pigment	Ceramic membrane	• Environmentally friendly approach for printing porous ceramic membranes • SLA technique for fabricating high-resolution ceramic membranes with proper precision • Controlling the thickness and surface roughness for better efficiency • Study on the effect of sintering to fabricate stable and porous membrane samples	• A large number of closed pores were formed in the range of 2–3 μm • Open pores with a range of 7.9–9.8 nm were observed • It was shown that the predicted theory of a greater proportion of the small-sized alumina particles could cause smaller pore sizes and the equal proportion of 0.3 and 50.0 μm caused relatively bigger pore sizes • The ceramic membrane was hydrophilic	Ray et al. (2020)

(*Continued*)

TABLE 9.6 (*Continued*)
Summary of Recent Studies on 3D Printed Membranes in Water and Wastewater Treatment Applications

3D Printing	Material	Membrane	Accomplishment	Performance	Reference
DIW	• Cellulose acetate • PVA • SiO_2 nanoparticles	Composite membrane	• CA/PVA/Si membrane was fabricated by DIW 3D printing • Superhydrophilic and superoleophobic performance • Proper oil-water separation • Good fouling-resistant and recyclability	• The 3D printed membrane sample with low water contact angle ~18.14° and a high underwater oil contact angle ~159.14°, demonstrated superhydrophilic and underwater superoleophobic characteristics • Superior mechanical durability was observed after 30 min sonication or 100 bending cycles • The oil-water separation efficiency of 99.0% was achieved using gravity as the driving force • Suitable antifouling characteristics and high oil-water separation efficiency after 50 cycles	Li et al. (2020)

(*Continued*)

TABLE 9.6 (*Continued*)

Summary of Recent Studies on 3D Printed Membranes in Water and Wastewater Treatment Applications

3D Printing	Material	Membrane	Accomplishment	Performance	Reference
FDM	• PMMA-g-PDMS • Polyamide 6	Hemodialysis membrane	• A combination of 3D printing and electrospinning technology was used to fabricate a prominent nanofibrous hemodialysis membrane • The surface area increased more than 2.5 times the surface area of a planar hemodialysis membrane • Esterification was used for the negatively charged hydrophilic modification • The proposed chemical modification could enhance the antifouling resistance and the hemodialysis performance	• The prominent hemodialysis nanofibrous membrane had pores < 0.1 μm • High surface area of 66 $m^2 \cdot g^{-1}$ was achieved • Excellent dialysis performance with a 0.072 μm nanofiber diameter and pore sizes in the range of 0.1–0.14 μm were achieved • Chemical modification could decrease the water contact angle from 82° to 50° • The flux recovery ratio was enhanced to 85% • The antifouling factor (Ft) improved to 0.112	Koh et al. (2020)
SLS	• Polyamide-12	Robust candle soot membrane	• Selective laser sintering was applied to fabricate a robust porous membrane • The fabricated membranes showed hydrophobic, superoleophilic, and underwater superoleophobic characteristics	• 3D printed membrane showed a high oil-water separation, i.e., more than 99% • Acceptable performance in the separation of immiscible organic mixtures • Highly solvent resistant 3D printed membrane	Yuan et al. (2020)

(*Continued*)

TABLE 9.6 (*Continued*)
Summary of Recent Studies on 3D Printed Membranes in Water and Wastewater Treatment Applications

3D Printing	Material	Membrane	Accomplishment	Performance	Reference
PolyJet	• Urethane acrylate oligomers • Paraffin wax	UF composite membranes	• 3D-printed flat and wavy supports • NIPS was used to form the selective layer	• Superior pure water permeance (PWP) • High permeance recovery ratio (87% vs 53%) after the first filtration cycle • 3D printed wavy membrane can maintain ~87% of its initial PWP after ten complete filtration cycles	Mazinani et al. (2019)
PolyJet	• Urethane acrylate oligomers • Paraffin wax	TFC	• ABS 3D printed flat and wavy structured supports • 3D printed membrane by depositing a thin PES selective layer	• The 3D printed wavy membrane has a 52% higher permeance recovery ratio • Wavy and flat membranes with oil rejection of ~96%	Al-Shimmery et al. (2019)
Drop-on-demand inkjet printing	• PVA • Yeast cells	Biocatalytic membrane	• BCMs were developed using cell surface-display enzymes • The enzymatic activity of BCMs was controlled by a simple inkjet printing technique • The stability and recyclability of BCMs were longer compared to free cells • BCMs showed competitive bioactivity in the degradation of emerging contaminants	• BCMs can be reused with high stability because they retained 76% of their initial activity • BCMs have been used promisingly in wastewater treatment • The proposed technology can be used as a single operation or in combination with other conventional treatment techniques	Chen et al. (2018a)

(Continued)

TABLE 9.6 (*Continued*)
Summary of Recent Studies on 3D Printed Membranes in Water and Wastewater Treatment Applications

3D Printing	Material	Membrane	Accomplishment	Performance	Reference
Stereolithography	• Diurethane dimethacrylate, • PEG diacrylate, • Dipentaerythritol penta-/hexa-acrylate • 4-vinylbenzyl chloride	AEMs	• Modification of homogeneous ion-exchange membranes (IEMs)	• The ion-exchange capacity increased with increasing number of functional groups • Selectivity decreased with increasing water content in AEMs • The resistance of AEMs increased with increasing membrane thickness • Patterned AEMs had less resistance than flat AEMs	Capparelli et al. (2018)
InkJet	• PES • Antimicrobial peptides	PES membrane	• The inkjet printer equipped with a UV lamp was used to modify the membrane surface • This system allowed the deposition of photoreactive peptides and immobilization on UF membranes	• Antimicrobial peptides coated UF membranes showed significant antibacterial ability • Biofilm growth decreased significantly	Mohanraj et al. (2018)
Binder jetting	• Kankara clay • Maltodextrin powder	Ceramic membrane	• Development of a low-cost technique for fabricating water treatment clay membranes • Useful for commercial and pilot scales	• COD and TSS decreased by 97.78% and 53.85%, respectively • Proper mechanical strength	Hwa et al. (2018)

AEMs, Anion-exchange membranes; BCMs, Biocatalytic membranes; DIW, Direct inject writing; DLP, Digital light processing; FDM, Fused deposition modeling; FFF, Fused filament fabrication; GO, Graphene oxide; NFES, Near-field electrospinning; NIPS, Non-solvent induced phase separation; PBI, Polybenzimidazole; PDMS, Poly(-dimethylsiloxane); PEG, Poly(ethylene glycol); PES, Polyethersulfone; PLA, Polylactic acid; PMMA, Poly(methyl methacrylate); PVA, Polyviny1alcohol; SCP, Solvent cast printing; SLA, Stereolithography; SLS, Selective laser sintering; TFC, Thin-film composite.

and components needed, as an all-in-one approach, using a wide range of materials for each component. Moreover, the 3D printing technique can provide more realistic alternatives and promising solutions compared to computer-based modeling (Soo et al., 2021).

One of the drawbacks of using 3D printing in module design is material selection. There are three possible steps for using polymer materials for 3D printing, including the melting step for heating the solid polymer for extrusion through a nozzle, the UV curing step for monomer polymerization, or the sintering step to fusing powder together. This highlights the importance of material selection for module design using 3D printing. Moreover, the use of non-toxic materials, such as cellulosic materials, can shed some light on the fabrication of cheaper and more sustainable modules (Lee et al., 2016; Chang et al., 2021).

In conclusion, there is a promising potential for the use of the 3D printing technique in the design of membrane module components, and in this regard, both material and resolution play a dominant role. Recently, however, a slow but positive trend has been seen in this regard.

9.4 CHALLENGES OF 3D PRINTING IN MEMBRANE ENGINEERING

Despite its many advantages, 3D printing technology still faces many challenging membrane-fabricating issues for water and wastewater treatment. Therefore, current challenges, such as resolution, accuracy, speed, and building size, must be considered to optimize the parameters and factors for fabricating the channel spacer, membrane, or the entire module.

9.4.1 Material Limitations

When using 3D printing in membrane engineering, there are issues with system and material compatibility. For example, extrusion-based 3D printers suffer from nozzle clogging as a result of material agglomeration and increased viscosity. Careful membrane design is necessary to prevent the embedding of unused materials in the membrane matrix. Materials trapped in the pore structure, either liquid or loose powder, should be noticed. It may be very difficult or even impossible in some cases. Wax refusal, as another challenging issue in the material jetting system, can be problematic, especially for cleaning small pores. However, the complete removal of wax from printed membrane pores has not yet been comprehensively investigated (Wang et al., 2019b; Cidonio et al., 2021).

Another challenging issue for materials in 3D printed spacers, membranes, or modules can be related to the stability of photopolymers in water- or organic-based solvents. Moreover, the thermal stability of 3D printed materials is often not suitable. In other words, the final printed object can experience a deviation from the original design. This is known as swelling or morphological change, which can occur when a membrane, spacer, or module is exposed to high temperatures or solvents. Mechanical strength can also be a challenge as well; 3D printed spacers bay become brittle after long-term soaking in water/solvent. This can be attributed to photoinitiators' stability in the polymer materials, swelling, or the polymer incompatibility with water. Moreover, inaccuracies in 3D printing can lead to the formation of microcracks that propagate when the membrane/spacer is exposed to solvents. Therefore, further development is needed to help move towards higher mechanical strength in 3D printing of spacers and membranes. Furthermore, more types of materials such as PVDF, PP, PTFE, polyimide (PI), polyamide (PA), etc., that are used for conventional membranes should be used for membrane printing (Zhang et al., 2019; Muravyev et al., 2019).

9.4.2 Process Limitations

Existing 3D printing systems can fabricate objects in a few hundred micrometers. However, the required membrane pore size is below the micrometer scale. Therefore, a 3D printer cannot print a functional membrane directly. In other words, existing 3D printers cannot provide the maximum achievable resolution for direct printing of water treatment membranes with desired pore sizes in the range of large-scale RO/NF membranes. The current achievable resolution range for membrane pore sizes is in the range of 0.1–5 μm, which means that 3D printing can only be used to fabricate MF membranes. Therefore, 3D printing technology still has a long way to go to fabricate RO/NF membranes (Tijing et al., 2020; Agrawal and Thompson, 2021).

9.4.3 Environmental Issues

Although 3D printing has been investigated as an evolving technology toward environmental sustainability, there are some environmental concerns that need to be addressed. For example, one of the problems associated with almost all 3D printing processes is the emission of particulates, either in the form of volatile organic compounds or in the form of ultrafine particles. According to some research results, the emission of ultrafine particles is higher when using ABS and PC fibers for printing compared to other fibers such as nylon and PLA. The emission rate of volatile organic compounds is measured as the highest for nylon, high-impact polystyrene (HIPS), and ABS fibers. Therefore, it is recommended to avoid 3D printing to fabricate membrane, spacer, and modules indoors, where it is not equipped with a ventilation system or particle filter. In addition, the use of toxic chemicals such as alcohols and propylene carbonate in the post-treatment step can also be harmful to the environment. Thus, life cycle assessment studies are needed to assess the environmental and health hazards as well as ecological side effects of 3D printing technology for membrane engineering. Almost high energy demand is another environmental concern associated with fabricating 3D printed membranes. In general, all of these issues can be minimized through further optimization and technology development for 3D printers (Aquino et al., 2018; Nachal et al., 2019).

9.4.4 Limitations and Costs of Scaling Up

3D printing technology uses complex types of machinery that make scalability complex, difficult and costly, especially for membrane engineering. Low printing speed can also cause the printing of a large membrane sheet with submicron pore size to take a long time. To overcome this challenge, a multi-nozzle and multi-laser source must be developed. It can provide a high-speed concurrent printer for fabricating high-resolution spacers, membranes, and modules. However, capital and operating costs can also increase (Low et al., 2017; Nunes et al., 2020).

The cost of materials, which depends on the type of system applied, is another limiting issue for the use of 3D printers in membrane engineering. It has been discussed in the literature that the cost of materials for desktop printers is lower than for large-scale systems. This can limit the fabrication of membranes, spacers, and modules using industrial 3D printers. Moreover, the quality of the materials used can affect its cost. Incorporating exotic materials or precious metals can further increase the overall cost. Due to the short and restricted lifespan of the laser source and UV lights that need to be replaced frequently, it can also increase the cost of operation. It is assumed that the overall cost of a fully 3D printed membrane module is likely to remain significantly higher than conventional approaches, even with improved technology. Therefore, development approaches should focus more on the application in specific separations rather than on production costs. In other words, 3D printed membrane modules should be used in targeted applications or niche membrane market, where custom-designed membranes for specific applications are required (Tijing et al., 2020; Nunes et al., 2020; Soo et al., 2021).

9.5 FUTURE PROSPECTS

Due to numerous advantages and potential applications of 3D printing technology in membrane fabrication, especially the unique architecture and morphology of the spacer/membrane that this fabrication technique can provide, the role of 3D printing technology could grow significantly in the near future. In fact, membrane engineering can enjoy the application of 3D printing technology when the materials used, resolution, and printing speed are sufficiently advanced and optimized. Once these bottlenecks and niches are removed, the 3D printing technique can be used to advance the production of the next generation of greener membranes in water and wastewater treatment. 3D printed membranes and modules can be fabricated using a wide range of organic/inorganic materials and the integration of polymers with organic/inorganic nanoparticles to fabricate a mixed matrix 3D printed membrane or spacer. The new membrane architecture is expected to increase selectivity and reduce fouling as well as reduce the effects of polarization. In addition, in terms of membrane fabrication, 3D printers can be used in innovative transport channels for highly selective separation. Figure 9.12 illustrates some of the specific applications of 3D printers in membrane engineering.

Since the desired resolution for membrane fabrication is not yet achievable with 3D printers, a multi-process strategy that combines 3D printing (to fabricate support or middle layer with unique architecture) and other conventional techniques (to prepare the top selective layer) should be investigated. For example, a combination of 3D printing and electrospinning, electrospraying, or solution-blow spinning can be used to fabricate composite membranes. Using this approach, a highly porous support layer can be formed by 3D printing, while a nanofibrous active layer can be electrospun on top. However, the proposed hybrid technique has not yet been used in the fabrication of water treatment membranes.

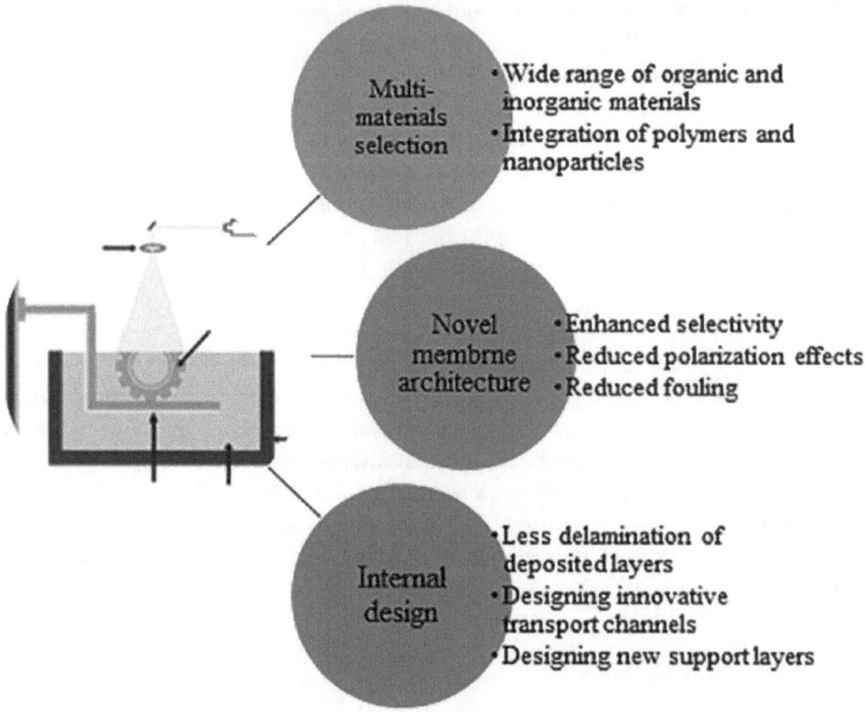

FIGURE 9.12 Prospects for the future of 3D printing technology for membrane engineering.

It is worth noting that almost all the 3D printing techniques studied are independent of time. However, 3D printing can be switched to 4D printing, as a future scenario in membrane technology, if the applied printer can support the time-dependent changes in the proposed printed pattern. This is a new concept that illustrates new horizons in the printing of spacers and membranes. In this regard, some 3D printing systems such as digital light processing, fused deposition modeling, inkjet printing, stereolithography, and photopolymerization can be pioneers.

REFERENCES

Abid H.S., D.J. Johnson, R. Hashaikeh, N. Hilal, A review of efforts to reduce membrane fouling by control of feed spacer characteristics, *Desalination* 420 (2017) 384–402. doi:10.1016/j.desal.2017.07.019.

Agrawal H., J.E. Thompson, Additive manufacturing (3D printing) for analytical chemistry, *Talanta Open 3* (2021) 100036. doi:10.1016/j.talo.2021.100036.

Ahmed F.E., R. Hashaikeh, N. Hilal, Hybrid technologies: The future of energy efficient desalination – A review, *Desalination* 495 (2020) 114659. doi:10.1016/j.desal.2020.114659.

Ali A., C.A. Quist-Jensen, M.K. Jorgensen, A. Siekierka, M.L. Christensen, M. Bryjak, C. Helix-Nielsen, E. Drioli, A review of membrane crystallization, forward osmosis and membrane capacitive deionization for liquid mining, *Resour. Conserv. Recycl.* 168 (2021) 105273. doi:10.1016/j.resconrec.2020.105273.

Al-Shimmery A., S. Mazinani, J. Ji, Y.M.J. Chew, D. Mattia, 3D printed composite membranes with enhanced anti-fouling behaviour, *J. Membr. Sci.* 574 (2019) 76–85. doi:10.1016/j.memsci.2018.12.058.

Aquino R.P., S. Barile, A. Grasso, M. Saviano, Envisioning smart and sustainable healthcare: 3D printing technologies for personalized medication, *Futures* 103 (2018) 35–50. doi:10.1016/j.futures.2018.03.002.

Aryanti N., A. Nafiunisa, T.D. Kusworo, D.H. Wardhani, Separation of reactive dyes using natural surfactant and micellar-enhanced ultrafiltration membrane, *J. Membr. Sci. Res.* 7 (2021) 20–28. doi:10.22079/JMSR.2020.120604.1335.

Azizi S., A. Hashemi, F.P. Shariati, H. Tayebati, A. Keramati, B. Bonakdarpour, M.M.A. Shirazi, Effect of different light-dark cycles on the membrane fouling, EPS and SMP production in a novel reciprocal membrane photobioreactor (RMPBR) by C. vulgaris species, *J. Water Proc. Eng.* 43 (2021) 102256. doi:10.1016/j.jwpe.2021.102256.

Balogun H.A., R. Sulaiman, S.S. Marzouk, A. Giwa, S.W. Hassan, 3D printing and surface imprinting technologies for water treatment: A review, *J. Water Proc. Eng.* 31 (2019) 100786. doi:10.1016/j.jwpe.2019.100786.

Balster J., I. Pünt, D.F. Stamatialis, M. Wessling, Multi-layer spacer geometries with improved mass transport, *J. Membr. Sci.* 282 (2006) 351–361. doi:10.1016/j.memsci.2006.05.039.

Bandehali S., H. Sanaeepur, A. Ebadi Amooghin, S. Shirazian, S. Ramakrishna, Biodegradable polymers for membrane separation, *Sep. Purif. Technol.* 269 (2021) 118731. doi:10.1016/j.seppur.2021.118731.

Bhattacharjee C., V.K. Saxena, S. Dutta, Fruit juice processing using membrane technology: A review, *Innov. Food Sci. Emerg. Technol.* 43 (2017) 136–153. doi:10.1016/j.ifset.2017.08.002.

Bucs S., N. Farhat, J.C. Kruithof, C. Picioreanu, M.C.M. van Loosdretch, J.S. Vrouwenvelder, Review on strategies for biofouling mitigation in spiral wound membrane systems, *Desalination* 434 (2018) 189–197. doi:10.1016/j.desal.2018.01.023.

Cao Y., X. Chen, S. S. Feng, Y. Wang, J. Luo, Nanofiltration for decolorization: Membrane fabrication, applications and challenges, *Ind. Eng. Chem. Res.* 59 (2020) 19858–19875. doi:10.1021/acs.iecr.0c04277.

Capparelli C., C.R. Fernandez Pulido, R.A. Wiencek, M.A. Hickner, Resistance and permselectivity of 3D-printed micropatterned anion-exchange membranes, *ACS Appl. Mater. Interfaces*, 29 (2018) 26298–26306. doi:10.1021/acsami.8b04177.

Cassano A., C. Conidi, E. Drioli, A comprehensive review of membrane distillation and osmotic distillation in agro-food applications, *J. Membr. Sci. Res.* 6 (2020) 304–318. doi:10.22079/JMSR.2020.122163.1349.

Castillo E.H.C., N. Thomas, O. Al-Ketan, R. Rowshan, R.K. Abu Al-Rub, L.D. Nghiem, S. Vigneswaran, H.A. Arafat, G. Naidu, 3D printed spacers for organic fouling mitigation in membrane distillation, *J. Membr. Sci.* 581 (2019) 331–343. doi:10.1016/j.memsci.2019.03.040.

Castro-Munoz R., G. Boczkaj, E. Gontarek, A. Cassano, V. Fila, Membrane technologies assisting plant-based and agro-food by-products processing: A comprehensive review, *Trends Food Sci. Technol.* 95 (2020) 219–232. doi:10.1016/j.tifs.2019.12.003.

Chang H., C.D. Ho, Y.H. Chen, L. Chen, T.H. Hsu, J.W. Lim, C.P. Chiou, P.H. Lin, Enhancing the permeate flux of direct contact membrane distillation modules with inserting 3D printing turbulence promoters, *Membranes* 11 (2021) 266. doi:10.3390/membranes11040266.

Chen Y., Gao P, Summe MJ, Phillip WA, Wei N, Biocatalytic membranes prepared by inkjet printing functionalized yeast cells onto microfiltration substrates, *J. Membr. Sci.* 550 (2018a) 91–100. doi:10.1016/j.memsci.2017.12.045.

Chen Z., Z. Li, J. Li, C. Liu, C. Lao, Y. Fu, C. Liu, Y. Li, P. Wang, Y. He, 3D printing of ceramics: A review, *J. Eur. Ceram. Soc.* 39 (2019) 661–687. doi:10.1016/j.jeurceramsoc.2018.11.013.

Chin J.Y., A.L. Ahmad, S.C. Low, Anti-wetting membrane distillation to treat high salinity wastewater: Review, *J. Membr. Sci. Res.* 6 (2020) 401–415. doi:10.22079/JMSR.2020.129954.1400.

Choi Y., G. Naidu, L.D. Nghiem, S. Lee, S. Vigneswaran, Membrane distillation crystallization for brine mining and zero liquid discharge: opportunities, challenges, and recent progress, Environ. *Sci.: Water Res. Technol.* 5 (2019) 1202–1221. doi:10.1039/C9EW00157C.

Choudhury M.R., J. Steffes, B.D. Huey, J.R. McCutcheon, 3D printed polyamide membranes for desalination, *Science* 361 (2018) 682–686. doi:10.1126/science.aar2122.

Cidonio G., M. Costantini, F. Pierini, C. Scognamiglio, T. Agrawal, A. Barbetta, 3D printing of biphasic inks: Beyond single-scale architectural control, *J. Mater. Chem. C* 9 (2021) 12489–12508. doi:10.1039/D1TC02117F.

Dang B.V., A.J. Charlton, Q. Li, Y.C. Kim, R.A. Taylor, P. L-Clech, T. Barber, Can 3D-printed spacers improve filtration at the microscale? *Sep. Purif. Tehnol.* 256 (2021) 117776. doi:10.1016/j.seppur.2020.117776.

Dehban A., A. Kargari, F. Zokaee Ashtiani, Preparation and optimization of antifouling PPSU/PES/SiO_2 nanocomposite ultrafiltration membranes by VIPS-NIPS technique, *J. Ind. Eng. Chem.* 88 (2020) 292–311. doi:10.1016/j.jiec.2020.04.028.

Del Borghi A., L. Moreschi, M. Gallo, Circular economy approach to reduce water-energy-food nexus, *Curr. Opin. Environ. Sci. Health* 13 (2020) 23–28. doi:10.1016/j.coesh.2019.10.002.

Dommati H., S.S. Ray, J.C. Wang, S.S. Chen, A comprehensive review of recent developments in 3D printing technique for ceramic membrane fabrication for water purification, *RSC Adv.* 9 (2019) 16869–16883. doi:10.1039/C9RA00872A.

Elsaid K., M. Kamil, E.T. Sayed, M.A. Abdelkareem, T. Wilberforce, A. Olabi, Environmental impact of desalination technologies: A review, *Sci. Total Environ.* 748 (2020) 141528. doi:10.1016/j.scitotenv.2020.141528.

Femmer T., A.J.C. Kuehne, J. Torres-Rendon, A. Walther, M. Wessling, Print your membrane: Rapid prototyping of complex 3D-PDMS membranes via a sacrificial resist, *J. Membr. Sci.* 478 (2015a) 12–18. doi:10.1016/j.memsci.2014.12.040.

Femmer T., A.J.C. Kuehne, M. Wessling, Estimation of the structure dependent performance of 3-D rapid prototyped membranes, *Chem. Eng. J.* 273 (2015b) 438–445. doi:10.1016/j.cej.2015.03.029.

Feng C., K.C. Khulbe, T. Matsuura, R. Farnood, A.F. Ismail, Recent progress in zeolite/zeotype membranes, *J. Membr. Sci. Res.* 1 (2015) 49–72. doi:10.22079/JMSR.2015.13530.

Figoli A., C. Ursino, S. Santoro, I. Ounifi, J. Chekir, A. Hafiane, E. Ferjani, Cellulose acetate nanofiltration membranes for cadmium remediation, *J. Membr. Sci. Res.* 6 (2020) 226–234. doi:10.22079/JMSR.2020.120669.1336.

Figoli A, T. Marino, S. Simone, E. Di Nicolo, X.M. Li, T. He, S. Tornaghi, E. Drioli, Towards non-toxic solvents for membrane preparation: a review, *Green Chem.* 16 (2014) 4034–4059. doi:10.1039/C4GC00613E.

Firouzjaei M.D., S.F. Seyedpour, S.A. Aktij, M. Giagnorio, N. Bazrafshan, A. Mollahosseini, F. Samadi, S. Ahmadalipour, F.D. Firouzjaei, M.R. Esfahani, A. Tiraferri, M. Elliot, M. Sangermano, A. Abdelrasoul, J.R. McCutcheon, M. Sadrzadeh, A.R. Esfahani, A. Rahimpour, Recent advances in functionalized polymer membranes for biofouling control and mitigation in forward osmosis, *J. Membr. Sci.* 596 (2020) 117604. doi:10.1016/j.memsci.2019.117604.

Frappa M., F. Macedonio, E. Drioli, Progress of membrane engineering for water treatment, *J. Membr. Sci. Res.* 6 (2020) 269–279. doi:10.22079/JMSR.2019.108451.1265.

Frappe M., F. Macedonio, E. Drioli, Progress of membrane engineering for water treatment, *J. Membr. Sci.* 6 (2020) 269–279. doi:10.22079/JMSR.2019.108451.1265.

Fritzmann C., M. Hausmann, M. Wiese, M. Wessling, T. Melin, Microstructured spacers for submerged membrane filtration systems, *J. Membr. Sci.* 446 (2013) 189–200. doi:10.1016/j.memsci.2013.06.033.

Fritzmann C., M. Wiese, T. Melin, M. Wessling, Helically microstructured spacers improve mass transfer and fractionation selectivity in ultrafiltration, *J. Membr. Sci.* 463 (2014) 41–48. doi:10.1016/j.memsci.2014.03.059.

Fuzil N.S., N.H. Othman, N.H. Alias, F. Marpani, M.H.D. Othman, A.F. Ismail, W.J. Lau, K. Li, T.D. Kusworo, I.Ichinose, M.M.A. Shirazi, A review on photothermal material and its usage in the development of photothermal membrane for sustainable clean water production, *Desalination* 517 (2021) 115259. doi:10.1016/j.desal.2021.115259.

Gao P., A. Hunter, M.J. Summe, W.A. Phillip, A method for the efficient fabrication of multifunctional mosaic membranes by inkjet printing, *ACS Appl. Mater. Interfaces* 8 (2016a) 19772–19779. doi:10.1021/acsami.6b06048.

Gao P., A. Hunter, S. Benavides, M.J. Summe, F. Gao, W.A. Phillip, Template synthesis of nanostructured polymeric membranes by inkjet printing, *ACS Appl. Mater. Interfaces* 8 (2016b) 3386–3395. doi:10.1021/acsami.5b11360.

Ghaffour N., T.M. Missimer, G.L. Amy, Technical review and evaluation of of the economics of water desalination: Current and future challenges for better water supply sustainability, *Desalination* 309 (2013) 197–207. doi:10.1016/j.desal.2012.10.015.

Goh P.S., T. Matsuura, A.F. Ismail, N. Hilal, Recent trends in membranes and membrane processes for desalination, *Desalination* 391 (2016) 43–60. doi:10.1016/j.desal.2015.12.016.

Hailemariam R.H., Y.C. Woo, M.M. Damtie, B.C. Kim, K.D. Park, J.S. Choi, Reverse osmosis membrane fabrication and modification technologies and future trends: A review, *Adv. Colloid Interface Sci.* 276 (2020) 102100. doi:10.1016/j.cis.2019.102100.

Homaeigohar S., M. Elbahri, Graphene membranes for water desalination, NPG Asia Mater. 9 (2017) e427. doi:10.1038/am.2017.135.

Hwa L.C., M.B. Uday, N. Ahmad, A.M. Noor, S. Rajoo, K.B. Zakaria, Integration and fabrication of the cheap ceramic membrane through 3D printing technology, *Mater Today Comm.* 1 (2018) 134–142. doi:10.1016/j.mtcomm.2018.02.029.

Issac M.N., B. Kandasubramanian, Review of manufacturing three-dimensional-printed membranes for water treatment, *Environ. Sci. Pollut. Res.* 27 (2020) 36091–36108. doi :10.1007/s11356-020-09452-2.

Jiang S., B.P. Ladewig, Green synthesis of polymeric membranes: Recent advances and future prospects, *Curr. Opin. Green Sustain. Chem.* 21 (2020) 1–8. doi:10.1016/j.cogsc.2019.07.002.

Joshi S.C., A.A. Sheikh, 3D printing in aerospace and its long-term sustainability, *Virtual Phys. Prototyp.* 10 (2015) 175–185. doi:10.1080/17452759.2015.1111519.

Kavitha J., M. Rajalakshmi, A.R. Phani, M. Padaki, Pretreatment processes for seawater reverse osmosis desalination systems-A review, *J. Water Proc. Eng.* 32 (2019) 100926. doi:10.1016/j.jwpe.2019.100926.

Kerdi S., A. Qamar, J.S. Vrouwenvelder, N. Ghaffour, Fouling resilient perforated feed spacers for membrane filtration, *Water Res.* 140 (2018) 211–219. doi:10.1016/j.watres.2018.04.049.

Khalil A., F.E. Ahmed, N. Hilal, The emerging role of 3D printing in water desalination, *Sci. Total Environ.* 790 (2021) 148238. doi:10.1016/j.scitotenv.2021.148238.

Koh E, Y.T. Lee, Development of an embossed nanofiber hemodialysis membrane for improving capacity and efficiency via 3D printing and electrospinning technology, *Sep. Purif. Technol.* 241 (2020) 116657. doi:10.1016/j.seppur.2020.116657.

Koo J.W., J.S. Ho, J. An, Y. Zhang, C.K. Chua, T.H. Chong, A review on spacers and membranes: Conventional or hybrid additive manufacturing? *Water Res.* 188 (2021) 116497. doi:10.1016/j.watres.2020.116497.

Lalia B.S., V. Kochkodan, R. Hashaikeh, N. Hilal, A review on membrane fabrication: Structure, properties and performance relationship, *Desalination* 326 (2013) 77–95. doi:10.1016/j.desal.2013.06.016.

Lee J.Y., W.S. Tan, J. An, C.K. Chua, C.Y. Tang, A.G. Fane, T.H. Chong, The potential to enhance membrane module design with 3D printing technology, *J. Membr. Sci.* 499 (2016) 480–490. doi:10.1016/j.memsci.2015.11.008.

Li F., W. Meindersma, A.B. de Haan, T. Reith, Novel spacers for mass transfer enhancement in membrane separations, *J. Membr. Sci.* 253 (2005) 1–12. doi:10.1016/j.memsci.2004.12.019.

Li Q., B. Lian, W. Zhong, A. Omar, A. Razmjou, P. Dai, J. Guan, G. Leslie, R.A. Taylor. Improving the performance of vacuum membrane distillation using a 3D-printed helical baffle and a superhydrophobic nanocomposite membrane, *Sep. Purif. Technol.* 248 (2021) 117072. doi:10.1016/j.seppur.2020.117072.

Li X., B. Zhu, J. Zhu, Graphene oxide based materials for desalination, *Carbon* 146 (2019) 320–328. doi:10.1016/j.carbon.2019.02.007.

Liang Y., J. Zhao, Q. Huang, P. Hu, C. Xiao, PVDF fiber membrane with ordered porous structure via 3D printing near field electrospinning, *J. Membr. Sci.* 618 (2021) 118709. doi:10.1016/j.memsci.2020.118709.

Lin W., Y. Zhang, D. Li, X.M. Wang, X. Huang, Roles and performance enhancement of feed spacer in spiral wound membrane modules for water treatment: A 20-year review on research evolvement, *Water Res.* 198 (2021) 117146. doi:10.1016/j.watres.2021.117146.

Liu J., A. Iranshahi, Y. Lou, G. Lipscomb, Static mixing spacers for spiral wound modules, *J. Membr. Sci.* 442 (2013) 140–148. doi:10.1016/j.memsci.2013.03.063.

Low Z.X., Y.T. Chua, B.M. Ray, D. Mattia, I.S. Metcalfe, D.A. Patterson, Perspective on 3D printing of separation membranes and comparison to related unconventional fabrication techniques, *J. Membr. Sci.* 523 (2017) 596–613. doi:10.1016/j.memsci.2016.10.006.

Low Z.X., Y.T. Chua, B.M. Ray, D. Mattia, I.S. Metcalfe, D.A. Patterson, Perspective on 3D printing of separation membranes and comparison to related unconventional fabrication techniques, *J. Membr. Sci.* 523 (2017) 596–613. doi:10.1016/j.memsci.2016.10.006.

Luelf T., D. Rall, D. Wypysek, M. Wiese, T. Femmer, C. Bremer, J.U. Michaelis, M. Wessling, 3D-printed rotating spinnerets create membranes with a twist, *J. Membr. Sci.* 555 (2018) 7–19. doi:10.1016/j.memsci.2018.03.026.

Lv J., Z. Gong, Z. He, J. Yang, Y. Chen, C. Tang, Y. Liu, M. Fan, W.M. Lau, 3D printing of a mechanically durable superhydrophobic porous membrane for oil-water separation, *J. Mater. Chem. A* 5 (2017) 12435–12444. doi:10.1039/C7TA02202F.

Madaeni S.S., M.E. Alami-Aleagha, P. Daraei, Preparation and characterization of metallic membrane using wire arc spraying, *J. Membr. Sci.* 320 (2008) 541–548. doi:10.1016/j.memsci.2008.04.051.

Mantihal S., R. Kobun, B.B. Lee, 3D food printing of as the new way of preparing food: A review, *Int. J. Gastron. Food Sci.* 22 (2020) 100260. doi:10.1016/j.ijgfs.2020.100260.

Matsuura T., Progress in membrane science and technology for seawater desalination – A review, *Desalination* 134 (2001) 47–54. doi:10.1016/S0011-9164(01)00114-X.

Mazinani S., A. Al-Simmery, Y.M.J. Chew, D. Mattia, 3D printed fouling-resistant composite membranes, *ACS Appl. Mater. Interfaces* 11 (2019) 26373–26383. doi:10.1021/acsami.9b07764.

Mikula K., D. Skrzypczak, G. Izydorczyk, J. Warchol, K. Moustakas, K. Chojnacka, A. Witek-Krowiak, 3D printing filament as a second life of waste plastics-a review, *Environ. Sci. Pollut. Res.* (2020) doi:10.1007/s11356-020-10657-8.

Mohanraj G., C. Mao, A. Armine, R. Kasher, C.J. Arnusch, Ink-jet printing-assisted modification on polyethersulfone membranes using a UV-reactive antimicrobial peptide for fouling-resistant surfaces, *ACS Omega* 3 (2018) 8752–8759. doi:10.1021/acsomega.8b00916.

Mozaffarikhah K., A. Kargari, M. Tabatabaei, H. Ghanavati, M.M.A. Shirazi, Membrane treatment of biodiesel wash-water: A sustainable solution for water recycling in biodiesel production process, *J. Water Process Eng.* 19 (2017) 331–337. doi:10.1016/j.jwpe.2017.09.007.

Muravyev N., K.A. Monogarov, U. Schaller, I.V. Fomenkov, A.N. Pivkina, Progress in additive manufacturing of energetic materials: Creating the reactive microstructures with high potential of applications, *Propellants Explos. Pyrotech.* 44 (2019) 941–969. doi:10.1002/prep.201900060.

Nachal N., J.A. Moses, P. Karthik, C. Anandharamakrishnan, Applications of 3D printing in food processing, *Food Eng. Rev.* 11 (2019) 123–141. doi:10.1007/s12393-019-09199-8.

Naseri Rad S, M.M.A. Shirazi, A. Kargari, R. Marzban, Application of membrane separation technology in downstream processing of Bacillus thuringiensis biopesticide: A review, *J. Membr. Sci. Res.* 2 (2016) 66–77. doi:10.22079/JMSR.2016.19154.

Nesan D., N.K. Rajantrakumar, D.J.C. Chan, Membrane filtration pretreatment and phytoremediation of fish farm wastewater, *J. Membr. Sci. Res.* 7 (2021) 38–44. doi:10.22079/JMSR.2020.120104.1324.

Ng V.H., C.H. Koo, W.C. Chong, J.Y. Tey, Progress of 3D printed feed spacers for membrane filtration, Mater. *Today: Proc.* 46 (2021b) 2070–2077.

Ng Z.C., W.J. Lau, T. Matsuura, A.F. Ismail, Thin film nanocomposite RO membranes: Review on fabrication techniques and impacts of nanofiller characteristics on membrane properties, *Chem. Eng. Res. Des.* 165 (2021a) 81–105. doi:10.1016/j.cherd.2020.10.003.

Ngo T.D., A. Kashani, G. Imbalzano, K.T.Q. Nguyen, D. Hui, Additive manufacturing (3D printing): A review of materials, methods, applications and challenges, *Compos. B. Eng.* 143 (2018) 172–196. doi:10.1016/j.compositesb.2018.02.012.

Niknejad A.S., S. Bazgir, M. Ardjmand, M.M.A. Shirazi, Spent caustic treatment using direct contact membrane distillation with electroblown styrene-acrylonitrile membrane, *Int. J. Environ. Sci.* Technol. 18 (2021a) 2283–2294. doi:10.1007/s13762-020-02972-x.

Nunes S.P., P.Z. Culfaz-Emecen, G.Z. Ramon, T. Visser, G.H. Koops, W. Jin, M. Ulbricht, Thinking the future of membranes: Perspectives for advanced and new membrane materials and manufacturing, *J. Membr. Sci.* 598 (2020) 117761. doi:10.1016/j.memsci.2019.117761.

Oladapo B.I., S.O.I. Ismail, T.D. Afolalu, D.B. Olawade, M. Zahedi, Review on 3D printing: Fight against COVID-19, *Mater. Chem. Phys.* 258 (2021) 123943. doi:10.1016/j.matchemphys.2020.123943.

Otitoju T.A., A.L. Ahmad, B.S. Ooi, Polyvinylidene fluoride (PVDF) membrane for oil rejection from oily wastewater: A performance review, *J. Water Proc. Eng.* 14 (2016) 41–59. doi:10.1016/j.jwpe.2016.10.011.

Parandoush P., D. Lin, A review on additive manufacturing of polymer-fiber composites, *Composite Structure* 182 (2017) 36–53. doi:10.1016/j.compstruct.2017.08.088.

Park S., Y.D. Jeong, J.H. Lee, J. Kim, K. Jeong, K.H. Cho, 3D printed honeycomb-shaped feed channel spacer for membrane fouling mitigation in nanofiltration, *J. Membr. Sci.* 620 (2021) 118665. doi:10.1016/j.memsci.2020.118665.

Rahmah W., A.K. Wardani, G. Lugito, I. Wenten, Membrane technology in deep seawater exploration: A mini review, *J. Membr. Sci. Res.* 6 (2020) 280–294. doi:10.22079/JMSR.2019.110529.1270.

Ravanchi M.T., T. Kaghazchi, A. Kargari, Application of membrane separation processes in petrochemical industry: A review, *Desalination* 235 (2009) 199–244. doi:10.1016/j.desal.2007.10.042.

Ray S.S., H. Dommati, J.C. Wang, S.S. Chen, Solvent based slurry stereolithography 3D printed hydrophilic ceramic membrane for ultrafiltration application, *Ceramics Inter* 46 (2020) 12480–12488. doi:10.1016/j.ceramint.2020.02.010.

Saleem H., S.J. Zaidi, Nanoparticles in reverse osmosis membranes for desalination: A state of the art review, *Desalination* 475 (2020) 114171. doi:10.1016/j.desal.2019.114171.

Samaei S.M., S. Gato-Trinidad, A. Altaee, The application of pressure-driven ceramic membrane technology for treatment of industrial wastewaters – A review, *Sep. Purif. Technol.* 200 (2018) 198–220. doi:10.1016/j.seppur.2018.02.041.

Sanaeepur H., A. Ebadi Amooghin, M.M.A. Shirazi, M. Pishnamazi, S. Shirazian, Water desalination and ion removal using mixed matrix electrospun nanofibrous membranes: A critical review, *Desalination* (2021) 115350. doi:10.1016/j.desal.2021.115350.

Seo J., D.I. Kushner, M.A. Hickner, 3D printing of micropatterned anion exchange membranes, *ACS Appl. Mater. Interf.* 6 (2016) 16656–16663. doi:10.1021/acsami.6b03455.

Shirazi M.M.A., A. Kargari, A.F. Ismail, T. Matsuura, Computational fluid dynamic (CFD) opportunities applied to the membrane distillation process: State-of-the-art and perspectives, *Desalination* 377 (2016) 73–90. doi:10.1016/j.desal.2015.09.010.

Shirazi M.M.A., A. Kargari, M. Tabatabaei, A.F. Ismail, T. Matsuura, Assessment of atomic force microscopy for characterization of PTFE membranes for membrane distillation (MD) process, *Desal. Water Treat.* 54 (2015a) 295–304. doi:10.1080/19443994.2014.883576.

Shirazi M.M.A., A. Kargari, M.J.A. Shirazi, Direct contact membrane distillation for seawater desalination, *Desal. Water Treat.* 49 (2012) 368–375. doi:10.1080/19443994.2012.719466.

Shirazi M.M.A., A. Kargari, S. Ramakrishna, J. Doyle, M. Rajendarian, P.R. Babu, Electrospun membranes for desalination and water/wastewater treatment: A comprehensive review, *J. Membr. Sci. Res.* 3 (2017a) 209–227. doi:10.22079/JMSR.2016.22349.

Shirazi M.M.A., S. Bazgir, F. Meshkani, A dual-layer, nanofibrous styrene-acrylonitrile membrane with hydrophobic/hydrophilic composite structure for treating the hot dyeing effluent by direct contact membrane distillation, *Chem. Eng. Res. Des.* 164 (2020) 125–146. doi:10.1016/j.cherd.2020.09.030.

Shrivastava A., S. Kumar, E.L. Cussler, Predicting the effect of membrane spacers on mass transfer, *J. Membr. Sci.* 323 (2008) 247. doi:10.1016/j.memsci.2008.05.060.

Shuit S.H., Y.T. Ong, K.T. Lee, B. Subhash, S.H. Tan, Membrane technology as a promising alternative in biodiesel production: A review, *Biotechnol. Adv.* 30 (2012) 1364–1380. doi:10.1016/j.biotechadv.2012.02.009.

Siddiqui A., N. Farhat, S.S. Bucs, R.V. Linares, C. Picioreanu, J.C. Kruithof, M.C.M. van Loosdrecht, J. Kidwell, J.S. Vrouwenvelder, *Water Res.* 91 (2016) 55–67. doi:10.1016/j.watres.2015.12.052.

Singh S., S. Ramakrishna, F. Berto, 3D printing of polymer composites: A short review, *Mater. Des. Process. Commun.* 2 (2020) e97. doi:10.1002/mdp2.97.

Singh S., S. Ramakrishna, R. Singh, Material issues in additive manufacturing: A review, *J. Manuf. Process.* 25 (2017) 185–200. doi:10.1016/j.jmapro.2016.11.006.

Soo A., S.M. Ali, H.K. Shon, 3D printing for membrane desalination: Challenges and future propects, *Desalination* 520 (2021) 115366. doi:10.1016/j.desal.2021.115366.

Sreedhar N., M. Kumar, S. Al Jitan, N. Thomas, G. Palmisano, H.A. Arafat, 3D printed photocatalytic feed spacers functionalized with β-FeOOH nanorods including pollutant degradation and membrane cleaning capabilities in water treatment, *Appl. Catal. B: Environ.* 300 (2022) 120318. doi:10.1016/j.apcatb.2021.120318.

Sreedhar N., N. Thomas, O. Al-Ketan, R. Rowshan, H. Hernandez, R.K. Abu Al-Rub, H.A. Arafat, 3D printed feed spacers based on triply predioc minimal surfaces for flux enhancement and biofouling mitigation in RO and UF, *Desalination* 425 (2018a) 12–21. doi:10.1016/j.desal.2017.10.010.

Sreedhar N., N. Thomas, O. Al-Ketan, R. Rowshan, H.H. Hernandez, R.K. Abu Al-Rub, H.A. Arafat, Mass transfer analysis of ultrafiltration using spacers based on triply periodic minimal surfaces: Effects of spacer design, directionaly and voidage, *J. Membr. Sci.* 561 (2018b) 89–98. doi:10.1016/j.memsci.2018.05.028.

Tabe S., A review of electrospun nanofiber membranes, *J. Membr. Sci. Res.* 3 (2017) 228–239. doi:10.22079/JMSR.2017.56718.1124.

Tan W.S., S.R. Suwarno, J. An, C.K. Chua, A.G. Fane, T.H. Chong, Comparison of solid, liquid and powder forms of 3D printing techniques in membrane spacer fabrication, *J. Membr. Sci.* 537 (2017) 283–296. doi:10.1016/j.memsci.2017.05.037.

Tan X.M., D. Rodrigue, A review on porous polymeric membrane preparation. Part I: Production techniques with polysulfone and poly (vinylidene fluoride), *Polymers* 11 (2019) 1160. doi:10.3390/polym11071160.

Tejo-Otero A., I. Buj-Corral, F. Fenollosa-Artes, 3D printing in medicine for preoperative surgicl planning: A review, *Ann. Biomed. Eng.* 48 (2020) 536–555. doi:10.1007/s10439-019-02411-0.

Thomas N., M. Kumar, G. Palsimano, R.K. Abu Al-Rub, R.Y. Alnuaimi, E. Alseinat, R. Rowshan, H.A. Arafat, Antiscaling 3D printed feed spacers via facile nanoparticle coating for membrane distillation, *Water Res.* 189 (2021) 116649. doi:10.1016/j.watres.2020.116649.

Thomas N., N. Sreedhar, O. Al-Ketan, R. Rowshan, R.K. Abu Al-Rub, H.A. Arafat, 3D printed spacers based on TPMS architectures for scaling control in membrane distillation, *J. Membr. Sci.* 581 (2019) 38–49. doi:10.1016/j.memsci.2019.03.039.

Thomas N., N. Sreedhar, O. Al-Ketan, R. Rowshan, R.K.A. Al-Rub, H.A. Arafat, 3D printed triply preriodic minimal surfaces as spacers for enhanced heat and mass transfer in membrane distillation, *Desalination* 443 (2018) 256–271. doi:10.1016/j.desal.2018.06.009.

Tijing L.D., J.R.C. Dizon, I. Ibrahim, A.R.N. Nisay, H.K. Shon, R.C. Advincula, 3D printing for membrane separation, desalination and water treatment, *Appl. Mater. Today* 18 (2020) 100486. doi:10.1016/j.apmt.2019.100486.

Wang H.H., J.T. Jung, J.F. Kim, S. Kim, E. Drioli, Y. Moo Lee, A novel green solvent alternative for polymeric membrane preparation via nonsolvent-induced phase inversion (NIPS), *J. Membr. Sci.* 574 (2019a) 44–54. doi:10.1016/j.memsci.2018.12.051.

Wang J, Y. Liu, Z. Fan, W. Wang, B. Wang, Z. Guo, Ink-based 3D printing technologies for grapheme-based materials: A review, *Adv. Compos. Hybrid Mater.* 2 (2019b) 1–33. doi:10.1007/s42114-018-0067-9.

Warsinger D.M., S. Chakraborty, E.W. Tow, M.H. Plumlee, C. Bellona, S. Loutatidou, L. Karimi, A.M. Mikelonis, A. Achilli, A. Ghassemi, L.P. Padhye, S.A. Snyder, S. Curcio, C.D. Vecitis, H.A. Arafat, J.H. Lienhard V, A review of polymeric membranes and processes for potable water reuse, *Prog. Polym. Sci.* 81 (2018) 209–237. doi:10.1016/j.progpolymsci.2018.01.004.

Werber J.R., A. Deshmukh, M. Elimelech, The critical need for increased selectivity, not increased water permeability, for desalination membranes, *Environ. Sci. Technol. Lett.* 3 (2016) 112–120. doi:10.1021/acs.estlett.6b00050.

Yadav P., N. Ismail, M. Essalhi, M. Tysklind, D. Athanassiadis, N. Tavajohi, Assessment of the environmental impact of polymeric membrane production, *J. Membr. Sci.* 622 (2021) 118987. doi:10.1016/j.memsci.2020.118987.

Yanar N., M. Son, H. Park, H. Choi, Bio-mimetically inspired 3D-printed honeycombed support (spacer) for the reduction of reverse solute flux and fouling of osmotic energy driven membranes, *J. Ind. Eng. Chem.* 83 (2020b) 343–350. doi:10.1016/j.jiec.2019.12.007.

Yanar N., P. Kallem, M. Son, H. Park, S. Kang, H. Choi, A new era of water treatment technologies: 3D printing for membranes, *J. Ind. Eng. Chem.* 91 (2020a) 1–14. doi:10.1016/j.jiec.2020.07.043.

Yazdi M.K., V. Vatanpour, A. Taghizadeh, M. Taghizadeh, M.R. Ganjali, M.T. Munir, S. Habibzadeh, M.R. Saeb, M. Ghaedi, Hydrogel membranes: A review, *Mater. Sci. Eng. C* 114 (2020) 111023. doi:10.1016/j.msec.2020.111023.

Yuan S., Strobbe D, Li X, Kruth JP, Van Puyvelde P, Van der Bruggen B, 3D printed chemically and mechanically robust membrane by selective laser sintering for separation of oil/water and immiscible organic mixtures, *Chem. Eng. J.* 385 (2020) 123816. doi:10.1016/j.cej.2019.123816.

Zhang Y.Z., Y. Wang, T. Cheng, L.Q. Yao, X. Li, W.Y. Lai, W. Huang, Printed supercapacitors: Materials, printing and applications, *Chem. Soc. Rev.* 48 (2019) 3229–3264. doi:10.1039/C7CS00819H.

Zhou D, L. Zhu, Y. Fu, M. Zhu, L. Xue, Development of lower cost seawater desalination processes using nanofiltration technologies – A review, *Desalination* 376 (2015) 109–116. doi:10.1016/j.desal.2015.08.020.

Zhou L.Y., J. Fu, Y. He, A review of 3D printing technologies for soft polymer materials, *Adv. Funct. Mater.* 30 (2020) 2000187. doi:10.1002/adfm.202000187.

Zhou X., C. Liu, Three-dimensional printing for catalytic applications: Current status and perspectives, *Adv. Funct. Mater.* 27 (2017) 1701134. doi:10.1002/adfm.201701134.

10 Nanohybrid Membrane for Natural Rubber Wastewater Treatment

Tutuk Djoko Kusworo and Dani Puji Utomo
Universitas Diponegoro

CONTENTS

DOI: 10.1201/9781003167327-12

10.1 INTRODUCTION

Agricultural industry derives wastewater in large volumes with a high level of organic contaminants. Mostly the agricultural wastewater is underutilized or untreated prior to its appropriate disposal (Emanuele Tarantino et al. 2017). The volume and characteristics of wastewater discharged from agriculture post-harvest process are highly variable, depending on the specific types of the material processing operation (Muro, Riera, and Carmen Diaz 2012). Wastewater with high organic contaminants is generated in large volumes during the processing of natural rubber (Jiang et al. 2018). Disposing the liquid waste into the surrounding environment would cause a serious problem, while only treating it to meet disposal standards is not economical. Taking into account its large quantity, natural rubber wastewater is potentially to be used as recycling water in the factory. One of the advanced processes that exhibit many advantages for wastewater treatment is membrane technology (Sulaiman, Ibrahim, and Abdullah 2010; Kusworo, Kumoro, Aryanti et al. 2021). The membrane process can perform a wide range of separations from particulate to molecular separation with high selectivity (Ezugbe and Rathilal 2020). The clean water can be recovered under mild process conditions and is easily recycled. Economic and environmental benefits are obtained when the generated wastewater stream is reused by reducing clean water demand, minimizing operational costs, as well as minimizing waste disposal quantity.

The implementation of membrane technology in wastewater treatment has faced issues, one of them is the fouling problem which causes a performance drop (Abdel-Karim et al. 2018; Kusworo, Aryanti, Qudratun, Tambunan et al. 2018). Many attempts have been developed in membrane fabrication to produce membrane material with antifouling properties. The incorporation of nanomaterial with specific properties into the polymeric membrane is called a nanohybrid membrane and is one of these attempts (Alaei Shahmirzadi and Kargari 2018; Kusworo, Soetrisnanto, Aryanti, Utomo, Qudratun et al. 2018a). In this chapter, the review of nanohybrid membrane development and experimental application in natural rubber wastewater is presented. The nanohybrid membrane fabrication and its characteristics are analyzed, and the permeability and selectivity performance in natural rubber wastewater filtration are also evaluated. Experimental data and detailed discussions are provided regarding concerns in reuse of water within natural rubber industry plants.

10.2 NATURAL RUBBER WASTEWATER QUANTITY AND CHARACTERISTICS

Agriculture industrial activities have been reported in contributing to environmental pollution through effluent wastewater disposal. Various levels and characteristics are released into the environment depending on the type of industrial sector. Table 10.1. presents approximate wastewater production generated by various agriculture industrial sectors. Among the agricultural industrial sectors, natural rubber is an important contributor to human life. 70%–80% of raw rubber in the world is natural rubber primarily produced in Thailand, Indonesia, and Malaysia ("FAOSTAT" 2020). In 2020, the natural rubber production only in Southeast Asia is expected to be 13.75 million metric tons ("South East Asia Rubber Markets in 2020" 2020). It means that around 275–480 million m^3 of wastewater is produced in a year. The rapid growth in the rubber industry that reaches 2.7% per year has brought some consequences; one of them is the significant increase of natural rubber waste including solid crumb rubber waste and contaminated wastewater. Generally, natural rubber wastewater has a high content of organic contaminants such as fatty acids, proteins, lipids, carbohydrates, carotenoids, uncoagulated natural latex, residual acid coagulants, and inorganic salts (Nashrullah 2017). Natural rubber wastewater exhibits typical characteristics such as 4–6 of pH (acid), 100–9,500 ppm of BOD_5, 120–15,000 ppm of COD, 30–525 ppm of TSS, 100–300 ppm of ammoniacal nitrogen ($N\text{-}NH_3$) from the preservative agent residue, and ~20 ppm of phosphorus compounds (Tanikawa et al. 2016; Sarengat and Setyorini 2015; Ngteni et al. 2020). Indeed, these contaminants levels potentially deteriorate the ecosystem and also may be harmful to the living organism without any proper wastewater treatments. Considering its abundant quantity, natural rubber wastewater also offers an opportunity for reuse purposes after being managed with the appropriate treatment methods.

Primary and secondary treatments that involve physical, chemical, and biological processes are often used to degrade the high organic contaminants in wastewater. The conventional treatments of natural rubber wastewater have covered general requirements for disposal based on the consent of avoiding pollution. Hence, the final treated wastewater quality has not met the requirement standard for reuse purpose, and one advanced process is required to produce recycle water level within a natural rubber processing plant.

Membrane-based technology offers a wide range of separation processes from molecular to suspended solid removal from wastewater. Pressure-driven membranes such as microfiltration (MF), ultrafiltration (UF), nanofiltration (NF), and reverse osmosis (RO) are usually applied in wastewater treatment either singly application or in combination (Ezugbe and Rathilal 2020). MF and UF are used to remove suspended solids in wastewater and are able to produce treated water with a turbidity of less than 0.1 NTU (Zhang et al. 2012). The application of MF and UF sometimes is aimed as pretreatment to nanofiltration and reverse osmosis. NF and RO are used typically for the removal of dissolved solids such as organics and salts. The combination of MF/UF-NF/RO is also used to improve the separation performance. Kusworo et al. (2018) developed double-stage membrane filtration comprising of UF-NF membrane for oily wastewater treatment. The first stage was UF nanohybrid membrane of PES-nano SiO_2 that was responsible for removing suspended solid and organic macromolecule. The second stage was NF membrane to reduce the salts' content in the wastewater. Other types of membrane contactors such as membrane distillation (MD) (Mokhtar et al. 2015), membrane electrodialysis (ED) (Dermentzis 2010), and membrane bioreactor (MBR) (Sulaiman, Ibrahim, and Abdullah 2010) are also

TABLE 10.1
Approximate Wastewater Production of Various Agriculture Industrial Sectors

Industry	Wastewater Production (m^3 ton^{-1}) of Product	Ref.
Natural rubber	20–35	Tanikawa et al. (2016)
processing	90–140	Urošević and Trivunac (2020)
Dairy	~14	H. Aziz et al. (2018)
Poultry	8–64	Food Northwest (2021)
Raw vegetables	18–180	Food Northwest (2021)
Fruits and canned fruits	2–5	Asgharnejad et al. (2021)
Grains to oils		

developed in agricultural wastewater treatment. However, these processes have several drawbacks such as relatively low permeate flux, high thermal consumption, low organic removal by ED, and membrane fouling. Therefore, pressure-driven membrane filtration is preferred for agricultural wastewater treatment.

10.3 NANOHYBRID MEMBRANES DEVELOPMENT

The application of polymeric membrane in wastewater treatment is restricted by the trade-off problem between permeability and selectivity. The effort to enhance the permeability of the membrane has the consequence in reduction of selectivity. The development of nanohybrid membrane is the current attractive method for improving the performance of MF, UF, NF, RO, and FO technologies for water/wastewater treatment and desalination. The nanohybrid membrane is the incorporation of nanoparticles into polymer matrix to encounter the trade-off problem and improve the membrane properties (Bet-Moushoul et al. 2016). Many researchers have reported that the presence of nanomaterials significantly enhances the membrane hydrophilicity, pore structure, membrane stability, thermal/chemical/mechanical strength, and charge density (Ayyaru and Ahn 2018; Kusworo, Qudratun, Utomo, Indriyanti, and Ramadhan 2018; Alhoshan et al. 2013). The application of nanohybrid membrane in high-organic-content wastewater treatment also showed better antifouling behavior (Kusworo, Aryanti, Qudratun, Tambunan et al. 2018). The photocatalytic and self-cleaning ability have been also reported by the incorporation of photoactive nanoparticles, for example, TiO_2, ZnO, SnO_2, and other semiconductor metal oxides. The nanohybrid membrane has great potential in resolving the drawbacks of the conventional pressure-driven membrane in its application for natural rubber wastewater treatment. There are some nanomaterials that have been reported as membrane nanofiller in the application of natural rubber wastewater treatment.

10.3.1 Nanosilica (SiO_2)

The incorporation of nanosilica in polyethersulfone (PES) membrane for natural rubber wastewater treatment has been studied in the previous study (Kusworo, Al-Aziz, and Utomo 2020). The addition of nanosilica in PES was intended to solve the trade-off between permeability and selectivity, improve the membrane hydrophilicity, and mechanical strength. Low concentration of nanosilica is added into PES along with polyethylene glycol (PEG) as pore-forming agent in N-methyl-2-pyrrolidone solvent. The nanohybrid membrane fabrication was followed by UV irradiation to modify the membrane surface (Kusworo, Qudratun, and Utomo 2017). The study showed that the PES-SiO_2 nanohybrid membrane had improved performance in terms of permeability and selectivity in natural rubber wastewater treatment.

10.3.2 Zinc Oxide Nanoparticles (ZnO NPs)

The study of ZnO incorporation into PES membrane has been investigated in the previous work (Kusworo et al. 2020a). ZnO is selected due to its strong hydrophilicity, high mechanical and chemical stability, antibacterial activity, and low-cost material. The embedment of ZnO NPs into the PES membrane is expected to improve the structural properties, antifouling, prevent the growth of biofouling, and enhance both permeability and selectivity in natural rubber wastewater treatment. The presence of ZnO nanoparticles contributed to reducing the surface negative charge, thus enhancing the removal of NH_4^+ ions in wastewater. Furthermore, the membrane antifouling property was significantly improved.

10.3.3 Titanium Dioxide Nanoparticles (TiO_2 NPs)

Titanium dioxide NPs are widely utilized in nanohybrid membrane fabrication due to their unique properties such as strong hydrophilicity, chemical stability, commercial availability, and photoactive behavior. The investigation on the effect of TiO_2 loading in polysulfone (PSf) membrane has been carried out in the previous study (Kusworo, Ariyanti, and Utomo 2020). The study reported that the TiO_2 NPs loading caused enlargement of the finger-like structure of the membrane and the localization of nanoparticles at the membrane surface leading to the increase of surface roughness. The surface hydrophilicity and mechanical strength were significantly enhanced by low-concentration loading of TiO_2. The performance evaluation showed that the incorporation of TiO_2 NPs significantly enhanced the pollutant rejection in rubber wastewater with a slight decline of permeate flux compared with a neat PES membrane. The TiO_2-blended PES membrane also exhibited lower fouling resistance in rubber wastewater treatment. A similar result was also reported where the introduction of TiO_2 into PES membrane decreases the irreversible fouling resistance in BSA filtration (Ayyaru and Ahn 2018; Yang et al. 2019). These results indicate that the presence of TiO_2 significantly improves the antifouling properties of the polymeric membrane.

10.3.4 Graphene Oxide/Reduced Graphene Oxide (GO/rGO)

Graphene oxide is two-dimensional carbon material with oxygen-containing groups that can be derived from graphite. rGO is the further process of graphene oxide by eliminating oxygen-containing groups through the reduction process, thus the characteristic of rGO is close to graphene (Das, Deoghare, and Maity 2021).

The incorporation of GO and rGO into polymeric membranes has been studied in previous work (Kusworo, Susanto et al. 2021; T.D. Kusworo and Wulandari 2021). The GO and rGO can be introduced to polymeric membranes alone or combined with other metal oxide nanomaterials. The mixture of TiO_2/GO in PSf membrane has been reported to have remarkable enhancement of permeate flux, and removal of organic and ammonia in natural rubber wastewater treatment (Kusworo, Susanto et al. 2021). The presence of GO in the matrix has been revealed in the contribution of improving nanoparticle dispersibility and photocatalytic activity of TiO_2 through the formation of reactive oxygen species (ROS). The blending of TiO_2/rGO in PSf membrane also exhibited higher pollutant rejection without sacrificing the permeate flux rate.

10.3.5 The Combination of Nanoparticles

The combination of two or more nanomaterials can be an alternative method to achieve the synergistic effect of each nanoparticle in a simple way. The combination can be carried out by simple mixing or through a chemical reaction to improve the compatibility or interaction with the polymer. In some cases, the introduction of another type of nanomaterial can help the dispersibility of nanoparticles that prevent aggregation. The mixing of different nanomaterials also can be used to reduce the fabrication cost, for example, the mixing of TiO_2 and ZnO can be selected to overcome the high-cost of TiO_2 without decreasing the performance. However, to obtain the desired performance of combined nanomaterials, optimization of the embedding process and loading concentration is required.

Despite the significant improvement by incorporating nanoparticles into the polymeric membrane, there are still some challenges encountered during the nanohybrid membrane fabrication. These challenges need to be solved for practical application in industrial scale. The common problems encountered in the nanohybrid membrane development are low compatibility of nanomaterials with polymer matrix, poor nanoparticles dispersibility, agglomeration of nanoparticles within the polymer matrix, weak chemical interaction between nanomaterials with polymer, and the possibility of nanomaterial leakage that may harm the environment.

10.4 EXPERIMENTAL

The study of nanohybrid membrane development for natural rubber wastewater treatment is carried out from upstream to downstream experiments. The upstream experiment is the selection of membrane materials, formulation of the dope solution, optimization of the fabrication method, and membrane characterization. The downstream experiment is the application of a fabricated membrane for natural rubber wastewater filtration using real wastewater obtained from the rubber processing plant. For more detail, the explanation of each work stage will be presented in the following sub-chapters.

10.4.1 Nanohybrid Membrane Fabrication

There are several methods that have been developed in polymeric membrane fabrication such as sintering, track-etching, stretching, template-leaching, and phase inversion. Among these methods, phase inversion is the most common method used in polymeric membrane preparation due to its simplicity, facile, and controllable process. The phase inversion method is performed by preparing the initial polymer solution then the solvent is separated via thermodynamic instability by exposing the polymer-rich region to polymer-poor region. Based on its media for polymer solidification, there are four types of phase inversion methods in membrane fabrication: (a) non-solvent induced phase separation (NIPS), this method was developed by Loeb-Sourirajan in 1960 by immersing the polymer solution into non-solvent (third phase in a ternary phase diagram) usually demineralized water (Rana et al. 2015). (b) Vapor induced phase separation (VIPS), the proto-membrane is exposed to the non-solvent vapor rich environment to achieve thermodynamic instability (Ismail et al. 2020). The absorption of non-solvent vapor initiates the de-mixing process. (c) Evaporation induced phase separation (EIPS), in this method, the polymer is dissolved in a volatile solvent, and the phase separation occurs through solvent evaporation (Pervin, Ghosh, and Basavaraj 2019). (d) thermally induced phase separation (TIPS), in this method the polymer is dissolved in the solvent at an elevated temperature (Figoli 2014). The polymer solution was then cast and allowed to solidify with the decrease of its temperature. These aforementioned methods have their respective advantages and disadvantages. However, the NIPS is the most widely used technique for preparing polymeric membranes including nanohybrid membranes.

As mentioned in the previous section, the incorporation of nanomaterial into host-polymer in membrane fabrication has several challenges such as nanoparticle aggregation, low dispersibility of nanoparticle, and incompatibility of nanoparticle with host-polymer. Regarding these challenges, previous researchers have developed several techniques to optimize to loading of nanoparticles into the polymer matrix as presented in Figure 10.1. (a) bulk blending is the simplest method in nanohybrid membrane fabrication. Nanoparticles are dispersed in a solvent together with polymer and other additives. The nanoparticle-containing polymer solution is then cast and phase inversed in non-solvent immersion. With this method, the distribution of nanoparticles is uncontrolled and there is a possibility of nanoparticle leaching during phase inversion. (b) in-situ interfacial polymerization is performed by introducing

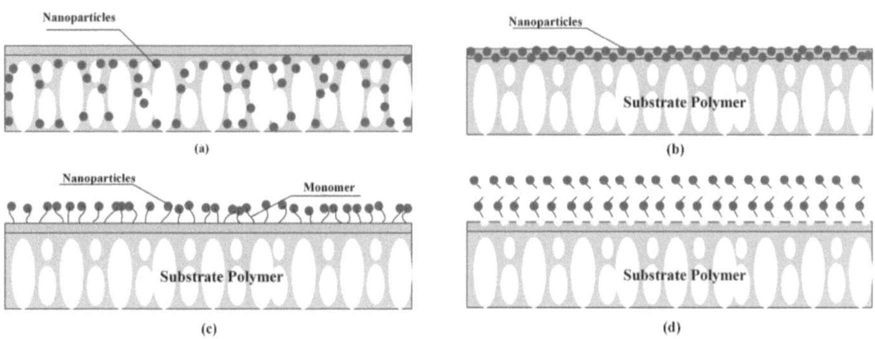

FIGURE 10.1 Scheme of nanohybrid membrane structure with different fabrication methods (a) bulk blending, (b) in-situ polymerization, (c) surface photo-grafting, (d) Layer-by-Layer (LbL) assembly.

the nanoparticle into an organic monomer and induced for polymerization under specific conditions. This method can immobilize the inorganic nanoparticle in a polymer matrix through chemical interaction with the monomer. This method will produce a nanohybrid membrane with nanoparticle distribution on the membrane surface. (c) the photo-grafting technique involves the incorporation of nanoparticles into polymer membrane surface through photo-initiated grafting process under UV light or irradiation exposure (Pejman et al. 2021). Photo-grafting of nanoparticle in nanohybrid membrane is intended to increase the immobility of nanoparticle in polymer matrix to prevent the nanoparticle from leaching. (d) layer-by-layer (LbL) assembly is carried out by creating a thin-layer of polyelectrolyte alternately with opposite charges on the polymer matrix (Saeki and Matsuyama 2017). This method can form an active layer for higher performance of NF, RO, and FO applications. The LbL assembly method has been reported to be a controllable method and produces a membrane with excellent thermal and chemical stability. In this study, the bulk blending method is selected due to its simplicity and low-cost technique. The pre/post-modifications are also involved in membrane preparation to improve the membrane properties.

The main issue in nanohybrid membrane preparation via bulk blending is nanoparticle aggregation/agglomeration. Therefore, the nanoparticles suspension and polymer solution are prepared separately and then blended in constant stirring as illustrated in Figure 10.2. An appropriate amount of nanoparticles (TiO_2, ZnO, SiO_2, SnO_2, GO, rGO, CeO_2) are dispersed in a solvent (NMP is the most common solvent used in membrane preparation). The nanoparticles suspension is then ultrasonicated for 30–60 minutes. The introduction of ultrasonic waves into suspension creates small vacuum bubbles in the liquid and then collapse as they reach saturation level in the high-pressure cycle (cavitation) (Kaboorani, Riedl, and Blanchet 2013). This cavitation process from sound energy will agitate the nanoparticles and prevent the aggregation of nanoparticles. Therefore, higher dispersibility can be achieved. At the same time, the polymer solution is also prepared by dissolving an appropriate amount of polymer (PES, PSf, PVDF) and pore-forming agent (PEG, PVP) in the solvent. The nanoparticles suspension and polymer solution are then mixed in continuous stirring for 8–12 hours until a homogeneous solution is obtained. The dope solution is degassed in a vacuum vessel to remove the air bubble and then cast onto a clean glass plate with a specific thickness. The proto-membrane is immersed in demineralized water for phase inversion.

Pretreatment modification is carried out to modify the surface properties as well as increase the interaction of nanoparticles with polymer via photo-initiated radical sites. The pretreatment can be performed by subjecting the proto-membrane to UV irradiation for a pre-determined time. UV light type-C with the wavelength of 200–280 nm is commonly used due to its high energy density that can initiate the formation of radical sites (Kusworo et al. 2017). The UV irradiated proto-membrane is then coagulated in the non-solvent for further process. The post-treatment of the membrane is conducted via thermal modification. The nanohybrid membrane is thermally cured at a temperature under its polymer-glass transition temperature (T_g) (Kusworo, Qudratun, Utomo, Indriyanti, and Ramadhan 2018). The thermal treatment of the membrane can initiate the mobility of the polymer chain to re-structure

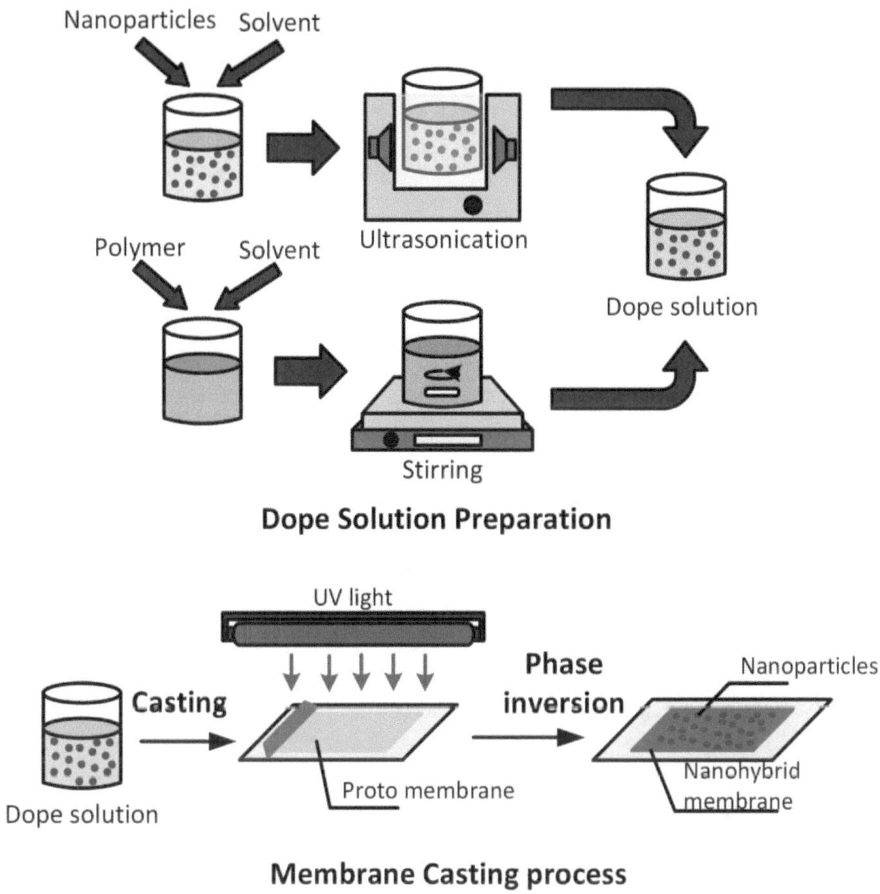

FIGURE 10.2 Schematic process diagram of nanohybrid membrane fabrication.

and increase the polymer packing density. The thermal treatment of polymeric membrane has been reported can improve the crystallinity property of the membrane. The different temperature ramp during thermal treatment also produces different membrane characteristics.

The nanoparticle incompatibility and poor distribution during the fabrication process can produce nanoparticle aggregation within the membrane matrix. The nanoparticles aggregate is presumed as the cause of nonselective gaps formation. Regarding this issue, Kusworo et al. applied crosslinked PVA on the nanohybrid membrane to overcome the problem of unselective gaps (Kusworo, Kumoro, Utomo et al. 2021). The experimental procedure of PVA coating on a nanohybrid membrane is schematically described in Figure 10.3. The nanohybrid membrane is surface dip-coated in a coating solution containing 2–3 wt.% PVA. The membrane is then dried in an oven at 50°C–60°C. PVA is a water-absorbing polymer due to its strong hydrophilicity, to prevent the swelling during the filtration process, the PVA layer is crosslinked using glutaraldehyde 5 wt.% with sulfuric acid as a catalyst. PVA

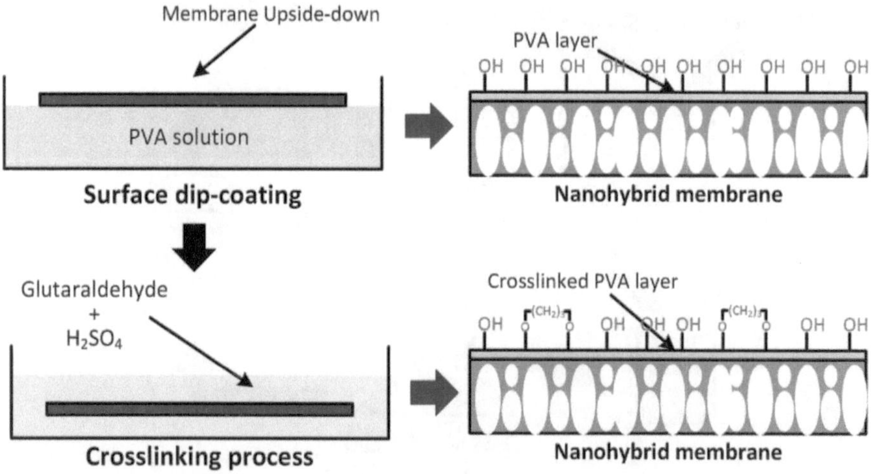

FIGURE 10.3 Illustration and schematic membrane structure of PVA coating process.

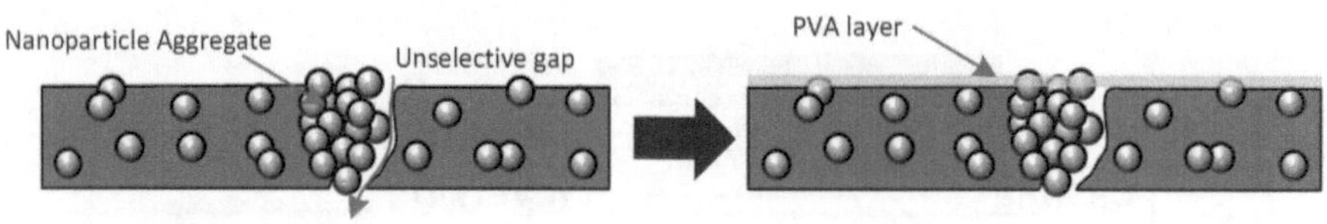

FIGURE 10.4 Illustration of unselective gaps removal by PVA coating.

layer can simultaneously improve surface hydrophilicity, removes unselective gaps, and alters surface electrostatic charge. The presumed mechanism of unselective gaps removal via PVA coating is presented in Figure 10.4.

10.4.2 Nanohybrid Membrane Characterization

The performance of the nanohybrid membrane in the filtration process strongly depends on the membrane characteristics. When modification treatment is applied on the conventional membrane either adding a new component into the matrix or reacting the membrane material with a specific compound through chemical reaction or physical interaction, the physicochemical structures of the fabricated membrane relatively change. Therefore, membrane characterization in developing novel modified membranes is strictly required. Membrane characterization provides a valuable source of information on the deep understanding of membrane permeability and selectivity behavior. There are several membrane characterization techniques that are normally used for characterizing the membrane prepared from polymer of composite materials. The common membrane characterization methods are spectroscopy methods, microscopy methods, physical and chemical characterization, mechanical properties characterization, mass transport-feed solution characterization methods.

10.4.2.1 Spectroscopy Methods for Membrane Characterization

Spectroscopy methods are used to obtain more fundamental information on membrane properties that help the effort on enhancing their performance. This fundamental characterization is applied to evaluate the chemical composition/structure of the membrane. the characterization instruments that are usually used in membrane development are infrared (IR) spectroscopy, Raman spectroscopy, nuclear magnetic resonance (NMR), X-ray photoelectron spectroscopy (XPS), and X-ray diffraction (XRD), electron paramagnetic resonance (EPR). Each of these instruments has its own analytical purpose. Among these different spectroscopy instruments, IR spectroscopy such as Fourier transform infrared (FTIR) is one of the most powerful tools for the evaluation of the chemical structure, functional group, and the possible molecular bond of the modified membrane. FTIR is very sensitive in detecting the single bond, double or triple bond, oxygen-containing groups, thus it can be used to determine whether specific functional groups exist within the material (Kusworo et al. 2020b). For the experimental procedure, sample preparation is a very crucial step in FTIR analysis. The solid sample of the membrane is crushed into finely ground solid powder. The sample powder is mixed with potassium bromide (KBr) with a ratio of about 1:100 and the mixture powder is compacted under high pressure for 1–2 min (Mohamed et al. 2017). The resultant sample-KBr pellet is inserted into a sample holder for FTIR analysis. The good sample preparation will result in clear IR spectra with no interfering bands. The key point in sample preparation is to make sure that the KBr pellet is finely ground and the sample is homogeneously mixed with KBr powder. The sample is usually observed by the mean of material absorbance in infrared spectra at wave numbers ranging from 4,000 to 400 cm^{-1}.

In following the FTIR analysis, XRD analysis for membrane characterization also provides important information on nanohybrid membrane. The membrane fabrication with involving nanoparticles incorporation, and the immersion process in a non-solvent bath is the crucial step. The improper procedure may lead to the nanoparticle release from the polymer matrix. Here, the XRD can be an alternative tool to confirm the presence of metal oxide nanoparticles since every pure metal oxide powder has a specific XRD diffraction pattern (Kusworo, Ariyanti, and Utomo 2020). The typical instrumental setting in XRD analysis is the usage of a monochromatic Cu Kα radiation source at 300 mA, 400 kV with a 2θ range of 10°–80°. The XRD analysis also can be used to evaluate the crystallinity structure of the polymeric membrane after applying the thermal treatment.

10.4.2.2 Microscopy Methods for Membrane Characterization

The microscopy method is a type of technique to observe the microstructural surface of the specimen directly. In the observation of either surface or cross-sectional area of the membrane, a scanning electron microscope (SEM) is the most common instrument used for this purpose. SEM has a higher magnification than a conventional light microscope that allow the researcher to observe the object in more detail. For evaluating the membrane morphology at higher magnification with a clear image, a field emission scanning electron microscope (FESEM) is usually used. In nanohybrid membrane fabrication, SEM/FESEM can be used to show the surface morphology and cross-sectional images of membrane specimens. The presence of nanoparticles can be observed with an SEM image with the naked eye. The formation of nanoparticle aggregation, unselective gaps, surface roughness, and nanoparticle dispersibility also can be evaluated using this technique. However, it requires further analysis for a deeper and more comprehensive understanding such as transmission electron microscope (TEM) at finer detail, atomic force microscopy (AFM) for three-dimensional topography, and energy-dispersive X-ray (EDX) for element mapping on the selected area (Kusworo, Aryanti, Qudratun, and Utomo 2018; Kusworo, Susanto et al. 2021). The sample preparation holds an important step in microscopy analysis. In observing the cross-sectional area of the membrane, the sample specimen should be immersed in liquid nitrogen and then subsequently fractured using tweezers. The defect of the membrane image will be obtained if the sample specimen is not well prepared.

10.4.2.3 Physical and Chemical Characterization Methods

The physical properties such as pore size, shape, connectivity, surface area, particle packing, porosity, and many more of the nanohybrid membrane are difficult to be determined directly. The physical characteristics of the membrane are an essential parameter that has to be determined extensively. Therefore, many researchers have developed methods to estimate the membrane porosity and pore size through the theoretical and empirical approaches. The membrane pore size can be estimated by calculating the molecular weight cut-off (MWCO) of the membrane. MWCO is defined as the molecular weight of the solutes that would be 90% rejected by the membrane (Yushkin et al. 2019). The MWCO is a popular parameter in UF membrane commercialization that can be used as a semiquantitative way of specifying the pore size of the membrane. The solutes that are commonly used for measuring the membrane MWCO are dextran, polyethylene glycol (PEG), and protein. After obtaining the MWCO value, the average pore size of the membrane can be estimated using the empirical Equation 10.1 developed by Panda and De (2013) for the correlation of PEG MWCO and average pore size.

$$r_m(cm) = 16.73 \times 10^{-10} \times (MWCO)^{0.557} \tag{10.1}$$

The correlation between average pore radius (in Å) and the MWCO value of dextran is expressed in the Equation 10.2 (Ren, Li, and Wong 2006).

$$r = 0.33(MWCO)^{0.46} \tag{10.2}$$

The theoretical approach was also applied to estimate the mean pore size of the fabricated membrane. The classic formula of the flow model in mesoporous medium developed by Guerout-Elford-Ferry is the most common formula used in the membrane pore size estimation due to its simplicity (Yuliwati et al. 2011). The most recent formula that has been developed for determining the mean pore radius are Pappenheimer-Ussing, Kedem-Katchalsky, and Mikulecky (Sarbolouki 1975). Where the three latter formulas introduced a more proper viscosity coefficient for the water inside the membrane and the latest formula is considering the extent of the membrane-water interaction. Equation 10.3 is the Guerout-Elford-Ferry formula for estimating the mean pore radius (r_m in m) of the membrane.

$$r_m = \sqrt{\frac{(2.9-1.75\varepsilon)8\eta lQ}{\varepsilon \times A \times \Delta P}} \tag{10.3}$$

Where ε is the membrane porosity, η is the pure water viscosity (8.9×10^{-4} Pa s^{-1}), l is the membrane thickness (m), Q is the permeate volume ($m^{-3}s^{-1}$), A is the effective membrane area (m^2), and ΔP is the transmembrane pressure (Pa). The measurement of membrane porosity (ε) is carried out using the fluid saturation method. The nanohybrid membrane sample is prepared with a certain area and conditioned in pure water for 24 hours then the wet membrane is weighed (M_{wet}). The membrane sample is then dried in a vacuum oven then dried (M_{dry}). The overall membrane porosity can be calculated using Equation 10.4.

$$\varepsilon(\%) = \left[\frac{(M_{wet} - M_{dry})}{\rho A l}\right] \times 100\% \tag{10.4}$$

where ρ is the density of pure water (0.998 g cm^{-3}), A is the total area of membrane (cm^2), and l is the membrane thickness (cm).

Membrane surface properties is important to be evaluated including the wetting properties. surface wetting properties of the membrane plays essential role on membrane performance during the filtration process such as permeate flux, rejection, and fouling growth behavior. Wetting properties is the interaction of membrane surface with a water molecule that is highly influenced by temperature, charge density, and surface energy. The surface energy of the membrane is related to surface hydrophilicity and surface roughness. Therefore, the evaluation of wetting properties and hydrophilicity are performed using the same technique. The most comment technique to determine the wettability and hydrophilicity of fabricated membrane is water contact angle measurement. Pure water is dripped on the membrane surface and the angle of the droplet is measured using a goniometer device. The lower contact angle indicates the hydrophilic nature of the tested material. It means that the material has a high affinity toward water molecules (M. Aziz, Arifin, and Lau 2019). The hydrophilicity properties of the membrane surface are provided by the presence of water-attracting/active polar functional groups (Huang et al. 2017). Higher membrane hydrophilicity and wettability are reported to have a significant influence on the enhancement of water flux, antifouling property, and self-life of the membrane.

10.4.2.4 Mechanical Properties and Characterization of Membrane

The improvement of permeate flux and solutes rejection rate must be accompanied by good mechanical durability to provide a long-life membrane in industrial applications. Therefore, the characterization of mechanical properties is required in nanohybrid membrane fabrication. Membrane mechanical properties are evaluated by measuring the material response to the applied force. The most common way to accomplish this purpose is by stretching the tested material at a controllable stretching rate (Batista et al. 2016; Wang et al. 2017). By measuring the force and the material deformation from the initial condition, a stress-strain behavior will be developed. The total applied force until the material fractured is termed as tensile strength (MPa) and the percentage of the material deformation from the initial size is termed as elongation break (%). Both tensile strength and elongation break are required to evaluate the membrane lifetime prediction and reliability.

10.4.2.5 Pure Water Permeability Characterization

The mass transfer in the pressure-driven membrane process is influenced by micro-hydrodynamics behavior and interfacial events occurring on the surface and inside the membrane. Pure water permeability or pure water flux (PWF) is defined as the volumetric flow through a certain membrane area at a certain pressure (Moradi et al. 2018). It is a simple technique to evaluate the permeability performance of the fabricated membrane. The PWF evaluation can be performed in either dead-end or cross-flow filtration with different operating conditions.

10.4.3 Nanohybrid Membrane Performance Evaluation

Membrane performance is evaluated by measuring the permeate water flux and pollutant rejection in natural rubber wastewater treatment. The natural rubber wastewater treatment used in the study is real wastewater obtained from primary effluent of waste treatment in the natural rubber processing plant. The typical characteristics of the natural rubber wastewater used in this work is presented in Table 10.2. The filtration process is performed in laboratory-scale cross-flow mode as shown in Figure 10.5. A tested nanohybrid membrane with an effective area of 15.9 cm^2 is applied on the membrane holder featured with a rubber seal and membrane mesh spacer. The membrane was then closed with a transparent resin cover tightly. Before the membrane is used to filter rubber wastewater, the demineralized water is fed into the membrane for 30 min for pre-conditioning and compaction. The wastewater filtration is performed at a transmembrane pressure of 5 bar and the permeate water was collected periodically. The permeate flux every 30 min is then calculated using Equation 10.5.

TABLE 10.2
Typical Characteristics of Natural Rubber Wastewater from Rubber Processing Plant

Parameter	Value
pH	6.8–8.0
Turbidity, NTU	25–30
Chemical oxygen demand (COD), mg/L	240–250
Biological oxygen demand (BOD_5), mg/L	85
Total dissolved solid (TDS), mg/L	330–350
Ammonia (NH_3), mg/L	18–175

Kusworo, Ariyanti, and Utomo (2020), Kusworo, Susanto et al. (2021), and Kusworo, Kumoro, Utomo et al. (2021).

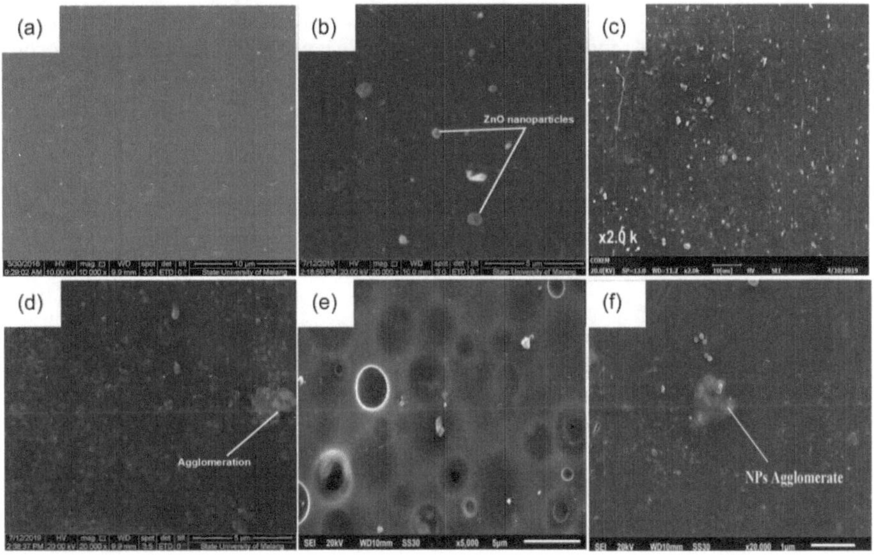

FIGURE 10.5 SEM images of membrane surface (a) neat PES membrane, (b) PES-ZnO nanohybrid membrane, (c) PSf-TiO_2 nanohybrid membrane, (d) PVA coated PES-ZnO nanohybrid membrane, (e) PSf-TiO_2/GO nanohybrid membrane, and (f) PES-ZnO/rGO nanohybrid membrane.

$$J = \frac{V}{A \cdot \Delta t} \tag{10.5}$$

Where J is the permeate water flux (L $m^{-2}h^{-1}$). V is the permeate volume collected every 30 min (L). A is the membrane effective area (0.00159 m^2). Δt is the interval time of the filtration process (h).

The selectivity performance of the membrane is evaluated by measuring the rejection rate of several pollutant components in natural rubber wastewater i.e., TDS, COD, NH_3, and turbidity. The rejection rate of each pollutant is calculated using Equation 10.6 as follows:

$$R(\%) = \left(1 - \frac{C_p}{C_f}\right) \times 100 \tag{10.6}$$

Where R is the rejection of pollutant (%), C_p is the concentration of pollutant in permeate (mg L^{-1}), and C_f is the concentration of pollutant in feed (mg L^{-1}).

The membrane stability in natural rubber wastewater treatment has been performed. The test procedure was similar with permeate flux evaluation with longer filtration time (8 hours). Flux profile of tested membrane was generated by calculating the permeate flux every 30 minutes. The flux recovery evaluations of sequential filtration-cleaning steps were also performed to investigate the membrane stability in terms of its potential for continuous applications. The fouling behavior in the nanohybrid membrane during natural rubber wastewater treatment was also investigated.

The fouling resistance of irreversible fouling (R_{ir}) and reversible fouling (R_r) are evaluated using a series model derived from Darcy's law of fluid flow through porous media (Kusworo, Kumoro, Aryanti et al. 2021). The R_{ir} and R_{ir} are calculated using Equations 10.7–10.9.

$$R_m = \frac{\Delta P}{\mu \times J_0} \tag{10.7}$$

$$R_{ir} = \frac{\Delta P}{\mu \times J_{ir}} - R_m \tag{10.8}$$

$$R_r = \frac{\Delta P}{\mu \times J_r} - R_m - R_{ir} \tag{10.9}$$

where J_0, J_{ir}, and J_r are the pure water flux (PWF) of a clean membrane, irreversible fouled membrane, and total fouled membrane ($m^3 s^{-1}$), respectively. μ is the viscosity of the pure water (Pa.s), and ΔP is the transmembrane pressure (Pa).

10.5 NANOHYBRID MEMBRANE CHARACTERISTICS

This chapter elaborates on the characteristics of the fabricated nanohybrid membranes from the previous studies which were used for treating natural rubber wastewater. Generally, the most important membrane characteristic is the membrane's ability to retain or filter specific particles or molecules. Where this selective behavior of the membrane is related to membrane physicochemical properties such as morphology, structure, porosity, thickness, surface wettability, electrostatic charge, polarity, and operating condition. Currently, the membrane properties can be changed by either pretreatment or post-treatment such as bulk modification, surface coating, surface modification, and nanoparticle incorporation. Therefore, several characterization techniques are required to evaluate the characteristics of the fabricated membranes i.e., membrane morphology using microscopy method, a chemical structure using FTIR analysis, material phase and component using XRD, physicochemical analysis, and mechanical test.

10.5.1 Nanohybrid Membranes Morphology

Several membrane performances such as particle retention, antifouling property, and water permeability are affected by the surface and internal structure of the membrane. In large pore size membrane (MF and UF), the particle exclusion is mostly affected by the pore size via the sieving mechanism, while in a narrower membrane (NF and RO), the rejection is also influenced by the electrostatic forces as well as the interaction between the membrane with solutes (Bitter 1991). In the nanohybrid membrane, the presence of nanoparticles plays important role in altering the membrane surface properties. However, it should be ensured that the nanoparticle has good distribution and compatibility with polymers. The processed image from

SEM analysis can be used to achieve this purpose. Figure 10.5 shows SEM images of nanohybrid membrane surfaces with different nanoparticle types and modification treatments. Figure 10.5a shows the neat PES membrane (Kusworo et al. 2016) with smooth and dense characteristics on the membrane surface. The SEM surface image analysis also can be used to observe the surface defect of the membrane. Figure 10.5b presents the PES membrane with ZnO incorporation via the simple blending method (Kusworo et al. 2020c). The membrane surface is smooth and white nodules are visually observed. The nanoparticles are slightly popped-up on the surface indicating that the nanoparticles are nearly leached out during the filtration process. This condition was reported to have increased surface roughness. The incorporation of 1.0 wt.% TiO_2 NPs into PSf membrane has a rougher surface as shown in Figure 10.5c (Kusworo, Ariyanti, and Utomo 2020). However, the NPs agglomerate and unselective void are not observed, which indicates that the nano TiO_2 has good compatibility with PSf polymer. Figure 10.5d is the PES-ZnO membrane with a mean of PVA coating on its surface (Kusworo et al. 2020c). It was observed the presence of NPs aggregation with no unselective was formed. The PVA coating may contribute to removing the unselective gaps. The incorporation of dual filler TiO_2 and GO in PSf membrane exhibits the most porous membrane surface as shown in Figure 10.5e. This porous surface formation could be due to the enhancement of hydrophilicity properties of the dope solution that accelerated the solvent exchange rate during phase inversion (Kusworo, Susanto et al. 2021). Another dual filler nanohybrid membrane SEM image is shown in Figure 10.5f, the membrane is rGO/ZnO incorporated PES membrane. The presence of rGO in the polymer matrix has helped the distribution of ZnO through synergistic enhancement of hydrophilicity and prevented the ZnO aggregation through intermolecular interaction between ZnO and rGO (Kusworo, Kumoro, Aryanti et al. 2021). This characterization shows that the surface properties are influenced by the concentration of nanoparticle, type of nanoparticle, and the dope solution preparation method.

With respect to the surface pore structure, not all membranes with more porous surfaces give the highest permeate flux in the filtration process. The internal porosity of the membrane also holds the important key in the transmembrane transfer process. The internal porosity of the nanohybrid membrane has been reported to be strongly affected by the addition of nanoparticles (Li et al. 2009). As shown in Figure 10.6, there are six cross-section SEM images of nanohybrid membranes. The letter A – F indicates the type of nanohybrid membranes which are neat PES membrane, PES with ZnO incorporation, PSf with TiO_2 loading, PES-ZnO with PVA coating, PSf membrane with dual filler TiO_2/GO, and PSf membrane with dual filler ZnO/rGO, respectively. The internal porous structure usually consists of extended finger-like and sponge-like micro-voids. The incorporation of hydrophilic nanoparticle such as ZnO and TiO_2 in the dope solution has been reported to have effect on altering the rheological properties and hydrophilicity of the dope solution (Ilyin, Kulichikhin, and Malkin 2015). The presence of nanoparticles also changes the thermodynamic instability during the phase inversion process that promotes a faster mass transfer rate of solvent with non-solvent resulting in a more porous membrane structure. Therefore, the nanohybrid membrane's internal structure can be modified by adjusting the type and concentration of the incorporated nanoparticles.

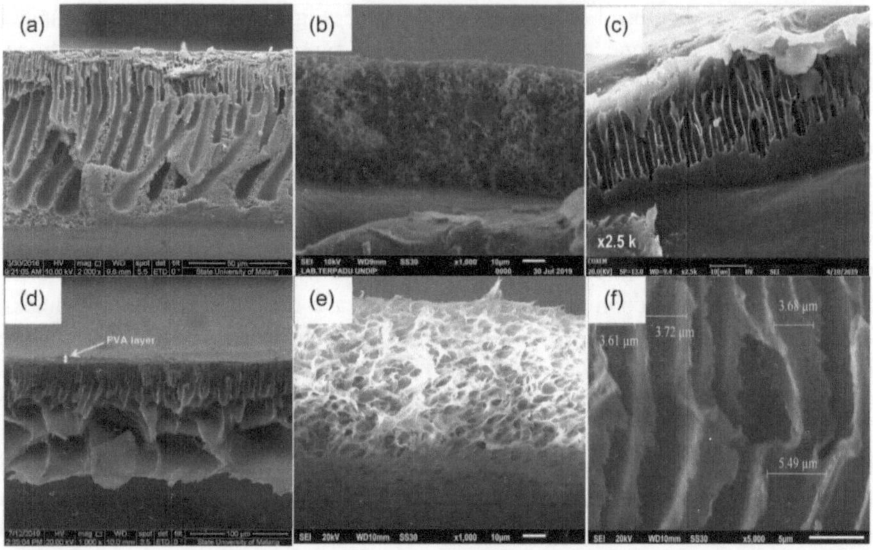

FIGURE 10.6 SEM images of membrane cross-section (a) neat PES membrane, (b) PES-ZnO nanohybrid membrane, (c) PSf-TiO_2 nanohybrid membrane, (d) PVA coated PES-ZnO nanohybrid membrane, (e) PSf-TiO_2/GO nanohybrid membrane, and (f) PES-ZnO/rGO nanohybrid membrane.

10.5.2 FTIR Analysis of Nanohybrid Membranes

The addition of nanoparticles in nanohybrid membrane fabrication may change the chemical structure of the membrane. This chemical alteration can be evaluated by analyzing the membrane FTIR spectra. The FTIR spectra can be used to investigate the effect of coating, UV irradiation, and surface modification on the membrane chemical functional groups. The FTIR spectra also can be an alternative method to evaluate the fouling deposition qualitatively. Figure 10.7 presents the utilization of FTIR spectra to evaluate the effect of nanoparticle addition (a), PVA coating on nanohybrid membrane (b), UV irradiation treatment (c) on the membrane chemical properties. The evaluation of fouling deposition qualitatively using FTIR is presented Figure 10.7d. The FTIR spectra of nanohybrid membranes and neat membrane have consistent peaks that indicate the signal of the main polymer molecule. For example, the peaks consistently appear at 1,240 and 1,585 cm^{-1} which are attributed to the IR absorption of O=S=O sulfone and C=C aromatic from PSf polymer (Jiang et al. 2019). The addition of metal oxide nanoparticles also doesn't show significant change on the FTIR spectra due to the IR absorption has low sensitivity of inorganic bond. Furthermore, the nanoparticle is incorporated in low concentration (less than 5 wt.%) into polymer. However, the addition of carbon-based nanomaterial such as GO has shown a new peak at ~1,690 cm^{-1} that belongs to the C=C bond of GO (Aher et al. 2017). The PVA coating on the nanohybrid membrane can be clearly identified using FTIR as shown in Figure 10.7b. A new strong peak at around 3,330 cm^{-1} appears as the presence of OH group in the PVA polymer chain. A peak at 1,727 cm^{-1}

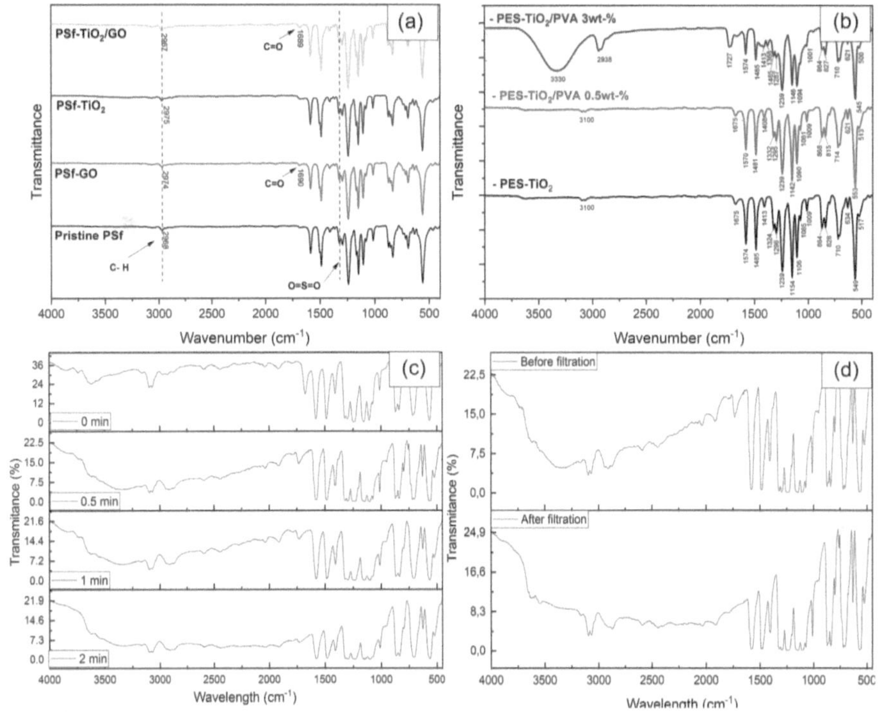

FIGURE 10.7 FTIR spectra of modified nanohybrid membranes (a) effect of TiO_2 loading, (b) effect of PVA coating on nanohybrid membrane, (c) effect of UV irradiation on nanohybrid membrane, and (d) before-after filtration spectra of nanohybrid membrane in natural rubber wastewater treatment.

also confirms the crosslinking of PVA with glutaraldehyde to form an acetal bond (Matty, Sultan, and Amine 2015). The chemical structure alteration as an effect of UV irradiation was also evaluated using FTIR as shown in Figure 10.7 (c) where the longer irradiation time, the higher IR absorption was around 3,500–3,000 cm^{-1}. This absorption improvement was reported due to the formation of hydroxyl and radical formation as the result of UV irradiation (Kusworo et al. 2020a; Abu Seman et al. 2010). These reported additional functional groups affected the membrane characteristics such as surface wettability, roughness, electrostatic charge, and affinity toward a particular molecule.

10.5.3 XRD Pattern of Nanohybrid Membranes

XRD analysis is a technique used for investigating the crystal structure of the membrane. in some studies, this technology is used to identify the presence of metal or metal oxide components in a composite material. Furthermore, the distance between the polymer chains can be evaluated. XRD patterns can identify the polymer structure of the membrane whether the structure is glassy or amorphous. The glassy polymer has a more ordered pattern and more crystallized structure than that of

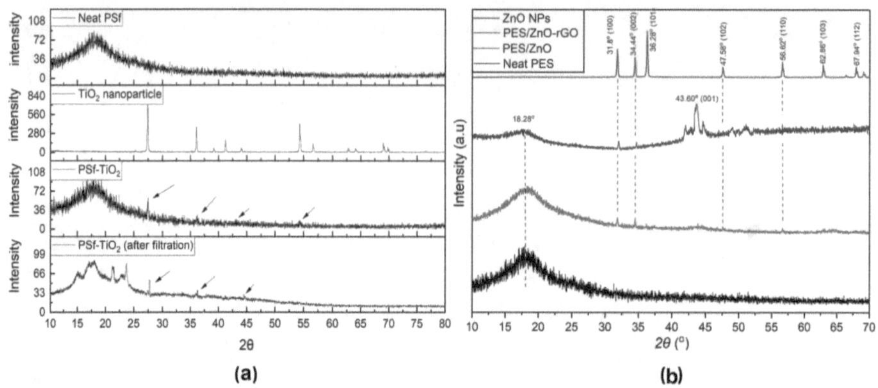

FIGURE 10.8 XRD pattern of nanohybrid membranes (a) identification of TiO_2 in PSf nanohybrid membrane matrix, and (b) identification of ZnO and rGO in the PES nanohybrid membrane matrix.

an amorphous polymer. More crystalline polymer is related to the higher mechanical properties than the amorphous membrane, thus the XRD is commonly used to investigate the modified polymeric membrane. Kusworo et al. (2017) investigated the crystal structure of the PES membrane after thermal treatment at a temperature nearly under its glass transition temperature (T_g), the XRD pattern showed a more ordered signal indicating the treated membrane has a higher crystalline structure (Aburideh et al. 2019). In the nanohybrid membrane development, the XRD can be utilized for the qualitative identification of nanoparticle filler. The filler's XRD pattern is supposed to be detected in the XRD peak along with the polymer. The XRD pattern of the filler is then compared to the peak standard of pure nanoparticle or identification using Joint Committee Powder on Diffraction Standards (JCPDS) database. For example, a nanohybrid membrane was fabricated by incorporating TiO_2 nanoparticles into PSf membrane (Kusworo, Ariyanti, and Utomo 2020). The XRD data as shown in Figure 10.8 (a) revealed the amorphous peak of PSf with additional crystalline peaks at 2θ that correspond to the crystalline peak of pure TiO_2. The membrane after being used for filtration was also analyzed in XRD to confirm that the nanoparticles were not leached out during the filtration process. The addition of nanoparticles into polymeric membranes also enhanced the overall crystallinity structure of the membrane Figure 10.8 (b). Previous studies also reported the relationship between crystallinity and water permeability and they concluded that the higher membrane crystallinity, the lower the water permeability. This phenomenon was answered by the limitation of water molecule diffusion through the increase of polymer packing density as the result of a more ordered polymer chain structure (Aburideh et al. 2019; Kusworo, Qudratun, Utomo, Indriyanti, and Ramadhan 2018).

10.5.4 Physicochemical Characteristics of Nanohybrid Membranes

The membrane transport and separation properties are provided by the structural properties of the membrane. Porosity and pore size are commonly used to assess the structural properties of the membrane. Porosity is defined as void space fraction to total membrane volume in a porous membrane. Membrane porosity may influence the transport properties while the pore size affects the separation properties. The measured porosity and pore size of the nanohybrid membrane for natural rubber wastewater treatment are presented in Table 10.3. The incorporation of nanoparticles seems to have a significant effect on membrane porosity. The previous study reported that the addition of a low concentration of nanoparticles in the nanohybrid membrane fabrication increased the porosity. However, the overloading of nanoparticles showed a slightly decline in porosity due to the high concentration of nanoparticles in the dope solution increasing the viscosity (Hong and He 2012) thus leading to a slow de-mixing process (Kusworo, Ariyanti, and Utomo 2020). The water contact angle (WCA) value as shown in Table 10.3 indicates that the addition of metal oxide nanoparticles (SiO_2, TiO_2, ZnO, etc) significantly enhanced the membrane hydrophilicity. This enhancement is provided by the oxygen molecule attached to the main metal atom that has water-attracting behavior (Kusworo, Kumoro, Aryanti et al. 2021; M. Aziz, Arifin, and Lau 2019; Huang et al. 2017). Furthermore, Yan et al. (2020) also obtained CNTs/TiO_2 composite membrane has presented smart wettability to activate on-demand separation for expected components. The higher hydrophilic properties of the nanohybrid membrane are beneficial in its application for

TABLE 10.3
Physicochemical Properties of Various Nanohybrid Membrane for Natural Rubber Wastewater Treatment

Nanohybrid Membrane	Polymer Conc.	Porosity (%)	Pore Size (nm)	WCA (°)	Ref.
Neat PSf	19 wt.%	22.79 ± 2.76	7.65 ± 0.54	61.83	Kusworo, Ariyanti, and Utomo (2020)
Neat PES	18 wt.%	33.5 ± 1.34	–	56.33	Kusworo et al. (2020a)
PES-SiO_2 0.5 wt.%	18 wt.%	35.42 ± 2.22	–	52 ± 2.65	Kusworo, Al-Aziz, and Utomo (2020)
PSf-TiO_2 1.0 wt.%	19 wt.%	36.78 ± 2.57	8.93 ± 0.62	44.17	Kusworo, Ariyanti, and Utomo (2020)
PES-ZnO 1.0 wt.%	18 wt.%	45.9 ± 1.84	–	56.35	Kusworo et al. (2020a)
PSf-GO 1 wt.%	16 wt.%	45.60 ± 3.85	4.33 ± 0.83	67.08	Kusworo, Susanto et al. (2021)
PSf-TiO_2/GO 1 wt.%	16 wt.%	59.84 ± 2.93	2.95 ± 0.39	65.25	Kusworo, Susanto et al. (2021)
PVA coated PES-ZnO	18 wt.%	45.8 ± 1.83	–	38.67	Kusworo et al. (2020c)

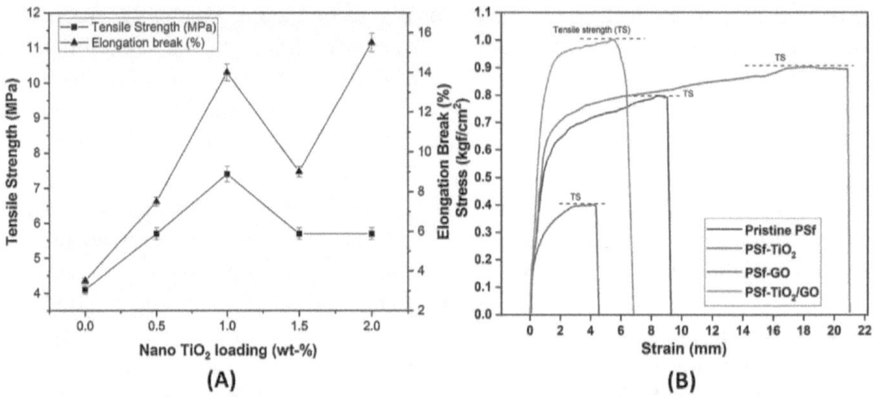

FIGURE 10.9 Mechanical properties evaluation of nanohybrid membrane (A) tensile strength and elongation break of TiO_2 loaded PSf membrane, and (B) stress-strain curve of the nanohybrid membranes.

natural rubber wastewater treatment as the hydrophilic membrane possesses higher antifouling property.

10.5.5 Mechanical Properties of Nanohybrid Membranes

The mechanical properties such as tensile strength and elongation break are important characteristics in membrane development, especially for pressure-driven membranes. The mechanical strength supports the filtration stability from the axial and tangential force of treated liquid (Tian et al. 2019). The structural characteristics such as pore topology, polymer chain order, and presence of nanofiller strongly influence the mechanical properties of the membrane. As shown in Figure 10.9, the addition of TiO_2 at various concentrations into PSf membrane shows a significant effect on membrane tensile strength and elongation break. The addition of certain concentration of nanoparticles improves the tensile strength due to the nanofiller helps in absorbing the energy applied to the membrane (Kusworo, Soetrisnanto, Aryanti, Utomo, Qudratun et al. 2018b). However, the nanoparticle loading is higher than the critical concentration, it reduces the tensile strength of the membrane as the result of agglomerate formation that disintegrates the polymer phase. The critical concentration of nanoparticles is different from one with others, it depends on the nanoparticle compatibility with polymer, surface energy, and physical properties. The improvement of membrane tensile strength was also correlated with the improvement of crystallinity structure consequently decreasing the elongation break of the membrane.

10.6 PERFORMANCE EVALUATION OF NANOHYBRID MEMBRANE FOR RUBBER WASTEWATER TREATMENT

Experimental studies of the nanohybrid membrane in the application of natural rubber wastewater treatment are focused on permeate flux performance, pollutant rejection, membrane stability, and fouling evaluation under different operating condition.

Previous studies have performed a systematic evaluation of various nanohybrid membranes with several modifications to optimize the trade-off between water flux and maximum pollutant rejection (Sulaiman, Ibrahim, and Abdullah 2010; Kusworo et al. 2020c; Jiang et al. 2018). There are many factors that affect the membrane separation performance such as membrane characteristics, hydrodynamic condition, membrane geometry, transmembrane pressure, temperature, and the feed conditions.

10.6.1 Permeate Water Flux and Pollutant Rejection Evaluation

Water permeability and selectivity of the membrane are the most important parameters in evaluating the performance. These performance parameters are influenced by membrane properties, operating conditions, and feed characteristics. The natural rubber wastewater has the characteristics of high organic pollutants and slightly acid conditions that might foul the membrane during application. The summary of performance evaluation in natural rubber wastewater treatment using a nanohybrid membrane with various types, concentrations, and modifications is presented in Table 10.4.

The experimental studies of membrane performance evaluation for natural rubber wastewater treatment as shown in Table 10.4 were performed in controlled operating condition. The wastewater was treated in a cross-flow membrane filtration at 5 bar transmembrane pressure. The study was focused on the effect of membrane modification on the performance parameters. With respect to membrane material, PSf and PES are the most common polymers used in polymeric membrane fabrication due to their high performance, robustness, and high chemical and thermal stability. As shown in Table 10.4, neat PES and PSf show moderate solutes rejection; however, the PSf reaches higher water flux due to its lower polymer concentration. The higher polymer concentration in the membrane fabrication resulted in a narrower pore size of the membrane. The incorporation of nanoparticles into the polymer matrix shows significant enhancement in the permeate flux. This flux increment has been explained as the effect of hydrophilicity enhancement by the addition of the nanoparticles and improves the membrane affinity toward water molecules (Mahmoudi et al. 2019). Moreover, the presence of nanoparticles provides a non-clogged water channel within the pores that enhances the water molecule diffusion. The pollutant rejection was also increased by the addition of nanoparticles. Nanoparticles could alter the membrane surface electrostatic charge that promoted Donnan's exclusion of charged solutes. For several nanoparticles, additional features were also reported in the membrane performance evaluation. TiO_2 shows photodegradation of organic molecules under UV irradiation thus leading to the lowering of absorbed foulant deposition. ZnO nanoparticles have been reported to show antibacterial activity that potentially mitigates the formation of biofouling in wastewater treatment. The addition of a dual filler could be an alternative way to gain a synergistic effect of each nanoparticle. For example, the addition of 1 wt.% of GO and TiO_2 into PSf polymer showed higher permeability and selectivity performance than those of either PSf-TiO_2 or PSf-GO. Surface modification via polymer coating on the membrane surface is the effective and simplest way to cover the unselective formation due to NPs aggregate formation. PVA is one of the most suitable polymers for coating purposes due to its

TABLE 10.4
Membrane Performance in Natural Rubber Wastewater Treatment

No.	Membrane	NPs Conc. (wt.%)	Average Permeate Flux (L $m^{-2}h^{-1}$)	Pollutant Removal (%)	Additional Features	Ref.
1	Neat PES (18 wt.%)	–	Initial: 6.3 Final: 1.3	TDS: 21% COD: 22% NH_3: 9.8% BOD_5: 23%	–	Kusworo et al. (2020c)
2	Neat PSf (16 wt.%)	–	Initial: 22 Final: 5	TDS: 10% COD: 82% NH_3: 30%	–	Kusworo, Susanto et al. (2021)
3	PES-SiO_2	0.5	17.5	TDS: 32% COD: 78% NH_3: 81%	–	Kusworo, Al-Aziz, and Utomo (2020)
4	PES-ZnO	1.0	Initial flux: 8.9 Final flux: 1.5	TDS: 22% COD: 24% NH_3: 11% BOD_5: 71%	The membrane has higher FRR than neat PES membrane	Kusworo et al. (2020a)
5	PSf-TiO_2 (PSf conc. 19 wt.%)	1.0	4.8	TDS: 14% COD: 88% NH_3: 89% Turbidity: 99%	Enhanced hydrophilicity and antifouling property	Kusworo, Ariyanti, and Utomo (2020)
6	PSf-TiO_2 (PSf conc. 16 wt.%)	1.0	17.5	TDS: 20% COD: 20% NH_3: 82%	Enhanced hydrophilicity and shows photodegradation	Kusworo, Susanto et al. (2021)
7	PSf-GO	1.0	9.8	TDS: 28% COD: 83% NH_3: 50%	Enhanced hydrophilicity	Kusworo, Susanto et al. (2021)
8	PSf-rGO/TiO_2	1.5	–	TDS: 79%	–	T.D. Kusworo and Wulandari (2021)
9	PSf-GO/TiO_2	1.0	Initial: 18.2 Final: 12	TDS: 40% COD: 60% NH_3: 90%	Enhanced hydrophilicity and provides photodegradation	Kusworo, Susanto et al. (2021)
10	PVA coated PES-TiO_2	1.0	28.5	TDS: 50% COD: 72% NH_3: 92%	Better antifouling properties, resistant to ozone exposure	Kusworo, Kumoro, Utomo et al. (2021)

strong hydrophilicity, high mechanical strength, chemical resistance, and relatively low-cost material. PVA coating on the membrane provides higher solutes rejection without deteriorating the water flux. The antifouling property of the membrane was also improved by PVA addition as well as the increasing the membrane resistant toward reactive chemicals.

10.6.2 Membrane Stability in Sequential Cleaning

Membrane filtration stability is an essential systematic study in membrane fabrication to predict long-term membrane application. There are several methods to evaluate the membrane stability such as flux stability in long-term filtration, flux stability under extreme operating conditions, flux recovery profile after harsh cleaning, and surface contact angle evaluation after chemical cleaning. Figure 10.10 shows the flux profile of ZnO and rGO incorporated PES membranes with sequential cleaning. The rapid flux decline could be attributed to the deposition of organic pollutants such as a peptide, lipids, fatty acids, latex, and residual coagulant in the rubber wastewater. After the cleaning procedure using pure water, the fluxes were recovered. The average flux recovery of ZnO incorporated membrane was higher than the neat PES membrane. One of the possible answers is the improvement of membrane porosity and hydrophilicity of the ZnO addition. The nanoparticle leaching during the filtration process is another issue that can be evaluated with this stability test. After the membrane is used for filtration and sequential cleaning, the presence of nanoparticles in the permeate stream was analyzed for the leaching possibility. Kusworo et al. (2021) reported that the presence of Zn in the permeate water was not detected. It indicates that the ZnO nanoparticles were not leached out during the filtration

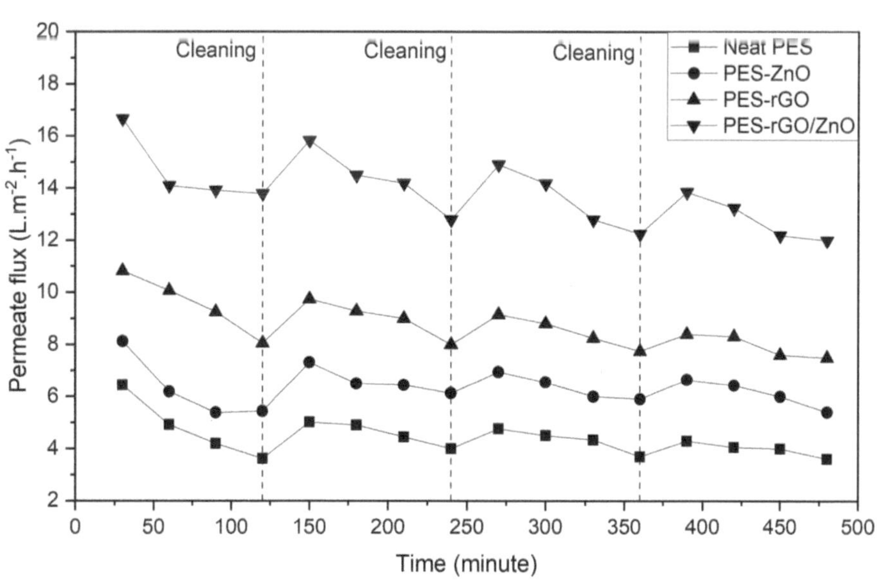

FIGURE 10.10 Filtration stability profile of nanohybrid membrane with sequential cleaning process (Kusworo, Kumoro, Aryanti 2021).

process. Therefore, the nanohybrid membrane has great potential for its application in wastewater treatment with excellent stability regarding water flux and membrane structure.

10.6.3 Fouling Evaluation

Fouling is the main obstacle in the membrane application for wastewater treatment, especially for a high organic load. Fouling in the membrane filtration is referred to the deposition or adsorption of particles, colloidal particles, or solute macromolecules on the membrane pores. While the deposition of growth and metabolism of bacteria cells is termed as biofouling. Both fouling and biofouling can cause severe flux decline and deteriorate the quality of water produced. Therefore, the fouling growth behavior of the fabricated membrane is important to be performed as it defines the functionality of the membrane. Several analytical methods have been developed by researchers for evaluating the fouling phenomenon in the membrane filtration process. The fouling resistances (Pertile et al. 2018), flux recovery ratio (FRR) (Samari et al. 2021), and flux decline ratio (FDR) (Mondal and De 2015) are common techniques in the membrane development study. Qualitative analysis via SEM imaging, FTIR spectra, and direct observation are sometimes used for membrane fouling autopsy. The fouling growth of the nanohybrid membrane in natural rubber wastewater has been reported. Table 10.5 shows the fouling resistances and FRR of nanohybrid membranes in their application for natural rubber wastewater treatment.

Reversible fouling is a type of fouling that is usually formed on the membrane surface. It can be removed by physical cleaning such as backwashing. Irreversible fouling refers to the internal fouling in the membrane pores which foulants are attached on the membrane pore walls. Irreversible fouling can be removed only by chemical cleaning. The incorporation of TiO_2 nanoparticles into PSf membrane significantly decreased

TABLE 10.5
Fouling Resistances and Flux Recovery Ratio (FRR) of Nanohybrid Membranes in Natural Rubber Wastewater Treatment

Membrane	Reversible Fouling Resistance	Irreversible Fouling Resistance	Flux Recovery Ratio (FRR)	Ref.
Neat PSf	$10.71 \times 10^{-16} m^{-1}$	$7.34 \times 10^{-16} m^{-1}$	–	Kusworo, Ariyanti, and Utomo (2020)
PSf-TiO_2 1.0 wt.%	$5.39 \times 10^{-16} m^{-1}$	$0.69 \times 10^{-16} m^{-1}$	–	Kusworo, Ariyanti, and Utomo (2020)
Neat PES	–	–	55%	T. Kusworo et al. (2020)
PES-ZnO	–	–	60%	T. Kusworo et al. (2020)
PES-rGO/ZnO	–	–	93%	Kusworo, Kumoro, Aryanti et al. (2021)

the foulant deposition within membrane pores. The irreversible fouling resistance of PSf membrane was decreased from 7.34×10^{-16} to $0.69 \times 10^{-16} m^{-1}$. The distribution of TiO_2 nanoparticles in the whole membrane pore walls prevents the attachment of fouling-causing organic solutes by providing a hydrophilic surface. ZnO and rGO nanoparticles also remarkably enhanced the FRR of the membrane up to 93% only with physical cleaning. These results have proven that the introduction of hydrophilic nanoparticles significantly improved the antifouling property of the nanohybrid membrane. This study result reveals the opportunity for nanohybrid membranes application in the treatment of rubber wastewater become techno-economically feasible.

10.7 INTENSIFICATION OF MEMBRANE SEPARATION

Intensification of membrane separation is defined as an improvement of process at membrane-based separation by integration of unit processes or operations, integration of functions, and enhancement of the performance parameters. The target of membrane process intensification could be productivity, product quality, energy consumption, environmental aspect, safety, and operational cost.

10.7.1 Membrane Integration with Other Processes

The intensification in natural rubber waste water treatment using nanohybrid membrane has been investigated. The membrane filtration was integrated with physical adsorption and ozonation processes. The adsorption and ozonation are expected to decrease the large particulate, organic macromolecules, and colloidal particles in the membrane feed. Thus, the product quality and quantity can be improved as well as the mitigation of fouling formation. The rubber wastewater treatment using nanohybrid PES-TiO_2 membrane involving sequential adsorption–ozonation processes showed organic pollutant rejection enhancement up to 96% (Kusworo, Kumoro, Utomo et al. 2021). The physical absorption using bentonite removes the large particulate and colloidal particles. While ozonation contributes to decreasing the organic contaminants and microorganisms. This proposed integrated process can minimize fouling and biofouling formation during application. Process intensification by involving activated carbon-bentonite adsorption and double stages membrane filtration was also developed in the polishing oily wastewater treatment (Kusworo, Aryanti, Qudratun, and Utomo 2018). Molinari et al. (2020) developed a hybrid membrane process coupled with the biological and chemical reactions for advanced wastewater treatment. Gaol et al. (2019) were intensifying the rubber wastewater treatment using a combination of sand filter, bentonite, and hybrid membrane (UF-RO). The study shows that the proposed process has reached remarkable metal ions removal of up to 96% and turbidity removal of up to 99% at a high feed flow rate (14 L min^{-1}). it reveals that there are many options for intensifying the treatment efficiency of rubber wastewater treatment using a nanohybrid membrane. The membrane can be combined such as UF-RO, UF-NF, or with other membrane types such as UF-MD, UF-PV, etc. The advanced oxidation processes (AOPs) also can be an alternative option in combination with membrane separation due to its high efficiency in removing organic contaminants (Azizah and Widiasa 2018; Boczkaj and Fernandes 2017).

10.7.2 Photocatalytic Membrane Filtration

The intensification via a functional enhancement in the membrane separation can be achieved by providing new capabilities such as photocatalytic degradation activity. The motive behind creating photocatalytic membrane separation is to catalyze the degradation reaction of organic contaminants thus the higher product quality can be achieved. The in-situ photocatalytic membrane process can be either adding the photocatalyst slurry into the feed or embedding photocatalyst nanoparticles into membrane material (Nasrollahi et al. 2021). The method of adding photocatalyst may risk pore-clogging during filtration. Therefore, the incorporation of nanoparticles with photocatalytic activity into the polymeric membrane is an alternative option. The semiconductor metal oxides such as TiO_2, ZnO, SnO_2, CeO_2, FeO, etc are the best material to be embedded in the polymer phase as nanocomposite membrane (Kusworo, Susanto et al. 2021; Regmi et al. 2021). The combination of TiO_2 with PSf polymer in enhancing pollutant removal in natural rubber wastewater treatment has shown remarkable results (Kusworo, Susanto et al. 2021). The filtration process under UV irradiation enhanced both permeate flux and organic removal efficiencies up to 33% and 200%, respectively. Moreover, the incorporation of TiO_2 and GO at once showed synergies between the capability of GO in binding specific contaminants and TiO_2 in catalyzing the degradation reaction. The utilization of ZnO as photocatalyst filler in PSf membrane was also evaluated, where the photocatalytic membrane has higher FRR during the cleaning process under UV irradiation (Kusworo, Kumoro, Aryanti et al. 2021). However, several problems are faced in developing photocatalytic membranes such as fouling, photocatalyst deactivation, and nanoparticle agglomeration. Therefore, an optimization of affecting variables such as loading concentration, dispersion degree, particle size, shape, and orientation geometry are required. Despite there are several limitations in the industrial-scale application, there are advantages of the photocatalytic membrane development for rubber wastewater treatment including the absence of pre- or post-treatment, stand-alone hybrid photocatalysis/filtration process, and an attractive feature for upscaling the technology.

10.8 FUTURE PROSPECT OF THE NANOHYBRID APPLICATION FOR NATURAL RUBBER WASTEWATER TREATMENT

The potency of the nanohybrid membrane application in natural rubber wastewater treatment is widely open. Several studies have confirmed that the incorporation of nanoparticles into polymeric membranes significantly improves the membrane characteristics. Rubber wastewater characteristics with high organic content have higher viscosity thus mechanical and chemical stable membrane is desired. Obviously, the nanohybrid membrane has answered this, where the mechanical properties of the membrane can be enhanced with nanoparticle loading. The major obstacle in membrane application is fouling also has been successfully mitigated by providing hydrophilic properties from the embedded nanoparticle. The pollutant rejection can be improved without decreasing the permeate flux by the addition of nanoparticles. Despite many limitations in the nanohybrid membrane application, there is however still room for improvement in many areas. The nanohybrid membrane can be integrated with other advanced processes for intensifying the treatment efficiency. The photocatalytic features of the membrane can be continuously studied to develop the fabrication method with excellent particle distribution, good compatibility, and long-life catalyst active site. The systematic and intensive study of optimizing the membrane characteristics, hydrodynamic parameters, operating conditions, and suitable membranes modules will help the nanohybrid membrane technology more viable for industrial large-scale applications. Wastewater treatment using membrane is also possible to produce higher water quality that can be reused as process water in the site plant. Thus, there are many useful aspects to the application of membrane technology in wastewater treatment including increasing treatment efficiency, environmental awareness, sustainability of clean water sources, and economical processes.

10.9 CONCLUSION

The natural rubber industry produced a vast amount of wastewater that is underutilized or untreated before its appropriate disposal. Natural rubber wastewater has characteristics of high organic content, slight acid pH, and high ammonia concentration. The treatment using conventional membrane treatment has been restricted by the formation of fouling that deteriorates selectivity, productivity, and membrane lifetime. The nanohybrid membrane which is a new class of composite membrane comprising nanomaterial and the polymeric membrane has been proven to have a significant influence in tailoring the membrane properties. The characterization methods show that an optimized nanoparticle loading into polymeric membrane improved the surface property, internal structure, mechanical strength, transport property as well as selectivity. The incorporation of nanoparticles into polymeric membrane material can also mitigate fouling formation by enhancing the hydrophilic surface property. The nanohybrid UF membrane can remove organic by decreasing the COD and BOD up to 88%. In the case of narrower pores membrane, the turbidity removal rate of 99% is possible to achieve. Membrane separation also widely opens the intensification process by integrating other processes such as adsorption, ozonation, advanced oxidation processes, and/or by providing attractive features such as photocatalytic capability. Therefore, the nanohybrid membrane in polishing high-organic wastewater has great potential for industrial large-scale application. Moreover, the membrane technology allows producing high-quality water than can be recycled as process water. Furthermore, it can be a major contribution to the continuous improvement plans in an environmental management system.

REFERENCES

Abdel-Karim, Ahmed, Sebastian Leaper, Monica Alberto, Aravind Vijayaraghavan, Xiaolei Fan, Stuart M. Holmes, Eglal R. Souaya, Mohamed I. Badawy, and Patricia Gorgojo. 2018. "High Flux and Fouling Resistant Flat Sheet Polyethersulfone Membranes Incorporated with Graphene Oxide for Ultrafiltration Applications." *Chemical Engineering Journal* 334 (February): 789–799. doi:10.1016/j.cej.2017.10.069.

Abu Seman, M.N., M. Khayet, Z.I. Bin Ali, and N. Hilal. 2010. "Reduction of Nanofiltration Membrane Fouling by UV-Initiated Graft Polymerization Technique." *Journal of Membrane Science* 355 (1–2): 133–141. doi:10.1016/j.memsci.2010.03.014.

Aburideh, Hanane, Nachida Kasbadji Merzouk, Mohamed Wahib Naceur, Zahia Tigrine, Djilali Tassalit, and Mohamed Abbas. 2019. "Thermal Annealing Effect on Morphology and Performance of Polysulfone-Cellulose Acetate Membranes: Application for Water Defluoridation." *Cellulose Chemistry and Technology* 53 (5–6): 1–15.

Aher, Ashish, Yuguang Cai, Mainak Majumder, and Dibakar Bhattacharyya. 2017. "Synthesis of Graphene Oxide Membranes and Their Behavior in Water and Isopropanol." *Carbon* 116 (May): 145–153. doi:10.1016/j.carbon.2017.01.086.

Alaei Shahmirzadi, Mohammad Amin, and Ali Kargari. 2018. "Nanocomposite Membranes." In *Emerging Technologies for Sustainable Desalination Handbook*, pp. 285–330. Elsevier. doi:10.1016/B978-0-12-815818-0.00009-6.

Alhoshan, Mansour, Javed Alam, Lawrence Arockiasamy Dass, and Nasser Al-Homaidi. 2013. "Fabrication of Polysulfone/ZnO Membrane: Influence of ZnO Nanoparticles on Membrane Characteristics." *Advances in Polymer Technology* 32 (4). doi:10.1002/adv.21369.

Asgharnejad, Hashem, Ehsan Khorshidi Nazloo, Maryam Madani Larijani, Nima Hajinajaf, and Hamidreza Rashidi. 2021. "Comprehensive Review of Water Management and Wastewater Treatment in Food Processing Industries in the Framework of Water-Food-Environment Nexus." *Comprehensive Reviews in Food Science and Food Safety* 20 (5): 4779–4815. doi:10.1111/1541-4337.12782.

Ayyaru, Sivasankaran, and Young-Ho Ahn. 2018. "Fabrication and Separation Performance of Polyethersulfone/Sulfonated TiO2 (PES–$STiO_2$) Ultrafiltration Membranes for Fouling Mitigation." *Journal of Industrial and Engineering Chemistry* 67 (November): 199–209. doi:10.1016/j.jiec.2018.06.030.

Aziz, Hamidi, Nur Puat, Motasem Alazaiza, and Yung-Tse Hung. 2018. "Poultry Slaughterhouse Wastewater Treatment Using Submerged Fibers in an Attached Growth Sequential Batch Reactor." *International Journal of Environmental Research and Public Health* 15 (8): 1734. doi:10.3390/ijerph15081734.

Aziz, Madzlan, Nur Fatihah Tajul Arifin, and Woei-Jye Lau. 2019. "Preparation and Characterization of Improved Hydrophilic Polyethersulfone/Reduced Graphene Oxide Membrane." *Malaysian Journal of Analytical Science* 23 (3): 479–487. doi:10.17576/mjas-2019-2303-12.

Azizah, Alif Nurul, and I. Nyoman Widiasa. 2018. "Advanced Oxidation Processes (AOPs) for Refinery Wastewater Treatment Contains High Phenol Concentration." *MATEC Web of Conferences* 156. EDP Sciences: 03012. doi:10.1051/matecconf/201815603012.

Batista, Natassia Lona, Philippe Olivier, Gérard Bernhart, Mirabel Cerqueira Rezende, and Edson Cocchieri Botelho. 2016. "Correlation between Degree of Crystallinity, Morphology and Mechanical Properties of PPS/Carbon Fiber Laminates." *Materials Research* 19 (1): 195–201. doi:10.1590/1980-5373-MR-2015-0453.

Bet-Moushoul, E., Y. Mansourpanah, Kh. Farhadi, and M. Tabatabaei. 2016. "TiO_2 Nanocomposite Based Polymeric Membranes: A Review on Performance Improvement for Various Applications in Chemical Engineering Processes." *Chemical Engineering Journal* 283 (January): 29–46. doi:10.1016/j.cej.2015.06.124.

Bitter, J. G. A. 1991. "Types of Membrane Separation Processes, Mechanisms of Separation." In *Transport Mechanisms in Membrane Separation Processes*, by J. G. A. Bitter, pp. 3–9. The Plenum Chemical Engineering Series. Boston, MA: Springer US. doi:10.1007/978-1-4615-3682-6_2.

Boczkaj, Grzegorz, and André Fernandes. 2017. "Wastewater Treatment by Means of Advanced Oxidation Processes at Basic PH Conditions: A Review." *Chemical Engineering Journal* 320 (July): 608–633. doi:10.1016/j.cej.2017.03.084.

Das, Papari, Ashish B. Deoghare, and Saikat Ranjan Maity. 2021. "A Novel Approach to Synthesize Reduced Graphene Oxide (RGO) at Low Thermal Conditions." *Arabian Journal for Science and Engineering* 46 (6): 5467–5475. doi:10.1007/s13369-020-04956-y.

Dermentzis, Konstantinos. 2010. "Removal of Nickel from Electroplating Rinse Waters Using Electrostatic Shielding Electrodialysis/Electrodeionization." *Journal of Hazardous Materials* 173 (1–3): 647–652. doi:10.1016/j.jhazmat.2009.08.133.

Ezugbe, Elorm Obotey, and Sudesh Rathilal. 2020. "Membrane Technologies in Wastewater Treatment: A Review." *Membranes* 10 (5): 89. doi:10.3390/membranes10050089.

"FAOSTAT." 2020. Accessed October 14. http://www.fao.org/faostat/en/#data/QC.

Figoli, Alberto. 2014. "Thermally Induced Phase Separation (TIPS) for Membrane Preparation." In *Encyclopedia of Membranes*, edited by E. Drioli and L. Giorno, 1–2. Berlin and Heidelberg: Springer. doi:10.1007/978-3-642-40872-4_1866-1.

Hong, Junming, and Yang He. 2012. "Effects of Nano Sized Zinc Oxide on the Performance of PVDF Microfiltration Membranes." *Desalination* 302 (September): 71–79. doi:10.1016/j.desal.2012.07.001.

Huang, Lijuan, Shuanglin Jing, Ou Zhuo, Xiangfeng Meng, and Xizhang Wang. 2017. "Surface Hydrophilicity and Antifungal Properties of TiO_2 Films Coated on a Co-Cr Substrate." *BioMed Research International* 2017: 1–7. doi:10.1155/2017/2054723.

Ilyin, S. O., V. G. Kulichikhin, and A. Ya. Malkin. 2015. "Rheological Properties of Emulsions Formed by Polymer Solutions and Modified by Nanoparticles." *Colloid and Polymer Science* 293 (6): 1647–1654. doi:10.1007/s00396-015-3543-6.

Ismail, Norafiqah, Antoine Venault, Jyri-Pekka Mikkola, Denis Bouyer, Enrico Drioli, and Naser Tavajohi Hassan Kiadeh. 2020. "Investigating the Potential of Membranes Formed by the Vapor Induced Phase Separation Process." *Journal of Membrane Science* 597 (March): 117601. doi:10.1016/j.memsci.2019.117601.

Jiang, Shi-Kuan, Gui-Mei Zhang, Li Yan, and Ying Wu. 2018. "Treatment of Natural Rubber Wastewater by Membrane Technologies for Water Reuse." *Membrane Water Treatment* 9 (1): 17–21. doi:10.12989/MWT.2018.9.1.017.

Jiang, Yi, Qingqing Zeng, Pratim Biswas, and John D. Fortner. 2019. "Graphene Oxides as Nanofillers in Polysulfone Ultrafiltration Membranes: Shape Matters." *Journal of Membrane Science* 581 (July): 453–461. doi:10.1016/j.memsci.2019.03.056.

Kaboorani, Alireza, Bernard Riedl, and Pierre Blanchet. 2013. "Ultrasonication Technique: A Method for Dispersing Nanoclay in Wood Adhesives." *Journal of Nanomaterials* 2013: 1–9. doi:10.1155/2013/341897.

Kusworo, Andri Cahyo Kumoro, Dani Puji Utomo, Faishal Maulana Kusumah, and Mardiyanti Dwi Pratiwi. 2021. "Performance of the Crosslinked PVA Coated PES-TiO2 Nano Hybrid Membrane for the Treatment of Pretreated Natural Rubber Wastewater Involving Sequential Adsorption – Ozonation Processes." *Journal of Environmental Chemical Engineering* 9 (2): 104855. doi:10.1016/j.jece.2020.104855.

Kusworo, Andri Cahyo Kumoro, Nita Aryanti, and Dani Puji Utomo. 2021. "Removal of Organic Pollutants from Rubber Wastewater Using Hydrophilic Nanocomposite RGO-ZnO/PES Hybrid Membranes." *Journal of Environmental Chemical Engineering* 9 (6): 106421. doi:10.1016/j.jece.2021.106421.

Kusworo, D. P. Utomo, N. Aryanti, and Qudratun. 2017. "Synergistic Effect of UV Irradiation and Thermal Annealing to Develop High Performance Polyethersulfone-Nano Silica Membrane for Produced Water Treatment." *Journal of Environmental Chemical Engineering* 5 (4): 3290–3301. doi:10.1016/j.jece.2017.06.035.

Kusworo, Danny Soetrisnanto, Nita Aryanti, Dani Puji Utomo, Qudratun, Via Dolorosa Tambunan, and Natalia Rosa Simanjuntak. 2018a. "Evaluation of Integrated Modified Nanohybrid Polyethersulfone-ZnO Membrane with Single Stage and Double Stage System for Produced Water Treatment into Clean Water." *Journal of Water Process Engineering* 23 (June): 239–249. doi:10.1016/j.jwpe.2018.04.002.

Kusworo, Habib Al-Aziz, and Dani Puji Utomo. 2020. "UV Irradiation and PEG Additive Effects on PES Hybrid Membranes Performance in Rubber Industry Wastewater Treatment." In, 050009. Semarang, Indonesia. doi:10.1063/1.5140921.

Kusworo, Heru Susanto, Nita Aryanti, Nur Rokhati, I. Nyoman Widiasa, Habib Al-Aziz, Dani Puji Utomo, Dewi Masithoh, and Andri Cahyo Kumoro. 2021. "Preparation and Characterization of Photocatalytic PSf-TiO_2/GO Nanohybrid Membrane for the Degradation of Organic Contaminants in Natural Rubber Wastewater." *Journal of Environmental Chemical Engineering* 9 (2): 105066. doi:10.1016/j.jece.2021.105066.

Kusworo, Nita Ariyanti, and Dani Puji Utomo. 2020. "Effect of Nano-TiO2 Loading in Polysulfone Membranes on the Removal of Pollutant Following Natural-Rubber Wastewater Treatment." *Journal of Water Process Engineering* 35 (June): 101190. doi:10.1016/j.jwpe.2020.101190.

Kusworo, Nita Aryanti, Enny Nurmalasari, and Dani Puji Utomo. 2020a. "Surface Modification of PES-Nano ZnO Membrane for Enhanced Performance in Rubber Wastewater Treatment." In, 050012. Semarang, Indonesia. doi:10.1063/1.5140924.

Kusworo, Nita Aryanti, Enny Nurmalasari, and Dani Puji Utomo. 2020c. "PVA Coated Nano Hybrid PES-ZnO Membrane for Natural Rubber Wastewater Treatment." *AIP Conference Proceedings* 2197 (1): 050013. doi:10.1063/1.5140925.

Kusworo, Nita Aryanti, Qudratun Qudratun, Via Dolorosa Tambunan, and Natalia Rosa Simanjuntak. 2018. "Development of Antifouling Polyethersulfone (PES)-Nano ZnO Membrane for Produced Water Treatment." *Jurnal Teknologi* 80 (3–2). doi:10.11113/jt.v80.12729.

Kusworo, Nita Aryanti, Qudratun, and Dani Puji Utomo. 2018. "Oilfield Produced Water Treatment to Clean Water Using Integrated Activated Carbon-Bentonite Adsorbent and Double Stages Membrane Process." *Chemical Engineering Journal* 347 (September): 462–471. doi:10.1016/j.cej.2018.04.136.

Kusworo, Nita Aryanti, R.A. Anggita, T.A.D. Setyorini, and D.P. Utomo. 2016. "Surface Modification and Performance Enhancement of Polyethersulfone (PES) Membrane Using Combination of Ultra Violet Irradiation and Thermal Annealing for Produced Water Treatment." *Journal of Environmental Science and Technology* 10 (1): 35–43. doi:10.3923/jest.2017.35.43.

Kusworo, Qudratun, and Dani Puji Utomo. 2017. "Performance Evaluation of Double Stage Process Using Nano Hybrid PES/SiO_2-PES Membrane and PES/ZnO-PES Membranes for Oily Waste Water Treatment to Clean Water." *Journal of Environmental Chemical Engineering* 5 (6): 6077–6086. doi:10.1016/j.jece.2017.11.044.

Kusworo, Qudratun, Dani Puji Utomo, Indriyanti, and Iqbal Ryan Ramadhan. 2018. "Enhancement of Separation Performance of Nano Hybrid PES –TiO_2 Membrane Using Three Combination Effects of Ultraviolet Irradiation, Ethanol-Acetone Immersion, and Thermal Annealing Process for CO_2 Removal." *Journal of Environmental Chemical Engineering* 6 (2): 2865–2873. doi:10.1016/j.jece.2018.04.023.

Kusworo, T.D., and Lutfi Mia Wulandari. 2021. "Fabrication of High Performance PSf-RGO/TiO_2 UF Membrane for Ruberry Wastewater Treatment." *IOP Conference Series: Materials Science and Engineering* 1053 (1): 012025. doi:10.1088/1757-899X/1053/1/012025.

Kusworo, Tutuk, Nita Aryanti, Dani Utomo, and Enny Nurmala. 2020. "Performance Evaluation of PES-ZnO Nanohybrid Using a Combination of UV Irradiation and Cross-Linking for Wastewater Treatment of the Rubber Industry to Clean Water." *Journal of Membrane Science and Research* 7 (1): 4–13. doi:10.22079/jmsr.2020.120490.1334.

Li, Jing-Feng, Zhen-Liang Xu, Hu Yang, Li-Yun Yu, and Min Liu. 2009. "Effect of TiO_2 Nanoparticles on the Surface Morphology and Performance of Microporous PES Membrane." *Applied Surface Science* 255 (9): 4725–4732. doi:10.1016/j.apsusc.2008.07.139.

Lumban Gaol, Elsa Rama, Subriyer Nasir, Hermansyah Hermansyah, and Agung Mataram. 2019. "Rubber Industry Wastewater Treatment Using Sand Filter, Bentonite and Hybrid Membrane (UF-RO)." *Sriwijaya Journal of Environment* 4 (1): 14–18. doi:10.22135/sje.2019.4.1.14.

Mahmoudi, Ebrahim, Law Yong Ng, Wei Lun Ang, Ying Tao Chung, Rosiah Rohani, and Abdul Wahab Mohammad. 2019. "Enhancing Morphology and Separation Performance of Polyamide 6,6 Membranes by Minimal Incorporation of Silver Decorated Graphene Oxide Nanoparticles." *Scientific Reports* 9 (1): 1216. doi:10.1038/s41598-018-38060-x.

Matty, Fadhel S., Maha T. Sultan, and Adiba Kh Amine. 2015. "Swelling Behavior of Cross-Link PVA with Glutaraldehyde." *IBN Al-Haitham Journal for Pure and Applied Science* 28 (2): 11. https://www.iasj.net/iasj?func=fulltext&aId=105121.

Mohamed, M.A., J. Jaafar, A.F. Ismail, M.H.D. Othman, and M.A. Rahman. 2017. "Fourier Transform Infrared (FTIR) Spectroscopy." In *Membrane Characterization*, pp. 3–29. Elsevier. doi:10.1016/B978-0-444-63776-5.00001-2.

Mokhtar, N. M., W. J. Lau, A. F. Ismail, and D. Veerasamy. 2015. "Membrane Distillation Technology for Treatment of Wastewater from Rubber Industry in Malaysia." *Procedia CIRP* 26: 792–796.

Molinari, Raffaele, Cristina Lavorato, and Pietro Argurio. 2020. "Application of Hybrid Membrane Processes Coupling Separation and Biological or Chemical Reaction in Advanced Wastewater Treatment." *Membranes* 10 (10): 281. doi:10.3390/membranes10100281.

Mondal, Mrinmoy, and Sirshendu De. 2015. "Characterization and Antifouling Properties of Polyethylene Glycol Doped PAN–CAP Blend Membrane." *RSC Advances* 5 (49): 38948–38963. doi:10.1039/C5RA02889B.

Moradi, Golshan, Sirus Zinadini, Laleh Rajabi, and Soheil Dadari. 2018. "Fabrication of High Flux and Antifouling Mixed Matrix Fumarate-Alumoxane/PAN Membranes via Electrospinning for Application in Membrane Bioreactors." *Applied Surface Science* 427 (January): 830–842. doi:10.1016/j.apsusc.2017.09.039.

Muro, Claudia, Francisco Riera, and Maria del Carmen Diaz. 2012. "Membrane Separation Process in Wastewater Treatment of Food Industry." In *Food Industrial Processes - Methods and Equipment*, edited by B. Valdez. InTech. doi:10.5772/31116.

Nashrullah, Syarif. 2017. "Pengolahan Limbah Karet dengan Fitoremidiasi Menggunakan Tanaman Typha angustifolia (Ruber Waste Treatment via Phytoremediation using Typha angustifolia Plant)." *Jurnal Teknologi Lingkungan Lahan Basah* 5 (1). doi:10.26418/jtllb.v5i1.18546

Nasrollahi, Nazanin, Leila Ghalamchi, Vahid Vatanpour, and Alireza Khataee. 2021. "Photocatalytic-Membrane Technology: A Critical Review for Membrane Fouling Mitigation." *Journal of Industrial and Engineering Chemistry* 93 (January): 101–116. doi:10.1016/j.jiec.2020.09.031.

Ngteni, Rahmat, Md. Sohrab Hossain, Mohd Omar Ab Kadir, Ahmad Jaril Asis, and Zulhafiz Tajudin. 2020. "Kinetics and Isotherm Modeling for the Treatment of Rubber Processing Effluent Using Iron (II) Sulphate Waste as a Coagulant." *Water* 12 (6): 1747. doi:10.3390/w12061747.

Panda, Swapna Rekha, and Sirshendu De. 2013. "Role of Polyethylene Glycol with Different Solvents for Tailor-Made Polysulfone Membranes." *Journal of Polymer Research* 20 (7): 179. doi:10.1007/s10965-013-0179-4.

Pejman, Mehdi, Mostafa Dadashi Firouzjaei, Sadegh Aghapour Aktij, Ehsan Zolghadr, Parnab Das, Mark Elliott, Mohtada Sadrzadeh, Marco Sangermano, Ahmad Rahimpour, and Alberto Tiraferri. 2021. "Effective Strategy for UV-Mediated Grafting of Biocidal Ag-MOFs on Polymeric Membranes Aimed at Enhanced Water Ultrafiltration." *Chemical Engineering Journal* 426 (December): 130704. doi:10.1016/j.cej.2021.130704.

Pertile, Carine, Márcia Zanini, Camila Baldasso, Mara Zeni Andrade, and Isabel Cristina Tessaro. 2018. "Evaluation of Membrane Microfiltration Fouling in Landfill Leachate Treatment." *Matéria (Rio de Janeiro)* 23 (1). doi:10.1590/s1517-707620170001.0297.

Pervin, Rumiaya, Pijush Ghosh, and Madivala G. Basavaraj. 2019. "Tailoring Pore Distribution in Polymer Films *via* Evaporation Induced Phase Separation." *RSC Advances* 9 (27): 15593–15605. doi:10.1039/C9RA01331H.

Rana, Dipak, T. Matsuura, Mohd Kassim, and A. Ismail. 2015. "Reverse Osmosis Membrane." In *Handbook of Membrane Separations*, edited by A. Pabby, S. Rizvi, and A. Requena, pp. 35–52. CRC Press. doi:10.1201/b18319-5.

Regmi, Chhabilal, Saeed Ashtiani, Zdeněk Sofer, Zdeněk Hrdlička, Filip Průša, Ondřej Vopička, and Karel Friess. 2021. "CeO_2-Blended Cellulose Triacetate Mixed-Matrix Membranes for Selective CO_2 Separation." *Membranes* 11 (8): 632. doi:10.3390/membranes11080632.

Ren, Jizhong, Zhansheng Li, and Fook-Sin Wong. 2006. "A New Method for the Prediction of Pore Size Distribution and MWCO of Ultrafiltration Membranes." *Journal of Membrane Science* 279 (1): 558–569. doi:10.1016/j.memsci.2005.12.052.

Saeki, Daisuke, and Hideto Matsuyama. 2017. "Ultrathin and Ordered Stacking of Silica Nanoparticles via Spin-Assisted Layer-by-Layer Assembly under Dehydrated Conditions for the Fabrication of Ultrafiltration Membranes." *Journal of Membrane Science* 523 (February): 60–67. doi:10.1016/j.memsci.2016.09.056.

Samari, Mahya, Sirus Zinadini, Ali Akbar Zinatizadeh, Mohammad Jafarzadeh, and Foad Gholami. 2021. "A New Fouling Resistance Polyethersulfone Ultrafiltration Membrane Embedded by Metformin-Modified FSM-16: Fabrication, Characterization and Performance Evaluation in Emulsified Oil-Water Separation." *Journal of Environmental Chemical Engineering* 9 (4): 105386. doi:10.1016/j.jece.2021.105386.

Sarbolouki, M. N. 1975. "Pore Flow Models and Their Applicability." *Ion Exchange and Membranes* 2 (2): 117–122.

Sarengat, Nursamsi, and Ike Setyorini. 2015. "Pengaruh Penggunaan Adsorben Terhadap Kandungan Amonia (NH3 –N) Pada Limbah Cair Industri Karet RSS (The Effect of Adsorben Utilization on the Ammonia (NH3-N) Content in RSS Rubber Industry Wastewater)." *National Conference Proceeding of Leather, Rubber, and Plastics* 4: 75–84.

"South East Asia Rubber Markets in 2020." 2020. *HQTS*. May 14. https://www.hqts.com/south-east-asia-rubber-markets-in-2020/.

Sulaiman, Nik Meriam Nik, Shaliza Ibrahim, and Sarah Lim Abdullah. 2010. "Membrane Bioreactor for the Treatment of Natural Rubber Wastewater." *International Journal of Environmental Engineering* 2 (1–3): 92. doi:10.1504/IJEE.2010.029823.

Tanikawa, D., K. Syutsubo, M. Hatamoto, M. Fukuda, M. Takahashi, P. K. Choeisai, and T. Yamaguchi. 2016. "Treatment of Natural Rubber Processing Wastewater Using a Combination System of a Two-Stage Up-Flow Anaerobic Sludge Blanket and down-Flow Hanging Sponge System." *Water Science and Technology* 73 (8): 1777–1784. doi:10.2166/wst.2016.019.

Tarantino, Emanuele, Grazia Disciglio, Giuseppe Gatta, Angela Libutti, Laura Frabboni, A. Gagliardi, and Annalisa Tarantino. 2017. "Agro-Industrial Treated Wastewater Reuse for Crop Irrigation: Implication in Soil Fertility." *Chemical Engineering Transactions* 58 (July): 679–684. doi:10.3303/CET1758114.

Tian, Xuyu, Shuntao Wu, Peiling He, Xiong Zhou, Zhaonian Bian, and Jinlin Hu. 2019. "Research on Membrane Structure Performance." In *E3S Web of Conferences*, by P. Zhou and Y. He. 79: 01014. doi:10.1051/e3sconf/20197901014.

Urošević, Tijana, and Katarina Trivunac. 2020. "Achievements in Low-Pressure Membrane Processes Microfiltration (MF) and Ultrafiltration (UF) for Wastewater and Water Treatment." In *Current Trends and Future Developments on (Bio-) Membranes*, pp. 67–107. Elsevier. doi:10.1016/B978-0-12-817378-7.00003-3.

Wang, K., A.A. Abdala, N. Hilal, and M.K. Khraisheh. 2017. "Mechanical Characterization of Membranes." In *Membrane Characterization*, pp. 259–306. Elsevier. doi:10.1016/B978-0-444-63776-5.00013-9.

"WaterandWastewaterUseintheFoodProcessingIndustry."2021.FoodNorthwest.AccessedJuly25. https://www.foodnorthwest.org/index.php?option=com_content&view=article&id=83%-3Awater-and-wastewater-use-in-the-food-processing-industry&catid=20%3Asite-content&showall=1&id=83:water-and-wastewater-use-in-the-food-processing-industry.

Yan, Luke, Chaohui Liu, Junyuan Xia, Min Chao, Wenqin Wang, Jincui Gu, and Tao Chen. 2020. "CNTs/TiO_2 Composite Membrane with Adaptable Wettability for on-Demand Oil/Water Separation." *Journal of Cleaner Production* 275 (December): 124011. doi:10.1016/j.jclepro.2020.124011.

Yang, Tao, Houfeng Xiong, Fen Liu, Qiyong Yang, Bingjie Xu, and Changchao Zhan. 2019. "Effect of UV/TiO_2 Pretreatment on Fouling Alleviation and Mechanisms of Fouling Development in a Cross-Flow Filtration Process Using a Ceramic UF Membrane." *Chemical Engineering Journal* 358 (February): 1583–1593. doi:10.1016/j.cej.2018.10.149.

Yuliwati, E., A.F. Ismail, T. Matsuura, M.A. Kassim, and M.S. Abdullah. 2011. "Characterization of Surface-Modified Porous PVDF Hollow Fibers for Refinery Wastewater Treatment Using Microscopic Observation." *Desalination* 283 (December): 206–213. doi:10.1016/j.desal.2011.02.037.

Yushkin, Alexey, Roman Borisov, Vladimir Volkov, and Alexey Volkov. 2019. "Improvement of MWCO Determination by Using Branched PEGs and MALDI Method." *Separation and Purification Technology* 211 (March): 108–116. doi:10.1016/j.seppur.2018.09.043.

Zhang, Tian C., Roa Y. Surampalli, Saravanamuthu Vigneswaran, R. D. Tyagi, Say Leong Ong, and C. M. Kao, eds. 2012. *Membrane Technology and Environmental Applications.* Reston, VA: American Society of Civil Engineers. doi:10.1061/9780784412275.

11 Mixed Matrix Membrane (MMM) in the Agriculture Industry

E. Yuliwati and S. Martini
Universitas Muhammadiyah Palembang

Ahmad Fauzi Ismail and Pei Sean Goh
Universiti Teknologi Malaysia

CONTENTS

DOI: 10.1201/9781003167327-13

11.1 INTRODUCTION

Water is indispensable for every human being, and water usage has been reported to be under great stress due to climate change, urbanization, industrialization, population growth, and food demands. These factors have asserted an extra discussion in the water purification industry. Many conventional and non-conventional technologies namely, adsorption, disinfection, coagulation, and flocculation have been developed to treat water and wastewater to reach the desired water quality for daily use [1]. Few technologies have failed to satisfy the level of water standards. Membrane technology has become the most viable option to overcome this issue [2]. Polymeric and ceramic membrane materials have been used extensively in the water and wastewater treatment field. The increasing awareness of keeping environmental sustainability has resulted in an increasing interest in creating effective membrane systems to purify contaminated water and wastewater. As these contaminated solutions derived from various industrial, household, and agricultural activities contain harmful pollutants giving negative impacts on the environment and human health, proper treatment then is crucial to improve water quality before final disposal [3].

Generally, membrane science research can be divided into seven major areas, that is, material selection, material characterization, membrane fabrication, membrane characterization and evaluation, transport phenomena, membrane module design, and process performance. Among these areas, material selection that is further used for membrane fabrication is the most important part of the membrane technology, and this phenomenon can be reflected by the significant technique [2–4]. Over the years, researchers have improved the performance of membranes [5–7]. They combined the effective features of polymeric and organics additives that are called mixed matrix membranes (MMMs). MMMs are comprising polymer matrix-containing fillers that can be an alternative to overcome the limitations of laminate membranes. MMMs can offer additional functions such as antifouling properties [8], enzyme mobilization [9], mechanical reinforcement, and removal of pollutants in the aqueous phase [10].

Several types of inorganic fillers such as silica [11], zeolite [12], TiO_2 [13,14], carbon nanotubes [15,16], multiwalled carbon nanotubes [17], and silver [18] have been widely used. Figure 11.1 shows various inorganic fillers utilized in preparing MMMs for water purification applications.

MMMs preparation takes days for homogenization that is caused by agglomeration tendency [19]. The incompatibility between inorganic fillers and polymer matrix provokes the formation of undesirable voids in the interface that are difficult to avoid. Meanwhile, MMMs can be defined as membranes that contain homogeneously dispersed fillers with a small composition of nanomaterials. MMMs can be also reused by adjusting pH, contributing to further reduction of harmful wastewater. They can simultaneously remove pollutants from an aqueous solution by adsorption and size exclusion [20,21]. Moreover, MMMs are more suitable for mass production of large-area membranes applicable to standard membrane modules in wastewater filtration.

Incorporating organic polymer with inorganic nanofillers has afforded a viable matrix membrane with overwhelming performance for liquid separation processes

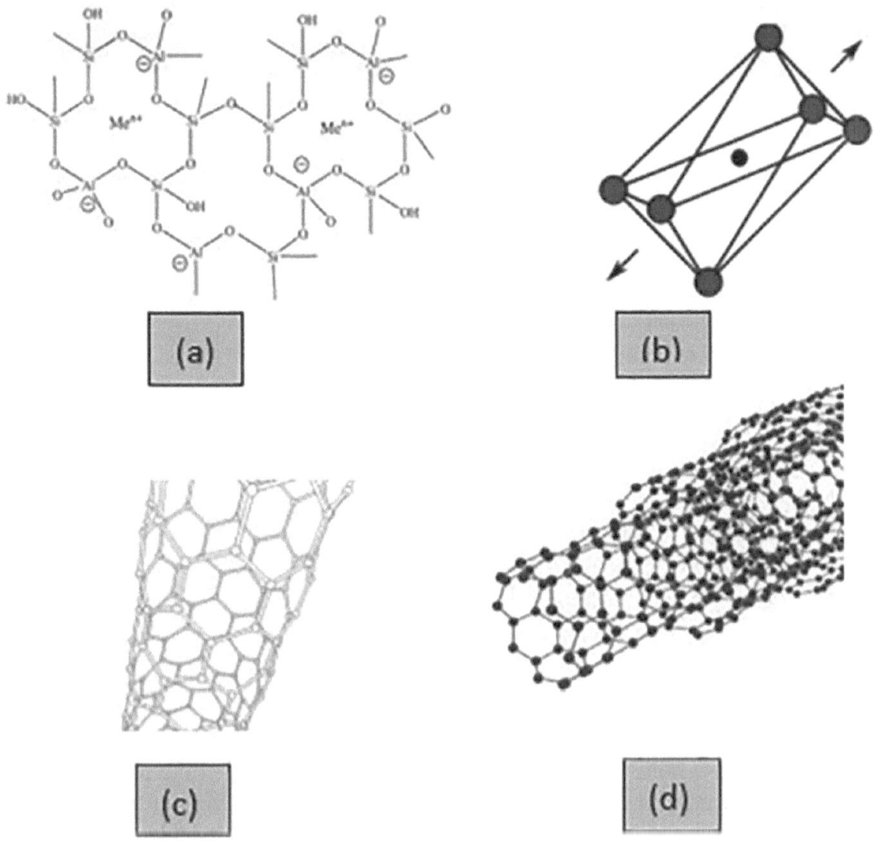

FIGURE 11.1 Various inorganic fillers in preparing MMMS for water purification: (a) zeolite, (b) metal oxide, (c) carbon nanotubes, and (d) multiwalled carbon nanotubes.

on industrial adoption. Table 11.1 summarizes the inorganic fillers for water treatment applications [17–21]. In membrane-based separation technology, Li et al. [19] considered using TiO_2 nanofiller in PA TFC membrane for the water treatment process. Their results revealed that the optimum membrane that contains 5 wt.% TiO_2 gives the better flux and selectivity. Further increase of TiO_2 content led to the significant reaction interference of nanofillers in the PA structure resulting in reduction in salt rejection due to a low degree of polymerization. Others researched zeolites that can be synthesized with SiO_2 higher or lower than in nature for the same framework type. Higher SiO_2 generally gives greater hydrothermal stability, stronger acid catalytic activity, and greater hydrophobicity as adsorbents. Conversely, lower SiO_2 gives greater cation exchange capacity and higher absorbance for polar molecules. Both natural and synthetic forms of the same zeolite are available in commercial quantity. The variable phase purity of the natural zeolite and chemical impurities are costly to remove. Zinc oxide is one of the metal oxides that has received significant attention due to its low cost, high surface area, photocatalytic activity, and

TABLE 11.1
Summary of Inorganic Fillers for Water Treatment Applications

Inorganic fillers	Properties	References
Iron-based	Highly reactive, larger surface area, detoxification of organic and inorganic pollutants, adsorption capacity, hydrophilic, fouling resistant, magnetic oscillation, hydraulic turbulence.	23, 26, 57
Silver-based	Anti bacterials, good transport facilitator, good selective barrier, high reactivity, low toxicity to humans	23, 24, 57, 59
Zeolite	Hydrophilic, fouling resistant, anti-adhesion to protein, effective sorbents, ion exchange media for metal ions	38, 39
Silica-based	Hydrophilic, fouling resistant, anti-adhesion to protein	18, 19
Titanium dioxide-based	Hydrophilic, fouling resistant, anti-adhesion to protein, photo catalytic, disinfection, anti-adhesion to protein, decomposition of organic compounds, reduced surface roughness, oxidative and reductive catalyst for organic and inorganic pollutant, killing bacteria.	10, 25, 30, 42
Carbon nanotube-based	Anti-microbial, hydrophilic, biofouling resistant, anti-adhesion to protein, selective sorbents for organic compound.	15, 16, 27, 28, 29

antibacteria properties [22]. Wang et al. [23] improved cellulose acetate membranes using Zeno-NPs (4 wt.%) which led to the enhancement of 111.1% of the flux compared to the pristine membranes. Previous researchers have found that the improvement in membrane permeability occurred due to the presence of zinc oxide (ZnO) in the PES membrane [24]. Additionally, this membrane has the highest fouling resistance during oleic acid filtration. However, at high polymer concentration, the mixed matrix membrane shows a reduction in the membrane permeability due to a drop in the dispersion rate. Silica nanofillers have also rapidly become focus of research on mixed matrix membranes due to their unique characteristics such as small size, strong surface energy, high scattered performance, and thermal resistance. Moreover, silica nanofiller has wider resources and lower price than that of TiO_2 nanofiller [25].

Zeolite has the chemical formula, $Mn_2OAl_2O_3.xSiO_2.yH_2O$, where the charge-balancing non-framework cation M has valence n, while x is 2.0 or more, and y is the moles of water in the voids. The Al and Si tetrahedral atoms in the forms of AlO_4 and SiO_4 tetrahedra are linked by shared oxygen ions. Synthetic zeolite can be more attractive for specific applications. Yellowtail et al. synthesized a nanocomposite membrane by blending zeolite in the polysulfide polymer, resulting in a membrane of better permeability and antifouling ability [25]. The result showed that zeolite raised the water flux up to 1.6 times. This resulted in permeation was attributed to increased hydrophilicity of nanocomposite. The zeolite also provided huge surface energy, which clustered small water molecules producing more polysulfone membrane. Table 11.1 tabulated the summary of inorganic fillers for water treatment applications [26]. The application of MMMs of adding inorganic zeolite in separation and purification processes was used to exclude molecules that are too large to enter the pores and admit smaller ones.

Carbon nanotubes (CNTs) have exceptional properties such as high mechanical and chemical stability and high electrical conductivity [16,27]. The properties of CNTs have made them attractive candidates for overcoming water scarcity and water pollution issues. Several authors have shown that the effect of CNTs in membrane polymer tends to enhance the hydrophilicity due to the decrease in contact angle, leading to greater water flux. The synthesis of polymer/CNT further improves membrane performances with respect to permeability, chlorine tolerance, thermal resistance, solvent stability, and fouling resistance [28] and the method for incorporating CNT to the mixed matrix membrane as in Figure 11.2.

Several authors demonstrate the functionalization of CNTs in a strong acid mixture as illustrated in Figure 11.3 [28]. They observed peaks of 1,680, 1,715, 2,875,

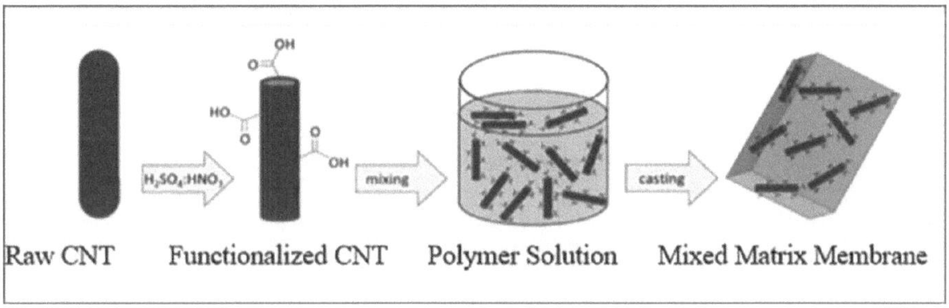

FIGURE 11.2 Schematic representation of MMMs with CNT-fillers.

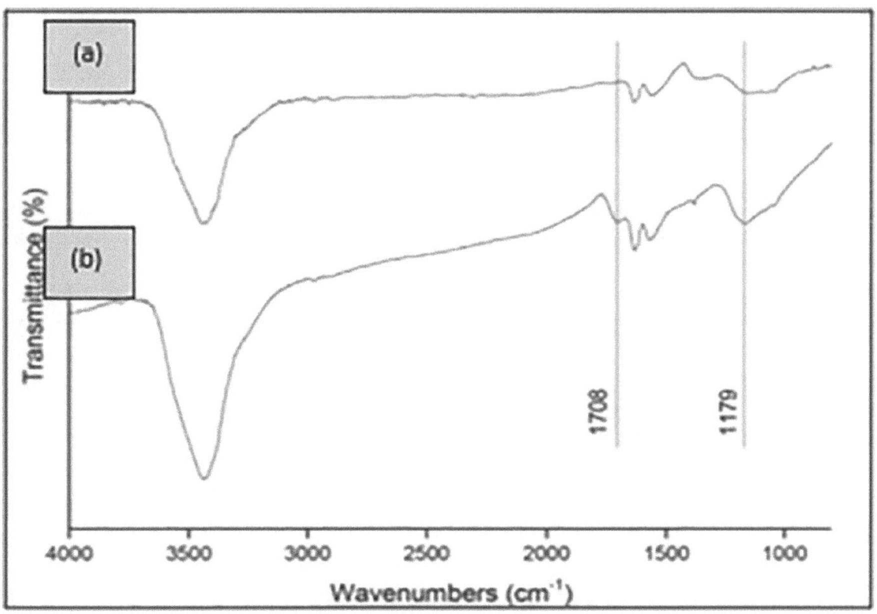

FIGURE 11.3 Functionalization of CNTs: (a) raw CNTs and (b) functionalized CNTs.

and 3,435 cm^{-1} that correspond to COOH, C–C, C–O, C=O, C–H, and –COOH, respectively, with Fourier Transform Infra-Red (FTIR) analysis. The length of CNTs was tens of micrometers before H_2SO_4:HNO_3 treatment, but reduced to hundreds of nanometers after H_2SO_4:HNO_3 treatment [28]. CNTs were broken into smaller CNTs, tips were open, and carboxylic groups were at the tips and defect sites of the CNTs. These results verify the successful functionalization of the CNTs.

The incorporation of multiwalled carbon nanotubes (MWCNTs) throughout the super selective thin-film layer was also explored as a facile approach to producing superior hydrophilic membrane. The unique molecular architecture of the tubes embedding in the membrane matrix has the potential to increase both permeability and selectivity. MWCNTs were not well dispersed in the non-polar solvent of the organic phase, but generally agreed that the rapid transport rates exist because the walls of nanotubes are much smoother (on atomic scales) than other materials leading to the increase in surface area of MMMs. This phenomenon tends to increase the rejection rate of MMMs [29].

11.2 MIXED MATRIX MEMBRANES

The combination of polymeric and inorganic/organic materials in one new material is called mixed matrix membranes (MMMs). MMMs can also be defined as incorporation of nanomaterials in solid–liquid phase or both that are dispersed or embedded in a continuous phase. These phases have been combined to have the effective features of both polymeric and filler. The sole purpose of developing this new material has been to associate the advantageous characteristics of two types of membranes boosting the overall process. Material advancement in membrane technology has made it possible to fine-tune the process efficiency and has successfully paved the way for MMMs in water treatment applications [29].

Figure 11.4 shows the schematic of an ideal MMMs that could offer the physicochemical stability of inorganic/organic material and polymeric materials while promising the desired morphology with higher values of permeate selectivity, hydrophilicity, fouling resistance, along with better thermal, mechanical, and chemical strength over a wider range of temperature and pH [28,29]. Polymer matrix plays a big role in permeability whereas the inorganic filler is a controlling factor for the

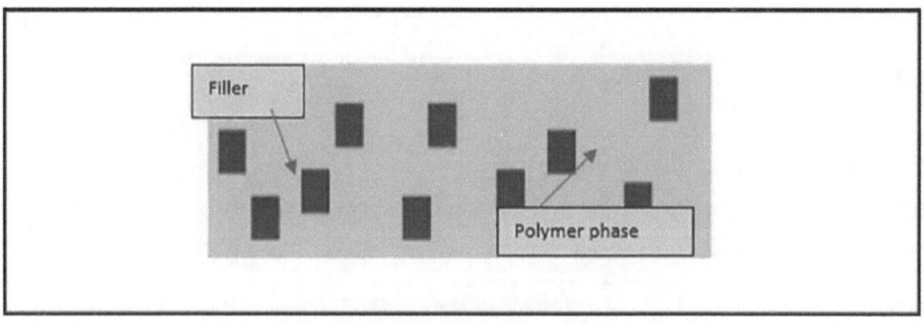

FIGURE 11.4 Schematic of ideal mixed matrix membranes.

selectivity of separation performance. The interfacial compatibility between the two phases is important to serve the desired purpose for such membranes [26]. The addition of fillers asserts their effect on the morphology of MMMs. The transport phenomenon determines the overall performance of newly developed membranes. The interfacial void formation, aggregation, and pore blockage are some of the key effects in resultant MMMs [29]. The presence of interfacial voids creates additional channels that allow the solvent to pass through the membranes [30]. Meanwhile, the mechanical strength and rejection rate are concerned by density. These features should be controlled or avoided by optimizing the process parameters of polymer concentration, filler concentration, casting technique, and coating technique [31–33]. Therefore, addition of these fillers in membrane synthesis could be challenging since controlling the placement of dispersion and its shedding/loss during the process is an important task that restricts its commercialization.

Goh and Ismail summarized their types of MMMs in four different types, namely conventional nanocomposite, thin-film composite with nano thin film, nanocomposite, thin-film composite with nanocomposite substrate, and surface-located nano composite [34]. These types of MMMs based on their corresponding filler types, such as inorganic filler-based MMMs, organic filler-based MMMs, biofilter-based MMMs, and hybrid filler-based MMMs.

11.2.1 Inorganic Filler-Based MMMs

An inorganic filler based membrane is an active MMM in which inorganic fillers attach themselves to support materials by covalent bonds, van der Waals forces, or hydrogen bonds. These inorganic fillers are prepared through processes such as sol–gel, inert gas condensation, pulsed laser ablation, spark discharge generation, ion sputtering, spray pyrolysis, photothermal synthesis, flame synthesis, low-temperature reactive synthesis, mechanical alloying, milling, and electrodeposition [35].

11.2.1.1 Zeolite Filler-Based MMMs

Recently, many reports demonstrated catalytic activity of polymer–zeolite MMM because the interaction of materials in the membrane matrix and the shape-selective catalytic properties of zeolites can improve permselective separations. The membrane also functions as a separator in the gas phase between different gaseous molecules. Thus, the membrane should be permeable enough to give efficient separation. For liquid-phase separation, metal-organic complexes and inorganic filler such as zeolite have been used [36]. It is well presented mostly that polydimethylsiloxane (PDMS) is incorporated as a polymer matrix because of its high permeability, an affinity for reagents, thermal, mechanical, and chemical stability [37]. Langhendries and Baron studied the catalytic activity of zeolite-filled poly(dimethylsiloxane) polymer membranes [38]. Catalyst performance was found to improve significantly as zeolite-encaged iron-phthalocyanine was incorporated into a dense hydrophobic polymer membrane.

SEM image of a zeolite MMMs showed a homogenous distribution of zeolite particles in the polymer matrix at different loads [35,39] (Figure 11.5). Zeolite mixed matrix membranes (zeolite MMMs) are used for sustainable engineering

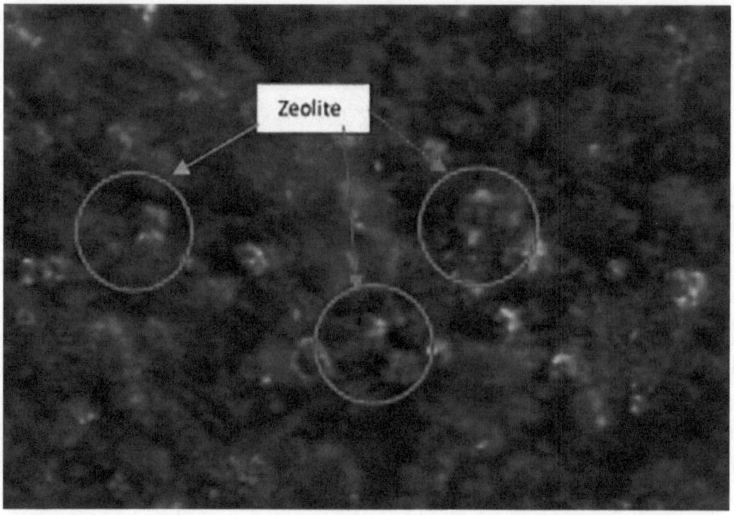

FIGURE 11.5 SEM images of zeolite MMMs.

improvement in catalyst performance. Both mathematical model and kinetics determined exact concentrations in polymer and catalyst, and subsequently, the resulting catalyst activity and selectivity. Their results also indicate that hydrophobic poly-(dimethylsiloxane) is an attractive polymer for the incorporation of the hydrophilic zeolite-encaged iron-phthalocyanine catalyst. As a result, diffusion through composite catalytic membranes can be predicted using the mass transfer coefficients of pure zeolite and pure polymer material, and a tortuosity factor based on the zeolite loading as a catalyst.

Figure 11.5 shows the dispersion of zeolite that is produced by synthesis of MMMs with homogenous mixing between polymer and zeolite. Meanwhile, Drioli and Giorno investigated the incorporation of polydimethylsiloxane (PDMS) into a polymer matrix and silicate for the permeation of various gases [40]. In their study, only a couple of very high-zeolite loadings were investigated, and they indicated that zeolite plays an important role as a molecular sieve in the membrane by facilitating the permeation of smaller molecules while it prevents the permeation of larger ones.

11.2.1.2 Titanium Dioxide Filler-Based MMMs

Titanium dioxide (TiO_2) nanofiller has a high specific area and hydrophilicity that affect the mass transfer during the membrane process. At lower TiO_2 concentration (< 2 wt.% of TiO_2), an increase in the amount of hydrophilic TiO_2 tends to draw more water into polymer dope, resulting in an increase in the length of finger-like macrovoids and a decrease in the thickness of the intermediate sponge-like layer [41]. Whereas at higher concentrations of the TiO_2 (3–10 wt.% of TiO_2), an increase in TiO_2 concentration would increase the viscosity of the polymer dope and decrease the rate of water intrusion into polymer dope. This phenomenon then results in shorter finger-like macrovoids and a thicker intermediate sponge-like layer.

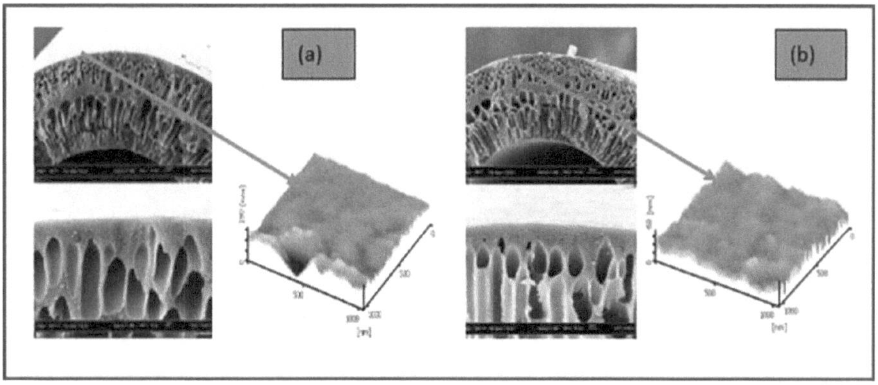

FIGURE 11.6 FESEM images of cross section and AFM images of the outer surface of MMMs with (a) lower TiO_2 and (b) higher TiO_2 concentrations.

A field emission scanning electron microscope (JEOL JSM-6700F) was used to examine the morphology of the PVDF hollow fiber membrane. Prior to analysis, the membrane samples were first immersed in liquid nitrogen and fractured carefully [42]. The samples were then coated with sputtering platinum before testing. The FESEM micrographs of the cross section and outer surface of the hollow fiber membranes were taken at various magnifications.

Figure 11.6 illustrates the AFM image of membrane surface that is not smooth. The nodule-like structure and nodule aggregates are formed at PVDF/TiO_2 MMMs' surface. Yuliwati et al. [42] reported the effect of TiO_2 on MMMs surface. They reported that the surface became smoother, and the nodules were separated from each other leading to a rougher MMMs surface than that of a neat PVDF membrane. This result may be attributed to PVDF/TiO_2 MMMs outer surfaces that experienced coalescence and orientation of polymer aggregates before gelatin in the external coagulation contained. The relaxation of the polymer occurs on the outer surfaces during relaxation, and the macromolecules tend to coil and entangle with each other, enhancing the fusion of nodular aggregates.

11.2.1.3 Carbon Nanotubes Filler-Based MMMs

Carbon nanotubes (CNTs) often refer to single-wall carbon nanotubes (SWCNTs) with diameters in the range of a nanometer. CNTs exhibit remarkable electrical conductivity, while others are semiconductors. They have exceptional tensile strength and thermal conductivity due to their nano structure and the strength of the bonds between carbon atoms. These properties are expected to be valuable in many areas of technology, such as electronics, optics, mixed matrix materials, and nanotechnology. Carbon nanotube membranes can be classified into different categories according to the fabrication methods; however, the two broad classes are: (a) freestanding CNT membranes, and (b) mixed (nanocomposite) CNT membranes. The two main types of freestanding CNT membranes, typically used in desalination and water treatment applications, are vertically aligned CNT (VACNT) membranes and bucky-paper membranes [43].

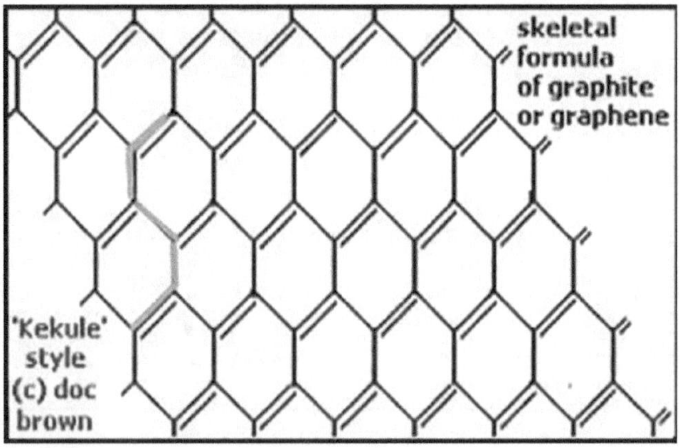

FIGURE 11.7 Structure molecule of carbon nanotubes (CNTs)-based membranes.

Figure 11.7 shows the zigzag and armchair configurations that are part of some structures of a single-walled nanotube. On some carbon nanotubes, there is a closed zigzag path that goes around the tube. The length of the carbon–carbon bonds is fairly fixed. There are constraints on the diameter of the cylinder and the arrangement of the atoms on it.

CNTs are the most used fillers in the development of MMMs. To employ CNTs as effective reinforcement in the polymeric matrix, proper dispersion, and suitable interfacial adhesion between CNTs and the polymer matrix have to be guaranteed. Figure 11.8 shows the application of CNTs in many industries, such as energy, biology, electronics, materials, agriculture, and tools. The current trend in polymeric MMMs is the incorporation of filler-like nanoparticles to improve the separation performance. Ismail et al. fabricated carbon nanotubes-mixed matrix membranes (CNT-MMMs) that offer a viable route to overcome the limitation demonstrated by the conventional polymeric and inorganics membranes. The excellent diffusivity properties of CNT have a promising outlook in wastewater separation processes.

11.2.2 Organic Filler-Based MMMs

Organic filler-based membranes are a modern type of MMMs in which organic fillers (such as cyclodextrin, polypyrrole, polyaniline (PANI), chitosan beads, wheat straw, yellow birch, pine, and rice husk) are introduced into substrate matrix, mostly through blending and phase inversion [44]. Organic fillers have the distinct advantage of having more functional groups attached to them, hence making them more adaptable than inorganic fillers. Their ability to attach themselves to a substrate through chemical reactions or binding themselves, especially with a hydrophobic surface makes them a better option for developing specialized (antifouling, highly hydrophilic, specific component rejection or higher porosity) membranes [45]. Zhao et al. synthesized a nanocomposite membrane by blending PANI nanofibers in polysulfone polymer, resulting in a membrane having better permeability and antifouling properties [46]. As a result, the water flux of PANI nanofibers increased up to 1.6

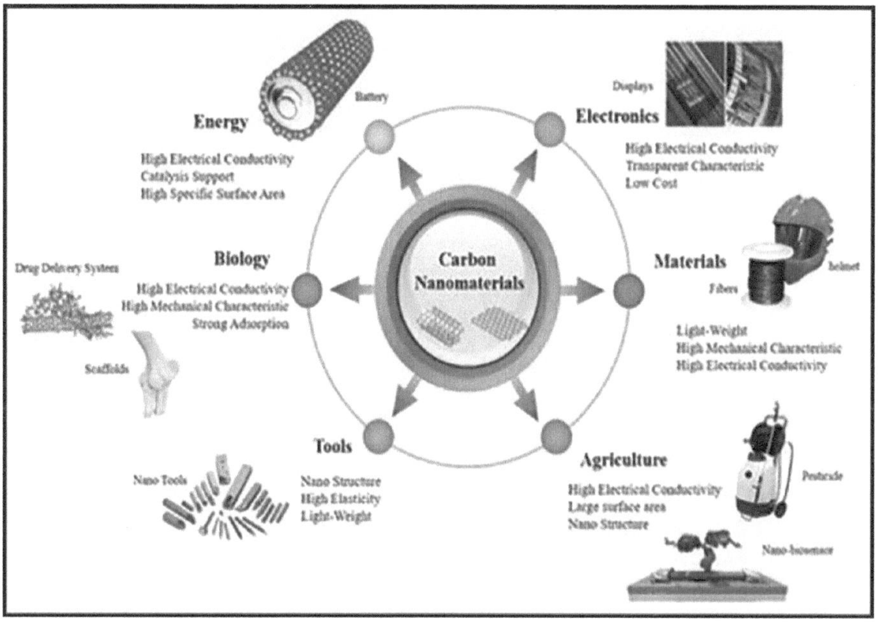

FIGURE 11.8 The applications of carbon nanotubes (CNTs) [28].

times. This improvement can be attributed to the increasing hydrophilicity of nanocomposite, since PANI (molecular structure as in Figure 11.9) fibers provided huge surface energy, which clustered the small water molecules, hence producing a more permeable polysulfone membrane .

Teli et al. also obtained the same results for PANI based nanocomposite membranes with added polyvinylpyrrolidone (PVP) [47]. They successfully enhanced the pure water flux, antifouling, separation efficiency, and mechanical strength of the resultant membranes with further PVP additions. Their study showed satisfying results because the addition of PVP (below 0.5 wt.%) with PANI organic filler in polysulfone matrix produced the aforementioned characteristics in the resulting membrane. BSA rejection through these membranes only occurs due to the hydrophilic nature of nanofillers (PANI) on the surface and the sieving mechanism due to the larger sizes of BSA molecules, though the addition of PVP may not affect the rejection significantly [47]. Zhong et al. blended a β-cyclodextrin polyurethane into a polysulfone matrix for removing Cd^{+2} ions from water [48]. The addition of β-cyclodextrin polyurethane increased the permeability of the MMM to 489 $Lm^2.h^{-1}$ by providing more wide pores on the surface, higher hydrophilicity, and better connectivity within finger-like pores.

11.2.3 Biomaterials-Based MMMs

Incorporation of biomaterials (biofillers) (such as aquaporin, amphiphilic, or lignin) into continuous matrix is an innovative technique to enhance the effectiveness of membrane technology. Biofiller-based MMMs deliver better permeability,

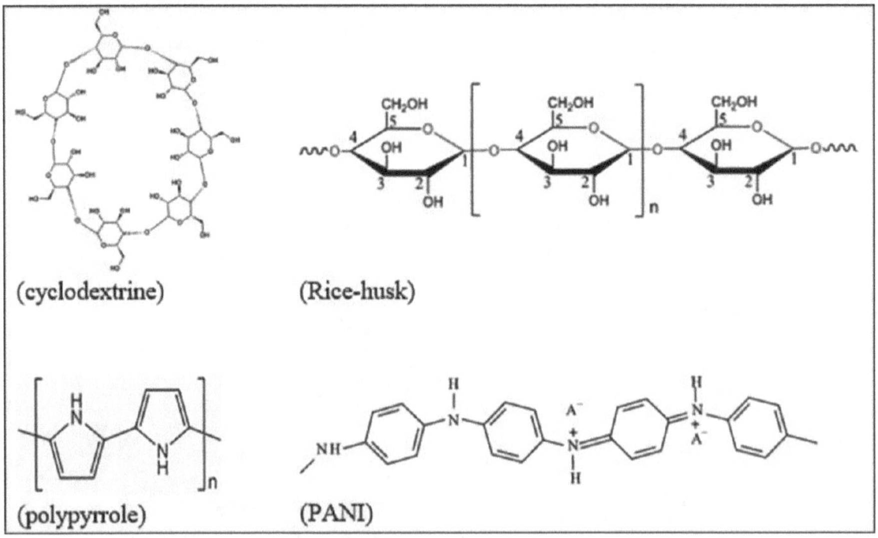

FIGURE 11.9 Various types of organic fillers.

antifouling ability, and certain functionalities such as mechanical reinforcement effect, which are either lacking or quantitatively low in the nascent membrane [49]. Two design strategies to synthesize these membranes are extensively reported in the literature. In the first strategy, aquaporin containing lipid bilayer is coated directly on the membrane substrate, while in the second strategy, vesicles or proteoliposomes (aquaporin incorporated in liposomes/ polysomes) are coated on the support surface [50].

Figure 11.10 presents a design of a vesicular membrane incorporated with a commonly used biofiller, such as aquaporin. Recent work by Wang et al. proposed that the introduction of aquaporin filler in amphiphilic triblock polymer vesicles (PMOXA15-PDMS10-PMOXA15) demonstrated an excellent performance related to permeability and driving force that was claimed to be 800-fold better than the simple polymeric membranes [51]. These newly developed MMMs also offered the unique ability to achieve a controlled permeability. They were found to be an excellent barrier toward urea, glucose, glycerol, and salt by reporting their relative reflection coefficient higher than unity. Nevertheless, the limiting concentration and incorporation method of biofillers in a polymer matrix have to be accounted properly since they could cause a significant decrease in membrane productivity under different biofiller concentrations [52].

Furthermore, Wang et al. used plant waste as biofiller in their study for cationic dye removal [53]. They added biofiller, e.g., banana peel, tea waste, and shaddock peel in polyethersulfone and reported the rejection of up to 95% of cationic dyes. The addition of such biofillers provided better charge interaction, hydrophobic interaction, and hydrogen bonding, hence improving the overall rejection of developed MMMs. Further improvement in cationic dye removal from wastewater is also suggested if the simple polymeric matrix is removed with biopolymers [53]. Other

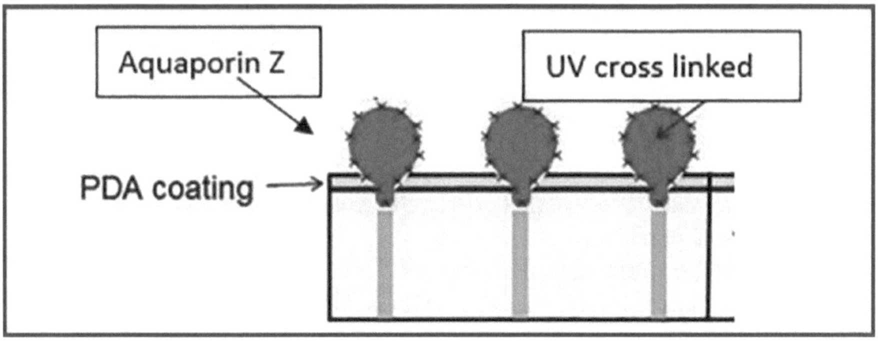

FIGURE 11.10 Design of biofiller mixed matrix membranes.

polymers have also been tested in various studies as Li et al. introduced aquaporin containing liposomes into polyamide imide (PAI) polymer matrix to synthesize a nanofiltration membrane with high permeability and higher rejection efficiency [54]. Their study reveals that at optimal composition (liquid-to-protein ratio of 200), the membrane showed the maximum pure water flux of 36.6 L $m^{-2}h^{-1}$, whereas the rejection of divalent salts was as high as 95%. The high permeability of the membrane was attributed to the availability of more passage for water molecules provided by aquaporin. On the other hand, high rejection of membrane was the direct result of the liposomes selective layer. Nevertheless, the method of incorporation of aquaporin in any matrix needs proper consideration since aquaporin placed near the top surface or exposed to the external environment would lose its activity.

Nevárez et al. synthesized gold and silver nanoparticles for producing a composite membrane for metal ions rejection [55]. They incorporated three different types of propinated lignin (kraft) in cellulose triacetate (CTA) through vapor-induced phase separation method. These showed that propinated kraft lignin (KL) improved the mechanical strength of the resultant KL/CTA membrane. Propination of kraft lignin increased the compatibility of the propyl group with cellulose acetate due to more London dispersion forces between the biofiller and the substrate polymer. However, propination adversely affected the mechanical characteristics of the developed membranes in the cases of organoslov (OL) and hydrolytic (HL) lignin. Propination in OL and HL lignin increased the particle sizes of resultant biofillers that diminished the adhesion between CTA and incorporated lignin nanoparticles, thus making the membranes less mechanically stable. Rejection measure for propinate Kl/OL/HL-based CTA MMMs showed a better rejection rate of arsenic ions for OL-based CTA MMMs (17%–22.8%), while the other two lignin suffered reduction due to antagonistic effects of divalent ions in the solution [56].

11.2.4 Hybrid Filler-Based MMMs

Hybrid fillers are a recent addition to the MMMs technology. This type of membrane contains two different fillers (independently or in composite form) added to the continuous phase. These hybrid materials are incorporated either to accomplish any targeted purposes or to improve the overall process effectiveness of the resultant membrane. A conceptual multifunctional membrane is depicted with hierarchal nanofillers where, on different layers, different types of nanofillers are introduced to achieve diverse functionalities. Mahmoudi et al. introduced the combination of iron (II, III) oxide and polyaniline into a polyethersulfone matrix to be able to accomplish 85% of Cu (II) removal from water [57]. The results showed that adsorption, in this case, was the dominating separation mode, otherwise this membrane could offer better reusability and durability.

A novel hybrid material chitosan-montmorillonite (CS-MMT) was dispersed in a polyethersulfone (PES) matrix by Saf et al. [58]. This novel hybrid filler CS-MMT raised the membrane antifouling ability due to its highly hydrophilic nature and increased the membrane mechanical strength by restricting the polymer chain mobility forming interrelated structures. They showed that a high flux recovery of up to 92% was achieved due to a loose active layer and the enhanced hydrophilic nature of the membrane [59]. Alpatova et al. also synthesized an antifouling MMM through inclusion of Fe_2O_3 nanoparticles and multiwalled carbon nanotube (MWCNT) in polyvinyldene fluoride (PVDF) [60]. The addition of this hybrid filler raised the degradation of fouling compounds such as cyclohexanoic acid and humic acid resulting in better antifouling behavior than that of the nascent one.

11.3 APPLICATION OF MMMs

Membrane-based technologies used for water and wastewater filtration can be a promising alternative to treat wastewater or water including both conventional and emerging pollutants and economic advantages over other water treatment processes. As aforementioned, MMMs comprising polymer matrix-containing fillers can be a viable alternative to overcome the limitation of laminate membranes. Only a small amount of inorganic nanomaterials is required to prepare MMMs as compared to the total weight, thereby minimizing the environmental and safety issues accompanied by the preparation of inorganic nanomaterials. MMMs can be reused by adjusting pH and simultaneously removing different types of pollutants from aqueous solution by adsorption and size exclusion. Moreover, MMMs are more suitable for mass production of large-area membranes used in water and wastewater treatment. In this chapter, we review the MMMs application in the agriculture industry, such as the polyvynilidene fluoride/titanium dioxide for virgin coconut oil purification and polysulfone/titanium dioxide fillers for palm oil wastewater.

11.3.1 MMMs on Purification of Virgin Coconut Oil

Virgin coconut oil (VCO) is obtained from the fresh and mature kernel (12 months old from pollination) of the coconut (Cocos nucifera L.) by mechanical or natural means with or without the application of heat, which does not lead to alteration of the nature of oil. VCO has not undergone chemical refining, bleaching, or deodorizing. VCO consists mainly of medium-chain triglycerides, which are resistant to peroxidation. VCO has been also acknowledged as the healthiest crop oil and can be extensively used in various fields such as food, beverage, medicinal, pharmaceutical,

TABLE 11.2
Essential Composition and Quality Factors of Virgin Coconut Oil (SNI 7381:2008)

Parameter	Standard, SNI
Moisture and impurities (%)	Max 0.1
Volatile matters at 120^{0}C (%)	Max 0.2
Free Fatty Acid (%)	Max 0.2
Peroxide Value meq/kg	Max 3
Relative density	0.915–0.920
Refractive Index at 40^{0}C	1.4480–1.4492
Insoluble impurities per cent by mass	Max 0.05
Saponification value (Mg KOH/g oil)	250–260 min
Color	Water clear
Odor and Taste	Natural fresh coconut scent, free of sediment, free from rancid odor and taste

nutraceutical, and cosmetics [61]. VCO is colorless, but multiple processes could change the color of VCO to bright yellow. This color change tends to decrease the VCO quality. The essential composition and quality factors of VCO have been tabulated in Table 11.2 [62].

High-quality VCO has some advantages such as it is odorless, colorless, and free of sediments. The odorless VCO can be produced by filtration through PVDF MMMs, which have porous fibers consisting mostly of titanium elements that are distributed on the MMMs' outer surface. Being covalently bonded and having a very large surface area were influenced by titanium dioxide fillers in PVDF polymer [40,61].

11.3.1.1 Characteristics and Quality of VCO

11.3.1.1.1 Dye Removal

Dyes are colored substances that establish chemical bonds with the substrates. It has been estimated that virgin coconut oil contains dyes contributing to the quality of VCO. The quality of VCO in Indonesia is standardized accordingly (SNI 7381:2008) [62].

11.3.1.1.2 VCO Analysis

The relative density of VCO samples was measured according to the AOAC method (AOAC, 2000) at a temperature of 30°C. The fatty acid profile of VCO samples was measured as fatty acid methyl esters (FAMEs). Prepared FAMEs were injected into the gas chromatography (Shimadzu, Kyoto, Japan) equipped with the flame ionization detector (FID) at a split ratio of 1:20. A fused silica capillary column (0.25 mm), coated with bonded polyglycol liquid phase, was used to analyze the fatty acids. The analytical conditions were an injection port temperature of 250°C and a detector temperature of 270°C. The oven temperature was set up within the range of 170°C–225°C at a rate of 1°C min^{-1} (no initial or final hold). The retention time of FAME standards was used to identify chromatographic peaks of the samples. Fatty acid content was calculated based on the peak ratio and expressed as g fatty acid/100g oil. The acid value of all VCO samples was measured by the AOCS method (AOCS, 2009), and FFA was analyzed by the following equation using the conversion factor of 2.81 for lauric acid. The results showed the characteristics of MMMs (PVDF/TiO_2 membrane) that identified four parameters namely odor, color, relative density, and free fatty acid content. Table 11.3 illustrates the composition of VCO after filtration.

11.3.1.2 Effect of Surface Morphology on Filtration of PVDF/TiO_2 MMMs

The results of the surface morphological analysis of the MMMs were obtained using a scanning electron microscope (SEM) as shown in Figure 11.11. The figure shows the surface morphology of bio-adsorbent obtained using an electron microscope scanning the field emission with magnification 374 times for each samples a and b. Figure 11.11a and b illustrates bio-adsorbent used by SEM with magnification of 8,740 and 8,750. The bio-adsorbent pores are spread evenly on the surface area.

The pore diameter can also be measured using the same tool, which is an average of 1.59 μm, as shown in Figure 11.11. The filtration process occurred when the VCO passed through the pores of PVDF/TiO_2 MMMs. The average pore diameter of the PVDF/TiO_2 can be analyzed through SEM. The density of permeate was analyzed

TABLE 11.3
Composition of Produced Virgin Coconut Oil After Filtration

Component	Unit	Total	Standard, SNI [62]
Odor	None	1	1–2
Color	None	1	1–2
Relative Density	gr/mL	0.917	0.915–0.917
Free Fatty Acid	%	0.057	0.03–0.09

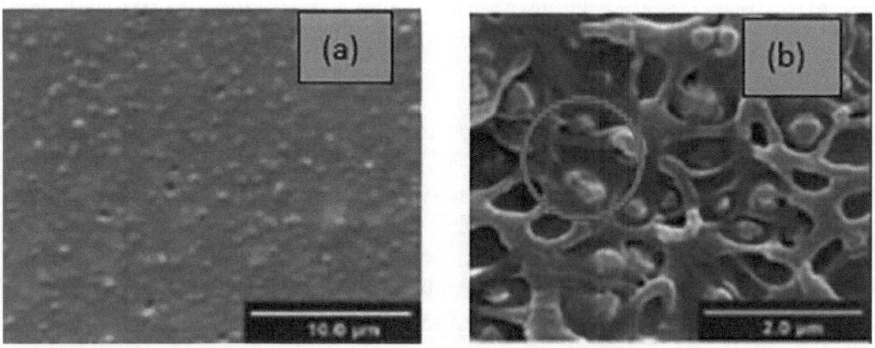

FIGURE 11.11 Surface morphology of MMMs using SEM (a) before and (b) after purification.

by a pycnometer with an average density of 0.917 gr mL^{-1}, free fatty acids (ALB) of 0.057%, as shown in Table 11.4. As can be seen, the MMMs which have a high total porosity and average pore size tend to increase the flux and remove more impurities significantly.

11.3.2 Application of Optimum Process Condition of PVDF/TiO_2 MMMs on Filtration of Palm Oil Wastewater

Wastewater streams typically contain many regulated inorganic and organic contaminants that can restrict their use or disposal thereof. Standards promulgated by state agency that regulate the maximum content of contaminants in wastewater streams disposed into publicly owned treatment works or discharged into waste injection wells have become increasingly stricter. Thus, processes for reducing the content of the inorganic and organic contaminants to an acceptable level in the wastewater streams have been employed to comply with these standards.

Palm oil is broadly produced in Indonesia and its production reached 48.68 million tons in 2018 [63]. Indonesians commonly choose palm oil as an alternative to other vegetable oils for daily needs. The current market showed a high demand of the palm oil industry leading to a higher production rate [64]. Meanwhile, the increase in palm oil production resulted in contaminants in its production wastes, such as solid waste and liquid waste. The contamination of water supplies by traces of the palm oil industry is a global issue causing environmental and health concerns resulting in an increasing demand for water remediation technologies. Nowadays, membrane system has grown up significantly and has received big attention from both academic and industry for answering the problems of water remediation technology. The usage of low-pressure membrane processes such as ultrafiltration has been increasing significantly for water and wastewater reclamation. The properties of wastewater also have a major impact on membrane fouling. Fouling is a major limitation for their implementation and also can affect the permeate quality and overall operation cost.

The palm oil industry mostly generates two types of waste, namely solid and liquid wastes. Liquid waste has been known as palm oil mill effluent, which is thick brownies viscous liquid waste, slurry, has high colloidal suspension, and has an unpleasant odor [65]. POME contains 95%–96% water, 0.6%–0.7% oil, and 4%–5% total solids including 2%–4% suspended solids [9,14]. Suspended solids consist mainly of debris from palm fruit mesocarp. POME was produced from sterilizer condensate, clarification condensate, and hydro-cyclone waste in the amount of 1,130 L of each ton of processed palm oil [64]. POME is a non-toxic liquid waste with an unpleasant odor and a high concentration of chemical oxygen demand (COD) and biological oxygen demand (BOD). This composition causes serious pollution and environmental problem to the water sources. Table 11.4 tabulates characteristic of POME and standard discharge of treated POME in Indonesia [65].

Palm oil mill effluent management treatment mostly applied conventional biological treatments of anaerobic or facultative digestion. The aerobic and facultative ponds rely on bacteria to break down the organic matter into simple end products of methane, carbon dioxide, hydrogen sulfide, and water [63,66]. It consists of a series

TABLE 11.4
The Standard Discharge of Treated POME

Parameters	Type of POME	Standard discharge, mg/L	Concentration of pollution loading, mg/L
BOD_5	25,000–29,000	100	225 (1.01)
COD	51,000–64,000	350	888 (0.25)
TSS	18,000–23,000	250	263 (1.02)
Oil and grease	6,000–7,000	25	63 (0.08)
Total nitrogen	750–1,200	50	21.5 (0.53)
pH	4..5	6–9	

of ponds connected to each other where each pond has its own purpose tending to increase the operational cost. Moreover, biological treatment can also produce biogas which is known as a corrosive and hazardous substance. These problems could be overcome by applying a membrane system. It could also treat POME with better consistency regardless of effluent variations allowing recycle process of the selected waste stream within a plant, so treated wastewater could be reused in the mill. The primary advantage of the membrane system is that it lowers the overall cost regarding supply of water and its further treatment namely, the elimination of the pollutant of POME. Moreover, the operational cost of the membrane system is less than conventional treatment. The increase of air gap length in membrane spinning affects the decrease of pure water permeation and wall thickness, the increase of pore size of outer surface skin layer thickness. TiO_2 particles loading possessed smaller pore size, and more apertures inside the membrane that increased the membrane hydrophilicity. Moreover, TiO_2 could degrade the color pigment contained in POME [66]. Some previous studies reported that permeate volume is enhanced with the use of aeration [67]. The combined liquid and gas flow has been shown to have more effect on fouling than liquid flow with higher velocity [68]. The bubble flow rate used for membrane filtration could provide oxygen to the biomass, maintain the solids suspension, and reduce the rate of the membrane fouling [69]. The objective of this current research is to investigate the performance of mixed matrix membrane in treating wastewater of palm oil mill effluent toward the different morphologies of used membrane based on varied air gap differences of 0, 3, and 5 cm. Moreover, the use of aeration was also studied on the size and velocities of bubbles. Herein, in this work, polivinylidene fluoride (PVDF), an inexpensive hydrophobic polymer, was chosen as a polymer for preparing PVDF-MMM through a non–solvent-induced phase inversion technique. This research also investigated the effect of suspended solids concentration, air bubbles flowrate (2.0, 3.0, and 4.0 mL min^{-1}), and size of bubble generation (4 and 8.5 μm) on flux and suspended solids removal.

The membrane is a promising method to be used in POME treatment due to its high packing density and the ease of module manufacture and operation. In this device, membranes are directly immersed in the feed reservoir with the withdrawal of permeate through the fibers by the application of vacuum on the outlet of the fiber lumen [70]. According to the reports, palm oil industry wastewater was characterized

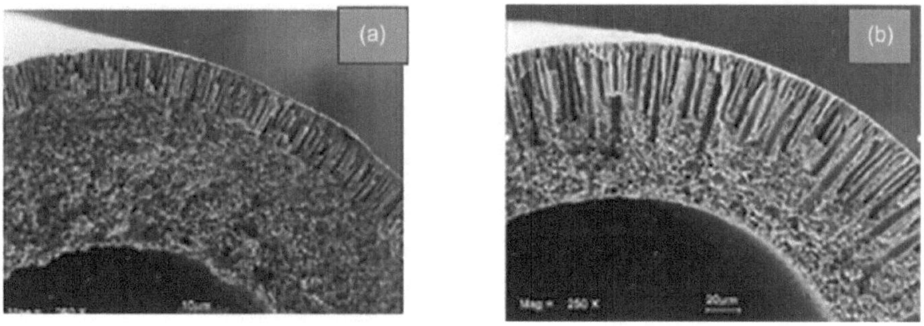

FIGURE 11.12 Cross-section image of PVDF membrane (a) without TiO_2 and (b) with TiO_2 4 wt.%.

by the presence of several organic and inorganic substances, namely, oil and grease, chemical oxygen demand (COD), total organic carbon (TOC), sulfide, free chlorine, ammonia nitrogen, and total suspended solids (TSS) [71]. The PVDF MMMs were produced by adding inorganic TiO_2 of 4 wt.% to the dope solution. TiO_2 concentration changed the membrane structure from sponge-like to finger-like, which significantly increased the water flux. It was caused by characteristic changing from hydrophobic to hydrophilic as shown in Figure 11.12.

The process efficiency in the membrane filtration is generally affected by many factors such as aeration flow rate, mixed liquor suspended solid (MLSS) concentration, pH, and hydraulic retention time (HRT). The optimization of these variables may significantly increase the process efficiency and the practicality of MLSS concentration. HRT, pH, and air bubble flow rate (ABFR) were considered as variables, while flux, total suspended solids (TSS), and ammonia nitrogen (NH_3-H) removal efficiencies were considered as parameters that have to be maximized due to the environmental regulation.

The properties of PVDF/TiO_2 mixed matrix membranes used in this work have been described in detail in the previous study[40,72]. As a semi-crystalline polymer, PVDF generally exhibits more complicated phase separation behavior than amorphous polymer. TiO_2 was added to the spinning dope to improve thermodynamic/kinetic relations during the phase inversion process in the preparation of PVDF-based membranes, to increase the surface hydrophilicity, and thus improve membrane water productivity [72]. The lab-scale experimental set-up shown in Figure 11.13 was used in this work. The membrane separation system consisted of a feed reservoir of 12 L volume, hollow fiber bundles, a peristaltic pump, a permeate flowmeter, and an permeate collector.

The filtration experiments were conducted at room temperature and under vacuum on the permeate side (0.5 bar abs) created using a peristaltic pump (Master flex model 7553-79, Cole Palmer) with the permeate being withdrawn from the open end of fibers and constant transmembrane pressure (TMP) of 0.5 bar was maintained to let water permeate from outside to the inside of the hollow fiber. The continuous aeration produced a turbulent flow which could decrease the cake layer thickness and the average particle size.

The permeation flux and rejection of modified PVDF (Merck) membranes with adding organic additive of rutile TiO_2 in varied concentrations (2 and 4 wt.%) and

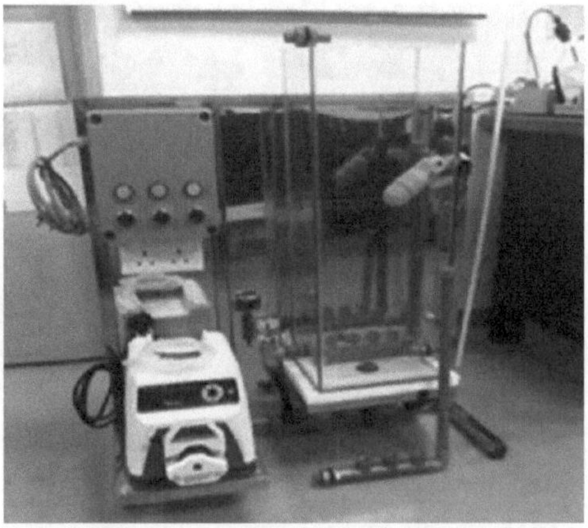

FIGURE 11.13 Membrane system.

dimethyl acetamide (DMAc, Merck) as a solvent for graywater were resulted by ultrafiltration experimental equipment as shown in Figure 11.13. Prepared modified module membrane with compositions of PVDF of 18 wt.% and titanium dioxide of 2–4 wt.% was submerged in the membrane reservoir. The suspension in a membrane reservoir with a volume of 9 L has been prepared with varying compositions of suspended solid within 3.0–6.0 mg L^{-1}. A cross-flow stream was produced by air bubbling generated by a diffuser situated underneath the membrane module for mechanical cleaning of the membrane module. The air bubbling flow rates per unit projection membrane area were set at 1.2–3.0 mL min^{-1} in order to produce proper turbulence. The filtration pressure was supplied by a vacuum pump and controlled by a needle valve and hydraulic retention time was set up at 180–240 minutes. Bubble size generation was adjusted with difference aerators that have a diameter of around 500 nm in order to have turbulence flow. Finally, permeate flow rates were continually recorded using flow meters, respectively.

11.3.2.1 Analytical Methods

The morphology of modified PVDF/TiO_2 MMMSs was analyzed using field emission scanning electron microscopy (FESEM; S-800M, Hitachi High Technology Co. Ltd., Tokyo, Japan). In order to observe the membrane cross-sections, membranes were first frozen in liquid nitrogen and then submitted to fracture. All samples of cross-sections were sputter-coated with a thin gold film prior to FESEM observation at a magnification of 8k.

11.3.2.2 Flux

Pure water permeation flux (J) was measured at reduced pressure (0.5 bar absolute) on the permeate side. Then, the flux (J) was calculated as follows:

$$J = \frac{Q}{A \cdot \Delta t} \tag{11.1}$$

where J is the flux (l/m^2h), V is the permeate volume (l), A is the membrane surface area (m^2), and t is the time (h).

11.3.2.3 Rejection Rate

The rejection rate was calculated according to the following equation:

$$R\% = \left(1 - \frac{C_p}{C_f}\right) \times 100 \tag{11.2}$$

where C_p and C_f are concentrations of permeate and feed solutions, respectively.

11.3.2.4 Total Suspended Solids and Ammonium Nitrogen Removal

Membrane performance of total suspended solids (TSS) and ammonium nitrogen (NH_3-N) concentrations were measured using a spectrophotometer (DR 5000, HACH) in accordance with the standard procedures of method 8006 (Photometric method) and method HR TNT 10031 (salicylate method), respectively. During the operation with high organic loading rates, the concentrations were evaluated daily and sampling was carried out three times a week. The TSS and NH_3-N removal efficiencies were calculated by Equations 11.3 and 11.4.

$$TSS\ removal\ (\%) = \frac{TSS_0 - TSS}{TSS_0} \times 100 \tag{11.3}$$

where TSSo and TSS are the initial TSS concentration of the feed synthetic greywater and permeate, respectively.

$$NH_3 - N\ removal\ (\%) = \frac{NH_3 - N_0 - NH_3 - N}{NH_3\ \ N_0} \times 100 \tag{11.4}$$

where NH_3-No and NH_3-N are the initial NH_3-N concentration of the feed graywater and permeate, respectively.

11.3.2.5 Morphology and Structural of Membrane

Field emission scanning electromagnetic microscopy (FESEM) of the modified membranes that has composition of PVDF of 18 wt.% and TiO_2 of 2 and 4 wt.% has the improvement of membrane morphology (Figure 11.13). It was observed that for addition of a small amount of TiO_2 nanoparticles the membrane tends to have a finger-like structure more than sponge-like macrovoids. TiO_2 nanoparticles have high specific areas and hydrophilicity that affect the mass transfer during the spinning process [40, 73]. The cross-sectional images for all hollow fibers consist of finger-like macrovoids extending from both the inner and outer wall of the hollow fiber, and an intermediate sponge-like layer. The thickness of the sponge-like layer decreases initially with an increase in TiO_2 concentration of 2–4 wt.% of the PVDF of 18 wt.% (Figure 11.14).

Based on Figure 11.14a and b, it can be assumed that there is a decreasing rejection rate (Table 11.5) with an increasing pore size of the outer surface leading to the increase in the air gap distance. This indicates that the solute transport may be governed by the pore size and pore size distribution of the external surfaces of the

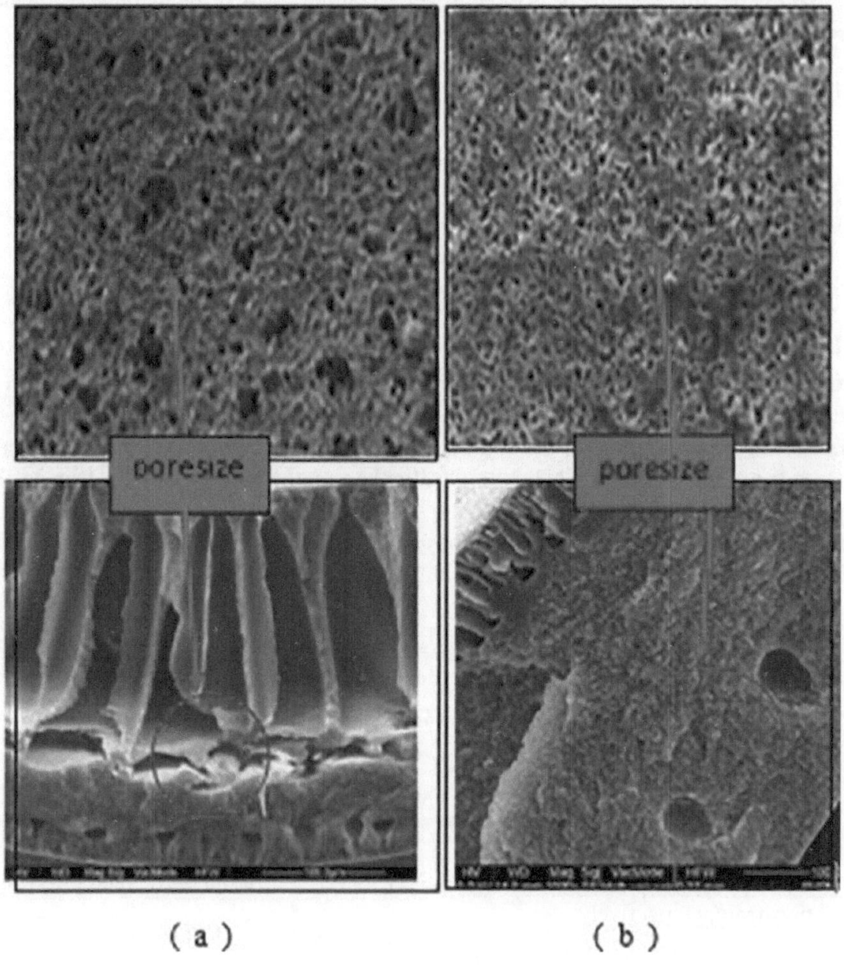

FIGURE 11.14 FESEM images of cross section of PVDF MMMs (a) with adding TiO_2 2 wt.% and (b) with adding TiO_2 4 wt.%.

TABLE 11.5
Structural and Morphology of PVDF/TiO_2 Mixed Matrix Membranes

Sample	Fluks, $L/m^2.h$	Avg poresize, nm	Porosity,%	Rejection Rate, %
PVDF		28.4		
PVDF + TiO_2 2 wt.%	30.00	10.00	65	80.25
PVDF + TiO_2 4 wt.%	87.50	38.20	86	98.90

membranes. In previous studies, it was found that lower solute separation values for hollow-MMMs have larger pore sizes [71–73]. These images showed that the membrane always used dimethyl acetamide (DMAc) as organic solvent. This solvent will change the structure of the membrane in wet conditions. The microstructural images of membranes used for liquid filtration normally performed under dry conditions. It can be proven that the structure of the membrane changes by adding the TiO_2 concentration of 2–4 wt.% where the pore size tends to be shorter and the diameter decreases. This phenomenon has been caused by swelling. This means that the microstructure images of the dry membrane could be changed significantly in wet condition, which is in the wet solvent that is trapped in membranes. These also illustrated the changes in average pore size in the membrane's outer surface with addition of TiO_2 concentration. Dzinun et al. (2020) studied mixed matrix membranes made from polyamide and reported that TiO_2 could induce an aggregate phenomenon and be absorbed into the substructure of PVDF membrane [73]. Yuliwati et al. (2011) also studied the effect of organic additives on membrane structure [40]. TiO_2 blocked the pores and caused a decrease in the average pore size of the membrane surface. The results were attributed to the porous structure and have correlated to average pore size, and porosity affecting permeate rate.

Table 11.5 shows the flux values of the used membrane containing TiO_2 of 4 wt.%. The porosity and average pore size information of the modified membrane are also tabulated in Table 11.5. All modified membranes possessed good porosity in the range of 65%–86%, which can be attributed to low polymer concentration in the membrane dope solution. The solubility of polymer and additives in the solvent plays a big role due to production of homogenous solutions before the fabrication of the hollow mixed matrix membranes. High porosity has been provided by addition of TiO_2 of 4 wt.% meaning the interaction and extrusion of TiO_2 in the porous structure and the aggregation of TiO_2 particles inside the pores. Vanneste et al. (2014) reported the interaction among active TiO_2 photocatalytic membranes [74]. They reported the possible influence of TiO_2 on the porosity of membranes. The small portion of TiO_2 in membrane dope solution could affect the interfacial stresses between polymer and TiO_2 particles. This would affect the formed pores of the organic phase during the remixing process.

The flux also increased with the increase in the concentration of TiO_2 from 2 to 4 wt.%. Table 11.5 demonstrates that the size of the average pore on the surface membrane is influenced by the value of flux.

11.3.2.6 Total Suspended Solids Removal

It can be concluded that the increase in TSS removal would occur with increasing ABFR and HRT and decreasing pH and MLSS concentrations. However, a further increase in ABFR resulted in a decrease in TSS removal. The ABFR increased from 1.2 to 2.1 ml min^{-1}, and TSS removal increased with an increase in ABFR because concentration polarization was reduced due to forceful turbulence. However, excessive aeration can cause size reduction of depositing particles due to shear-induced diffusion and inertial lift forces resulting in more severe pore blockage. Thus, there is a critical value beyond the increase of ABFR that has virtually no effect on the fouling resistance. Moreover, it could have a detrimental effect [71–73]. The turbulent flow may consume transmembrane pressure of the system, causing weaker hydraulic and attachability factors that lead to the decline of the suspended solids removal [74].

The TSS removal was highly dependent on the pH of the feed solution. The TSS removal under various pH values was affected not only by the characteristics of the membrane but also by the properties of the solute (droplet). The size of the emulsion droplet was not uniform, and the micelles carried charges due to the reaction of surfactants. At a low pH level, the contribution of the charge neutralization predominated for micro-floc formation of relatively large sizes that were sufficient for steric hindrance. This is consistent with the conclusion made by Dzinun et al. [75]. Hence, despite the coagulated suspension that accumulated densely in the form of a cake layer at the membrane surface, the extent of suspended solid rejection was improved. The increase in suspended solid aggregation and in turn TSS removal reduction was caused by the formation of a thicker suspended solid deposit. This is likely due to the reduction of electrostatic repulsion.

11.3.2.7 Ammonium Nitrogen Removal

NH_3-N removal increased with an increase in ABFR from 1.0 to 2.5 mL min^{-1}, and then decreased with a further increase in ABFR at pH 8.00. The reason for the presence of a critical ABFR value has already been given while discussing its effect on TSS removal. ABFR must be carefully controlled to maintain adequate expansion and liquid–liquid mass transfer while minimizing shear effects. Lai et al. (2018) also mentioned that the smaller particle size in aerated submerged ultrafiltration was mainly due to the violent turbulence that aeration produced under membrane bundles [74].

It can also be concluded that NH_3-N removal increased with an increase in HRT and pH and a decrease in MLSS concentration. ABFR must be carefully controlled to maintain adequate expansion and mass transfer while minimizing shear effects near the optimum value. Lower NH_3-N removal at high MLSS concentration is due to serious membrane fouling such as membrane adsorption and pore plugging that occurs on the membrane surface. The concentration polarization on the membrane surface was also one of the factors, as has been observed at low ABFR [75]. It should however be noted that there are other reasons for the high NH_3-N removal observed. The nitrogen compounds are adsorbed to the deposited matters that are retained by the membrane in the filtration process. Besides, the biomass also assimilates organic nitrogen. This high removal value was also possible due to nitrification reactions that occurred in the reservoir where ammonium was highly soluble in water. The ammonium ions formed can be readily reduced to nitrite and nitrate. The optimal process parameters in suspended solids and ammonia nitrogen removal from palm oil wastewater were very challenging.

11.4 FUTURE CHALLENGES

Recently, novel MMMs have attracted great attention in membrane technology, due to their excellent advantages, such as some improvement in several parameters like permeability, selectivity, thermal and chemical stability, and mechanical strength of a polymeric membrane. Furthermore, the recent development demonstrated that gas separation, as well as water treatment, has obtained significant benefit from membrane technology advancements enabling its further application in wider real industrial aspects. However, a comprehensive understanding of organic–inorganic interfaces is in a great need. MMMs performance suffers from defects caused by poor contact at the molecular sieve/polymer interface, the complexity of the synthesis process, high cost, identification of compatible inorganic particles, agglomeration, zeolite, and its applications, inorganic particle concentration, phase separation, control of morphology, and structural defects. Moreover, some MMMs for water purification application are considered to be of the potential hazard to humans and environment, and also need more research to determine the hazardous character of these nanoparticles and the mechanism of nanoparticles embedded membrane fouling in industrially water purification in the future.

One of the many difficulties associated with membrane technology is the fouling phenomenon. Although several strategies such as the incorporation of antifouling nanoparticles, and surface modification have been used to overcome this problem, intensive investigations are needed to stop the regeneration of microbial colonies on the membrane surface and to reduce the leaching of filler. The next generation MMMs should be developed by producing nano-size fillers without aggregation to improve their separation properties for the membrane industry, especially MMMs. There are several reasons to produce nano-size fillers, especially zeolite fillers such as more polymer/particle interfacial area and enhanced polymer–filler interface contact by smaller particles. The potential of incorporating fillers such as titanium dioxide particles has not been attained up to the expectation of MMMs performance, due to the smaller sizes, homogeneous distribution, agglomeration, price, availability, compatibility with polymer interface, their relationship with water chemistry, better interfacial contact, and stability.

There are limitations to developing novel materials due to costly synthesis processes. The molecular dynamic simulations (MD) of mixed matrix materials could be an effective approach to predicting the diffusive performance of MMM, especially zeolite MMMs, and to provide experimental guidelines for tuning the membrane permeability at the molecular level without high costs. Although there are previously predicted models for predicting the processes contributing to membrane separations, however, studies in MMMs showed inadequate suitable models. Therefore, MD will be essential and effective to predict the morphology and intrinsic properties of these fillers and their interaction with the polymeric matrix.

Ultimately, membrane morphology could change the properties of membranes, and subsequently, it will influence the membrane performance. Therefore, improving membrane performance in real conditions such as high temperature, high pressure, and incorporating a plasticizer into the polymer solution would be possible and essential in order to provide better thermally and chemically MMMs at different operating conditions for sustainable engineering.

11.5 CONCLUSIONS

Mixed matrix membranes with zeolite fillers have attracted a lot of attention in membrane technology research due to their excellent advantages, such as high permeability and improved selectivity. Zeolite MMMs could be considered an ideal candidate for the purification industry since they combine the properties of polymeric matric and zeolite inorganic fillers. The application and fabrication techniques of zeolite-reinforced polymeric membranes have been comprehensively reviewed in this chapter to optimize interfacial interaction between the zeolite and the polymeric matrix. Compatibility between zeolite and polymer matrix can be improved with several methods, such as by applying high processing temperature during membrane formation, the silane modification, and priming on the particle's surface, annealing that can relax the stress imposed on hollow fiber and result in higher packing density of polymer chains, and the introduction of an LMWA agent between the polymer matrix and inorganic particles. There have been numerous implementations to incorporate zeolite particles in polymer matrices in water purification applications and for gas separation due to its superior separation properties and size exclusion. Applications of zeolite MMMs were re-evaluated for a variety of industrial processes, including water purification, medical industry, catalytic, and gas separation.

However, despite its advantages, there are still issues and difficulties associated with zeolite MMMs that have restricted their wider applications. It can be concluded that the advancements in the application and fabrication of zeolite MMM need further intensive investigations. Future research should be conducted to develop new techniques that provide a better understanding of zeolite incorporation into polymer structures. New materials should also be considered as a way of reducing fouling concerns. Additional study is necessary for an improved understanding of the basic transport mechanism occurring through the MMMs. The next generation MMMs must be developed with nano-size fillers and without aggregation to improve their separation properties severely needed in the membrane industry. Some results indicate that the nanosized zeolite particles incorporated in MMMs offer better performance in comparison with micron size particles. New additives and modification agents should be produced to improve adhesion between polymer and inorganic fillers. In conclusion, despite all the identified problems, MMM technology with zeolites could be considered a strong candidate for the modern purification industry due to the remarkable properties of polymeric and inorganic zeolite materials.

REFERENCES

1. Bokhary A., Tikka A., Leitch M., Liao B. 2018. Membrane fouling prevention and control strategies in pulp and paper industry application: A review. *J. Membr. Sci. Res.*, 4, 181–197.
2. Andrade L.H., Mendes F.D.S., Espindola J.C., Amaral M.S.C. 2014. Nanofiltration as tertiary treatment for the reuse of dairy wastewater treated by membrane bioreactor. *Sep. and Purif. Technol.*, 126, 21–29.
3. Delcolle R., Gimenes M.L., Fortulan C.A., Moreira W.M., Martins N.D., Pereira N.C. 2017. A comparison between coagulation and ultrafiltration processes for biodiesel wastewater treatment. *Chem. Eng. Trans.*, 57, 271–276.
4. Boutin O., Rowland D.J., Carpenter. 2016. Evaluating trivalent chromium toxicity on wild terrestrial and wetland plants. *Chemosphere*, 162, 355–366.
5. Eternadi H., Qazvini H. 2020. Investigation of alumina nanoparticles role on the critical flux and performance of polyvinyl chloride membrane in a submerged membrane system for the removal of humic acid. *Poly. Bull.* doi:10.1007/s00289-020-03234-z.

6. Firdayati M., Indiyani A., Prihandrijanti M., Otterpohl R. 2015. Greywater in Indonesia: Characteristic and treatment systems. *Jurnal Teknik Lingkungan.*, 21(2), 98–114.
7. Frioui R, Oumeddour R., 2008. Investment and production costs of desalination plants by semi-empirical method, *Desalination*, 223, 457–463.
8. Indriani T., Herumurti W. 2010. Grey Water Treatment Using ABR-AF Reactor. Bachelor thesis. Surabaya: Institut Teknologi Sepuluh Nopember Surabaya.
9. Leal R.H., Temmink H. 2010. Comparison of three systems for biological greywater treatment. Water J., 2, 155–169. ISSN 2073-4441. Water.
10. Herdiansyah M.I., Yuliwati E., Mahyudin, Ismail A.F. 2017. Mathematical model of optimum composition on membrane fabrication parameters for treating batik palembang wastewater. *J. Eng. Appl. Sci.*, 12(4), 797–802.
11. Khouni I., Marrot B., Amar R.B. 2010. Decolourization of the reconstituted dye bath effluent by commercial laccase treatment: Optimization through response surface methodology. *Chem. Eng. J.*, 156, 121–133.
12. Ptel R., Park J.T., Hong H.P., Kim J.H., Min B.R. 2011. Use the block copolymer as compatibilizer in polyimide/zeolite composite membranes. *Poly. Adv. Tech.*, 22(5), 768–722.
13. Chen L.C., Huang C.M., Hsiao M.C., Tsai F.R. 2010. Mixture design optimization of the composition of S, C, SnO_2-codoped TiO_2 for degradation of phenol under visible light. *Chem. Eng. J.*, 165, 482–489.
14. Lai G.S., Yusob M.H., Lau W.J., Gohari R.J., Emadzadeh D., Ismail A.F. 2017. Novel mixed matrix membranes incorporated with dual-nanofillers for enhanced oil-water separation. *Sep. Purif. Technol.*, 178, 113–121.
15. Hebbar R.S., Isloor A.M., Inamuddin, Asiri A.M. 2017. Carbon nanotube-and graphene-based advanced membrane materials for desalination. *Environ. Chem. Lett.*, 15(6430), 643–671.
16. Liao Z.L, Chen H, Zhu B.R, Li H.Z. 2015. Combination of powdered activated carbon and powdered zeolite for enhancing ammonium removal in micro-polluted raw water. *Chemosphere*, 134, 127–132.
17. Daraei P., Madaeni S.S., Ghaemi N., Ahmadi Monfared H., Khavidi M.A. 2013. Fabrication of PES nanofiltration membrane by simultaneous use of multi-walled carbon nanotube and surface graft polymerization method: Comparison of MWCNT and PAA modified MXCNT. *Sep. Purif. Technol.*, 104, 32–44.
18. Kobya M., Demirbas E., Bayramoglu M., Sensoy M.T. 2010. Optimization of electrocoagulation process for the treatment of metal cutting wastewaters with response surface methodology. *Water Air Soil Poll.*, 215, 399–410.
19. Li Y., Su Y, Dong Y., Zhao X., Jiang Z., Zhang R., Zhao J., 2014. Separation performance of thin film composite nanofiltration membrane through interfacial polymerization using different amine monomers. *Desalination*, 333, 59–65.
20. Madaria P.R., Nagarajan M., Rajapagol C., Garg B.S. 2005. Removal of chromium from aqueous solutions treatment with carbon aerogel electrodes using response surface methodology. *Ind. Eng. Chem. Res.*, 44, 6549–6559.
21. Malisie, A. 2008. Sustainability assessment on sanitation system for low income urban areas in Indonesia. Dissertation for Doctoral Degree. Hamburg, Germany: TUHH.
22. Maghami M., Abdelrasoul A. 2018. Zeolite mixed matrix membranes (Zeolite-MMMs) sor sustainable engineering. Chapter 7 Interchopen.
23. Misdan N., Lau W.J., Ismail A.F., Matsuura T. 2013. Formation of thin film composite nanofiltration membrane: Effect of polysulfone substrate characteristics. *Desalination*, 329, 9–18.
24. Ahmad A.A., Abdulkarim A.A., Ismail S., Seng O.B. 2016. Optimization of PES/ZnO mixed matrix membrane preparation using response surface methodology for humid acid removal. *Korean J. Chem. Eng.*, 33(3), 997–1007.

25. Yuliwati E., Ismail A.F., Matsuura T., Kassim M.A., Abdullah M.S. 2011. Effect of modified PVDF hollow fiber submerged ultrafiltration membrane for refinery wastewater treatment. *Desalination*, 283, 214–220.
26. Qadir D., Mukhtar H., Keong L.K. 2017. Mixed matrix membranes for water purification applications. *Sep. Purif. Rev.*, 46, 62–80.
27. Mohammad A.W., Teow Y.H., Ang W.L, Chung Y.T., Oatley-Radcliffe D.L., Hilal N. 2015. Nanofiltration membranes review: Recent advances and future prospects, *Desalination*, 356, 226–254.
28. Celik E., Par H., Choi H. 2011. Carbon nanotube blended polyethersulfone membranes for fouling control in water treatment. *Water Res.*, 45, 274–280.
29. Vatanpour V., Madaeni S.S., Moradian R., Zinadini S., Astinchap B. 2012. Novel antibifouling nanofiltration polyethersulfone membrane fabricated from embedding TiO_2 coated multi-walled carbon nanotubes. *Sep. Purif. Technol.*, 90, 211–221.
30. Moustakes N.G., Katsaros F.K., Kontos A.G., Ramanos G. E., Dionysiou D.D., Falaras P. 2014. Visible light active TiO_2 photocatalytic filtration membranes with improved permeability and low energy consumption. *Catal. Today*, 224, 56–69.
31. Lau W.J., Ismail A.F. 2009. Polymeric nanofiltration membranes for textile dye wastewater treatment: Preparation, performance evaluation, transport modelling, and fouling control- a review, *Desalination*, 245, 321–348.
32. Lau W.J., Ismail A.F. 2010. Application of response surface methodology in PES/SPEEK blend NF membrane for dyeing solution treatment. *Membr. Water Treat.*, 1, 49–60.
33. Goh P.S., Ismail A.F., Hilal N. 2016. Nano-enabled membranes technology: Sustainable and revolutionary solution for membrane desalination? *Desalination*, 380, 100–104.
34. Goh P.S, Ismail F. 2014. Review: Is interplay between nanomaterial and membrane technology the way forward for desalination? *J. Chem. Technol. Biotechnol.*, 99, 971–980.
35. Pressdee J.R., Veerapaneni S., Shorne-Darby H.L., Clement J.A., Van der Hoek J.P. 2006. Integration of membrane filtration into water treatment systems, American Water Works Association (AWWA) research foundation, USA.
36. Silva E.M., Rogez H., Larondelle Y. 2007. Optimization of extraction of phenolics from Inga edulis leaves using response surface methodology. *Sep. Purif. Technol.*, 55, 381–387.
37. Vanneste J. Peumans W.J., Van Damme E.J.M., Darvishmanesh S., Bernaderts K., Geuns J.M.C., Van der Bruggen, B. 2014. Novel natural and biomimetic ligands to enhance selectivity of membrane processes for solute–solute separations: Beyond nature's logistic legacy. *J. Chem. Tech. Biotech.*, 89, 354–371.
38. Langhendries G., Baron G.V. 1998. Mass transfer in composite polymer-zeolite catalytic membranes. *J. Membr. Sci.*, 141, 265–275.
39. Khin M.M., Nair A.S., Babu V.J., Murugan R., Ramakrishna S. 2012. A review on nanomaterials for environmental remediation. *Energy Environ. Sci.*, 5, 8075–8109.
40. Drioli E, Giorno L. 2010. *Comprehensive Membrane Science and Engineering*, Elsevier, Amsterdam, Netherland.
41. Yuliwati E., Mohruni A.S., Mataram A. 2018. Green technology contribution in development of coolant wastewater filtration. *Sriwijaya. J. Env.*, 3(2), 74–79.
42. Yuliwati E., Porawati, H., Elfidiah, Melani, A. 2019. Performance of composite membrane for palmoil wastewater treatment. *J. Appl. Memb. Sci. Yech.*, 23(2), 1–10.
43. Zhu J., Tian M., Zhang Y., Zhang H., Liu J. 2015. Fabrication of a novel "loose" nanofiltration membrane by facile blending with Chitosan-Montmorillonite nanosheets for dyes purification. *Chem. Eng. J.*, 265, 184–193.
44. Wang R., Jiang X., He A., Xiang T., Zhao C. 2015. An in situ crosslinking approach towards chitosan-based semi IPN hybrid particles for versatile adsorption of toxin. *RSC Adv.*, 5, 51631–51641.

45. Chung T.S., Jiang L.Y., Li Y., Kulprathipanja S. 2007. Mixed matrix membranes (MMMs) comprising organic polymers with dispersed inorganic fillers for gas separation. *Prog. Polym. Sci.*, 32, 483–507.
46. Zhao Q., Hou J., Shen J., Liu J., Zhang Y. 2015. Long-lasting antibacterial behavior of a novel mixed matrix water purification membrane. *J. Mat. Chem. A.*, 3, 18696–18705.
47. Teli S.B., Molina S., Calvo E.G., Lozano A.E., de Abajo J. 2012. Preparation, characterization and antifouling property of polyethersulfone-PANI/PMA ultrafiltration membranes. *Desalination*, 299, 113–122.
48. Zhong P.S., Chung T.S., Jeyaseelan K., Armugam A. 2012. Aquaporin-embedded biometric membranes for nanofiltration. *J. Membr. Sci.*, 407–408, 27–33.
49. Shen Y.X., Sabo P.O., Sines I.T., Erbakan M., Kumar M. 2014. Biomimetric membranes: A review. *J.Membr. Sci.*, 454, 359–381.
50. Tang C, Wang Z, Petrinic L, Fane A.G, Halix-Nielson C. 2015. Biomimetic aquaporin membranes coming of age. *Desalination*, 368, 89–105.
51. Wang M, Wang Z, Wang X, Wang S, Ding W, Gao C. 2015. Layer-by-layer assembly of Aquaporin Z-incorporated biomimetic membranes for water purification. *Environ. Sci. Technol.*, 49, 3761–3768.
52. Theron J., Walker J.A., Cloete T.E. 2008. Nanotechnology and water treatment: Applications and emerging opportunities. *Crit. Rev. Microbiol.*, 34, 43–69.
53. Wang H., Chung T.S., Tong Y.W., Jeyaseelan K., Armugam A., Chen Z., Hong M., Meier W. 2012. Highly permeable and selective pore-spanning biomimetic membrane embedded with Aquaporin Z. *Small*, 8, 1185–1190.
54. Li X., Wang R., Tang C., Vararattanavech A., Zhao Y., Torres J., Fane T. 2012. Preparation of supported lipid membranes for aquaporin Z incorporation. *Coll. Surf. B Biointerf.*, 94, 333–340.
55. Nevárez L.M., Casarrubias L.B., Canto O.S., Celzard A., Fierro V., Gómez R.I., Sánchez G.G. 2011. Biopolymers-based nanocomposites: Membranes from propionated lignin and cellulose for water purification. *Carbohydrate Poly.*, 86(2), 732–741.
56. Tesh S.J., Scott T. 2014. Nano-composites for water remediation: A Review. *Adv. Mat.*, 26, 6056–6068.
57. Mahmoudi E., Ng L.Y., Ba-abbad M.M., Mohmmad W. 2015. Novel nanohybrid polysulfone membrane embedded with silver nanoparticles on graphene oxide nanoplates. *Chem. Eng. J.*, 277, 1–10.
58. Saf A.O., Akin I., Zor E., Bingol H. 2015. Preparation of a novel PSf membrane containing rGO/PTh and its physical properties and membrane performance. *RSC Adv.*, 5, 42422–42429.
59. Nasir R., Mukhtar H., Man Z., Mohshim D.F. 2013. Material advancements in fabrication of mixed matrix membranes. *Chem. Eng. Technol.*, 36, 717–727.
60. Alpatova A., Meshref M., McPhedran, K.N., Gamal El-Din M. 2015. Composite polyninylidene fluoride (PVDF) membrane impregnated with Fe_2O_3 nanoparticles and multiwalled carbon nanotube for catalytic degradation of organic contaminants. *J. Membr. Sci.*, 490, 227–235.
61. Biswas A.K., Md Islam R., Choudhury Z.S., Mostafa A., Kadir M.F. 2014. Nanotechnology based approaches in cancer therapeutics. *Adv. Nat. Sci Nanosci. Nanotechnol.*, 5, 043001–043012.
62. Essential composition and Quality Factors of Virgin Coconut Oil (SNI 7381: 2008), 2008.
63. Geise G.M., Paul D.R., Freeman B.D. 2014. Fundamental water and salt transport properties of polymeric materials. *Prog. Polym. Sci.*, 39, 1–42.
64. Mukherjee R., Sharma R., Saini P., De S. 2015. Nanostructured polyaniline incorporated ultrafiltration membrane for desalination of brackish water. *Environ. Sci. Water Res. Technol.*, 1, 893–904.

65.
66. Standard Regulation of
67. Lukina A.O., Boutin C., Rowland O., Carpenter D.J. 2016. Evaluating trivalent chromium toxicity on wild terrestrial and wetland plants. *Chemosphere*, 162, 355–366.
68. Bokhary A., Tikka A., Leitch M., Liao B. 2018. Membrane fouling prevention and control strategies in pulp and paper industry application: A review. *J. Membr. Sci. Res.*, 4, 181–197.
69. Eternadi H., Qazvini H. 2020. Investigation of alumina nanoparticles role on the critical flux and performance of polyvinyl chloride membrane in a submerged membrane system for the removal of humic acid. Polymer Bulletin, Springer Verlag GmbH.
70. Moustakes N.G., Katsaros F.K., Kontos A.G., Ramanos G.E., Dionysiou D.D., Falaras P. 2014. Visible light active TiO_2 photocatalytic filtration membranes with improved permeability and low energy consumption. *Catal. Today*, 224, 56–69.
71. Fane A.G., Yeo A., Law A., Parameshwaran K., Wicaksana F., Chen V. 2005. Low pressure membrane processes- doing more with less energy. *Desalination*, 185, 159–165.
72. Razmjou A., Arifin E., Dong G., Mansouri J., Chen V. 2012. Superhydrophobic modification of TiO_2 nanocomposite PVDF membranes for applications in membrane distillation. *J. Membr. Sci.*, 415–416, 850–863.
73. Dzinun H., Ichikawa Y., Honda M., Zhang O. 2020. Efficient immobilised TiO_2 in polyvinylidene fluoride (PVDF) membrane for photocatalytic degradation of methylene blue. *J. Membr. Sci. Res.*, 6, 188–195.
74. Lai G.S., Lau W.J., Goh P.S., Ismail A.F., Tan Y.H., Chong C.Y. 2018. Tailor-made thin film nanocomposite membrane incorporated with graphene oxide using novel interfacial polymerization technique for enhanced water separation. *Chem. Eng J.*, 344, 524–534.
75. Vanneste J., Peumans W.J., Van Damme E.J.M., Darvishmanesh S., Bernaderts K., Geuns J.M.C., Van der Bruggen B. 2014. Novel natural and biomimetic ligands to enhance selectivity of membrane processes for solute–solute separations: Beyond nature's logistic legacy. *J. Chem. Tech. Biotech.*, 89, 354–371.

12 Water Filtration and Organo-Silica Membrane Application for Peat Water Treatment and Wetland Saline Water Desalination

Muthia Elma, Fitri Ria Mustalifah, Aliah, Nurul Huda, Erdina Lulu Atika Rampun, and Aulia Rahma
Lambung Mangkurat University

CONTENTS

DOI: 10.1201/9781003167327-14

12.1 INTRODUCTION

Various technologies have been developed for treating a peat water treatment such as filtration (Parabi et al. 2019), membrane (Mahmud, Abdi, and Mu'min 2013), adsorbent (Zulfikar et al. 2013), and ozone (Ali, Lestari, and Putri 2019). However, advanced technology is necessary to treat peat water into clean water due to the high level of dissolved organic compounds, low pH, and dark color. Several methods have been applied to treat peat water as well as wetland saline water, such as:

- Filtration: It is strongly recommended to build a grid-independent water filtration system in order to access clean water according to water quality standards (biological, chemical, and physical). It has been known that peat water has organic matter and other pollutants.
- Coagulation-pervaporation process for wetland saline water desalination via pervaporation: This combination method has been reported by Rahma et al. (2019a) and shown to be an excellent approach to treating peat water and wetland saline water.
- Membranes ultrafiltration: Membranes ultrafiltration is also known as low-pressure membrane process with pore sizes ranging around 2–100 nm (Raman, Cheryna, and Rajagopalan 1994). In the application of wetland water treatment, previous research reported that ultrafiltration could also treat peat water with high natural organic matter (NOM). It is also found that two adsorptions may increase the ability of rejection and reduce the adsorbent to one single adsorption (Mahmud et al. 2020).
- Adsorbent: Activated carbon (AC) is an amorphous carbonaceous material that can be fabricated in any carbonaceous material, and it has a high surface area and porosity. Recently, anthracite and bituminous are the most carbonaceous material sources. AC is often applied in wastewater treatment and emissions as adsorbents (Saleem et al. 2019).
- Ozone: In principle, ozone (O_3) damages bacteria by passing oxygen through high-voltage electrodes as shown in Figure 12.1. The benefits of using ozone

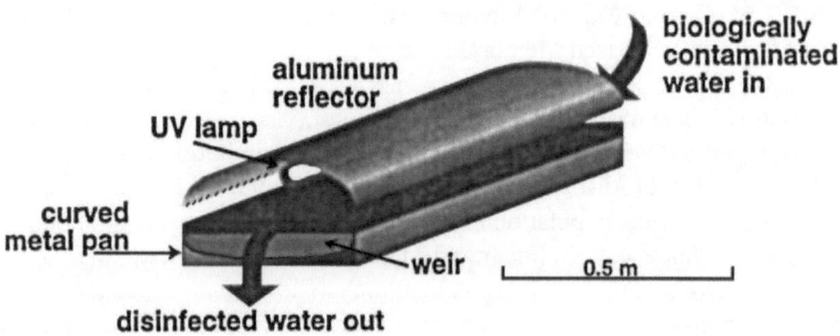

FIGURE 12.1 Water disinfection with UV lamp is placed below aluminum reflector, above water flow in tray (Gadgil 1998).

in water treatments are that it is easy to operate, is environmental friendly, and no residue is released during its application (Remondino and Valdenassi 2018).

Constructed Wetland. In terms of usage, the wetland is not only supposed to be treated, but it can also act as a technology named constructed wetlands (CWs). Most developing countries have warm tropical and subtropical climates. It is also generally acknowledged that CWs are more suitable for wastewater treatment in tropical regions than in temperate regions (Kivaisi 2001; Denny 1997; Haberl 1999). Typical solar disinfection is SODIS (solar water disinfection) as shown in Figure 12.2. It is a simple method that is applied by filling water in the transparent polyethylene terephthalate (PET) bottles or clear plastic bags in a batch method (Gutiérrez-Alfaro et al. 2017) and disinfecting it using direct sunlight for 6 hours to kill the pathogens.

12.2 LOW-COST WATER FILTRATION SET-UP FOR PEATLAND AND WETLAND SALINE WATER TREATMENT

Water, in general, is used for living or habitat, transportation, fisheries, and recreation purposes. The high content of natural organic matter in the water makes swamp water brown and unsuitable for use. In addition, seawater entrusts that occur at maximum tide into the swamp aquifer cause the swamp water to become salty which is known as salt marsh water. In terms of quality, swamp water is not suitable for use because it has a relatively low pH of 3–5, it is brown in color, and has a high organic

FIGURE 12.2 SODIS batch reactor for treating 100 NTU turbid water (Ubomba-Jaswa et al. 2010).

matter content of 38–280 mg L^{-1} $KMnO_4$ (Rahma et al. 2019b). Wetland systems can be environmentally and economically sustainable viable options for upgrading existing secondary wastewater treatment systems to achieve tertiary level wastewater with minimum operational input and energy (Park, Craggs, and Tanner 2018).

One simple way of ensuring good water quality is to purify it with a household ceramic water filter. As reported, the manufacturing and supply of water filters suitable for removing suspended solids, pathogenic bacteria, and other toxins from drinking water are very important. Microporous ceramic water filter with micron-sized pores is developed using a traditional slip casting process. This locally produced filter has the advantage of utilizing less raw materials, costs, labor, energy, and expertise and is becoming more effective and efficient compared to other low-cost production filters.

12.3 ORGANO-SILICA MEMBRANES FOR WETLAND SALINE WATER DESALINATION

Organo-silica membranes are fabricated from pure silica membranes by adding organic materials as a template. Pure silica membranes are classified as inorganic membranes employing silica as the main precursor. The most visible characteristic of this material is that the hardness level is quite high. Therefore, this material is utilized as a membrane coating with the potential that silica can filter out substances or small particles of impurities. One of the silica precursors is tetraethyl orthosilicate (TEOS). This material has good mechanical properties in the SiO_2 layer precipitation process, is easy to do hydrolysis reaction, and is easy for –OH group replacement. The silanol (Si-OH) groups react with the non-hydrolyzed alkoxide groups to form siloxane (Si-O-Si) bonds and then silica networks will be easily formed. Figure 12.3 shows the structure of the TEOS.

Pure silica membranes are obtained from the sol–gel process. The synthesis of sol silica is often known as the sol–gel method. Generally, the fabrication of silica membranes employs acid–base as the catalyst. During the fabrication process of silica sols, three reactions take place, namely the hydrolysis reaction, alcohol condensation, and water condensation.

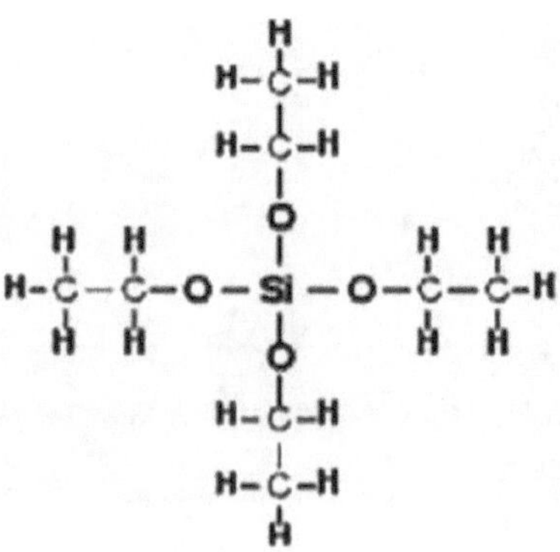

FIGURE 12.3 Structure of TEOS (Elma, Mujiyanti, and Amalia 2019).

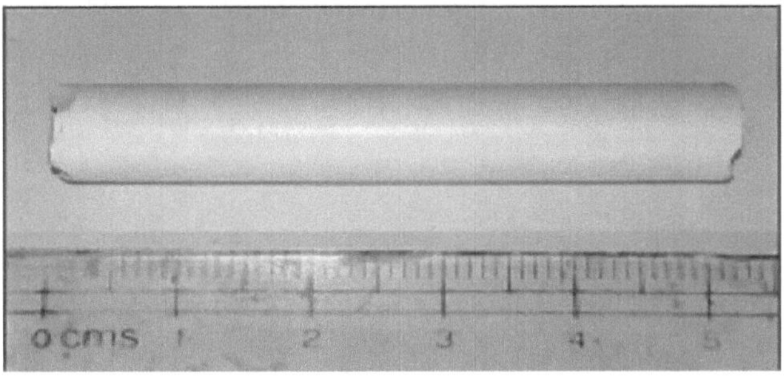

FIGURE 12.4 Pure silica membrane (Elma and Assyaifi 2018).

Recently, research on silica membrane pores with micropore and mesopore pore sizes in water desalination have been well developed (Duke, Mee, and da Costa 2007; Balistreri et al. 2016). Figure 12.4 shows a pure silica membrane.

During fabrication, this membrane is conducted by two main membrane calcination techniques: (a) CTP (Conventional Thermal Process) method and (b) RTP (Rapid Thermal Processing) method. The CTP technique requires a longer time to produce each layer of the silica membrane (~4 hours for each layer), while the RTP calcination technique requires a shorter time (1 hour for each layer). Unfortunately, the RTP technique produces a thicker layer membrane than the CTP technique. Initially, the silica membranes were manufactured with three interlayers of membranes for the filtering process, but now innovations have emerged by using pure silica membranes without any interlayers. This gives advantages of lower production prices and shorter execution times. The performance of the pure silica membrane applied for water desalination can be determined via the pervaporation process.

The strength of silica membrane can be seen in long-term performance. The membrane structure may collapse over time. This is due to low hydrostability when applied for water desalination. It happened due to the high concentration of silanol (Si-OH) groups in the silica network. The hydrophilic nature of silanol reacts with water molecules causing a decrease in membrane performance. Efforts have been developed to increase the hydrostability of membranes by adding carbon to pure silica membrane materials, for example using natural carbon from pectin (Rampun et al. 2019) and adding P123 (Elma et al. 2018b) carbon to pure silica membranes (Sumardi, Elma, Rampun, et al. 2021; Pradhana et al. 2021; Muthia Elma 2021).

12.3.1 Si-P123 Membranes

The latest innovation in the silica-P123 membrane application is to increase the hydrostability of the silica membrane applied for wetland saline water desalination. The silica-P123 membrane is a pure silica sol with the addition of organic material type P123 (a non-ligand triblock copolymer P123) as a template. The P123 material is a longer carbon chain, and the addition of P123 concentration to the silica membrane causesthe membrane structure to have a mesopore pore size (Elma 2020b). Membrane silica with triblock copolymer P123 template is shown in Figure 12.5.

FIGURE 12.5 Silica-P123 tubular membrane (Elma et al. 2018a).

This wetland desalination application shows that the addition of carbon from P123 to the silica membrane with a mesopore pore size (2.2 nm) produces a high enough permeate flux with portable water quality standards. The resulting permeate flux reaches 2.6–4.5 kg $m^{-2}h^{-1}$ and salt rejection reaches >98% (Rahma et al. 2020). Long-term in the application with over 400 hours shows an average permeate flux of 2.6 kg $m^{-2}h^{-1}$, with salt rejection in the solution still reaching 99% (Rahma et al. 2020). Another application of silica membranes using P123 (Syauqiyah et al. 2019) explains that there was an increase in membrane hydrostability. The water flux is also excellent (8.5 kg $m^{-2}h^{-1}$) and salt rejection is >99.5%. This result shows a higher performance as compared to pure silica membranes. However, the addition of P123 material as carbon sources into the silica matrix increases the operational cost if applied on a large scale (Yang et al. 2017).

12.3.2 Si-P (Silica-Pectin) Membranes

Pectin is an alternative templating material for the silica membrane. Pectin is a biodegradable polysaccharide that is mostly found in fruits and other plants, such as

FIGURE 12.6 Silica-pectin membrane has a 4 cm length (Elma, Mustalifah, et al. 2020).

apple peels, jackfruit skins, banana peels, and cherry tomatoes. The pectin content in apple skin is around 24.5% of the apple plant part. Meanwhile, banana peels account for 14% of the weight of banana peels (Tuhuloula, Budiyarti, and Fitriana 2013). Pectin is a substitute group for –OH in the C atom. Carbon chain bonds in silica membranes create a stronger silica network and affect the excellent mechanical and thermal stability of membrane performance (penelitian Elma and Wang). But, the calcination temperature may affect the carbon chain bonds in the silica-pectin material. This is because carbon is an organic and flammable material (Pratiwi et al. 2019). Silica-pectin membrane is shown in Figure 12.6.

Desalination of wetland saline water using a silica-pectin membrane has been carried out and provides a breakthrough in the silica membrane fabrication applied for the water desalination process. The presence of a high permeate flux in desalination reached 5.73 kg $m^{-2}h^{-1}$ compared to the permeate flux of silica-P123, which ranged from 1.9 to 3.7 kg $m^{-2}h^{-1}$ (Syauqiyah et al. 2019). The performance of the silica-pectin membrane in wetland saline water desalination shows that the addition of pectin gives great effects on water flux. The increase in water flux was 19 times higher than pure silica membranes with also great salt rejection (>99%) (Rampun et al. 2019). In another study, it was shown that the performance of silica membranes using a carbon template from pectin showed an increase in water flux and high salt rejection (>99%) (Elma et al. 2019). The thickness of the silica-pectin membrane (~2 μm) is thicker than membranes fabricated using CTP techniques (30–50 nm) (Elma et al. 2012). Although it is slightly thicker than other carbon templates, the performance of the membrane is still excellent. An excellent membrane performance was investigated and showed high water fluxes (8.3 kg $m^{-2}h^{-1}$) and salt rejection (99.4%) (Elma,

Pratiwi, et al. 2020). As a result, silica-pectin membranes can offer reinforcement to silica structures. From an economic point of view, non-ionic micelles formed by pectin silica-pectin membrane are more economical than P123 material (Syauqiyah et al. 2019).

12.3.3 Si-Glucose Membranes

Glucose can replace the role of a triblock copolymer (P123) (Elma, Wang, Yacou, and Diniz da Costa 2015) as an organic compound to form a hybrid material that is embedded in the silica matrix. The effect of the concentration of glucose addition on the fabrication of silica membranes for desalination of wetland water resulted in a greater ratio of siloxane/Si-C. On the contrary, the values of silanol/siloxane and silanol/Si-C are much lower along with the higher concentration of added glucose(Elma, Wang, Yacou, Motuzas, and Diniz da Costa 2015). The performance of the membrane glucose-silica in water desalination using NaCl solution (7.5%wt.) as a feed shows fluxes ranging from 0.22 to 2.28 kg $m^{-2}h^{-1}$ with salt rejection >93%.

Figure 12.7A represents the image of silica-glucose. PtNP was well dispersed in CS solution. Particle sizes were observed in the range of about 3–4 nm, indicating that the amino groups play an important role in stabilizing PtNP in the CS solution. No nanoparticle aggregation was observed after 2 months. When CNTs were functionalized with carboxyl groups and MTOS were added and sonicated, nanohybrids and sol–gels were formed. PtNP and CNT depicted in Figure 12.7B were dispersed in the sol–gel.

Table 12.1 summarizes the performance of organo-silica membranes for different feeds desalination prepared with RTP calcination. Pure silica has the lowest flux and salt rejection compared to organo-silica membranes. It proves that the carbon addition in silica could improve the membrane performance in removing salt particles even at high salt concentrations.

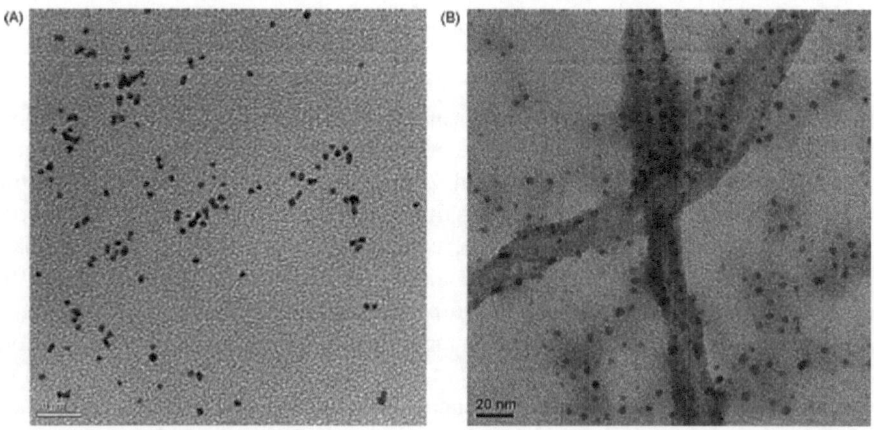

FIGURE 12.7 Transmission electron microscopy (TEM) images of PtNP–CS solution (A) and CNT–PtNP–CS–MTOS sol–gel (B) of the glucose (Kang et al. 2008).

TABLE 12.1
Various Organo-Silica Membrane Performance Applied for Water Desalination

Various Organo-Silica Membrane	Calcination Technique	Feed Water	Water Flux (kg m²h⁻¹)	Salt Rejection	Ref.
Pure silica	RTP	NaCl 3.5 wt.%	0.61–1.19	84.95 %	Elma and Assyaifi (2018)
Silica-P123 membrane	RTP	Wetland saline water	2.6–4.5	>98%	Elma (2020b)
Silica-pectin membrane	RTP	Wetland saline water	3.4 (pectin 0.5%wt calcined 300°C)	±98%	Rampun et al. (2019)
Silica-glucose membrane	RTP	7.5% NaCl	1.8, 2.2 and 4.8	>99%	Mujiyanti, Elma, and Amalia (2019)

12.3.4 Single-Catalyst Membranes

The presence of inorganic particles that can be completely dispersed in the polymer matrix has been further developed to enhance the membrane performance (Elma, Mujiyanti, and Amalia 2019). Many new types of organic–inorganic hybrid materials have the potential to obtain the desired hybrid material properties, improving the mechanical and thermal properties of inorganic compounds with flexibility. Currently, organic–inorganic hybrid composites have attracted many researchers because of the properties possessed by them. These properties include relatively easy control such as thermal, optical, and mechanical resistance. Organic–inorganic hybrid materials have found their application as optical materials, sensors, catalysts, adsorbents, and thin layers as membrane coating (Zulfikar et al. 2006).

Inorganic acid catalysts, including HNO_3 (Zhang et al. 2018), are often employed in silica xerogel fabrication. This acid catalyst in the hydrolysis reaction produces silanol groups generally with small pore sizes. On the other hand, a base catalyst produces siloxane groups with a relatively large pore size during condensation reaction (Rezki et al. 2020). Thus, in the application of membrane for water desalination, the pore size should be controlled (slightly in mesoporous). For that, an organic acid catalyst is suitable for fabricating silica matrix due to the ability to form mesoporous materials as well as act as an organic carbon source. Citric acid has the chemical formula $C_6H_8O_7$ and therefore has several carbon bonds, in addition to being organic in nature, and thus it is able to increase the silica material's hydrophobicity (Pratiwi et al. 2019). Silica xerogel formulated with the addition of a single catalyst (citric acid) exhibited siloxane, silanol, and silica carbon peaks at 1050, 964, and 800 cm^{-1}, respectively, during FTIR measurement (Sumardi, Elma, Lestari, et al. 2021).

12.3.5 Organo-Catalyst Membranes

Organo-catalyst is the reaction carried out by a number of sub-stoichiometric organic compounds that do not contain even a small amount of enzyme or inorganic element (Raj and Singh 2009). Organo-catalysts have become very important because of their green chemistry perspective. The most important constituent in the reaction is the solvent. One of the solvents that satisfy this criterion of green chemistry is "water," which is a suitable solvent in various biosynthetic reactions responsible for sustaining life. Therefore, water has the potential to actas a reaction medium in synthetic organic chemistry (Heine et al. 2001). Other than that, water possesses some unique physical properties such as high surface tension, hydrogen bonding capability, and polarity. All these properties have an important role to play in the ultimate influence that water exerts on organic reactions (Raj and Singh 2009; Elma and Setyawan 2018).

Initially, water is not considered a suitable solvent for organic reactions. This is because the functional groups present in the reactants can react on their own with water molecules. Further, most organic reactants are non-polar, so they do not dissolve in water. In addition, water can interfere with the transition state that is formed between the organo-catalysts molecule and the substrate. This can interfere with hydrogen bonding and other polar interactions, thereby impairing catalytic activity and stereo control. Therefore, it is assumed that the reaction in water causes a slow reaction rate and lower yield of the desired product (Klijn and Engberts 2005).

Although water does not dissolve organic reagents, it still accelerates the rate of many reactions (Pirrung 2006). Now, the reaction takes place in a concentrated organic phase away from water molecules, resulting in an increase in the rate (Heine et al. 2001). Organo-catalysts are also more stable and can tolerate the presence of moisture and air. Therefore, considerable efforts have been made in developing water-compatible small organic molecules. Organo-catalyst has also evolved as the primary way of catalyzing the wide variety of organic reactions, besides enzymes and metal catalysis.

Silica as an inorganic material has several advantages such as high chemical, thermal, and mechanical stabilities (Benfer et al. 2001). In order to reduce the production cost and efficiency of silica as an inorganic material, organic material is an excellent way to be used as a catalyst and a carbon source to form an organo-silica network (Elma 2020). The addition of carbon from an organic catalyst can create strength and change the structure of the silica matrix (Elma, Wang, Yacou, Motuzas, and da Costa 2015). Configuring Si-C in silica networks is a new potential strategy to solve a major problem in hydrostability. Sol xerogel silica was successfully fabricated using a double-base acid catalyst (citric acid and ammonia) using low calcination (200°C and 250°C). And, the carbon functional groups are lost due to high temperatures; however, the functional groups of siloxanes and silanol still remain. Based on FTIR measurement, organo-catalyst also enhances the hydrostability and robustness of the silica network (Elma 2020).

12.3.6 TEVS-Based Membranes

Mesoporous silica is often applied for membrane fabrication in both water and gas separation. Triethoxyvinylsilane (TEVS) is often used to produce microporous silica membranes on interlayer porous substrates (Elma, Wang, Yacou, and da Costa 2015). This is because TEVS included in the vinyl group act as a silica ligand pendant. The methyl silica ligands pendant group is known for producing high-quality micro silica membrane (Elma, Rampun, et al. 2020).

12.3.7 ES40-Based Membranes

Recently, the interlayer-free ES40 and TEOS-derived silica membranes were successfully developed for the desalination of seawater by rapid thermal treatment (RTP) (Wang, Wang, et al. 2016). ES40 as a partially condensed silica precursor to TEOS can be derived from strong silica structures that are more resistant to pressure and thermal expansion. The chemical effect of P123 as the carbon content is implanted into the silica network with ES40 as a precursor obtained from the silica membrane through the sol–gel method. Furthermore, the optimized P123 synthesis is selected to be implanted in the manufacturing of an organo-silica hybrid membrane (ES40-P123) (Rahman et al. 2020). According to FTIR result, the application of copolymer P123 given into silica sols by sol–gel method is a prominent way to produce a combination of silica and carbon bonded together in a silica matrix to produce xerogel as a material for making membrane coating applied for water desalination. P123 functions as a carbon content to replace the silanol (Si-OH) groups with silica carbon (Si-C). Hydrophobicity and loss of pore volume of organo-silica (ES40-P123) and pure ES40 materials depend on the ratio of the peak of vibration of silanol or siloxane. The presence of carbon chains in the silica network may improve the mechanical properties of the membrane and result in the best membrane desalination performance (Rahman et al. 2020).

12.4 FOULING EFFECT ON WETLAND WATER TREATMENT

Wetland water characteristic is similar to other surface water sources that contains a high concentration of natural organic matters (NOMs) (Wenten et al. 2020). The most problematic membrane foulant during the membrane process for wetland water treatment was initiated by NOMs and extent the membrane fouling, in particular irreversible fouling. Mainly, NOMs are composed of proteins, humic acids, lipids, and polysaccharides (Wang et al. 2020). Thus, the effect of the wetland water treatment process on membrane fouling is required to investigate more. Due to this, it is necessary to decide the best way to treat wetland water implemented by membrane technology.

Fouling is also a crucial problem that is limiting the membrane performance. The fouling of pressure-driven membranes is commonly reconciled to the deposition, accumulation, or adsorption of foulants onto the membrane surface and/or on the interior of the membrane pores. The basic functions of the membrane separation technique could be molested, such as dwindling of over filtration time, consisting permeate flow, pressure drop, and solute removal efficiency (Goh et al. 2018). The fouling mechanism depends on membrane type classification which is divided into dense and porous membranes. Frequently, the dense membrane-like RO majorily has detected the surface fouling mechanism. Nonetheless, the RO fouling mechanism can be classified into scaling, chemical oxidation, biological fouling, and organic fouling, whereas the porous membranes are associated with pore blocking, cake formation, concentration polarization, and organic adsorption (Guo, Ngo, and Li 2012).

Organic fouling is the combination of deposition, reactions, and interactions of high-molecular-weight organic molecules (e.g., NOMs on the surface of the membrane). The existence of NOMs, such as fulvic and humic acids in the seawater is due to the degradation and decomposition of living organisms (Goh et al. 2018). On the other hand, the presence of NOMs in wetland water originated from its wetland or peat soil that contains organic substances (Rahma et al. 2020). NOMs in wetland or peat water are the main parameters of water quality that affect the process of drinking water treatment (Elma, Rahma, et al. 2020; Rahma et al. 2019b). The presence of NOMs in wetland water is contributed to cause tea-color, odor, chromatization, and regrowth of biological in the water distribution network and causing the formation of by-products from disinfection in the form of carcinogen compounds (Tang et al. 2014; Zularisam et al. 2009).

Figure 12.8 displays the schematic of the fouling mechanism by attacking NOMs for wetland saline water treatment by pervaporation. A recent study reported that a mesoporous silica membrane was applied for wetland saline water desalination (Rahma et al. 2020; Elma, Riskawati, and Marhamah 2018). At the beginning of pervaporation time, the performance of the mesoporous silica-based membrane exhibited high water fluxes. However, the water flux gradually declines in line with pervaporation time, because of NOMs attacking within the membrane pores.

Figure 12.9 shows the fouling mechanism on the ultrafiltration membrane for peat water treatment. The permeable UF membrane exhibited a slightly rapid flux decline at the early stages of peat water filtration. The important point of the UF membrane during NOMs fouling is pore size. At the initial stage of ultrafiltration, pore blocking in the membrane pores is a dominant factor that causes fouling problems. Then, this fouling mechanism starts to form pore blocking to create cake formation (Mahmud

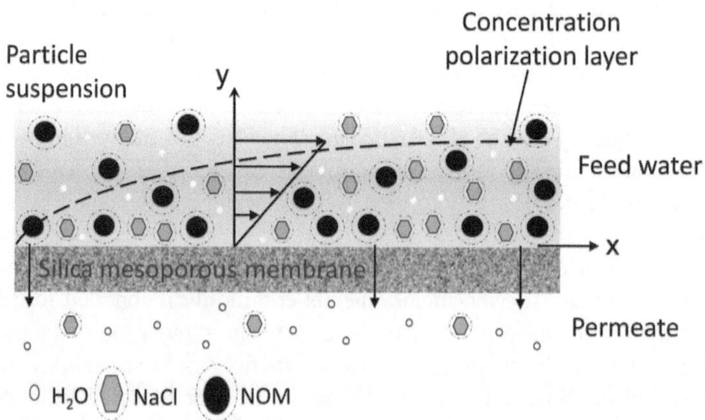

FIGURE 12.8 Schematic of fouling mechanism (concentration polarization). NOM in wetland saline water was attacked after it flowed through the silica mesoporous membrane. The clean permeate was obtained.

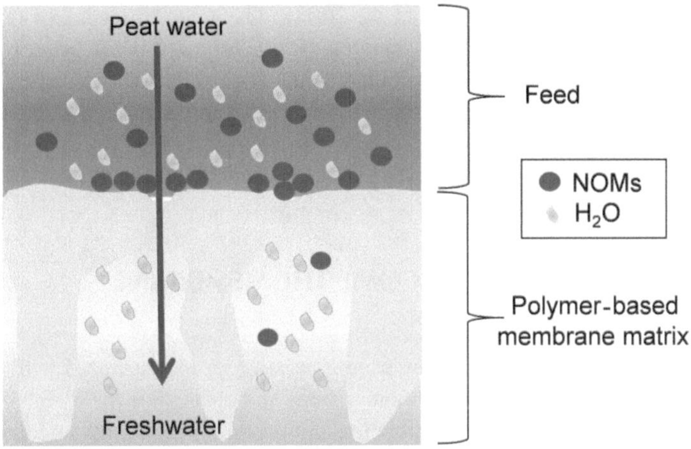

FIGURE 12.9 Schematic of fouling mechanism (deposition particles) on ultrafiltration membrane for treating peat water. The NOMs attached to the membrane surface and block the membrane pore.

et al. 2020; Rahma, Mahmud, and Abdi 2018; Rosadi, Mahmud, and Abdi 2017). NOMs (majority colloidal) block the pores on the permeable membranes. In addition, the structure of the fouling layer on the surface of the membrane is related to the operating condition (Costa, de Pinho, and Elimelech 2006).

Both organic and dissolved inorganic compunds can result in membrane fouling. Dissolved organic matter (DOM) is prevalent in wastewater and surface water, which can be categorized according to their origins into three different sections: (a) refractory NOMs derived from drinking water sources; (b) synthetic organic compounds (SOC) added by consumers and disinfection by-products (DBPs) produced during disinfection processes of water and wastewater treatment; and (c) soluble microbial products (SMP) generated during the biological treatment processes due to decomposition of organic compounds. NOMs have been identified as a major foulant of polymeric membranes in drinking water applications.

NOMs are a complicated heterogeneous mixture of compounds formed as a result of the decomposition of plant and animal materials in the environment, particularly in wetland systems. The heterogeneous compounds (MW ranges from 1,000 to 100,000 Da) can be divided into humic acids, fulvic acids, and humin according to their solubility in acidic (pH) solutions. The amount of NOMs in peat water includes humic acid (Mahmud et al. 2020). Generally, NOMs are divided into three fractions of organic substances: (a) hydrophobic (dissolved organic carbon with larger molecular weight), (b) hydrophilic (smaller molecular weight such as polysaccharides, amino acid, protein, etc.), and (c) transphilic (molecular weight between hydrophobic and hydrophilic) (Zularisam et al. 2007). Hence, it is required to treat peat water before processing with membrane filtration to avoid membrane fouling (Zularisam, Ismail, and Salim 2006a). Previous study shows that NOMs may cause fouling in UF membranes. Although hydrophilic fractions have less impact on water quality, studies showed large-size molecules of hydrophilic NOMs have greatly contributed to fouling (Zularisam et al. 2007). Therefore, in order to mitigate or retard the fouling of the membrane, pre-treatment is required before treating wetland water. Pre-treatment is normally executed to the feed water before passing through the membrane system to diminish membrane fouling. In addition, this pre-treatment procedure is applied to eliminate foulant, extend membrane life, optimize recovery, and system productivity. Other than this, pre-treatment also hinders the physical damage to the membranes matrix.

12.5 STRATEGIES TO OVERCOME THE MEMBRANE FOULING LIMITATION

Fouling is a major challenge in membrane operation. It is associated with the accumulation of substances on the membrane surface and also inside the membrane pores. This phenomenon not only leads to decreasing productivity but also needs a supply of additional energy to maintain the membrane performance consistently. Membrane fouling by NOMs is not easy to be recovered only through physical washing or scrubbing using water and air flow (Sun et al. 2013). Frequent membrane cleaning will increase the operational cost as well as will decrease membrane module performance and operation efficiency (Shan et al. 2016). To settle this problem, various strategies have been proposed which include pre-treatment optimization, controlling operation conditions, and developing membrane with antifouling properties. In addition, the complex chemical cleaning process or the pre-treatment is required to maintain the long-term stable operation of a membrane. There are several strategies that have been reported to face the fouling phenomenon for wetland and peat water treatment. The strategies are more specific on the membrane process, both ultrafiltration and pervaporation.

Ultrafiltration and pervaporation membranes have relatively mesoporous to microporous structures, which leads to undesirable removal efficiency of NOMs. Meanwhile, pollutants in raw water can cause severe flux decline of the membrane. Therefore, pre-treatment such as adsorption, coagulation, and/or photocatalytic is necessary to couple with ultrafiltration or pervaporation to lessen fouling membrane.

12.5.1 Hybrid Adsorption Pre-Treatment Membranes

Adsorption is the combination process that uses a low-pressure membrane-like UF. Powdered activated carbon (PAC) is commonly used to remove NOMs in the peat water treatment (Mahmud et al. 2020; Rahma et al. 2019b; Mahmud et al. 2012). Dosing PAC can improve the removals of pollutants in the water by adsorption. Membrane filtration resistance may decline by the addiction of PAC, due to changing the structure of the cake layer instead of eliminating pollutants. PAC was able to decrease membrane filtration resistance and improve filtration flux to a limited level because PAC could only adsorb a small part of soluble organics with low molecular weight while the adsorption of macromolecular organic that affected filtration flux was poor (Sun et al. 2013).

Several studies conducted by Kim, Cai, and Benjamin (2008); Kang and Choo (2010); and Wang and Benjamin (2016) state that the adsorption-UF process can reduce fouling of the membrane. The results of the study by Tomaszewska and Mozia (2002) stated that the utilization of PAC as an adsorbent in the UF system is effective in setting aside organic matter with small and large molecular weights. In contrast, Sun et al. (2017) mentioned that the use of PAC has an influence to decrease membrane flux because addition of PAC forms a *layer* that is capable to withstand organic substances and plays a role in the occurrence of fouling. The interaction between PAC particles and NOMs molecules has a significant effect on fouling and can reduce the occurrence of irreversible fouling on membranes.

Research on the influence of PAC on the UF process has been conducted by Tomaszewska and Mozia (2002) to find out UF performance, flux value, and *backwashing*. The type of membrane used is made from polyvinylidene fluoride to set aside humic acid and phenol, which are NOMs in water. In addition, PAC-UF is very effective in setting aside humic acid and phenols in water samples. Mahmud et al. (2020) investigated the polysulfone UF membrane by the addition of PAC via two-stage adsorption process. The two-stage adsorption process is operated by the distributed adsorbent capability to transfer adsorbate to the adsorbent surface itself. This step would optimize the removal and demand more low dosage rather than conventional adsorption. The two-stage adsorptions dosing by PAC as pre-treatment is carried out to investigate its effect on polysulfone UF membrane performance for peat water treatment. Flat sheets of UF membranes were varied at 1–3 bar. Further, $KMnO_4$ and UV_{254} rejections are determined to represent organic and NOMs.

During blank-feed pass through the polysulfone ultrafiltration membrane, it delivers 54% lower than demineralized water fluxes. It is expected that the PAC might contribute to the decline in the water flux. This case is also in line with the research reported by Tomaszewska and Mozia (2002), however, the opposite result was concluded by Lin, Lin, and Hao (2000) where the PAC does not affect to sustain hydrophilic ultrafiltration membrane's water flux (Mozia, Tomaszewska, and Morawski 2005; Li and Chen 2004).

The two-stage adsorption/UF demonstrated the highest water fluxes of peat water (Mahmud et al. 2020). The water flux significantly drops until 73% and 58% of feed demineralize water and blank at three bar conditions, respectively. Two-stage adsorption doses of 3/4:1/4 are, subsequently, used as the feed of UF. Overall, the water fluxes were gradually decreased at each operating pressure. This phenomenon has occurred due to polarization concentration, which is some particles were accumulated on the membrane surface and generated a layer onto it. Thus, it can reduce the UF membrane performance. Also, NOM was attached to the top of the membrane surface known as the cake layer. The cake layer remains tightly compact and contributes to worse water flux. Two-stage adsorption-polysulfone ultrafiltration obtained 93% and 88% of removal efficiency for organic substances ($KMnO_4$) and UV_{254}, respectively. NOM cannot be removed only by using UF alone (Mozia, Tomaszewska, and Morawski 2005). By operating UF, lower pressure applied is good for removal. It is because the smaller feed velocity makes particles easy to filter. These results agree with other studies reported (Maddah et al. 2018).

To overcome NOM-PAC deposition, pre-treatment and backwashing are mostly used (Zularisam, Ismail, and Salim 2006b). Several parameters such as the structural properties of PAC and membrane type (Mozia, Tomaszewska, and Morawski 2005;

Li and Chen 2004) may influence NOM removal and fouling mitigation. Hence, they need to be considered first. Many methods have been reported to minimize fouling such as photocatalytic (Singh et al. 2016), oxidation and ozonation (Hutagalung et al. 2014), coagulation-flocculation (Elma, Rahma, et al. 2020; Rahma et al. 2019b), and adsorption (Shimabuku et al. 2014). Among them, adsorption using activated carbon is good to remove hydrophilic NOM (Uyak et al. 2014).

12.5.2 Hybrid Coagulation Pre-Treatment Membranes

Coagulation, as a physicochemical pre-treatment process, is most widely and successfully used due to its low cost and relatively easy to operate characteristics. Coagulation combined with UF is a promising process with respect to the removal of contaminants, the maintenance of a high membrane performance, and the reduction of subsequent formation of disinfection by-product. Coagulants are abiding between organic macromolecules and inorganic salts (Guo, Ngo, and Li 2012). The coagulation/sediment is mostly utilized for removing turbidity, nevertheless, its capability to remove organic matter is restricted. In general, NOMs with high molecular weight and insolubility organic are effectively removed by the coagulation process, meanwhile, NOMs with low molecular weight and large solubility organic inefficiency are removed (Sun et al. 2013).

Bing-zhi et al. (2007) asserted that online coagulation could form a cake layer on the membrane surface, which was positive for the removal of NOMs to improve effluent quality and decrease membrane fouling. Rosadi, Mahmud, and Abdi (2017) concluded based on the series of experiments that were compared to ultrafiltration. When it is investigated, the controlling irreversible membrane fouling of polysulfone ultrafiltration membrane done by pre-coagulation/sedimentation predicted that the pre-coagulation/sedimentation could not eliminate macromolecular organics perfectly, including protein and polysaccharides, which were the major contributor to irreversible membrane fouling. Considering pre-treatment coagulation for improving filtration efficiency of membrane filtration depends on choosing proper coagulants and control coagulation conditions, which had the best coagulation-ultrafiltration treatment efficiency based on wetland water qualities.

Elma, Rahma, et al. (2020) also investigated wetland saline water desalination via pervaporation and multiple feeds of the sample with pre-coagulation and without pre-coagulation. Coupling processes of coagulation as pre-treatment and silica-pectin membranes exhibit high water flux at 60°C of 12.2 kg $m^{-2}h^{-1}$. It also conducted UV_{254} rejection of feed water after coagulation/sedimentation displays 43% higher than the membrane pervaporation process only. Overall, salt rejection for all processes was to be demonstrated over 99%. The application of coagulation as pre-treatment process is promising to enhance the performance of the silica-pectin membrane for wetland water desalination.

Performance of silica-pectin membrane pervaporation sorted the water flux from highest to smallest after coagulation is better than without coagulation at multiple feed temperatures. These results are influenced by the presence of salt and NOMs in the feed. Wetland saline water without pre-treatment consists of high NOMs, which lead to membrane fouling. Fouling plays a role in blocking membrane pores and affecting the membrane's performance, resulting in the decrease of water flux (Goh et al. 2018). Pre-treatment of coagulation increase the membrane's water flux in treating water with high NOMs content (Sillanpää et al. 2018). The coagulation process plays a role in reducing NOMs in water to ease the work of membranes (Bing-zhi et al. 2007). It can be concluded that the coagulation pre-treatment greatly improves the performance of silica-pectin membranes for desalination of wetland saline water. The small amount of UV_{254} rejection for desalination of these feeds, due to membrane fouling, might occur. The presence of high NOMs in water contributes to the occurrence of organic fouling on the membrane (Goh et al. 2018).

Rahma et al. (2019b) also demonstrated a combination of coagulation/pervaporation processes for wetland saline water desalination carried out at room temperature (~25°C). Measurement of water flux and salt rejection of organic matter (UV_{254}) have been done on wetland saline water as *feed* either without pre-treatment or with pre-coagulation treatment. The water flux of wetland saline water with pre-treatment coagulation at optimum doses of 30 mg L^{-1} has a higher value than without coagulation (5.4 kg $m^{-2}h^{-1}$) as shown in Figure 12.10. The presence of high organic matter in water contributes to the cause of organic fouling of the membrane (Goh et al. 2018). Fouling can lower the value of water flux.

Removal of UV_{254} in wetland saline water without pre-treatment is relatively low at 58.2%. Meanwhile, the removal of UV_{254} after the coagulation process decreases to 88.8% (Figure 12.10). The small refection of UV_{254} for the desalination of wetland saline water is caused by membrane fouling. Organic fouling of

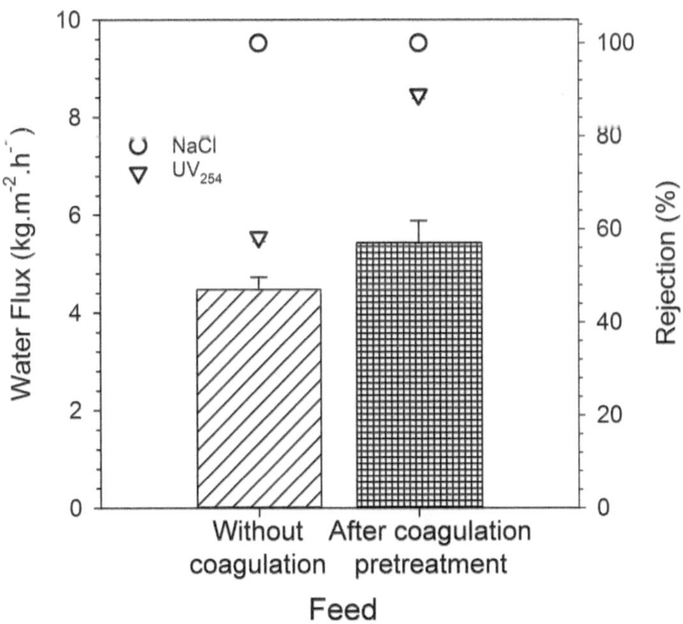

FIGURE 12.10 Silica-pectin membrane pervaporation in variated feed wetland saline water. A bar chart for water flux is shown on the left side and a line plot for (NaCl and UV_{254}) rejections on the right side (Rahma et al. 2019b).

FIGURE 12.11 Image of wetland saline water before processing (raw water), after coagulation pre-treatment and after coagulation-pervaporation (Rahma et al. 2019b)

silica-pectin membranes causes refection to decrease (Goh et al. 2018). In addition, the pre-treatment process of coagulation has succeeded in lowering the relatively high content of organic matter. A comparison of wetland saline water results by pre-treatment of coagulation and a combination of coagulation-pervaporation processes can be seen in Figure 12.11.

12.5.3 Photocatalytic Integrated Membrane

Photocatalytic method can be used as purification and degrades organic compounds, such as humic acid in peat water (Stiadi 2013; Wang, Elma, et al. 2016; Zhang et al. 2018). Photocatalysis method using a titanium oxide catalyst (TiO_2) can degrade organic matter and a reduction in the organic material in peat water (Jayadi, Destiarti, and Sitorus 2014). This indicates the potentialof photocatalytic in the decomposition of organic materials in the wetland water into non-complex compounds. On the other hand, the utilization of illumination using UV rays can diminish 99% of pathogenic bacteria and 99% of viruses (Abbaszadegan et al. 1997) which can increase the quality of wetland saline water.

The research has an idea to improve the results of water flux and salt rejection by combining photocatalytic use of ultraviolet rays and TiO_2 catalysts. Photocatalytic application-membrane enhances membrane performance (Song et al. 2017; Wang et al. 2017). TiO_2 in this regard is often used as a photocatalytic catalyst because it is environmentally friendly (Molinari, Argurio, and Palmisano 2015). The appropriate selection of membrane materials is also an important factor in reducing process costs without compromising membrane capability. The mechanism of NOMs removal by combination process photocatalytic and membrane is shown in Figure 12.12.

When the TiO_2 catalyst is charged with UV rays the excitation of electrons that cause interaction with TiO_2 occurs. Interactions that occur form •OH radicals due to illumination. •OH radicals interact with organic matter in the degradation process (Ramadhani, Destiarti, and Syahbanu 2017).

12.6 PERSPECTIVE ON FUTURE TREND

According to the previous research findings, filtration has been applied in most low-middle income nations since it can be set up at home and uses inexpensive materials.

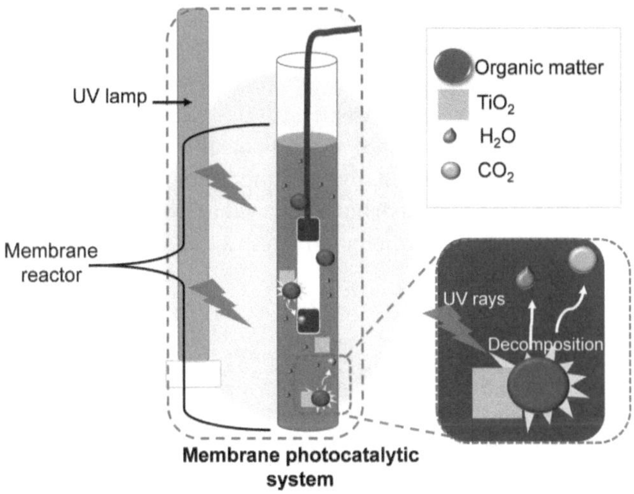

FIGURE 12.12 Photocatalytic-membrane processes with TiO_2 addition leading to NOM decomposition in a peat water.

Yet, in large production, the filtration required complexity in infrastructure where the water can be easily poured; it is also difficult to clean the filter materials in a scale-up plant after a long operation. The filtration operator should be aware of the formation of a biofilm layer because after the filtration is cleaned, it takes 10–20 days for the filter to create the biofilm layer again (Clark et al. 2012). This brings up another problem about the sustainability of filtration. Aside from filtering, coagulation treatment has long been regarded as a simple method capable of removing NOM in surface water. But, disinfection by-product (DBP) might be present after chlorination in raw water where NOM exists (Ghernaout, Ghernaout, and Kellil 2009).

Among all treatment methods, the membrane technology has a huge potential performance to treat the wetland water because of the high rejection and water flux production. In addition, the membrane does not need a large area. However, a previous study found that a pervaporation process had a low rejection in NOMs after treating a wetland with saline water. Nonetheless, in terms of salt rejection, pervaporation has a high salt rejection rate of >99% (Rahma et al. 2019a; Rampun et al. 2019). To accomplish the high salt rejection as well as NOMs rejection, the pre-treatment for pervaporation is needed to be adjusted. Unfortunately, there are still a few studies that reported on the ability of pervaporation in NOM removal. Other studies have reported on the effectiveness of ultrafiltration (UF) membrane in reducing DOC and UV_{254} (Mahmud, Abdi, and Mu'min 2013; Wan et al. 2021; Yuan et al. 2021; Gibert et al. 2015). Nevertheless, a cake layer attached to the top of the membrane becomes a major concern in UF when it is applied to natural water. The UF has been coupled with a coagulant, adsorbent, pre-treatment membrane technologies (nanofiltration or reverse osmosis), or membrane bioreactors (MBR) to reduce the fouling caused by NOMs. Based on the recent study, photocatalytic-coagulation-ultrafiltration has been proposed (Wang et al. 2020). The photocatalytic greatly decreases the NOMs content in natural water.

Despite all the advantages, the largest challenge to run the membrane is energy demand. The future researches trend likely not only focus on the membrane performance, but also on cost energy consumption. The cost can be minimized by integrating the membrane with renewable energies such as solar, wind, water, and geothermal (Song et al. 2019; Ali et al. 2018; Esmaeilion 2020; Gude and Fthenakis 2020; Mehmood and Ren 2021). Finally, it is important to note that the implementation strategy, maintenance, and sustainability need to be considered before selecting the wetland water treatment, especially in remote areas.

ACKNOWLEDGMENTS

The authors would like to thank the facilities provided by M^{2}ReG Lambung Mangkurat University (ULM). Muthia Elma thanks to Basic Research Grant 2021, Applied Research Universities Grant 2021–2023, World Class Research Grant 2021. The Ministry of Research and Technology/National Research and Innovation Agency, The Ministry of Education and Culture of the Republic of Indonesia.

REFERENCES

Abbaszadegan, Morteza, Michaela N Hasan, Charles P Gerba, Peter F Roessler, Barth R Wilson, Roy Kuennen, and Eric Van Dellen. 1997. 'The disinfection efficacy of a point-of-use water treatment system against bacterial, viral and protozoan waterborne pathogens.' *Water Research*, 31: 574–582.

Ali, Aamer, Ramato Ashu Tufa, Francesca Macedonio, Efrem Curcio, and Enrico Drioli. 2018. 'Membrane technology in renewable-energy-driven desalination.' *Renewable Sustainable Energy Reviews*, 81: 1–21.

Ali, Firdaus, Dwi Lintang Lestari, and Marsya Dyasthi Putri. 2019. 'Application of ozone plasma technology for treating peat water into drinking water.' In *IOP Conference Series: Earth and Environmental Science*, 012056. IOP Publishing.

Balistreri, Noemie, Dorian Gaboriau, Claude Jolivalt, and Franck Launay. 2016. 'Covalent immobilization of glucose oxidase on mesocellular silica foams: Characterization and stability towards temperature and organic solvents.' *Journal of Molecular Catalysis B: Enzymatic*, 127: 26–33.

Benfer, S, U Popp, H Richter, C Siewert, and G Tomandl. 2001. 'Development and characterization of ceramic nanofiltration membranes.' *Separation and Purification Technology*, 22: 231–237.

Bing-Zhi, Dong, Chen Yan, Gao Nai-Yun, and Fan Jin-Chu. 2007. 'Effect of coagulation pretreatment on the fouling of ultrafiltration membrane.' *Journal of Environmental Science*, 19: 278–283.

Clark, Peter A., Catalina Arango Pinedo, Matthew Fadus, and Stephen Capuzzi. 2012. 'Slow-sand water filter: Design, implementation, accessibility and sustainability in developing countries.' *Medical Science Monitor: International Medical Journal of Experimental and Clinical Research*, 18: RA105-17.

Costa, Ana Rita, Maria Norberta de Pinho, and Menachem Elimelech. 2006. 'Mechanisms of colloidal natural organic matter fouling in ultrafiltration.' *Journal of Membrane Science*, 281: 716–725.

Denny, Patrick. 1997. 'Implementation of constructed wetlands in developing countries.' *Water Science and Technology*, 35: 27–34.

Duke, MC, S Mee, and JC Diniz da Costa. 2007. 'Performance of porous inorganic membranes in non-osmotic desalination.' *Water Research*, 41: 3998–4004.

Elma, Muthia. 2020. 'Functionalization of Si-C using TEOS (Tetra Ethyl Ortho Silica) as precursor and organic catalyst'. *E3S Web Conf*, 148: 07008.

Elma, Muthia, and Zaini Lambri Assyaifi. 2018. 'Desalination process via pervaporation of wetland saline water.' In *IOP Conference Series: Earth and Environmental Science*, 012009. IOP Publishing.

Elma, Muthia, Arief Rakhman, and Rahmi Hidayati. 2018. 'Silica P123 membranes for desalination of wetland saline water in south kalimantan.' In *IOP Conference Series: Earth and Environmental Science*, 012007. IOP Publishing.

Elma, Muthia, Mahmud Mahmud, Noni Handayani, Via Susetia Putri, and Aulia Rahmah. 2019. 'Performance of interlayer-free silica pectin for seawater desalination on temperature variations of 25°C dan 40°C with membrane calcination temperature of 300°C.' In *Prosiding Seminar Nasional Lingkungan Lahan Basah*, 279–284.

Elma, Muthia, Dwi Rasy Mujiyanti, and Mufidah Nur Amalia. 2019. 'The characterization of silica-organic xerogel from glucose.' *Konversi*, 8: 17–22.

Elma, Muthia, Fitri Ria Mustalifah, Lilis Suryani, Erdina Lulu Atika Rampun, and Aulia Rahma. 2020. 'Wetland saline water and acid mine drainage desalination by interlayefree silica pectin membrane from banana peels.' *Nusantara Science and Technology Proceedings*, 2020:271–279.

Elma, Muthia, Amalia Enggar Pratiwi, Aulia Rahma, Erdina Lulu Atika Rampun, and Noni Handayani. 2020. 'The performance of membranes interlayer-free silica-pectin templated for seawater desalination via pervaporation operated at high temperature of feed solution.' In *Materials Science Forum*, 349–355. Trans Tech Publications Ltd.

Elma, Muthia, Aulia Rahma, Amalia E Pratiwi, and Erdina LA Rampun. 2020. 'Coagulation as pretreatment for membrane-based wetland saline water desalination.' *Asia-Pacific Journal of Chemical Engineering*, 15: e2461.

Elma, Muthia, Erdina LA Rampun, Aulia Rahma, Zaini L Assyaifi, Anna Sumardi, Aptar E Lestari, Gesit S Saputro, Muhammad Roil Bilad, and Adi Darmawan. 2020. 'Carbon templated strategies of mesoporous silica applied for water desalination: A review.' *Journal of Water Process Engineering*, 38: 101520.

Elma, Muthia, Nur Riskawati, and Marhamah. 2018. 'Silica membranes for wetland saline water desalination: Performance and long term stability.' In *IOP Conference Series: Earth and Environmental Science*, 175: 012006.

Elma, Muthia, and Heru Setyawan. 2018. 'Synthesis of silica xerogels obtained in organic catalyst via sol gel route.' In *IOP Conference Series: Earth and Environmental Science*, 012008.

Elma, Muthia, David K Wang, Christelle Yacou, and João C Diniz da Costa. 2015. 'Interlayer-Free Hybrid organo-silica membranes based teos and tevs for water desalination.' In *International Conference on Oleo and Petrochemical Engineering*, Pekanbaru, 1–9.

Elma, Muthia, David K Wang, Christelle Yacou, Julius Motuzas, and João C Diniz da Costa. 2015. 'High performance interlayer-free mesoporous cobalt oxide silica membranes for desalination applications.' *Desalination*, 365: 308–315.

Elma, Muthia, David K Wang, Christelle Yacou, and João C Diniz da Costa. 2015. 'Interlayer-free P123 carbonised template silica membranes for desalination with reduced salt concentration polarization.' *Journal of Membrane Science*, 475: 376–383.

Elma, Muthia, David K Wang, Christelle Yacou, Julius Motuzas, and João C Diniz da Costa. 2015. 'High performance interlayer-free mesoporous cobalt oxide silica membranes for desalination applications.' *Desalination*, 365: 308–315.

Elma, Muthia, Christelle Yacou, David K Wang, Simon Smart, and João C Diniz da Costa. 2012. 'Microporous silica based membranes for desalination.' *Water*, 4: 629–649.

Esmaeilion, Farbod. 2020. 'Hybrid renewable energy systems for desalination.' *Applied Water Science*, 10: 84.

Gadgil, Ashok. 1998. 'Drinking water in developing countries.' *Annual Review of Energy and the Environment*, 23: 253–286.

Ghernaout, Djamel, Badiaa Ghernaout, and Amara Kellil. 2009. 'Natural organic matter removal and enhanced coagulation as a link between coagulation and electrocoagulation.' *Desalination and Water Treatment*, 2: 203–222.

Gibert, Oriol, Benoît Lefèvre, Albert Teuler, Xavier Bernat, and Joana Tobella. 2015. 'Distribution of dissolved organic matter fractions along several stages of a drinking water treatment plant.' *Journal of Water Process Engineering*, 6: 64–71.

Goh, PS, WJ Lau, MHD Othman, and AF Ismail. 2018. 'Membrane fouling in desalination and its mitigation strategies.' *Desalination*, 425: 130–155.

Gude, Veera Gnaneswar, and Vasilis Fthenakis. 2020. 'Energy efficiency and renewable energy utilization in desalination systems.' *Progress in Energy*, 2: 022003.

Guo, Wenshan, Huu-Hao Ngo, and Jianxin Li. 2012. 'A mini-review on membrane fouling.' *Bioresource Technology*, 122: 27–34.

Gutiérrez-Alfaro, Sergio, Asunción Acevedo, Manuel Figueredo, Matthias Saladin, and Manuel A Manzano. 2017. 'Accelerating the process of solar disinfection (SODIS) by using polymer bags.' *Journal of Chemical Technology & Biotechnology*, 92: 298–304.

Haberl, R. 1999. 'Constructed wetlands: A chance to solve wastewater problems in developing countries.' *Water Science and Technology*, 40: 11–17.

Heine, Andreas, Grace DeSantis, John G Luz, Michael Mitchell, Chi-Huey Wong, and Ian A Wilson. 2001. 'Observation of covalent intermediates in an enzyme mechanism at atomic resolution.' *Science*, 294: 369–374.

Hutagalung, Sutrisno Salomo, Imamul Muchlis, Bambang Herlambang, and Arjon Turnip. 2014. 'Removal of chemical and biological contaminants on peat water by ozone-based advanced oxidation processes with reverse osmosis.' In *2014 2nd International Conference on Technology, Informatics, Management, Engineering & Environment*, 288–291. IEEE.

Jayadi, Sony Fajar, Lia Destiarti, and Berlian Sitorus. 2014. 'P Photocatalyst reactor fabrication and its application for degradation of organic peat water using TiO_2 catalyst.' *Journal Kimia Khatulistiwa*, 3.

Kang, S.K., and K.H. Choo. 2010. 'Why does a mineral oxide adsorbent control fouling better than powdered activated carbon in hybrid ultrafiltration water treatment?' *Journal of Membrane Science*, 355: 69–77.

Kang, Xinhuang, Zhibin Mai, Xiaoyong Zou, Peixiang Cai, and Jinyuan Mo. 2008. 'Glucose biosensors based on platinum nanoparticles-deposited carbon nanotubes in sol–gel chitosan/silica hybrid.' *Talanta*, 74: 879–886.

Kim, Jaeshin, Zhenxiao Cai, and Mark M Benjamin. 2008. 'Effects of adsorbents on membrane fouling by natural organic matter.' *Journal of Membrane Science*, 310: 356–364.

Kivaisi, Amelia K. 2001. 'The potential for constructed wetlands for wastewater treatment and reuse in developing countries: A review.' *Ecological Engineering*, 16: 545–560.

Klijn, Jaap E, and Jan BFN Engberts. 2005. 'Fast reactions 'on water'.' *Nature*, 435: 746–747.

Li, Chi-Wang, and Yi-Shiou Chen. 2004. 'Fouling of UF membrane by humic substance: Effects of molecular weight and powder-activated carbon (PAC) pre-treatment.' *Desalination*, 170: 59–67.

Lin, Cheng-Fang, Tze-Yao Lin, and Oliver J Hao. 2000. 'Effects of humic substance characteristics on UF performance.' *Water Research*, 34: 1097–1106.

Maddah, Hisham A, Abdulazez S Alzhrani, M Bassyouni, MH Abdel-Aziz, Mohamed Zoromba, and Ahmed M Almalki. 2018. 'Evaluation of various membrane filtration modules for the treatment of seawater.' *Applied Water Science*, 8: 1–13.

Mahmud, Chairul Abdi, and Badarudddin Mu'min. 2013. 'Removal natural organic matter (nom) in peat water from wetland area by coagulation-ultrafiltration hybrid process with pretreatment two stage coagulation.' *Journal of Wetlands Enviromental Management*, 1: 42–49.

Mahmud, Muthia Elma, Erdina Lulu Atika Rampun, Aulia Rahma, Amalia Enggar Pratiwi, Chairul Abdi, and Raissa Rosadi. 2020. 'Effect of two stages adsorption as pre-treatment of natural organic matter removal in ultrafiltration process for peat water treatment.' *Materials Science Forum*, 988: 114–121.

Mahmud, Suprihanto Notodarmojo, Tri Padmi, and Prayatni Soewondo. 2012. 'Adsorption of natural organic matter (NOM) in peat water on natural and activated peat clay: Equilibrium study of adsorption isotherms and kinetics.' *Info-Teknik*, 13: 28–37.

Mehmood, Aamir, and Jingzheng Ren. 2021. 'Chapter 11- Renewable energy-driven desalination for more water and less carbon.' In Jingzheng Ren (ed.), *Renewable-Energy-Driven Future* (Academic Press).

Molinari, R, P Argurio, and L Palmisano. 2015. '7- Photocatalytic membrane reactors for water treatment.' In Angelo Basile, Alfredo Cassano and Navin K Rastogi (eds.), *Advances in Membrane Technologies for Water Treatment* (Woodhead Publishing: Oxford).

Mozia, Sylwia, Maria Tomaszewska, and Antoni W Morawski. 2005. 'Studies on the effect of humic acids and phenol on adsorption–ultrafiltration process performance.' *Water Research*, 39: 501–509.

Mujiyanti, Dwi Rasy, Muthia Elma, and Mufidah Amalia. 2019. 'Interlayer-free glucose carbonised template silica membranes for brine water desalination.' In *MATEC Web of Conferences*, 03010. EDP Sciences.

Muthia Elma, Mahmud Mahmud, Nurul Huda, Zaini L Assyaifi, Elsa Nadia Pratiwi, Mita Riani Rezki, Dewi Puspita Sari, Erdina Lulu Atika Rampun, Aulia Rahma. 2021. 'PVDF-TiO_2 hollow fibre membrane for water desalination.' *Journal Riset Teknologi Pencegahan Pencemaran Industri*, 12: 1–6.

Parabi, A, R Christiana, D Octaviani, M Zalviwan, E Noerhartati, and LT Muharlisiani. 2019. 'Neutralization, coagulation and filtration process in peat water.' In *Journal of Physics: Conference Series*, 033019. IOP Publishing.

Park, Jason BK, Rupert J Craggs, and Chris C Tanner. 2018. 'Eco-friendly and low-cost Enhanced Pond and Wetland (EPW) system for the treatment of secondary wastewater effluent.' *Ecological Engineering*, 120: 170–179.

Muthia Elma, Mahmud Mahmud, Aulia Rahmah, Noni Handayani, Via Susetia Putri. 'The effect of adding pectin as a silica-pectin-free interlayer membrane coating through salt water desalination process.' In *Environmental Wetland National Conference Proceeding*.

Pirrung, Michael C. 2006. 'Acceleration of organic reactions through aqueous solvent effects.' *Chemistry–A European Journal*, 12: 1312–1317.

Pradhana, EA, M Elma, MHD Othman, N Huda, MD Ul-haq, Erdina LA Rampun, and A Rahma. 2021. 'The functionalization study of PVDF/TiO2 hollow fibre membranes under vacuum calcination exposure.' In *Journal of Physics: Conference Series*, 012035. IOP Publishing.

Pratiwi, Amalia Enggar, Muthia Elma, Aulia Rahma, Erdina LA Rampun, and Gesit Satriaji Saputro. 2019. 'Deconvolution of pectin carbonised template silica thin-film: synthesis and characterisation.' *Membrane Technology*, 2019: 5–8.

Rahma, Aulia, Muthia Elma, Mahmud Mahmud, Chairul Irawan, Amalia Enggar Pratiwi, and Erdina Lulu Atika Rampun. 2019a. 'Removal of natural organic matter for wetland saline water desalination by coagulation-pervaporation.' *Journal of Scientific & Applied Chemistry*, 22: 8.

Rahma, Aulia, Muthia Elma, Erdina Lulu Atika Rampun, Amalia Enggar Pratiwi, Arief Rakhman, and Fitriani. 2020. 'Rapid thermal processing and long term stability of interlayer-free silica-P123 membranes for wetland saline water desalination.' *Advanced Research in Fluid Mechanics and Thermal Sciences*, 71: 1–9.

Rahma, Aulia, Mahmud Mahmud, and Chairul Abdi. 2018. 'The effect of active carbon adsorption pre-treatment on polysulphone (Uf-Psf) ultrafiltration membrane fouling on removal of natural organic materials (Boa) peat water.' *Jernih: Thesis Mahasiswa Program Studi Teknik Lingkungan*, 1.

Rahman, Sazila Karina, Aulia Rahma, Isna Syauqiah, and Muthia Elma. 2020. 'Functionalization of hybrid organosilica based membranes for water desalination–Preparation using Ethyl Silicate 40 and P123.' *Materials Today: Proceedings*, 31: 60–64.

Raj, Monika, and Vinod K Singh. 2009. 'Organocatalytic reactions in water.' *Chemical Communications*: 6687–6703.

Ramadhani, Safitri Ulfah, Lia Destiarti, and Intan Syahbanu. 2017. 'Degradation of organic matter in peat water with thin layer TiO_2 photocatalyst.' *Jurnal Kimia Khatulistiwa*, 6.

Raman, Lakshminarayan P, M Cheryna, and Nandakishore Rajagopalan. 1994. 'Consider nanofiltration for membrane separations.' *Chemical Engineering Progress Rajagopalan*, 90.

Rampun, Erdina Lulu Atika, Muthia Elma, Isna Syauqiah, Meilana Dharma Putra, Aulia Rahma, and Amalia Enggar Pratiwi. 2019. 'I Interlayer-free Silica Pectin Membrane for Wetland Saline Water via Pervaporation.' *Jurnal Kimia Sains dan Aplikasi*, 22: 99–104.

Remondino, Marco, and Luigi Valdenassi. 2018. 'Different Uses of Ozone: Environmental and Corporate Sustainability.' *Literature Review and Case Study*, 10: 4783.

Rezki, Mita Riani, Muthia Elma, Mahmud Mahmud, Sunardi Sunardi, Elsa Nadia Pratiwi, Era Nandita Radiya Oktaviana, Siti Fatimah, and Aulia Rahma. 2020. 'Silica templated carbon membranes from nipa palm (Nypa fruticans) carbon for wetland saline water desalination applications'.

Rosadi, Raissa, Mahmud, and Chairul Abdi. 2017. 'Effect of two-stage coagulation hybrid process and polysulfone ultrafiltration membrane on removal of natural organic materials from peat water.' *Jukung Jurnal Teknik Lingkungan*, 3: 55–69.

Saleem, Junaid, Usman Bin Shahid, Mouhammad Hijab, Hamish Mackey, and Gordon McKay. 2019. 'Production and applications of activated carbons as adsorbents from olive stones.' *Biomass Conversion and Biorefinery*, 9: 775–802.

Shan, Linglong, Hongwei Fan, Hongxia Guo, Shulan Ji, and Guojun Zhang. 2016. 'Natural organic matter fouling behaviors on superwetting nanofiltration membranes.' *Water Research*, 93: 121–132.

Shimabuku, Kyle K, Hyukjin Cho, Eli B Townsend, Fernando L Rosario-Ortiz, and R Scott Summers. 2014. 'Modeling nonequilibrium adsorption of MIB and sulfamethoxazole by powdered activated carbon and the role of dissolved organic matter competition.' *Environmental Science & Technology*, 48: 13735–13742.

Sillanpää, Mika, Mohamed Chaker Ncibi, Anu Matilainen, and Mikko Vepsäläinen. 2018. 'Removal of natural organic matter in drinking water treatment by coagulation: A comprehensive review.' *Chemosphere*, 190: 54–71.

Singh, Pardeep, MC Vishnu, Karan Kumar Sharma, Rishikesh Singh, Sughosh Madhav, Dhanesh Tiwary, and Pradeep Kumar Mishra. 2016. 'Comparative study of dye degradation using TiO_2-activated carbon nanocomposites as catalysts in photocatalytic, sonocatalytic, and photosonocatalytic reactor.' *Desalination and Water Treatment*, 57: 20552–20564.

Song, L, B Zhu, S Gray, M Duke, and S Muthukumaran. 2017. 'Performance of hybrid photocatalytic-ceramic membrane system for the treatment of secondary effluent.' *Membranes (Basel)*, 7: 20.

Song, Xiangju, Yuchao Wang, Canzhu Wang, Minghua Huang, Saeed Gul, and Heqing Jiang. 2019. 'Solar-intensified ultrafiltration system based on porous photothermal membrane for efficient water treatment.' *ACS Sustainable Chemistry Engineering*, 7: 4889–4896.

Stiadi, Yeni. 2013. 'Fotokatalis komposit magnetik TiO_2-$MnFe_2O_4$.' *Prosiding Semirata*, 2013, 1.

Sumardi, Anna, Muthia Elma, Aptar Eka Lestari, Zaini Lambri Assyaifi, Adi Darmawan, Isna Syauqiah, Erdina Lulu Atika Rampun, Yanti Mawaddah, and Linda Suci Wati. 2021. 'Deconvolution of TEOS/TEVS xerogel by single or dual organic catalyst addition.' *Journal Kimia Valensi*, 6: 208–214.

Sumardi, Anna, Muthia Elma, Erdina Lulu Atika Rampun, Aptar Eka Lestari, Zaini Lambri Assyaifi, Adi Darmawan, Dede Heri Yuli Yanto, Isna Syauqiah, Yanti Mawaddah, and Linda Suci Wati. 2021. 'Designing a mesoporous hybrid organo-silica thin film prepared from an organic catalyst.' *Membrane Technology*, 2021: 5–8.

Sun, Lihua, Ning He, Tianmin Yu, Xi Duan, Cuimin Feng, and Yajun Zhang. 2017. 'The removal of typical pollutants in secondary effluent by the combine process of PAC-UF.' *Water Science & Technology*, 75: 1485–1493.

Sun, Wen, Junxia Liu, Huaqiang Chu, and Bingzhi Dong. 2013. 'Pretreatment and membrane hydrophilic modification to reduce membrane fouling.' *Membranes*, 3: 226–241.

Syauqiyah, Isna, Muthia Elma, Meilana D Putra, Aulia Rahma, Amalia E Pratiwi, and Erdina LA Rampun. 2019. 'Interlayer-free silica-carbon template membranes from pectin and p123 for water desalination.' In *MATEC Web of Conferences*, 03017. EDP Sciences.

Tang, Congcong, Zhangwei He, Fangbo Zhao, Xiaoyang Liang, and Zhanshuang Li. 2014. 'Effects of cations on the formation of ultrafiltration membrane fouling layers when filtering fulvic acid.' *Desalination*, 352: 174–180.

Tomaszewska, Maria, and Sylwia Mozia. 2002. 'Removal of organic matter from water by PAC/UF system.' *Water Research*, 36: 4137–4143.

Tuhuloula, Abubakar, Lestari Budiyarti, and Etha Nur Fitriana. 2013. 'Pectin characterization by utilizing banana peel waste using the extraction method.' *Konversi*, 2: 21–27.

Ubomba-Jaswa, Eunice, Pilar Fernández-Ibáñez, Christian Navntoft, M Inmaculada Polo-López, Kevin G McGuigan. 2010. 'Investigating the microbial inactivation efficiency of a 25 L batch solar disinfection (SODIS) reactor enhanced with a compound parabolic collector (CPC) for household use.' *Journal of Chemical Technology McGuigan, and Biotechnology*, 85: 1028–1037.

Uyak, Vedat, Muge Akdagli, Mehmet Cakmakci, and Ismail Koyuncu. 2014. 'Natural organic matter removal and fouling in a low pressure hybrid membrane systems.' *The Scientific World Journal*, 2014: 1–11

Wan, Ying, Pengchao Xie, Zongping Wang, Jingwen Wang, Jiaqi Ding, Raf Dewil, and Bart Van der Bruggen. 2021. 'Application of UV/chlorine pretreatment for controlling ultrafiltration (UF) membrane fouling caused by different natural organic fractions.' *Chemosphere*, 263: 127993.

Wang, David K, Muthia Elma, Julius Motuzas, Wen-Che Hou, Diego Ruben Schmeda-Lopez, Tianlong Zhang, and Xiwang Zhang. 2016. 'Physicochemical and photocatalytic properties of carbonaceous char and titania composite hollow fibers for wastewater treatment.' *Carbon*, 109: 182–191.

Wang, David K, Muthia Elma, Julius Motuzas, Wen-Che Hou, Fengwei Xie, and Xiwang Zhang. 2017. 'Rational design and synthesis of molecular-sieving, photocatalytic, hollow fiber membranes for advanced water treatment applications.' *Journal of Membrane Science*, 524: 163–173.

Wang, Long-Fei, and Mark Benjamin. 2016. 'A multi-spectral approach to differentiate the effects of adsorbent pretreatment on the characteristics of NOM and membrane fouling.' *Water Research*, 98: 56–63.

Wang, Nan, Xing Li, Yanling Yang, Zhiwei Zhou, Yi Shang, and Xiaoxuan Zhuang. 2020. 'Photocatalysis-coagulation to control ultrafiltration membrane fouling caused by natural organic matter.' *Journal of Cleaner Production*, 265: 121790.

Wang, Shengnan, David K Wang, Julius Motuzas, Simon Smart, and João C Diniz da Costa. 2016. 'Rapid thermal treatment of interlayer-free ethyl silicate 40 derived membranes for desalination.' *Journal of Membrane Science*, 516: 94–103.

Wenten, IG., K Khoiruddin, AK Wardani, PTP Aryanti, DI Astuti, and AAIAS Komaladewi. 2020. 'Preparation of antifouling polypropylene/ZnO composite hollow fiber membrane by dip-coating method for peat water treatment.' *Journal of Water Process Engineering*, 34: 101158.

Yang, Hong, Muthia Elma, David K Wang, Julius Motuzas, and João C Diniz da Costa. 2017. 'Interlayer-free hybrid carbon-silica membranes for processing brackish to brine salt solutions by pervaporation.' *Journal of Membrane Science*, 523: 197–204.

Yuan, Xiao-Tong, Lei Wu, Hong-Zhang Geng, Luda Wang, Wenyi Wang, Xue-Shuang Yuan, Benqiao He, Yi-Xuan Jiang, Yu-jie Ning, and Ze-Ru Zhu. 2021. 'Polyaniline/polysulfone ultrafiltration membranes with improved permeability and anti-fouling behavior.' *Journal of Water Process Engineering*, 40: 101903.

Zhang, Tianlong, Muthia Elma, Fengwei Xie, Julius Motuzas, Xiwang Zhang, and David K Wang. 2018. 'Rapid thermally processed hierarchical titania-based hollow fibres with tunable physicochemical and photocatalytic properties.' *Separation and Purification Technology*, 206: 99–106.

Zularisam, A, A Ismail, M Salim, M Sakinah, and H Ozaki. 2007. 'The Effect of Natural Organic Matter (NOM) Fractions on Fouling Characteristics and Flux Recovery of Ultrafiltration Membranes.' *Desalination*, 212: 191–208.

Zularisam, AW, AF Ismail, and Razman Salim. 2006a. 'Behaviours of Natural Organic Matter in Membrane Filtration for Surface Water Treatment-A Review.' *Desalination*, 194: 211–231.

Zularisam, AW, AF Ismail, MR Salim, Mimi Sakinah, and T Matsuura. 2009. 'Application of coagulation-ultrafiltration hybrid process for drinking water treatment: Optimization of operating conditions using experimental design.' *Separation and Purification Technology*, 65: 193–210.

Zularisam, AW, AF Ismail, and Razman Salim. 2006b. 'Behaviours of natural organic matter in membrane filtration for surface water treatment—A review.' *Desalination*, 194: 211–231.

Zulfikar, MA, E Novita, R Hertadi, and SD Djajanti. 2013. 'Removal of humic acid from peat water using untreated powdered eggshell as a low cost adsorbent.' *International Journal of Environmental Science Technology*, 10: 1357–1366.

Zulfikar, Muhammad Ali, A Wahab Mohammad, Abdul Amir Kadhum, and Nidal Hilal. 2006. 'Poly (methyl methacrylate)/SiO2 hybrid membranes: Effect of solvents on structural and thermal properties.' *Journal of Applied Polymer Science*, 99: 3163–3171.

13 Eco-Friendly Dye Degradation Approaches for Doped Metal Oxides

Muhammad Ikram
Government College University Lahore

Jahanzeb Hassan
Riphah International University

Anwar Ul-Hamid
King Fahd University of Petroleum & Minerals

Salamat Ali
Riphah International University

CONTENTS

DOI: 10.1201/9781003167327-15

13.1 INTRODUCTION

Wastewater contamination is regarded as a worldwide concern due to its negative implications for human and aquatic life [1,2]. According to statistical analysis, more than one billion people worldwide will face a scarcity of hygienic and portable water until 2025 [3–5]. Domestic and industrial pollution containing organic solvents, pharmaceuticals, common household chemicals, and pesticides are the leading causes of polluted water. These emissions cause a slew of dreadful diseases that endanger the population's health [6,7]. In this regard, wastewater decontamination is needed to address these serious issues. To decontaminate wastewater, various methods such as oxidation, membrane filtration, photocatalytic degradation, reverse osmosis, adsorption, liquid–liquid extraction, biological degradation, nano-filtrations, and UV irradiation are used [8–10]. As a result, numerous methods for wastewater management have been developed, with photocatalysis being one of the most realistic and cost-effective approaches for the finest degradation of water contaminants [11–13]. The term photocatalysis is a combination of two words: photo, which means photon, and catalysis, which requires the use of a nanocatalyst to carry out the chemical reaction that degrades pollutants. This fundamental method for mineralizing wastewater pollutants by light absorption in the presence of nanocatalyst initiates and accelerates the subsequent chemical reactions. The basic idea of a catalyst is to increase the rate of a chemical reaction. The standard photocatalytic process equation is shown in Equation 13.1 [14,15]:

$$\text{Organic pollutant} + O_2 \xrightarrow[\text{presence of UV/solar light}]{\text{Semiconductor in the}} CO_2 + H_2 + \text{mineral acids} \qquad (13.1)$$

When compared to other traditional degradation methods, photocatalysis has several advantages. Primarily, the degradation path is environmentally friendly, producing no harmful chemicals or by-products. Second, conditions/states of reaction such as room temperature, normal atmospheric pressure, and sunlight are needed, all of which are voluntarily available. A photocatalyst is a fundamental component of the photocatalytic technique, and a variety of materials can be used as a photocatalyst. As a consequence, photocatalysis is an entirely reasonable, cost-effective, environmentally friendly, and adaptable method for removing harmful pollutants from wastewater in order to provide a sustainable environment [11,16–18]. Because of its effective use of limitless and naturally occurring solar energy, the semiconductor photocatalytic method has been recognized as the most important green approach to reporting environmental and global energy-related problems, making it the best choice to remove organic contaminants from wastewater [15,19,20].

Metal oxide (MO) based heterostructures are closely packed structures formed by generating coordination bonds between metal ions and oxides. Because of their peculiar physicochemical properties, these heterostructures have a wide variety of promising favorable applications [11]. MO has been thoroughly studied in order to comprehend its physical properties, such as magnetism, transparency, and electron transport, as well as its chemical properties, such as gas sensing and, most notably, photocatalysis. In addition to their stoichiometric diversity and compositional ease, these properties result from strong electron-electron interactions [21]. Furthermore, MO can be modified by doping of heteroatoms (fabrication of nanocomposites), as well as chemical and structural changes, allowing them to be extensively studied [8]. Moreover, they exhibit outstanding stability and a significant recycling rate making them economical for a variety of applications [17,19]. Therefore, MO has been utilized in various applications including poisonous gas sensing, biomedical applications, exclusion of heavy metals, and significantly photocatalytic route for the degradation of contaminants.

The MO and its composites demonstrate tremendous photocatalytic progress for wastewater management. Further, MO-based nanomaterials are well established with controlled structural, non-toxicity, crystalline surface potentials, and are extremely stable in water [8]. Materials for photocatalytic progress should be inexpensive, non-toxic, and abundant on Earth for long-term sustainability in many environmental and energy-related aspects; hence, using MO as photocatalysts resulted in well-aligned these characteristics [17]. Titanium dioxide or titania (TiO_2), copper oxide (CuO), Zinc oxide (ZnO), nickel oxide (NiO), tungsten trioxide (WO_3), and tin dioxide (SnO_2) are the prominent examples of MO. Among these, TiO_2 and ZnO are close to being nominated as ideal photocatalysts because of their exceptional photocatalytic performance [11]. Controlling the morphology, increasing the surface-to-volume ratio, and doping with specific materials can all help to improve photocatalytic reactions. Nanomaterials have a higher surface-to-volume ratio, which promotes photocatalysis by providing more surface area for oxidation-reduction reactions. As compared to bulk materials, the optical and chemical properties of nanoscale-based materials are superior. As an example, typical bulk gold doesn't express any reactivity whereas gold nanoparticles are chemically reactive [22–24]. Numerous routes had been employed to produce nano-metal oxides as described in the literature. These routes comprise co-precipitation, sol–gel, electrostatic spinning, magnetron sputtering, spray drying, sonochemical, vapor transport, hydrothermal, and solution combustion [25–28].

The mechanisms of photocatalysis and the various techniques used to synthesize doped metal oxide nanostructures, as well as their properties and applications, will be covered in this chapter. This chapter, in particular, compares the photocatalytic efficiency of doped and undoped TiO_2, ZnO, SnO_2, and CuO.

13.2 METAL OXIDES-BASED NANOMATERIALS

13.2.1 Characteristics and Properties

MOs have played an important role in a variety of disciplines, including material science, physics, and chemistry, in recent decades. Essentially, the distinct physicochemical properties of MO nanostructures bind them to a wide range of applications. Among the many applications of MO are poisonous gas sensing, biomedical applications, heavy metal removal, textile coating (for wearable electronic devices), and the most important, which is the focus of this chapter, photocatalytic process for organic pollutant degradation [11,29,30].

Transporting materials at the nanoscale has many advantages, since it contains smaller particle sizes, resulting in a higher fraction of atoms on the surface of nanoparticles, which increases surface area. The increased surface area in nanostructured MO contributes to greater reactivity as compared to normal crystals because they can respond to a greater extent. As a result, more molecules of MO nanoparticles are available for reaction with molecules arriving on the surface. Nanoscale oxides have excellent surface chemistry, which is now well understood [31]. Currently, MO nanomaterials for photocatalytic progress are recognized as front-liners for environmental remediation, particularly for the degradation of wastewater organic pollutants [8]. These MOs are stable, economical, and have a high response to visible-light-driven photocatalytic progress which enables them as encouraging materials for environmental remediation [17].

13.2.2 Synthesis of Doped Metal Oxide Nanoparticles

Various methods are employed for the fabrication of MO nanoparticles. In our study, we talk over repeatedly used and substantial techniques for their synthesis.

13.2.2.1 Co-Precipitation Approach

The co-precipitation method is used to create MO photocatalysts; the synthesis process is usually carried out at room temperature. In this scheme, a combination of precipitating fluid and a salty precursor, such as a metallic salt of nitrate chloride or nitrate, is used to precipitate oxo-hydroxide from the solvent (normally water). By combining a base and a compound, metallic oxo-hydroxides are precipitated in water. Typically, precipitating materials $(C_2H_5)_4NOH$, KOH, and NaOH are used in this technique, with the structure and properties of these MO being highly dependent on the characteristics and pH level of the alkaline mixture. The mode and concentration of adding these alkali solutions into the reaction mixture must be carefully considered to achieve the necessary MO properties using this technique. Surfactants are used to critically optimize the surface properties of MOs. Surfactants can reduce the size and range of fabricated nanomaterials. The best part is that the synthesis process is inexpensive and can be carried out under normal conditions, such as at room temperature [31,32]. The ZnO was prepared by K_2CO_3 and $Zn(NO_3)_3$ in place of precursors using this technique [33].

13.2.2.2 Hydrothermal Synthesis

The hydrothermal synthesis route is solvothermal in nature, but we normally use water as a solvent, so the process is referred to as hydrothermal. As previously mentioned, the thermal decomposition of metal subjects is accomplished by boiling the material in an autoclave or an inert atmosphere. Agglomeration is also possible, so an appropriate capping or stabilizing agent is used in the reaction at the appropriate time to obstruct growth production. The capping agent also aids in the dissolution of particles in solvents [31].

13.2.2.3 Sol–Gel Process

The sol–gel method is usually used to fabricate the MOs treated by the metal reactive predecessors (normally consists of alkoxides in an alcoholic mixture). In this way, hydroxide is formed, which further leads to fabrication of the metal hydroxide polymer which appears as a densely packed porous gel. The formation of target size powder crystal of MO was done by the process of heating and drying the gel. This fabricated MO may appear as the nanoparticle, bulk, or oxygen insufficient material which depends on the heat treatment. This procedure was used to synthesize the CuO nanoparticle with a mixture of copper acetate in ethanol as its predecessor [34].

13.2.2.4 Green Synthesis

The traditional processes formed in laboratory set up in the presence of various limitations and devastating factors such as the possibility of random error, bio accretion, unreliability, reusability, etc. A new mode of synthesis procedures has been introduced to tackle these major factors. The green synthesis is used to prevent the fabrication of irrelevant and devastating by-products by retreating to reliable, durable, and eco-friendly procedures of synthesis [35].

13.2.2.5 Sonochemical Synthesis

A beam of ultrasonic waves i.e. 20kHz to 10MHz is directed to the solution of the initial material that splits the bond of the compound. Stankie et al. [32] reported that the beam of ultrasonic waves develops the fabrication, growth, and collision of bubbles in a mixture, named acoustic cavitation. A large amount of energy that leads to the heating of the solution is gained due to the collision of million bubbles in the solution. The shape and morphology of MO crystal depend on the duration of the solution cooling process. This synthesis is beneficial for producing uniform size, high surface area, and purity besides, it requires less time for the fabrication [31]. This synthesis was used to fabricate the TiO_2 nanoparticle with increased photocatalytic properties which are used for the cavitation by using an ultrasonic beam. This process produces a distinguished environment for hydrolysis of titanium alkoxide and the fabrication of seed nuclei to develop the growth of TiO_2 particles [36].

13.3 METAL OXIDE AS PHOTOCATALYST

Different MOs explicitly TiO_2, ZnO, SnO_2, and CuO have been broadly utilized as photocatalysts and follow general components of photocatalytic movement. Photocatalytic measurement begins with the treatment of light that isolated charges to create hole openings equipped for oxidizing substrates [37–40]. The scattering of light initiates MOs and electrons, which become energized and transfer from lower energy VB to CB, for example, excitons pair. These excitons initiate the oxidation or reduction of substrates/reactants adsorbed on the surface of photocatalysts. The photocatalytic capability of MOs is attributed to the formation of OH radicals and oxygen radicals caused by Goodness anion oxidation and oxygen decrease, respectively. The generation of radicals and anion, which degrade the contaminants, will produce less dangerous by-products [37,40]. Previously mentioned MOs have been utilized as heterogeneous photocatalysts since the most recent couple of many years being biocompatible, steadier alongside their capacity to create charge continuous light enlightenment. Different attributes of MO like electronic arrangement, light assimilation limit, capacity to ship charges, and their lifetime made MOs the most demanding contender as such photocatalysts [40,41]. Inferable from their high contamination remediation limit, different MO photocatalysts like TiO_2, ZnO, SnO_2 and CeO_2 have been broadly used for the transformation of destructive poisons to CO_2 and H_2O [37–40,42–46].

The photocatalysis based on MO provides fascinating features because it can be utilized as photovoltaics, inhibit glass fogging, and is capable of producing O_2 and H_2through the process of water splitting. This mechanical significance connotes their job in different fields like natural sciences, remediation of contaminations, hardware, and hydrogen stockpiling [37,39,42]. The method of photocatalysts based on heterogeneous is used in plant water decontamination and can also be used for water reusing by removing hazardous and poisonous natural mixtures, contaminants, and bacterial microorganisms. The removal of hazardous chemicals from photocatalyst surfaces results in the exchange of water and carbon dioxide [38,47,48]. The electrons generated as a result of the reaction of photocatalyst in the presence of O_2 are named catalysts [38,39,49,50]. Photocatalysis based on induction of visible light reports another notable instrument utilized for remediation of natural toxins and harmful synthetic chemicals [44–46,51,52]. Physiochemical qualities of photocatalyst assume a critical part in photocatalysis and rely upon the size, morphology, shape, and arrangement of MO impetus [53,54]. Different strategies utilized for the arrangement of materials with the wanted size and morphology may work with the development of photoactive material as a powder or thin film with improved photocatalytic execution [55,56].

13.3.1 Photocatalytic Potential of Zinc Oxide

Zinc Oxide (ZnO) is a water-insoluble mineral that is mined in white powder form from the mineral Zincite using synthetic methods. It is a member of the II–IV group of the periodic table and belongs to the family of n-type semiconductors with large bandgap energy. It is a cost-effective material with chemical stability and non-toxicity that is widely used in a variety of industries including ceramics, plastics, glass, cement, and rubber. ZnO holds enormous radiation hardness, significant optical absorption (i.e. in the UV region), efficacious transparency, and exceptional thermal properties. These wonderful characteristics make ZnO remarkable in numerous applications containing solar cells, antibacterial agents, medicine, optoelectronics, catalytic activity, and finally photocatalysis.

Pawar et al. [57] produced the microstructure of the polar surface of Cu-doped ZnO via a single-stage chemical approach. Field Emission Scanning Electron Microscopy (FESEM) micrographs disclosed the rods nature of ZnO into spheres whereas the incorporation of doping disk-like structure was detected. Enormous absorption of Cu above ZnO stemmed from the enhancement of the photocatalytic process for the degradation of both rhodamine B (RhB), and methylene blue (MB) dyes. Additionally, the presence of polar surfaces, production of CuO, and the introduction of O_2 vacancies were the factors for improved photocatalysis. Yin et al. [58] prepared hierarchical Ni-doped ZnO nanostructures via a solvothermal route. Transforming of solid into a hollow spherical structure was done through Ostwald ripening strategy. The degradation progress of RhB dye was attained by utilizing 1 mol% doping of Ni in the ZnO nanostructure. Utilizing 10 mol% Ni-doping causes to reduce energy and enhances the recombination of the photo-induced electrons hence resulting in the dominance of photocatalysis.

Štengl and colleagues [59] prepared the bismuth (Bi^{3+})-doped ZnO nano-rods by adopting thermal hydrolysis of zinc-peroxo complex. FESEM micrographs displayed a greater concentration of Bi^{3+} and the introduction of doping occasioned to alter morphology from nano-rods to flower. Results demonstrate that the prepared product (45.4 wt.% Bi_2O_3) boosted the photocatalytic progress to degrade orange II dye. Moreover, improvement of the photocatalytic process in the visible portion was noticed by annealing the prepared samples at 400°C. Yildirim et al. [60] had produced cobalt (Co)-doped ZnO thin films using glass substrate by means of sol–gel technique in which doping percentage of Co was 0–5 at. %. X-ray diffraction (XRD) analysis signifies the shifting of peaks in the direction of a lower diffraction angle with the introduction of doping. This shifting represents the contraction of the lattice parameter along the C-axis. Photocatalysis of the prepared product was executed for MB degradation which illustrates the finest value of Co was 3 at.% for the remarkable photocatalytic performance of Co-doped ZnO films.

Sadiq et al. [61] manufactured mixed nanostructures of ZnO and Ni-tungstate (WO_4) with reduced graphene oxide (rGO) that was N_2-doped (NrGO/ZnO/NiWO_4) through microwave irradiation process. Sol–gel strategy was engaged to synthesize Al-doped ZnO comprehended by polymerization of polyaniline (PANI) in order to synthesize hybrid polyaniline-ZnO doped via (Al) (PZA) by Mitra and his co-workers. The photocatalytic progress was assessed for the degradation of MB in which a 2.5% sample (NrGO/ZnO/NiWO_4) exhibited nine times superior degradation effectiveness as compared to bare $NiWO_4$. In photocatalytic reactions, the active species are photogenerated holes which were verified through type-II configuration via band edge location and trapping experiment. This reaction undergoes via prepared photocatalyst in the existence of $NaBH_4$ which turns into the remarkable reduction of 4-amino as well as 4-nitro phenol, respectively [62]. Morphological analysis (via FESEM and TEM) publicized the development of Al/ZnO and PANI nano-rods owing to an average particle size of 74.62 nm. Figure 13.1 illustrates the degradation of MO and rose bengal (RB) dyes, respectively, using definite time intervals; (a) in the absence of photocatalyst in dark (b) in the occurrence of photocatalyst using irradiation, (c–i) ZnO, AlZnO, PANI, PAZ 1, PAZ 2, PAZ 3 and PAZ 4 hybrid in light. For Al/ZnO hybrid (22 wt.%), percentage degradation of ~98% and 92.5% were respectively noticed for RB and MO within 150 minutes.

Lavand et al. [63] fabricated the carbon (C) and iron (Fe) adapted ZnO nanostructures through a microemulsion routine. Photoabsorption capability of ZnO was upgraded in the visible portion by changing Fe concentration. Degradation of 2,4,6-trichlorophenol (TCP) using photocatalytic strategy was greater for C and Fe co-doped with ZnO in comparison with all other prepared photocatalysts such as pristine, Fe-doped ZnO, and C-doped ZnO. Furthermore, 2.07 wt.% C and Fe adapted ZnO display superior photo-activity compare to the rest of all doped samples. Low excitons recombination rate is achieved through the synergetic outcomes of C and Fe which causes boosted degradation efficiency. Ardekani et al. [64] utilized an ultrasonic spray pyrolysis process (one step) to accumulate nanocomposite of co-doped N and sulfur (S) with ZnO-CeO_2 (NSZC) having a size of ~20 nm via

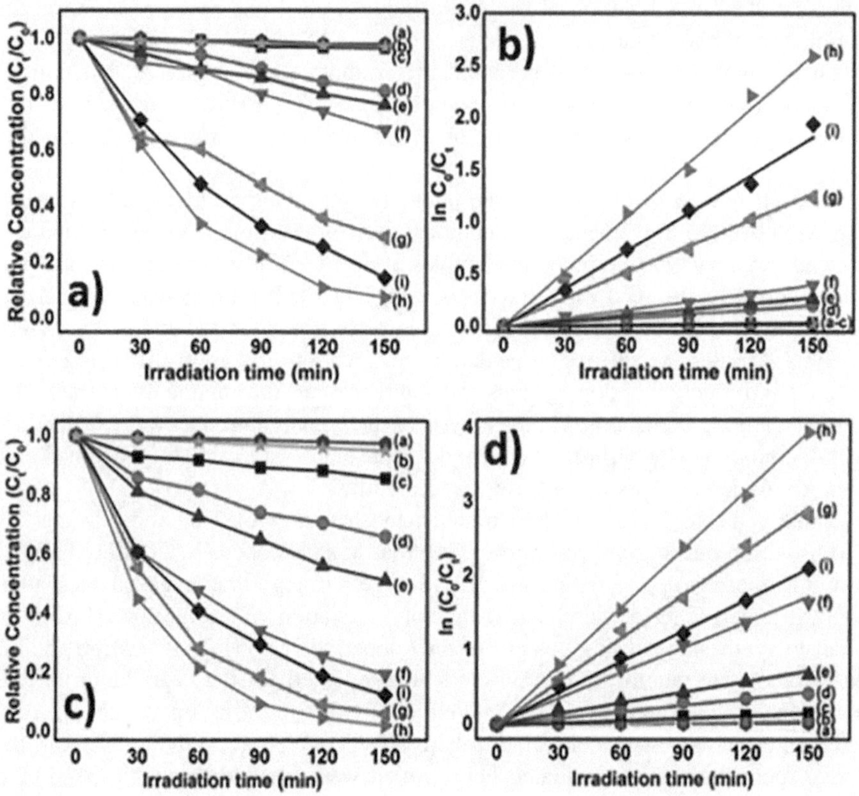

FIGURE 13.1 (a and c) Reaction profile of MO and RB degradation in an aqueous solution respectively against specific time intervals under various conditions. (b and d) First-order kinetic plot of ln (C_o/C_t) vs. irradiation time of MO and RB degradation for (a) without catalyst in dark, (b) without catalyst in light and with catalysts, (c) ZnO, (d) AlZnO, (e) PANI, (f) PAZ 1, (g) PAZ 2, (h) PAZ 3, and (i) PAZ 4 hybrid in light. (Reproduced with permission from Ref. [62] Copyright 2017 Elsevier B.V.)

the surface of FTO substrate, as presented in Figure 13.2a. A variety of techniques was adopted to characterize the prepared samples. FESEM examination reveals that crumbled shape particles were uniformly distributed over the surface of the substrate; visualized in Figure 13.2b. Furthermore, crystallite size ranged from 15 to 25 nm (Figure 13.2c). Absorption spectra were acquired from methyl orange (MO) in the attendance of NSZC nanocomposite presented in Figure 13.2d. Reduction in the intensity of the characteristic peak of dye point outs the successful degradation of dye. In the case of doped photocatalyst, ~100% degradation efficiency was detected for 180 min. The photocatalytic mechanism of NSZC nanocomposite is displayed in Figure 13.2e in which transfer of an electron from CeO_2 headed for CB of ZnO while hole excited from ZnO headed for CeO_2 VB. Also, holes generate •OH, whereas electrons produced superoxide radicals that constructively suggest the degradation of dye.

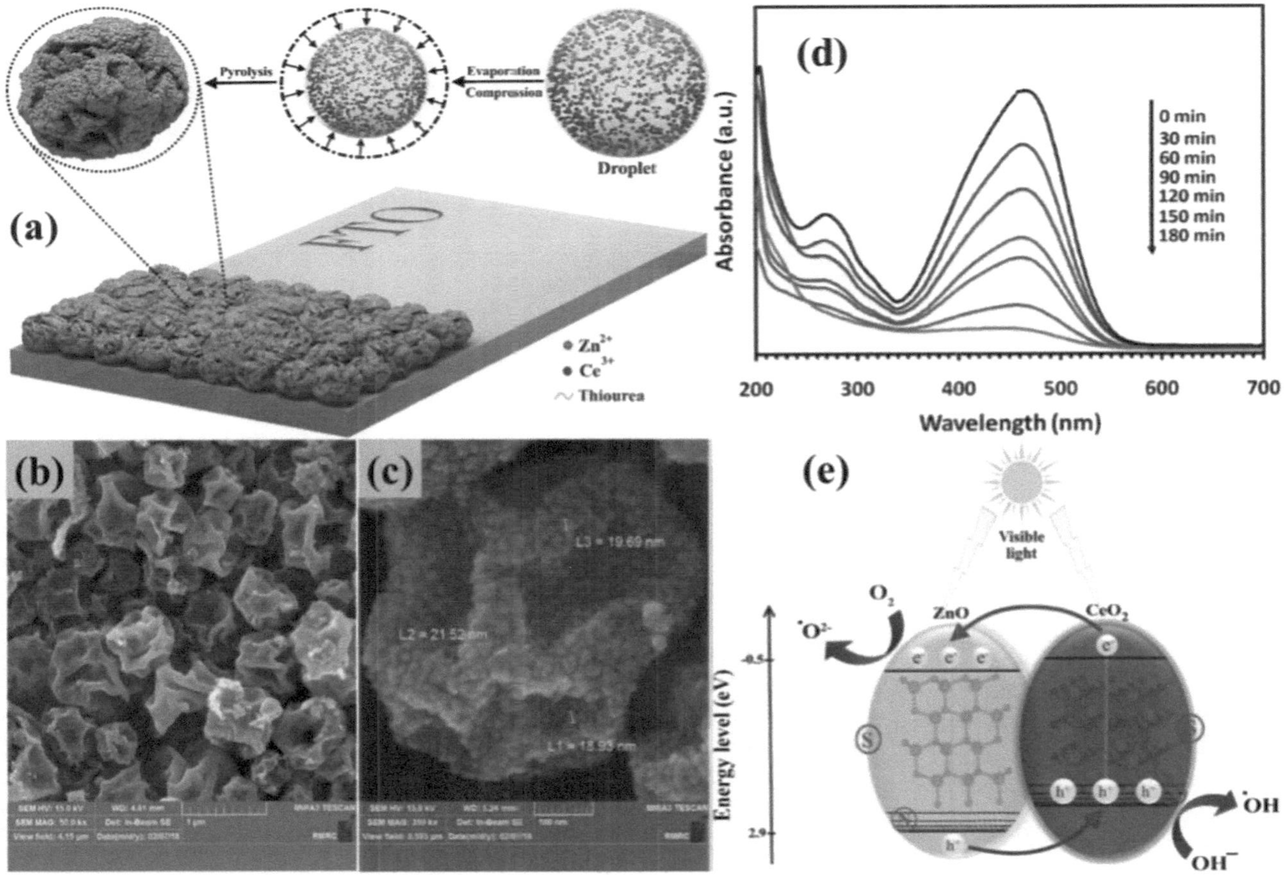

FIGURE 13.2 (a) Schematic illustration of the synthesis of NSZC nanocomposites using USP, (b and c) FESEM images of NSZC nanocomposites, and (d and e) UV–Vis spectra obtained from MO solution in the presence of NSZC nanocomposite and schematic illustration of PC mechanism of NSZC nanocomposite respectively, under visible light irradiation. (Reproduced with permission from Ref. [64] Copyright 2019 Elsevier B.V.)

13.3.2 Photocatalytic Potential of Titanium Dioxide

Titanium dioxide (TiO_2) is also known as titanium (IV) oxide, and titania is a remarkable photocatalyst due to its long physiochemical stability and, more significantly, its low cost. It corresponds to the bandgap energy of the order of 3.0–3.2 eV, which is in the UV portion of the spectrum [65,66]. There exist three polymorphs of TiO_2 which are nominated as anatase, brookite, and rutile; among these crystallographic phases, the rutile phase is recognized as the highest stable whereas the remaining two belong to the metastable state of stability. Other phases can be transformed into the most stable rutile phase at prominent temperatures irreversibly. Additionally, titania photocatalyst has significant merits in antibacterial, lithium-ion batteries, sensors, anticancer, and prominently in catalysis as well as wastewater treatment. Owing to these parameters, the synthesis route of titania photocatalyst, its transformation from anatase to rutile, and the photocatalytic process are described in the following sections

13.3.2.1 Doping with Metal and Non-Metal Elements

A distinction in crystal structure and surface composition are the main features that can tune the characteristics and utilization of nanomaterials by choice. Modification in optical characteristics of nanomaterials can be achieved by the addition of electronically active material inside the crystal structure may drop the recombination rate [67].

Hou et al. [68] examined the influence of different ratios of deionized water (DIW) in the solutions and Aqueous-NH_3 on the morphological features of N-doped TiO_2 nanotubes that were synthesized via a hydrothermal process. The N-doped TiO_2 nanotubes liquefied in NH_3 yield advanced adsorption capacity as compared to water. Further, N-dopant causes to diminish the bandgap energy from 3.2 to 2.84 eV, hence enhancing the degradation response of MO dye as displayed in Figure 13.3A. Correspondingly, McManamon examined the impact of S-doping inside the crystal structure of TiO_2 which turns into a remarkable drop of bandgap energy from 3.2 to 1.7 eV as visualized in Figure 13.3B. The incorporation of S-dopant managed to amplify photocatalytic progress for the degradation of malachite green [69].

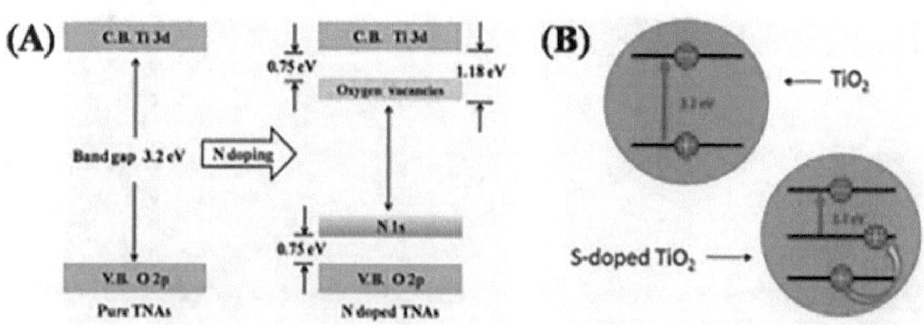

FIGURE 13.3 Schematic illustration of the energy band structures: (A) TiO_2 and N-doped TiO_2 and (B) TiO_2 and S-doped TiO_2. (Reproduced with permission from Refs. [68] and [69], Copyright 2014 and 2015, Elsevier B.V.)

Later, Bakar et al. [70] reported the template-free oxygen peroxide methodology followed by hydrothermal route for crystallization and synthesis of S-doped TiO_2. In this strategy, oxygen atoms were replaced with an S anion that extends the optical absorption spectra from UV to visible portion and lowers bandgap energy as shown in Figure 13.4A. Moreover, S-dopant produces oxygen vacancies on the surface that support to trap electrons and fall the recombination rate of excitons which surely lifted the degradation efficiency of methyl orange dye. Wu et al. examined the C-doped TiO_2 via a solvothermal process which displays similar features compare to S-doped TiO_2 with superior photocatalytic response for methyl orange dye and nitric oxide oxidation. Schematic visualization of bandgap energy difference is displayed in Figure 13.4b [71].

Sood et al. [72] prepared Fe-doped TiO_2 using an ultrasonic-assisted hydrothermal strategy in which Ti is replaced by Fe instead of O which turns into diminishing bandgap energy in the range of 3.2–2.9 eV as a consequence enhances the photocatalytic progress toward the degradation of para nitrophenol and its mechanism is presented in Figure 13.5A. Diverse behavior was detected for Cu-doped TiO_2 which was synthesized via a hydrothermal route. The d-orbital splitting explained the formation of intermediate states among the VB and CB for Cu-doped TiO_2. Sub-orbitals

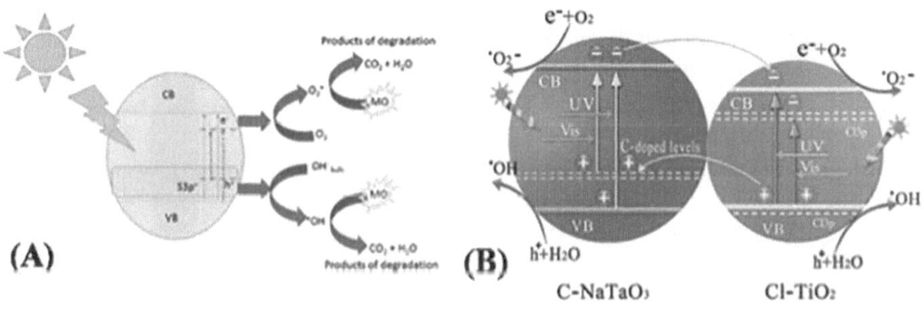

FIGURE 13.4 Schematic illustration of the band gaps and photocatalytic processes on S-doped TiO_2 (A) and C-doped TiO_2 (B). (Reproduced with permission from Refs. [70] and [71] Copyright 2016 Royal Society of Chemistry and 2013 Elsevier B.V.)

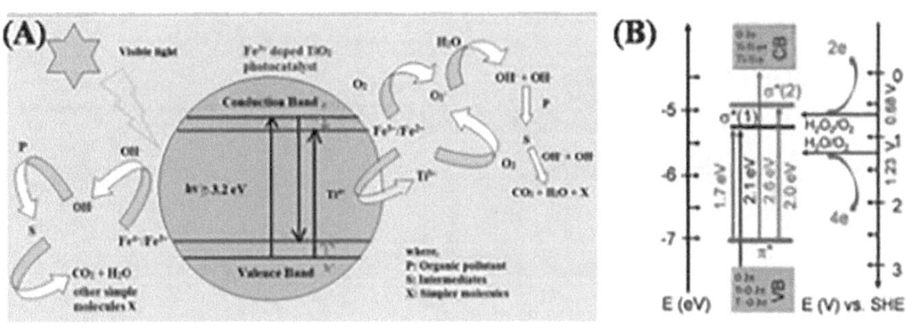

FIGURE 13.5 Schematic illustration of the band gaps structure and photocatalytic processes (A) in Fe-doped TiO_2 and (B) Cu-doped TiO_2. (Reproduced with permission from Refs. [72] and [73], Copyright 2015, Elsevier B.V. and 2014, American Chemical Society.)

involved in valence band are Ti–O 2σ, Ti–O 2π, and O 2π, for conduction band (bottom): Ti–Ti σ, Ti–Ti σ*, Ti–Ti π, Ti–Ti π*, and Ti–O π* while Ti –O σ* for conduction band top as shown in Figure 13.5B. Furthermore, an atomic number of transition metal rises owing to the fluctuation of π* orbitals energy level to CMB that diminishes bandgap energy between σ* (1) and σ*(2). These orbitals were located adjacent to peroxide oxidation sites and H_2O hence, they have a trend to respond to σ orbitals of transitional species and drop potential activity [73].

Sol–gel synthesis of indium (In)-doped TiO_2 reveals significant prospects including transferring of absorption band edge headed for lower wavelength, inferior recombination rate, rise in surface active sites, and large bandgap energy [74]. These features cause to produce CH_4 in higher concentrations by choosing In-doped TiO_2 to elevate photocatalytic progress for the reduction of CO_2. Additionally, the photocatalytic mechanism of bare and In-doped TiO_2 is related in Figure 13.6A and B, where electronic band structures of semiconducting materials are visible. An electron (e^-) in the valence b and VB is excited by photo-radiation to an unoccupied conduction band-CB, which is separated by forbidden bandgap, leaving a positive hole (h^+) in VB. The formation of excitons (e^-–h^+ pair) initiates oxidation and reduction reactions within synthetic compounds, which are then absorbed by the surface of the photocatalyst. The limitation mentioned above does not hold for electronic band structure of photocatalysts. As an example, isolated chemical species that do not possess aforesaid band structure in solid form may be considered photocatalysts. Even with the use of bulk material, photoabsorption and the resulting reaction may proceed on localized sites of photocatalyst.

Kőrösi et al. [75] synthesized PF co-doped anatase TiO_2 powder through a hydrothermal route by adopting hexafluorophosphoric acid (HPF_6) as a dopant by choosing

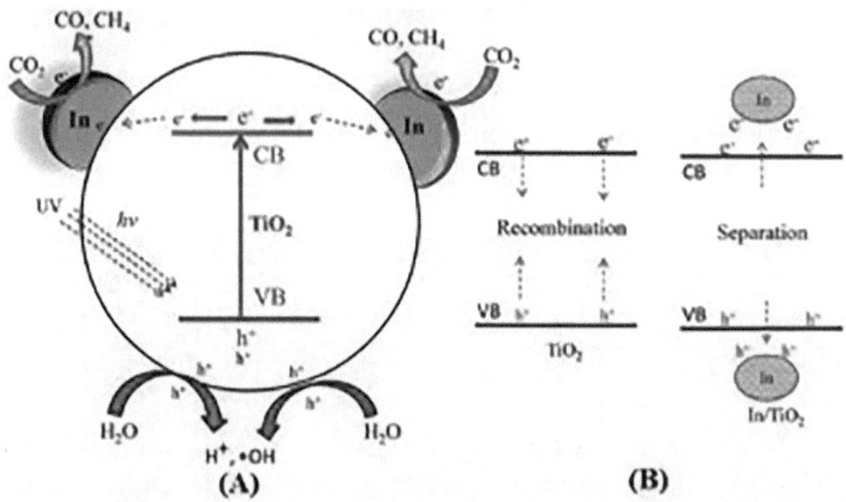

FIGURE 13.6 Schematic illustration for the possible reaction mechanism of photocatalytic CO_2 reduction with H_2O: (A) photocatalytic CO_2 conversion and redox reactions over In-doped TiO_2 and (B) photogenerated charge carrier's recombination and separation. (Reproduced with permission from Ref. [74], Copyright 2015, Elsevier B.V.)

several PF concentrations. The morphological features of nanoparticles were influenced by the concentration of dopants. Higher concentrations of HPF_6 stemmed into round-shaped particles whereas at lower concentration polymeric nanoparticles in spherical and rod-like shapes were detected. PF co-doped TiO_2 proved exceptional photocatalytic progress compare to bear, and doped TiO_2 (F-TiO_2, P-TiO_2). To pay critical significance to surface configuration, PF ratio must be modified effectively to attain maximum progress of photocatalysis using PF-TiO_2.

Li and his group [76] fabricated chains of high dimensional, mono-dispersed, and homo-genetic nanosized crystals of N-doped mesoporous TiO_2 spheres that were synthesized via reflux routine. Figure 13.7a is a graphical depiction of synthesis which indicates that TiO_2 seeds were developed in TiO_2 nanoparticles and self-assembled in TiO_2 sphere precursor. Ammonia was incorporated as a foundation of N by hydrothermal procedure, which assists to spread it homogeneously for calcination. Significant characteristics such as intense purity of anatase phase, a higher concentration of N-dopant, reasonably reduced crystallite size, and extraordinary specific area were attained efficaciously. These remarkable characteristics presented a wide range of UV absorption and the material was successfully hired for the photo-reduction of RhB dye. The influence of N-doping particularly 0.5/N/TiO_2 powerfully affects the

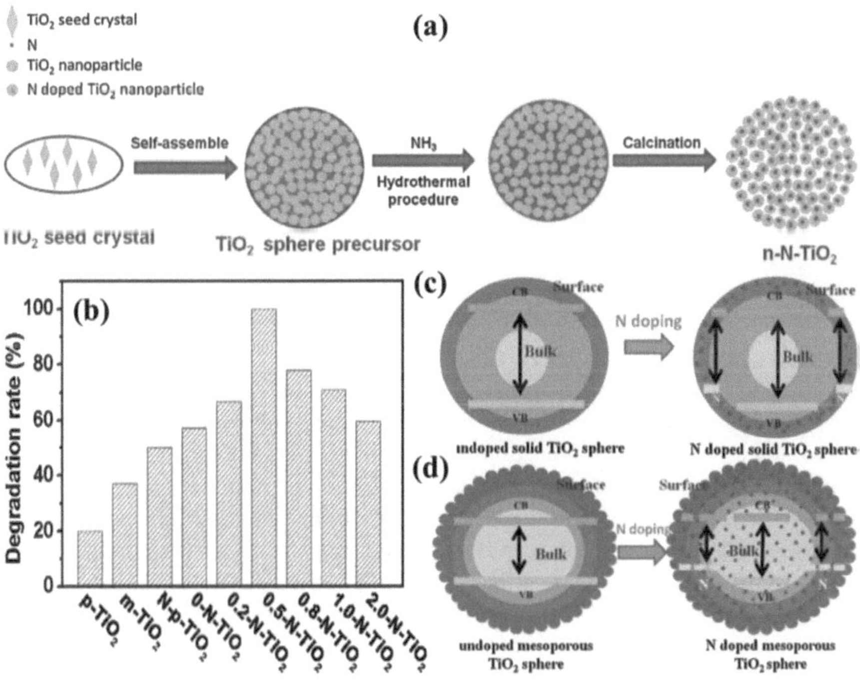

FIGURE 13.7 (a) Schematic diagram of the formation of N-doped mesoporous TiO_2 spheres, (b) PC degradation rate of RhB of undoped and N-doped TiO_2 spheres under visible light irradiation, Schematic diagram of the band structure of undoped and N-doped solid (c) and mesoporous (d) TiO_2 spheres. (Reproduced with permission from Ref. [76] Copyright 2015 Elsevier B.V.)

working as verified in Figure 13.7b. The enhancement in degradation rate with the incorporation of N-dopant into solid and mesoporous TiO_2 compared to pristine material. An important feature is structural features that contribute a key role in photocatalytic progress such as TiO_2 nanocrystals are fixed together to produce a solid TiO_2 sphere and impurity level is dispersed on the subsurface of the TiO_2 sphere surface upon N-dopant as presented in Figure 13.7c. Further, N-doped TiO_2 was homogeneously dispersed from inside out linked with pristine TiO_2 and formed energy bands from bulk to surface as communicated in Figure 13.7d. Formation of impurity levels by N-dopant seems to be accountable for boosted photocatalytic progress.

Liu et al. [77] adopted the hydrothermal technique for the synthesis of Ga-doped TiO_2 series. Time and temperature (10 hours) for synthesis (150°C) were set with no further post-heat treatment. The presence of the anatase phase was exposed by XRD analysis with an average grain size of 10–15 nm. The morphological assessment was done via TEM analysis that well attributes with XRD and absorption spectra (UV–Vis) authorize the redshift of Ga-TiO_2. Better surface area reduced grain size, the useful reaction of the visible region, and separation of photo-excited carriers are the characteristics features that were caused to enhanced photocatalytic efficiency of 0.6 mol% Ga–TiO_2.

13.3.3 Photocatalytic Potential of Tin (IV) Oxide

The chemical formula of tin (IV) oxide or stannic oxide is SnO_2 which is ordinarily considered as an oxygen-wanting n-type semiconductor that possesses extensive bandgap energy of 3.6 eV with significant characteristics such as amphoteric, diamagnetic, and a colored solid. It has been extensively exploited in rechargeable lithium-ion batteries, fuel sensitized solar cells, gas sensors, supercapacitors, and significantly for the degradation of organic dyes using the photocatalytic approach. These applications are due to their extraordinary physicochemical stability, greater photosensitivity, exceptional electrical properties, and easy availability. Unfortunately, the degradation efficiency of pristine SnO_2 is disappointing due to its large bandgap energy and high exciton recombination rate, which limits its usage in industrial processes. To overcome these challenges, a variety of modification schemes such as elemental doping, morphology modification, hetero-junction growth, and co-doping with noble metals have been investigated to improve the photocatalytic progress of SnO_2, as mentioned below. Al-Hamdi et al. [78] constructed bare and iodine (I)-doped SnO_2 nanoparticles using a sol–gel track with several ratios of iodine. Increasing the ratio of dopant material causes to reduce grain size as affirmed via XRD, Scanning electron microscope (SEM), and High-resolution transmission electron microscopy (HRTEM) analysis. Photocatalytic progress of SnO_2:I nano-photocatalyst was inspected via mineralization and degradation of phenol in water under direct contact with sunlight and UV light treatment. Nanoparticles' effectiveness was set up with 1.0 wt.% SnO_2:I, which was greater as compared to bare SnO_2. No considerable rise in phenol degradation was detected with the mounting of I-dopant. However, I-dopant enhanced the degradation of pollutants by altering oxygen vacancy concentrations in the SnO_2 crystal lattice.

Abdelkader et al. [79] utilize a solid-state approach for the fabrication of novel bismuth cuprites ($CuBi_2O_4$)-doped SnO_2 with different mass ratios to attain p–n hetero-junction semiconductors. Various characterization routes (XRD, UV–Vis, and SEM) were adopted to examine the respective properties of prepared samples. Photocatalytic progress was of p-$CuBi_2O_4$/n-SnO_2 photocatalysts was inspected for the degradation of congo red (CR) dye via probe reaction under UVA light treatment. Experimental outcomes disclosed the alteration in phase composition, morphology, and optical properties of the nanocomposites with medium concentrations of pH and mass. Compared with bare SnO_2, and $CuBi_2O_4$, higher photocatalytic efficiency of p-$CuBi_2O_4$/n-SnO_2 was detected. Maximum photo-degradation (58.6%) was detected due to successful excitons separation with 5 wt.% p-$CuBi_2O_4$ by 100 minutes of light treatment at 25°C and pH = 8.

In another examination, Zhou et al. [80] calculated the photocatalytic progress of hollow SnO_2 nano-fibers (H-SNFs) and hollow core–shell TiO_2-SnO_2 nano-fibers (TH-SNFs) that were synthesized via facile single-needle electro-spinning. During the calcination stage, the Kirkendall effect plays a vital part in molding the hollow structure as well as hollow core–shell authorization, which was infrequently exposed in a complex two-step protocol. SEM micrographs verified the development of pure and doped SnO_2 nano-fibers, as displayed in Figure 13.8a and b, respectively. Furthermore, Figure 13.8c shows the assessment in the degradation of MB using H-SNFs and TH-SNFs, which attained 79% MB photo-degradation within 10 minutes as compared to hollow SnO_2 nano-fibers (60%). Inset in Figure 13.8c denotes the first order kinetic rate constant.

13.3.4 Photocatalytic Potential of CuO

Tenorite mineral is a black solid inorganic substance with the formula CuO. It is also known as cupric oxide or copper (II) oxide. CuO is a p-type semiconductor with antiferromagnetic properties, low bandgap energy of about 1.2 eV, and low electrical resistance. The electrical conductivity drops when it is exposed to species of reduction gas. It owns a $3d^9$ electronic configuration, and it also corresponds to a monoclinic crystal system. CuO is a significant material owing to its applications in gas sensing, heterogeneous catalysis, solar cells, and lithium electrode materials. Various synthesis approaches can be utilized to synthesize CuO and its literature-related photocatalytic process is discussed below.

Ardekani et al. [81] deposited thin nanofilms of N-doped ZnO-CuO nanocomposite on the glass substrate through the spray pyrolysis method. Ammonium acetate, zinc acetate, and copper acetate were used as precursors in various molar ratios. FESEM examination reveals nano-dimensioned particles of crumpled shape. The elemental assessment was inspected via EDX exploration which verified the presence of copper, oxygen, nitrogen, and zinc. Moreover, XRD studies guaranteed the existence of crystalline phases of CuO and ZnO with a grain size of ~18 nm. Degradation response of MO dye signifies high progress of nanocomposite. Altered materials can be used as dopants with MO via such a technique.

Rajendaran et al. [82] applied natural herb extract to synthesize the biogenically nanoparticles to produce economical and non-toxic material. They employed

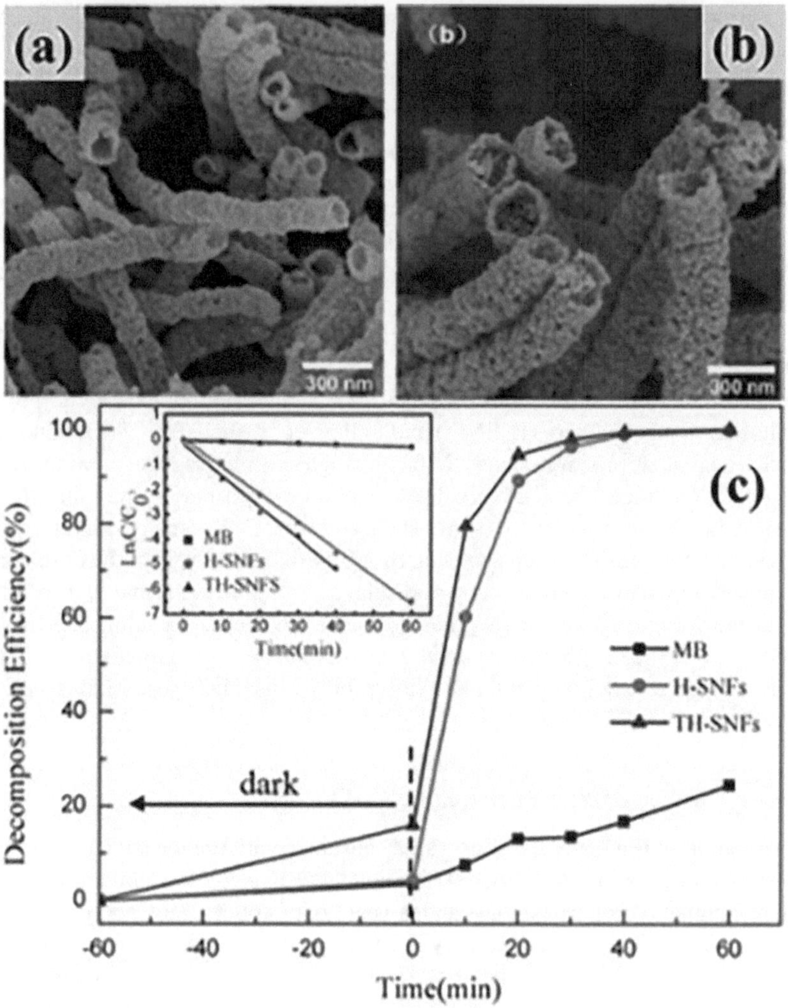

FIGURE 13.8 SEM images of (a) pure and (b) doped SnO_2 nano-fibers, (c) Comparison of decomposition efficiency of H-SNFs and TH-SNFs for the degradation of MB. The inset in (c) is the first-order kinetic rate constant data associated with MB PC degradation in the presence of H-SNFs and TH-SNFs. (Reproduced with permission from Ref. [80] Copyright 2016 Elsevier B.V.)

the biosynthesis method to produce CuO (CO), *Azadirachta indica* extract modified CuO (AzI-CO), molybdenum-doped *Azadirachta indica* extract modified CuO (AzI-MCO), Ag-doped *Azadirachta indica* extract modified CuO (AzI-ACO) and Ag, molybdenum with *Azadirachta indica* extract-doped CuO (AzI-MACO) nanoparticles by employing *Azadirachta indica* leaf extracts as a surfactant. The photocatalytic progress of the nanoparticles was assessed via MB degradation. Photocatalytic results of AzI-MACO were exhibits higher progress (99%) than that of CO (39%), AzI-CO (52%), AzI-ACO (74%), and AzI-MCO (88%) and followed pseudo-first-order

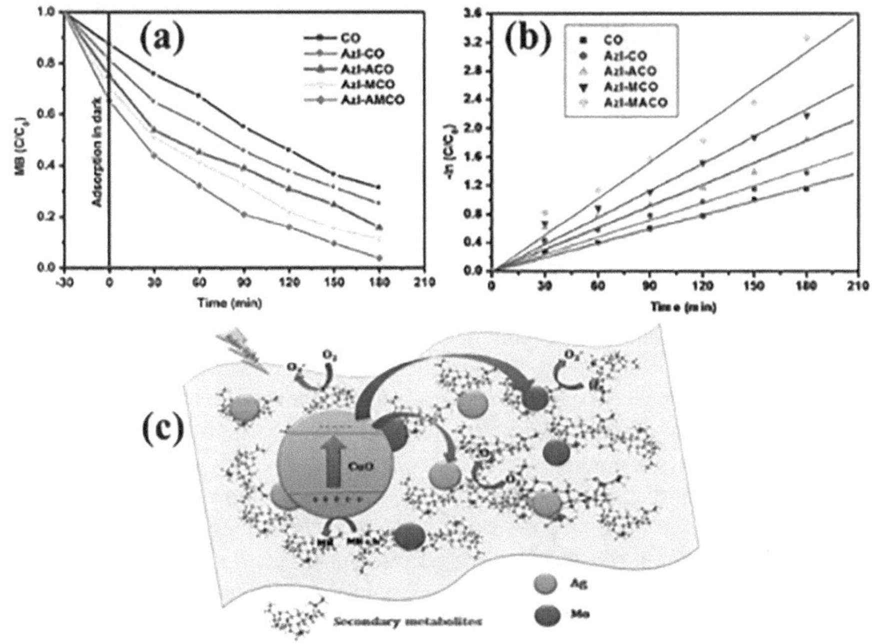

FIGURE 13.9 (a) Photo-degradation curve of MB using CO, AzI-CO, AzI-ACO, AzI-MCO and AzI-AMCO, (b) Kinetic plot of -ln(C/C_0) versus irradiation time for photo-degradation of MB, and (c) Schematic diagram of electron transfer in AzI-MACO under solar-light irradiation. (Reproduced with permission from Ref. [82] Copyright 2019 Elsevier B.V.)

kinetics (Figure 13.9a and b). An increase in efficiency was ascribed to the suppression of excitons as displayed in Figure 13.9c.

Reddy et al. [83] described the one-pot fabrication of a nanocomposite Cu-TiO_2/CuO photocatalyst via surfactant-aided ultra-sonication process. Various characterization approaches were employed for the evolution of various properties that indicate crystallite size of TiO_2 in the range of 20–25 nm which point out enhanced optical and surface properties. Cu-TiO_2/CuO photocatalyst displayed an increased H_2 production rate of catalyst (10,453 μmol $h^{-1} g^{-1}$) which was 21 times greater than bare TiO_2 nanoparticles. The degradation efficiency of MB was higher (90%) that reduced 55% of toxicity. Thus, the photocatalytic progress was considerably improved due to CuO capacity to separate excitons. The one-pot procedure was reflected as a time-efficient and easy process for the synthesis of Cu-TiO_2/CuO-nanomaterials on a large scale. In another research, Gadah et al. [84] utilized sol–gel process to fabricate CuO-silicon dioxide (CuO-SiO_2) nanoparticles and numerous ratios (0.5, 1.0, 3.0 and 5.0 wt.%) of Pt was incorporated to spread/decorate these nanoparticles through the photo-assisted deposition. The decoration of CuO-SiO_2 nanoparticles with Pt exposed a reduction in exciton recombination rate and bandgap energy from ranges from 2.38 to 1.91 eV. Photocatalytic progress of Pt@CuO-SiO_2 nanoparticles was evaluated with visible light for the degradation of Acridine orange (AO) dye. Maximum degradation (100%)

of AO dye was reached with a 1.2 g L^{-1} dose of 3.0 wt.% Pt@CuO-SiO_2 after 30 minutes reaction time, which PC stability for five times [85,86].

13.4 CONCLUSIONS AND SUMMARY

Many toxins are dumped into the atmosphere freely and inadvertently as a result of global urbanization. Environmental remediation research is currently receiving a lot of attention in order to counteract these negative effects and create a healthy environment through technical innovation. MOs have been studied extensively for decades due to their physicochemical properties, and they are now considered leading candidates for environmental remediation, especially in the photocatalytic degradation of wastewater pollutants. The most widely used photocatalysts, TiO_2, ZnO, SnO_2, and CuO, are discussed in this chapter. Chemical and structural modifications, such as morphology control, doping of heteroatoms or ionic structures, and the preparation of nanocomposites, are also possible ways to modify MOs. These changes successfully addressed the inherent shortcomings of the aforementioned photocatalysts and improved their photocatalytic performance. To maximize the usage of these MOs in a number of applications, such as heavy metal removal, textile coating for wearable electronic devices, biomedical applications, poisonous gas sensing, and the aforementioned photocatalytic degradation of organic contaminants, crucial feedback, and cutting-edge concepts are needed.

REFERENCES

1. Yu, Yichang, Chengjun Li, Shoushuang Huang, Zhangjun Hu, Zhiwen Chen, and Hongwen Gao, BiOBr hybrids for organic pollutant removal by the combined treatments of adsorption and photocatalysis, *RSC Advances*, 8 (2018) 32368–32376.
2. Raza, A., M. Ikram, M. Aqeel, M. Imran, A. Ul-Hamid, K.N. Riaz, and S. Ali, Enhanced industrial dye degradation using Co doped in chemically exfoliated MoS_2 nanosheets, *Applied Nanoscience*, 10 (2020) 1535–1544.
3. Gleick, Peter H., A look at twenty-first century water resources development, *Water International*, 25 (2000) 127–138.
4. Pi, Yunhong, Xiyi Li, Qibin Xia, Junliang Wu, Yingwei Li, Jing Xiao, and Zhong Li, Adsorptive and photocatalytic removal of Persistent Organic Pollutants (POPs) in water by metal-organic frameworks (MOFs), *Chemical Engineering*, 337 (2018) 351–371.
5. Ikram, M., A. Raza, M. Imran, A. Ul-Hamid, A. Shahbaz, and S. Ali, Hydrothermal synthesis of silver decorated reduced graphene oxide (rGO) nanoflakes with effective photocatalytic activity for wastewater treatment, *Nanoscale Research Letters*, 15 (2020) 95.
6. Mousavi, Mitra, Aziz Habibi-Yangjeh, and Shima Rahim Pouran, Review on magnetically separable graphitic carbon nitride-based nanocomposites as promising visible-light-driven photocatalysts, *Materials Science: Materials in Electronics*, 29 (2018) 1719–1747.
7. Wang, Puze, Jiping Yao, Guoqiang Wang, Fanghua Hao, Sangam Shrestha, Baolin Xue, Gang Xie, and Yanbo Peng, Exploring the application of artificial intelligence technology for identification of water pollution characteristics and tracing the source of water quality pollutants, *Science of the Total Environment*, 693 (2019) 133440.
8. Gusain, Rashi, Kanika Gupta, Pratiksha Joshi, and Om P. Khatri, Adsorptive removal and photocatalytic degradation of organic pollutants using metal oxides and their composites: A comprehensive review, *Advances in Colloid and Interface Science*, 272 (2019) 102009.

9. Ikram, M., M.I. Khan, A. Raza, M. Imran, A. Ul-Hamid, and S. Ali, Outstanding performance of silver-decorated MoS2 nanopetals used as nanocatalyst for synthetic dye degradation, *Physica E: Low-dimensional Systems and Nanostructures*, 124 (2020) 114246.
10. Ikram, M., I. Jahan, A. Haider, J. Hassan, A. Ul-Hamid, M. Imran, J. Haider, A. Shahzadi, A. Shahbaz, and S. Ali, Bactericidal behavior of chemically exfoliated boron nitride nanosheets doped with zirconium, *Applied Nanoscience*, 10 (2020) 2339–2349.
11. Gautam, S., H. Agrawal, M. Thakur, A. Akbari, H. Sharda, R. Kaur, and M. Amini, Metal oxides and metal organic frameworks for the photocatalytic degradation: A review, *Journal of Environmental Chemical Engineering*, 8 (2020) 103726.
12. J. Hassan, M. Ikram, A. Ul-Hamid, M. Imran, M. Aqeel, and S. Ali, Application of chemically exfoliated boron nitride nanosheets doped with co to remove organic pollutants rapidly from textile water, *Nanoscale Research Letters*, 15 (2020) 75.
13. Ikram, M., I. Hussain, J. Hassan, A. Haider, M. Imran, M. Aqeel, A. Ul-Hamid, and S. Ali, Evaluation of antibacterial and catalytic potential of copper-doped chemically exfoliated boron nitride nanosheets, *Ceramics International*, 46 (2020) 21073–21083.
14. Hitam, C.N.C., and A.A. Jalil, A review on exploration of Fe_2O_3 photocatalyst towards degradation of dyes and organic contaminants, *Journal of Environmental Management*, 258 (2020) 110050.
15. Danish, M.S., L.L. Estrella, I.M.A. Alemaida, A. Lisin, N. Moiseev, M. Ahmadi, M. Nazari, M. Wali, H. Zaheb, and T. Senjyu, Photocatalytic applications of metal oxides for sustainable environmental remediation, *Metals*, 11 (2021) 80.
16. Li, Ruixiang, Tian Li, and Qixing Zhou, Impact of titanium dioxide (TiO_2) modification on its application to pollution treatment: A review, *Catalysts*, 10 (2020) 804.
17. J. Theerthagiri, S. Chandrasekaran, S. Salla, V. Elakkiya, R.A. Senthil, P. Nithyadharseni, T. Maiyalagan, K. Micheal, A. Ayeshamariam, M.V. Arasu, N.A. Al-Dhabi, and H.-S. Kim, Recent developments of metal oxide based heterostructures for photocatalytic applications towards environmental remediation, *Journal of Solid State Chemistry*, 267 (2018) 35–52.
18. Ikram, M., R. Tabassum, U. Qumar, S. Ali, A. Ul-Hamid, A. Haider, A. Raza, M. Imran, and S. Ali, Promising performance of chemically exfoliated Zr-doped MoS2 nanosheets for catalytic and antibacterial applications, *RSC Advances*, 10 (2020) 20559–20571.
19. Ikram, M., J. Hassan, M. Imran, J. Haider, A. Ul-Hamid, I. Shahzadi, M. Ikram, A. Raza, U. Qumar, and S. Ali, 2D chemically exfoliated hexagonal boron nitride (hBN) nanosheets doped with Ni: synthesis, properties and catalytic application for the treatment of industrial wastewater, *Applied Nanoscience*, 10 (2020) 3525–3528.
20. Raza, A., U. Qumar, J. Hassan, M. Ikram, A. Ul-Hamid, J. Haider, M. Imran, and S. Ali, A comparative study of dirac 2D materials, TMDCs and 2D insulators with regard to their structures and photocatalytic/sonophotocatalytic behavior, *Applied Nanoscience*, 10 (2020) 3875–3899.
21. Kiriakidis, G., and V. Binas, Metal oxide semiconductors as visible light photocatalysts, *Journal of the Korean Physical Society*, 65 (2014) 297–302.
22. Manzoor, M., A. Rafiq, M. Ikram, M. Nafees, and S. Ali, Structural, optical, and magnetic study of Ni-doped TiO_2 nanoparticles synthesized by sol–gel method, *International Nano Letters* 8, no. 1 (2018) 1–8.
23. Rafiq, A., M. Imran, M. Aqeel, M. Naz, M. Ikram, and S. Ali, Study of transition metal ion doped CdS nanoparticles for removal of dye from textile wastewater, *Journal of Inorganic and Organometallic Polymers and Materials*, 6 (2019) 1915–1923.
24. Turaeva, N., and H.J.C. Krueger, Wolkenstein's model of size effects in CO oxidation by gold nanoparticles, *Catalysts*, 10 (2020) 288.
25. Qin, Ming, Di Lan, Jiaolong Liu, Hongsheng Liang, Limin Zhang, Hui Xing, Tingting Xu, and Hongjing Wu, Synthesis of single-component metal oxides with controllable multi-shelled structure and their morphology – Related applications, *The Chemical Record*, 20 (2020) 102–119.

26. Atanasova, G., A. Og Dikovska, T. Dilova, B. Georgieva, G. V. Avdeev, P. Stefanov, and N. N. Nedyalkov, Metal-oxide nanostructures produced by PLD in open air for gas sensor applications, *Applied Surface Science*, 470 (2019) 861–869.
27. Feng, Yining, Xiaodong Jiang, Ehsan Ghafari, Bahadir Kucukgok, Chaoyi Zhang, Ian Ferguson, and Na Lu, Metal oxides for thermoelectric power generation and beyond, *Hybrid Materials*, 1 (2018) 114–126.
28. Golestanbagh, M., M. Parvini, and A. Pendashteh. Preparation, characterization and photocatalytic properties of visible-light-driven $CuO/SnO_2/TiO_2$ photocatalyst, *Catalysis Letters*, 148 (2018) 2162–2178.
29. Khin, M.M., A.S. Nair, V.J. Babu, R. Murugan, and S. Ramakrishna, A review on nanomaterials for environmental remediation, *Energy & Environmental Science*, 5 (2012) 8075–8109.
30. Qumar, U., J. Hassan, S. Naz, A. Haider, A. Raza, A. Ul-Hamid, J. Haider, I. Shahzadi, I. Ahmad, and D.M. Ikram, Silver decorated 2D nanosheets of GO and MoS_2 serve as nanocatalyst for water treatment and antimicrobial applications as ascertained with molecular docking evaluation, *Nanotechnology* (2021). 10.1088/1361-6528/abe43c
31. Rodríguez, José A., and Marcos Fernández-García, eds. *Synthesis, Properties, and Applications of Oxide Nanomaterials*. John Wiley & Sons, 2007.
32. Stankic, S., S. Suman, F. Haque, and J. Vidic, Pure and multi metal oxide nanoparticles: synthesis, antibacterial and cytotoxic properties, *Journal of Nanobiotechnology*, 14 (2016) 73.
33. Farahmandjou, Majid, and Farzaneh Soflaee, Synthesis of iron oxide nanoparticles using borohydride reduction, *International Journal of Bio-Inorganic Hybrid Nanomaterials*, 3 (2014) 203–206.
34. Armelao, L., D. Barreca, M. Bertapelle, G. Bottaro, C. Sada, and E. Tondello, A sol–gel approach to nanophasic copper oxide thin films, *Thin Solid Films*, 442 (2003) 48–52.
35. Singh, J., T. Dutta, K.-H. Kim, M. Rawat, P. Samddar, and P. Kumar, 'Green' synthesis of metals and their oxide nanoparticles: Applications for environmental remediation, *Journal of Nanobiotechnology*, 16 (2018) 84.
36. Jimmy, C. Yu, Jiaguo Yu, Wingkei Ho, and Lizhi Zhang, Preparation of highly photocatalytic active nano-sized TiO_2 particles via ultrasonic irradiation, *Chemical Communications*, 19 (2001) 1942–1943.
37. Khan, M.M., S.F. Adil, and A. Al-Mayouf, Metal oxides as photocatalysts, *Journal of Saudi Chemical Society*, 19 (2015) 462–464.
38. Pelizzetti, E., and C. Minero, Metal oxides as photocatalysts for environmental detoxification, *Comments on Inorganic Chemistry*, 15 (1994) 297–337.
39. Hisatomi, T., J. Kubota, and K. Domen, Recent advances in semiconductors for photocatalytic and photoelectrochemical water splitting, *Chemical Society Reviews*, 43 (2014) 7520–7535.
40. Hisatomi, Takashi, Jun Kubota, and Kazunari Domen, Recent advances in semiconductors for photocatalytic and photoelectrochemical water splitting, *Chemical Society Reviews* 22 (2014) 7520–7535.
41. Fujishima, Akira, and Kenichi Honda, Electrochemical photolysis of water at a semiconductor electrode, *Nature* 238 (1972) 37–38.
42. Djurišić, Aleksandra B., Yu Hang Leung, and Alan Man Ching Ng. Strategies for improving the efficiency of semiconductor metal oxide photocatalysis. *Materials Horizons* 1, no. 4 (2014): 400–410.
43. Chen, Haihan, Charith E. Nanayakkara, and Vicki H. Grassian. Titanium dioxide photocatalysis in atmospheric chemistry. *Chemical Reviews* 112, no. 11 (2012): 5919–5948.

44. Kalathil, Shafeer, Mohammad Mansoob Khan, Sajid Ali Ansari, Jintae Lee, and Moo Hwan Cho. Band gap narrowing of titanium dioxide (TiO 2) nanocrystals by electrochemically active biofilms and their visible light activity. *Nanoscale* 5, no. 14 (2013): 6323–6326.
45. Wang, Junpeng, Zeyan Wang, Baibiao Huang, Yandong Ma, Yuanyuan Liu, Xiaoyan Qin, Xiaoyang Zhang, and Ying Dai. Oxygen vacancy induced band-gap narrowing and enhanced visible light photocatalytic activity of ZnO. *ACS Applied Materials & Interfaces* 4, no. 8 (2012): 4024–4030.
46. Ansari, Sajid Ali, Mohammad Mansoob Khan, Mohd Omaish Ansari, Shafeer Kalathil, Jintae Lee, and Moo Hwan Cho. Band gap engineering of CeO_2 nanostructure using an electrochemically active biofilm for visible light applications. *Rsc Advances* 4, no. 32 (2014): 16782–16791.
47. Fu, Xianzhi, Louis A. Clark, Qing Yang, and Marc A. Anderson. Enhanced photocatalytic performance of titania-based binary metal oxides: TiO_2/SiO_2 and TiO_2/ZrO_2. *Environmental Science & Technology* 30, no. 2 (1996): 647–653.
48. Sano, T., N. Negishi, K. Uchino, J. Tanaka, S. Matsuzawa, and K. Takeuchi. Photocatalytic degradation of gaseous acetaldehyde on TiO2 with photodeposited metals and metal oxides. *Journal of photochemistry and photobiology A: Chemistry* 160, no. 1–2 (2003): 93–98.
49. Wang, Huanli, Lisha Zhang, Zhigang Chen, Junqing Hu, Shijie Li, Zhaohui Wang, Jianshe Liu, and Xinchen Wang. Semiconductor heterojunction photocatalysts: Design, construction, and photocatalytic performances. *Chemical Society Reviews* 43, no. 15 (2014): 5234–5244.
50. Pelaez, Miguel, Nicholas T. Nolan, Suresh C. Pillai, Michael K. Seery, Polycarpos Falaras, Athanassios G. Kontos, Patrick S.M. Dunlop, et al. A review on the visible light active titanium dioxide photocatalysts for environmental applications. *Applied Catalysis B: Environmental* 125 (2012): 331–349.
51. Khan, Mohammad Mansoob, Sajid A. Ansari, D. Pradhan, M. Omaish Ansari, Jintac Lee, and Moo Hwan Cho. Band gap engineered TiO_2 nanoparticles for visible light induced photoelectrochemical and photocatalytic studies. *Journal of Materials Chemistry A* 2, no. 3 (2014): 637–644.
52. Khan, Mohammad Mansoob, Sajid Ali Ansari, Debabrata Pradhan, Do Hung Han, Jintae Lee, and Moo Hwan Cho. Defect-induced band gap narrowed CeO_2 nanostructures for visible light activities. *Industrial & Engineering Chemistry Research* 53, no. 23 (2014): 9754–9763.
53. Wang, Hongkang, and Andrey L. Rogach. Hierarchical SnO_2 nanostructures: Recent advances in design, synthesis, and applications. *Chemistry of Materials* 26, no. 1 (2014): 123–133.
54. Sun, Chunwen, Hong Li, and Liquan Chen. Nanostructured ceria-based materials: Synthesis, properties, and applications. *Energy & Environmental Science* 5, no. 9 (2012): 8475–8505.
55. Yu, Jiaguo, Jingxiang Low, Wei Xiao, Peng Zhou, and Mietek Jaroniec. Enhanced photocatalytic CO_2-reduction activity of anatase TiO_2 by coexposed {001} and {101} facets. *Journal of the American Chemical Society* 136, no. 25 (2014): 8839–8842.
56. Zhou, Peng, Jiaguo Yu, and Mietek Jaroniec. All-solid-state Z-scheme photocatalytic systems. *Advanced Materials* 26, no. 29 (2014): 4920–4935.
57. Pawar, Rajendra C., Da-Hyun Choi, Jai-Sung Lee, and Caroline S. Lee. Formation of polar surfaces in microstructured ZnO by doping with Cu and applications in photocatalysis using visible light. *Materials Chemistry and Physics* 151 (2015): 167–180.

58. Yin, Qiaoqiao, Ru Qiao, Zhengquan Li, Xiao Li Zhang, and Lanlan Zhu. Hierarchical nanostructures of nickel-doped zinc oxide: Morphology controlled synthesis and enhanced visible-light photocatalytic activity. *Journal of Alloys and Compounds* 618 (2015): 318–325.
59. Štengl, Václav, Jiří Henych, Michaela Slušná, Jakub Tolasz, and Kateřina Zetková. ZnO/Bi_2O_3 nanowire composites as a new family of photocatalysts. *Powder Technology* 270 (2015): 83–91.
60. Yildirim, Ozlem Altintas, Hanife Arslan, and Savaş Sönmezoğlu. Facile synthesis of cobalt-doped zinc oxide thin films for highly efficient visible light photocatalysts. *Applied Surface Science* 390 (2016): 111–121.
61. Sadiq, M. Mohamed Jaffer, U. Sandhya Shenoy, and D. Krishna Bhat. NiWO4-ZnO-NRGO ternary nanocomposite as an efficient photocatalyst for degradation of methylene blue and reduction of 4-nitro phenol. *Journal of Physics and Chemistry of Solids* 109 (2017): 124–133.
62. Mitra, Mousumi, Amrita Ghosh, Anup Mondal, Kajari Kargupta, Saibal Ganguly, and Dipali Banerjee. Facile synthesis of aluminium doped zinc oxide-polyaniline hybrids for photoluminescence and enhanced visible-light assisted photo-degradation of organic contaminants. *Applied Surface Science* 402 (2017): 418–428.
63. Lavand, Atul B., and Yuvraj S. Malghe. Synthesis, characterization and visible light photocatalytic activity of carbon and iron modified ZnO. *Journal of King Saud University-Science* 30, no. 1 (2018): 65–74.
64. Ardekani, Saeed Rahemi, Alireza Sabour Rouh Aghdam, Mojtaba Nazari, Amir Bayat, Esmaiel Saievar-Iranizad, and Mehrdad Najafi Liavali. Synthesis and characterization of photocatalytically active crumpled-shape nanocomposites of nitrogen and sulfur co-doped $ZnO–CeO_2$. *Solar Energy Materials and Solar Cells* 203 (2019): 110195.
65. Ikram, M., E. Umar, A. Raza, A. Haider, S. Naz, A. Ul-Hamid, J. Haider, I. Shahzadi, J. Hassan, S. Ali, Dye degradation performance, bactericidal behavior and molecular docking analysis of Cu-doped TiO_2 nanoparticles, *RSC Advances*, 10 (2020) 24215–24233.
66. Ikram, M., J. Hassan, A. Raza, A. Haider, S. Naz, A. Ul-Hamid, J. Haider, I. Shahzadi, U. Qamar, and S. Ali, Photocatalytic and bactericidal properties and molecular docking analysis of TiO_2 nanoparticles conjugated with Zr for environmental remediation, *RSC Advances*, 10 (2020) 30007–30024.
67. Humayun, M., F. Raziq, A. Khan, and W. Luo. Modification strategies of TiO2 for potential applications in photocatalysis: A critical review. *Green Chemistry Letters and Reviews*, 11, no. 2 (2018): 86–102.
68. Hou, Xian, Cheng-Wei Wang, Wei-Dong Zhu, Xiang-Qian Wang, Yan Li, Jian Wang, Jian-Biao Chen, Tian Gan, Hai-Yuan Hu, and Feng Zhou. Preparation of nitrogen-doped anatase TiO2 nanoworm/nanotube hierarchical structures and its photocatalytic effect. *Solid State Sciences* 29 (2014): 27–33.
69. McManamon, Colm, John O'Connell, Paul Delaney, Sozaraj Rasappa, Justin D. Holmes, and Michael A. Morris. A facile route to synthesis of S-doped TiO2 nanoparticles for photocatalytic activity. *Journal of Molecular Catalysis A: Chemical* 406 (2015): 51–57.
70. Bakar, Shahzad Abu, and Caue Ribeiro. Rapid and morphology controlled synthesis of anionic S-doped TiO_2 photocatalysts for the visible-light-driven photodegradation of organic pollutants. *RSC Advances* 6, no. 43 (2016): 36516–36527.
71. Wu, Xiaoyong, Shu Yin, Qiang Dong, and Tsugio Sato. Preparation and visible light induced photocatalytic activity of $C-NaTaO_3$ and $C-NaTaO_3–Cl-TiO_2$ composite. *Physical Chemistry Chemical Physics* 15, no. 47 (2013): 20633–20640.
72. Sood, Swati, Ahmad Umar, Surinder Kumar Mehta, and Sushil Kumar Kansal. Highly effective Fe-doped TiO_2 nanoparticles photocatalysts for visible-light driven photocatalytic degradation of toxic organic compounds. *Journal of Colloid and Interface Science* 450 (2015): 213–223.

73. Roy, Nitish, Youngku Sohn, Kam Tong Leung, and Debabrata Pradhan. Engineered electronic states of transition metal doped TiO_2 nanocrystals for low overpotential oxygen evolution reaction. *The Journal of Physical Chemistry C* 118, no. 51 (2014): 29499–29506.
74. Tahir, Muhammad, and NorAishah Saidina Amin. Indium-doped TiO_2 nanoparticles for photocatalytic CO_2 reduction with H_2O vapors to CH_4. *Applied Catalysis B: Environmental* 162 (2015): 98–109.
75. Kőrösi, László, Mirko Prato, Alice Scarpellini, Andreas Riedinger, János Kovács, Monika Kus, Vera Meynen, and Szilvia Papp. Hydrothermal synthesis, structure and photocatalytic activity of PF-co-doped TiO_2. *Materials Science in Semiconductor Processing* 30 (2015): 442–450.
76. Li, Xiao, Pengwei Liu, Yu Mao, Mingyang Xing, and Jinlong Zhang. Preparation of homogeneous nitrogen-doped mesoporous TiO_2 spheres with enhanced visible-light photocatalysis. *Applied Catalysis B: Environmental* 164 (2015): 352–359.
77. Liu, Xuemei, Matiullah Khan, Wenxiu Liu, Wei Xiang, Ming Guan, Peng Jiang, and Wenbin Cao. Synthesis of nanocrystalline Ga–TiO_2 powders by mild hydrothermal method and their visible light photoactivity. *Ceramics International* 41, no. 2 (2015): 3075–3080.
78. Al-Hamdi, A.M., M. Sillanpää, and J. Dutta, Photocatalytic degradation of phenol by iodine doped tin oxide nanoparticles under UV and sunlight irradiation, *Journal of Alloys and Compounds* 618 (2015) 366–371.
79. Abdelkader, E., L. Nadjia, and B. Ahmed, Preparation and characterization of novel CuBi2O4/SnO2 p–n heterojunction with enhanced photocatalytic performance under UVA light irradiation, *Journal of King Saud University – Science* 27 (2015) 76–91.
80. Zhou, Huimin, Zhiyong Li, Xiao Niu, Xin Xia, and Qufu Wei. The enhanced gas-sensing and photocatalytic performance of hollow and hollow core–shell SnO2-based nanofibers induced by the Kirkendall effect. *Ceramics International* 42, no. 1 (2016): 1817–1826.
81. Ardekani, Saeed Rahemi, Alireza Sabour Rouhaghdam, and Mojtaba Nazari. N-doped ZnO-CuO nanocomposite prepared by one-step ultrasonic spray pyrolysis and its photocatalytic activity. *Chemical Physics Letters* 705 (2018). 19–22.
82. Rajendaran, Karthiga, Rajarajan Muthuramalingam, and Suganthi Ayyadurai. Green synthesis of Ag-Mo/CuO nanoparticles using Azadirachta indica leaf extracts to study its solar photocatalytic and antimicrobial activities. *Materials Science in Semiconductor Processing* 91 (2019): 230–238.
83. Reddy, N. Lakshmana, M. V. Shankar, S. C. Sharma, and G. Nagaraju. One-pot synthesis of Cu–TiO_2/CuO nanocomposite: Application to photocatalysis for enhanced H_2 production, dye degradation & detoxification of Cr (VI). *International Journal of Hydrogen Energy* 45, no. 13 (2020): 7813–7828.
84. Gadah, R. H., and A. S. Basaleh. Influence of doped platinum nanoparticles on photocatalytic performance of CuO–SiO_2 for degradation of Acridine orange dye. *Ceramics International* 46, no. 2 (2020): 1690–1696.
85. Ikram, M., A. Raza, J. Z. Hassan, A. A. Rafi, A. Rafiq, S. Altaf, and A. Ashfaq. Rational design and advance applications of transition metal oxides, IntechOpen, London, ISBN: 978-1-83968-049-6, 2021. doi:10.5772/intechopen.96568.
86. Ikram, M., A. Raza, K. Shahzad, A. Haider, J. Haider, M. Ikram, *Advanced Carbon Materials: Base of 21st Century Scientific Innovations in Chemical, Polymer, Sensing & Energy Engineering*, IntechOpen, London, ISBN: 978-1-78985-924-9, 2021. doi:10.5772/intechopen.95869.

Index